relative atomic masses (approximate values)

aluminium	27	holmium	165	rhenium	186
antimony	122	hydrogen	1	rhodium	103
argon	40	indium	115	rubidium	85
arsenic	75	iodine	127	ruthenium	101
barium	137	iridium	192	samarium	150
beryllium	9	iron	56	scandium	45
bismuth	209	krypton	84	selenium	79
boron	11	lanthanum	139	silicon	28
bromine	80	lead	207	silver	108
cadmium	112	lithium	7	sodium	23
caesium	133	lutetium	175	strontium	88
calcium	40	magnesium	24	sulphur	32
carbon	12	manganese	55	tantalum	181
cerium	140	mercury	201	tellurium	128
chlorine	35·5	molybdenum	96	terbium	159
chromium	52	neodymium	144	thallium	204
cobalt	59	neon	20	thorium	232
columbium	93	nickel	59	thulium	169
copper	63·5	nitrogen	14	tin	119
dysprosium	162·5	osmium	190	titanium	48
erbium	167	oxygen	16	tungsten	184
europium	152	palladium	106	uranium	238
fluorine	19	phosphorus	31	vanadium	51
gadolinium	157	platinum	195	xenon	131
gallium	70	potassium	39	ytterbium	173
germanium	73	praseodymium	141	yttrium	89
gold	197	radium	226	zinc	65
hafnium	179	radon	222	zirconium	91
helium	4				

CHEMISTRY
for modern courses

F. Bennett, M.Sc., A.R.I.C., A.M.C.T.
Senior Chemistry Master, Farnworth Grammar School

C. Dobson, B.Sc.
Head Master, King Edward VI School, Southampton

J. Dyson, M.Sc., Ph.D.
Senior Chemistry Master, Harrogate Grammar School

J. G. A. Raffan, B.Sc.
Head of Science, The Netherhall School, Cambridge

D. E. Wilson, M.A., Ph.D.
Senior Chemistry Master, Manchester Grammar School

F. Winterbottom, B.Sc.
formerly Science Master, Manchester Grammar School

Editor: **C. Dobson, B.Sc.**

The English Universities Press Ltd

ISBN 0 340 16118 3

First printed 1975

The English Universities Press Ltd
St Paul's House, Warwick Lane
London EC4P 4AH

Printed in Great Britain by
Butler & Tanner Ltd, Frome and London

contents

preface

When the authors of this book were members of the Chemistry Department of Manchester Grammar School, they collaborated in designing a fully structured experimentally based course to meet the requirements of a new Ordinary Level syllabus for the General Certificate of Education. The course aimed to combine the best features of traditional subject-matter and teaching method with the new insights into the process of understanding and learning chemistry at school as developed primarily, though not solely, by the Nuffield Science Teaching Project. Having been tested for some years in different schools and revised in detail, the course has remained unaltered in its fundamental conception and has proved to be both an excellent foundation for further study in the subject and equally valuable for those who will not pursue their studies beyond this level.

In common with many other teachers, the authors felt the need for a textbook to act as a companion reference to experimental work; using the Manchester teaching course as a basis, they have broadened the scope of the subject-matter and altered the level of treatment slightly in order to produce this book, which it is hoped will meet the main requirements of most modern Ordinary Level syllabuses. The emphasis throughout, even in the theoretical sections, is on deductions drawn from experimental observations. The significance of energy changes in chemical reactions is a constantly recurring theme–this is the starting-point for the study of chemistry as represented here. It is not a course-book and, though it might be used as such, it will serve a better purpose as a companion to a detailed experimental course. Many teachers will wish to amend the order in which the topics are treated, but considerable thought has been given to order of presentation. Theoretical concepts are introduced gradually and are spaced from each other: properties interpreted in terms of molecules are treated first; later the constitution of molecules is introduced; the treatment of ions and electrons is delayed until sufficient evidence for their existence and their role in reactions has been gathered. It has been found valuable to study the properties of three 'families' of elements (noble gases, halogens, alkali metals), in some detail but at an elementary level, as a prelude to the full treatment of the Periodic Table and its basis in atomic structure. Similarly, acids and alkalis are studied qualitatively some time before a modern definition and explanation of acid character is attempted.

Use of mass spectrometer results may be found an original feature at this level; the authors have found it to be perfectly acceptable (and often exciting) to thirteen- and fourteen-year-old pupils, contributing notably to an understanding of the important idea of mass, as well as removing much of the mystery and difficulty from 'formulae'. The authors have not been afraid to simplify and must apologise if in places they have over-simplified.

At the present time the use of agreed nomenclature and units and the method of formulation of equations in school courses presents a problem. SI Units are used throughout. The authors have followed the recommendations of the Association for Science Education in the Report 'Chemical Nomenclature, Symbols and Terminology for Use in School Science' (1972), the earlier Report of this body, 'SI Units, Signs, Symbols and Abbreviations', and that of the Royal Society (1969), though they acknowledge that at the time of going to press these recommendations are by no means universally accepted. One important divergence should be noted: in this book a formula is used to represent one mole of a substance–indeed the authors believe this to be the most useful of the several meanings which may be attached to a formula. Nomenclature and units used in examination questions have been amended to conform with the practice used in the text and to minimise confusion, but this should not be taken as being indicative of the current practice of the examining boards.

It has been claimed, with some justification, that the use of a text-book is incompatible with a proper investigational approach to the study of chemistry. It is the authors' humble hope that this book may help to fill a practical need without detracting from the stimulus and excitement of the laboratory bench.

C. Dobson

acknowledgments

The authors wish to express their thanks to the following for providing information and photographic material:

Aerofilms; Aluminium Development Association; Associated Press; Burndy Library; Camera Press; Copper Development Association; D. Stubbs; Esso Petroleum Co. Ltd; Ever Ready; Fisons Ltd; Imperial Chemical Industries Ltd; Imperial War Museum; Mansell Collection; Mullard Ltd; Phillips; Pilkington Brothers Ltd; Science Museum; Shell; Tate & Lyle; United Kingdom Atomic Energy Authority.

Permission to publish questions from recent examination papers, given by the following examining boards, is gratefully acknowledged:

Associated Examining Board (AEB)
University of Cambridge Local Examinations Syndicate (C)
Eire, The Department of Education Secondary Education Branch (E)
Joint Matriculation Branch (JMB)
University of London School Examinations Council (L)
Nuffield Course Papers as published by the University of London (Nf)
Oxford Delegacy of Local Examinations (O)
Oxford and Cambridge Schools Examination Board (O&C)
Scottish Certificate of Education Examination Board (Sc)
Southern Universities Joint Board (S)
Welsh Joint Education Committee (W)

The authors are indebted to many people for help in the preparation of this book: to the authors whose standard works of reference we have consulted; to Mr R. Stone, formerly Senior Science Master and Second Master of Manchester Grammar School, for his valuable advice and criticism; to our colleagues and pupils for their assistance; and finally to our wives for their patient and forbearing encouragement.

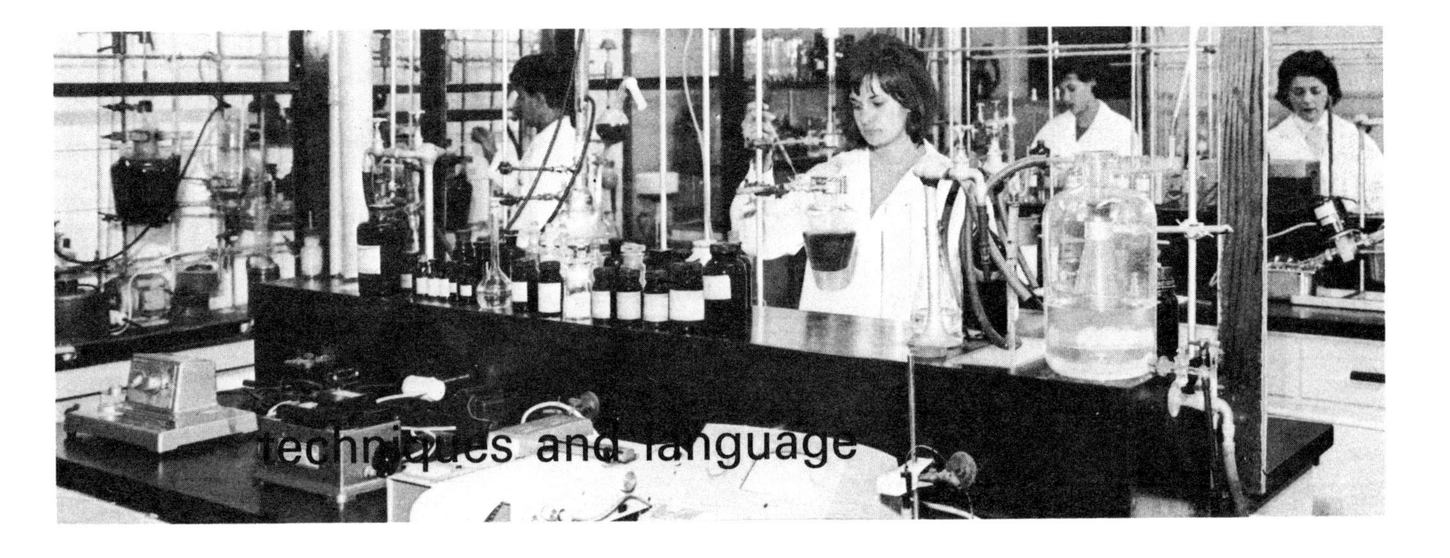

1.1 scope of the study of chemistry

This is a chemistry book – what sort of a study is chemistry? You may answer: chemistry is about test-tubes and flasks and beakers, acids and alkalis, atoms and molecules, heating things to 'see what happens'. If you asked a similar question 'What sort of a boy is Tom Smith?' similar answers might be 'He is a boy who wears glasses and a grey suit, he is a third year boy at my school, he has ginger hair and large feet, if I hit him he will hit me back'. These statements may all be true, but they do not describe Tom Smith fully. The study of chemistry is as difficult to describe fully as is Tom Smith.

Chemistry is a branch of science, and science is concerned with asking questions about the things around us, in such a way that clear and useful answers can be formed. The truth of the answers should be capable of proof by experiment and the answers often lead to further questions. In this way, scientific activity slowly builds up an understanding of the way things behave. Only such an understanding of the way Tom Smith behaves answers fully and usefully as to what sort of a boy he is.

Chemistry is particularly concerned with the way in which substances interact with each other to form different substances, how the new substances behave and what use they may have. The growth of a tree is studied by the biologist; the densities of different woods and friction at a wooden surface by the physicist; the chemist is interested in the products formed when wood is heated in air, the different products formed when it is heated in the absence of air, and how to account for the differences. To do this he must find answers to the questions 'What is wood made of?' and 'What is air?'

The question 'How does a tree grow?' is answerable within the terms of the science of biology; the question 'Why does a tree grow?' is not. 'What sort of a boy is Tom Smith?' can be given some kind of useful answer in scientific terms; 'Why is Tom Smith that sort of boy?' cannot. It is very important indeed to know how to frame the questions which you ask in your study of science.

It is neither possible nor desirable to check every single statement by experiment. To be absolutely certain of the truth of the statement 'all specimens of water have the formula H_2O', all specimens of water would have to be analysed – and this would leave us with no water to use. Nevertheless, we need to accept the statement as true in order to proceed with our study of chemistry and we do so because of the evidence of past experiments on many different samples of the substance. We must, however, be quite sure of what we mean by 'water' – the language of science must mean the same thing to all who use it. Of course we accept many statements which we hear or read about the nature of substances on the authority of those qualified to make them; we rely on the honesty of the experimenter and on the precision of his language in expressing his results. 'The elements in water are combined in the ratio two to one' is not a sufficiently accurate statement to be useful or even meaningful.

This book attempts to emphasise the experimental basis for the study of chemistry and how the study reveals the development of patterns and principles describing the behaviour of substances. Good experimental technique and clear, unambiguous language (including mathematical language, the clearest and least ambiguous of all) are the essential starting-points which are dealt with in this Chapter.

1.2 experimental techniques

1. use of the Bunsen burner

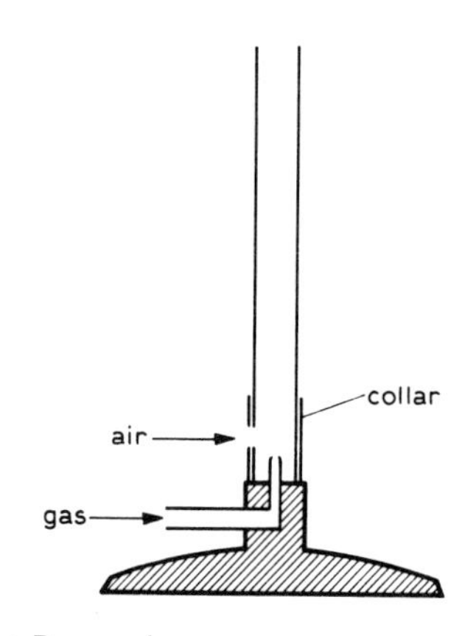

Fig. 1.1 a Bunsen burner

Many chemical operations require a source of heat and the Bunsen burner (Fig. 1.1) is an important piece of laboratory equipment. It enables pure gas or a mixture of gas and air to be burned in air. The air inlet is controlled by a rotating collar; this should always be closed before the burner is lit, giving a luminous, smoky flame. With the air inlet partially open, a silent blue flame is obtained which is the correct flame to use for most purposes. If high temperatures are required, the air inlet should be opened fully; a roaring flame is obtained in which the hottest part is just above the tip of the bright blue cone. When the burner is not being used, it should be adjusted to give a low, luminous flame, or extinguished.

A Bunsen burner should never be left burning under an empty tripod and gauze; the apparatus to be heated should first be placed on the cold gauze, then the burner lit and placed below it. When heating is over, the first operation should always be to remove the burner.

2. glassware

Small hard-glass test-tubes or ignition tubes held by a clothes-peg are best for heating solids; when high temperatures are employed a crucible should be used. Test-tubes of normal size are used for liquids and suspensions and should not normally be more than one-third full. Beakers and flasks (flat-bottomed and conical) are used for liquids, while gases may be handled in gas jars, syringes, or (on the small scale) in corked test-tubes or boiling tubes. The commonest error in early laboratory work is to over-fill vessels–wherever possible choose your apparatus so that it is never more than half-full.

3. heating solids and liquids

Most accidents occur during this procedure–always wear safety spectacles and never look into the mouth of a tube or flask during heating to see if anything has happened. A silent blue flame should be employed at first and this may subsequently be changed to the hotter, roaring flame–though this should rarely, if ever, be necessary when heating liquids. Liquids in test-tubes should be heated near the surface at first, with the tube held at an angle, mouth away from any persons nearby, and kept in constant gentle movement in the flame.

4. transferring liquids

When pouring from a reagent bottle, the label should be kept uppermost. It should not be necessary to put the stopper on the bench; if the receiving vessel is held between the thumb and first two fingers of the left hand, the third and fourth fingers can be used to withdraw and hold the stopper from the reagent bottle–though this may need a little practice. A liquid should never be poured direct from a reagent bottle into a jar of gas–the gas will assuredly get into the bottle and pollute the reagent.

Small quantities of liquids are best transferred using a dropping pipette with a rubber bulb. The glass tube of the pipette must be long enough to permit it to be held firmly by the second, third and fourth fingers, while the bulb is squeezed between the first finger and thumb. A dropping pipette should never be held by the rubber bulb alone: the resulting lack of control is messy and dangerous. One or two drops of liquid can be picked up and dispensed from a dropping pipette by bending the bulb over with the thumb and not squeezing it at all.

5. filtering and centrifuging

These techniques are used to separate a suspension into a solid precipitate and a clear liquid filtrate. When filtering, the paper should be folded in half, then again just short of a quarter (Fig. 1.2) and opened to give a cone having the wider of the two possible angles. The paper should be moistened with a little solvent (usually water) when it is fitted into the funnel. Pouring the suspension down a glass rod (Fig. 1.3) helps accuracy, while the stem of the funnel should be in contact with the side of the beaker receiving the filtrate, as shown. Suspensions filter more rapidly hot than cold.

Fig. 1.2 folding a filter paper

A centrifuge should always be charged with two tubes, one containing the suspension and another containing an equal bulk of water. The centrifuge should always have a shield, which should not be opened until the centrifuge has come to rest. Thus the temptation to hurry the slowing-down process by pressing the finger on top of the spindle–a dangerous malpractice–should never arise. After centrifuging, the filtrate can easily be withdrawn using a dropping pipette. Recovery of the precipitate from the bottom of the tube is much more difficult and the technique is most useful when the filtrate is required rather than the precipitate.

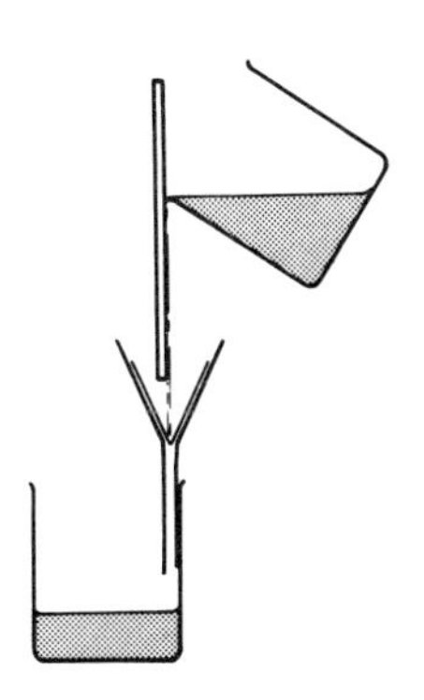

Fig. 1.3 filtration

6. evaporation

Liquids should be evaporated in evaporating basins; beakers, test-tubes and flasks are not adequate. If a solution is to be evaporated to dryness, the evaporation should be completed with the basin sitting on top of a beaker of boiling water. This, and squirting air from an empty dropping pipette over the surface, will minimise 'spitting' of pasty solid.

7. safety precautions

Wear safety spectacles. Wash hands thoroughly with soap and hot water immediately after any practical work, whether you think the substances which you have been handling are poisonous or not. Never leave cases or satchels on the laboratory floor. Never run in a laboratory. Never carry glass rods or tubes in the pockets. If any substances get on your hands, rinse them immediately with water–remember that the amount of material on a chemist's hands is inversely proportional to his ability. When testing a gas for smell, sniff it, don't inhale it. Keep reagents and apparatus towards the back of a bench, not the front. Above all, apply common sense to all laboratory work.

1.3 the balance: mass and weight

So many types of chemical balance are now available that instruction as to their use must be given in the laboratory and will depend on the balance used. All measure the gravitational force exerted on the object to be 'weighed'; a spring balance measures this force, while a beam-balance compares two such forces acting in close proximity (the opposite ends of the beam) and therefore effectively gives a measure of mass.

Mass is a property of a single object; **weight** is a force of attraction between two. The mass of a cricket ball measures its resistance to a change in its motion–it requires more energy to accelerate it than does a tennis ball; once moving, it is more difficult to stop. Its mass does not alter with a change in its surroundings; a blow on the head from a cricket ball will be just as painful in outer space as it is on the surface of the earth. The weight of the ball is the force of attraction between the ball and the earth–in outer space this disappears.

In chemistry we are more concerned with the mass than with the weight of an object and we use a balance to determine masses. The phrase 'to weigh an object' is so commonly used that it has not been avoided in this book, but it should be recognised that the true meaning of the phrase is 'to determine the mass of an object'.

When quoting a mass of five grams, 5 g means $5(\pm 0{\cdot}5)$g, 5·0 g means $5(\pm 0{\cdot}05)$g and 5·00 g means $5(\pm 0{\cdot}005)$g. In other words, 5 means 'nearer 5 than 6 or 4', 5·0 means 'nearer 5·0 than 5·1 or 4·9' and so on. It will save tedious correction and misunderstanding later if this convention is adopted from the outset. It is also wrong to report the result of a calculation as '12·237 g' if the balance employed in the experimental work is capable of measuring mass to an accuracy of only 0·1 g.

1.4 simple mathematics needed by chemists

This section contains some 'tricks of the trade' for chemists; it is not meant to replace a proper understanding of the mathematical ideas quoted. Two aspects of mathematics which can cause difficulty and which are essential to work in chemistry are indices and graphs.

1. indices

Scientists often deal with very large or very small numbers and it is inconvenient to write these out in full. For example: the speed of light in a vacuum is 300 000 000 m s^{-1}, but it is easier to write and remember 3×10^8 m s^{-1}.

The mass of 1 atom of hydrogen is 0·000 000 000 000 000 000 000 0016 g, but it is easier to remember $1{\cdot}6 \times 10^{-24}$ g.

$\mathbf{3 \times 10^8}$ means 3 multiplied by 10, eight times in succession:

$3 \times 10^1 = 30$; $3 \times 10^2 = 300$; $3 \times 10^3 = 3000$; $3 \times 10^8 = 300000000$

$\mathbf{1{\cdot}6 \times 10^{-24}}$ means 1·6 divided by 10, twenty-four times in succession:

$1{\cdot}6 \times 10^{-1} = 0{\cdot}16$; $1{\cdot}6 \times 10^{-2} = 0{\cdot}016$; $1{\cdot}6 \times 10^{-3} = 0{\cdot}0016$; etc.

To multiply 3×10^8 by $1{\cdot}6 \times 10^{-24}$, multiply the figures, but add the indices:

$$(3 \times 10^8) \times (1{\cdot}6 \times 10^{-24}) = (3 \times 1{\cdot}6) \times (10^{-24+8}) = 4{\cdot}8 \times 10^{-16}$$

To divide $1{\cdot}6 \times 10^{-24}$ by 3×10^8, divide 1·6 by 3 and subtract the indices:

$$(1{\cdot}6 \times 10^{-24}) \div (3 \times 10^8) = \frac{1{\cdot}6}{3} \times 10^{-24-8} = 0{\cdot}53 \times 10^{-32}$$

If you use indices, always leave one figure in front of the decimal point, thus:

for $0{\cdot}53 \times 10^{-32}$, write $5{\cdot}3 \times 10^{-33}$
for $12{\cdot}6 \times 10^{20}$, write $1{\cdot}26 \times 10^{21}$

There is no mention of fractions here; decimals are always used in scientific work.

2. *graphs*

It is best in scientific graphs to mark the points with a cross thus +, not as a point or a cross ×.

The best straight line or smooth curve is drawn so that the points are left evenly balanced about the line; rarely, if ever, is it correct to join points in a zig-zag fashion (Fig. 1.4).

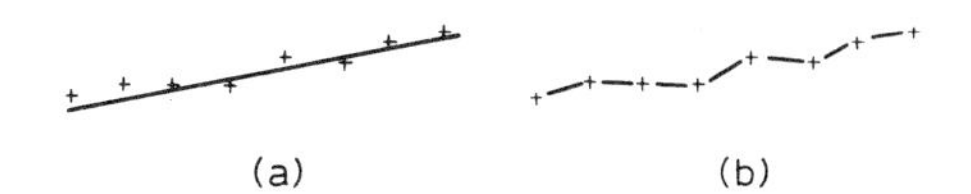

Fig. 1.4 (a) correct, (b) incorrect methods of drawing a 'straight-line' graph

The scale of the axes should be chosen to use as much of the graph paper as possible. Fig. 1.5 shows some correct and incorrect scales.

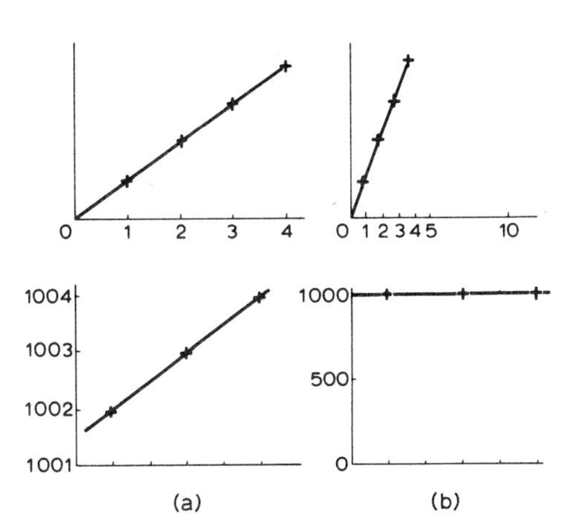

Fig. 1.5 (a) correct, (b) incorrect choices of scale

The interpretation of a straight line graph is often of importance in science. The equation $y = kx$ can be represented by a straight line passing through the origin (0, 0); the equation $y = kx + c$ can be represented by a straight line which meets the y-axis at point $(0, c)$. Examples of these two types of graph are given in Fig. 1.6.

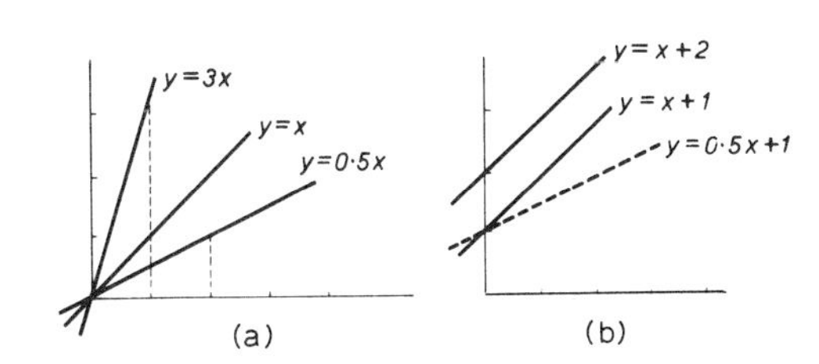

Fig. 1.6 graphs of the form
(a) $y = kx$
(b) $y = kx + c$

The converse argument is true: if a graph is a straight line and passes through the origin, it can be represented by an equation of the form $y = kx$.

In scientific work we use graphs to discover what equations can be written to connect two sets of observations. For example, the volume of a fixed mass of gas is found to increase with temperature. If the volumes occupied by the gas at different temperatures are measured and plotted graphically as shown in Fig. 1.7, a clear relationship emerges. The graph is a straight line passing through the origin and its equation is of the form

$$y = kx$$

Hence
$$V = kT$$

The volume of a fixed mass of gas is directly proportional to its absolute temperature.

The axes of a graph should always be labelled and **each axis should represent a pure number.**

$$\text{a physical quantity} = \text{a number} \times \text{a unit}$$

e.g.
$$V \quad = \quad 5 \quad m^3$$

It follows that

$$\frac{\text{a physical quantity}}{\text{a unit}} = \text{a number}$$

e.g.
$$\frac{V}{m^3} = 5$$

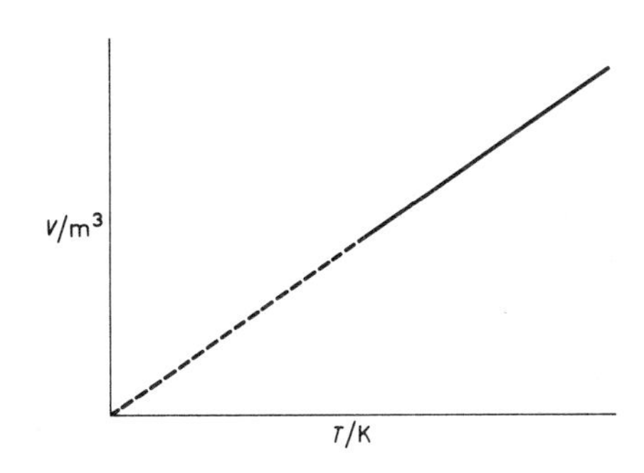

Fig. 1.7 graph showing how the volume (V) of a fixed mass of gas varies with the temperature (T)

If the axis of a graph is labelled by the physical quantity divided by its unit, as shown in Fig. 1.7, the axis will represent pure numbers. This convention is followed in this book.

1.5 units

Scientists have recently agreed to use a consistent system of units for measurement. The system used is called 'Système Internationale' or SI and it is used throughout this book.

Its basic units are the metre (length), kilogram (mass), second (time), ampere (electric current), kelvin (temperature) and mole (amount of substance). All other units are dependent on these – thus the newton (the unit of force) is defined as the force required to accelerate a mass of one kilogram at the rate of one metre per second every second.

Commonly encountered physical quantities and their symbols, together with the units in which they are expressed and the symbols for the units, are shown in Appendix 1. They are introduced throughout the book as they become necessary. However, you should be prepared for some abbreviations. The metre is a convenient unit for expressing the measurement of rooms and gardens, but it is less useful for measuring large or small distances. The prefix kilo is used to mean 'a thousand': the distance from Edinburgh to London is expressed as 650 kilometres rather than as 650 000 metres, and it is written in abbreviated form as 650 km. The width of a matchstick is about 0·0015 metre: here the prefix milli is used, meaning 'the thousandth part of' and the width is expressed as 1·5 millimetre or 1·5 mm. Such prefixes, with their symbols of abbreviation, are shown in Fig. B on page 306; they may be used with any unit.

An important form of abbreviation is used in place of phrases like 'per second' and 'per gram'. If you are walking fairly slowly, you may cover a distance of one metre every second: this is written as $1\,m\,s^{-1}$. The index ($^{-1}$) means 'every so many' or 'per'. Other indices can be used: a volume of two cubic metres is written as $2\,m^3$; a density of 8000 kilogram per cubic metre (about that of iron) is written as $8000\,kg\,m^{-3}$.

Inevitably, sources of information will for some time continue to use units other than those listed in Appendix 1 and a list of some useful conversion factors is printed at the beginning of this book.

1.6 nomenclature

The naming of chemical substances often seems complex, but the systematic method of naming, used in this book, avoids confusion by giving a clear and unambiguous name to each substance. The names are directly connected with the structures of substances in terms of the particles of which they are made; older systems of naming substances could not do this. You will soon realise that the apparently unwieldy name copper(II) sulphate(VI)-5-water conveys much exact and valuable information about the substance formerly known as 'blue vitriol'.

The derivation of systematic names is given in sections 10.14 to 10.17 and section 14.3. A full list of systematic and common names is given in Appendix 2.

1.7 symbols and equations

The authors are not in agreement with the recommendation that a symbol, e.g. S, should be used to represent one atom of sulphur or as a shorthand form meaning merely 'some sulphur'. The practice adopted in this book is that **the symbol S may represent one atom of sulphur or one mole of sulphur**; it will mean the latter when written in an equation. A symbol is not used as a shorthand form meaning 'some of the substance', except as an expedient in cases of dire necessity (e.g. as an abbreviation in a condensed table or when the full name would be unnecessarily unwieldy).

The same practice is used for formulae, so that H_2 may mean one molecule of hydrogen or one mole of hydrogen molecules; it is not used to represent 'some hydrogen'.

Thus— H means 1 mole of hydrogen atoms,
$\frac{1}{2}H_2$ means 0·5 mole of hydrogen molecules.

Fractions are therefore used in equations when it is desirable for clarity.

The arrow ($\rightarrow$) is used in equations rather than the equality sign (=) to denote a chemical change. The physical states of substances are represented in equations as (s) for solid, (l) for liquid, (g) for gas or vapour and (aq) to mean dissolved in water (quantitatively, at infinite dilution). Ionic solids are usually represented simply as, e.g., $Na_2CO_3(s)$; where it seems desirable to emphasise the ionic nature of the solid, the representation $(2Na^+ + CO_3^{2-})(s)$ is used. When only one ionic species in an ionic solid is involved in a reaction, the equation is written in the form

$$CO_3^{2-}(s) + 2H^+(aq) \rightarrow H_2O(l) + CO_2(g)$$

The hydrated hydrogen ion is written in equations as $H^+(aq)$ rather than H_3O^+ or $H_3O^+(aq)$, except for cases in which it is desirable to emphasise that the water is acting as a base.

2.1 the meaning of 'energy'

We use the terms 'energy' and 'energetic' casually in everyday speech. A cereal contains 'wheat, sugar and honey to give real energy'. An 'energetic' person might involve himself in a lot of activities. A footballer at the end of a Cup Final might well be 'drained of energy'. Astronauts are ordered to take rests, in spite of their excitement, to 'conserve their energy'.

In all of these cases, the presence (or absence) of energy is shown by what it can or cannot do; energy cannot be seen, or felt, or weighed. The cereal packet suggests that eating the cereal will enable you to undertake arduous physical exercise like climbing mountains. The footballer, having lost his energy, is unable to run up and down the pitch any more. The astronaut can carry out his allotted programme only if he has the energy to do so.

The scientific meaning of energy is closely related to the everyday sense. Energy is defined in terms of what it can do: **energy is the capacity to cause a change in the condition of the surroundings.** Thus the cereal will change your condition from 'unable-to-climb-mountains' to 'able-to-climb-mountains'. The footballer with no energy cannot change the condition of the ball; he has no energy to kick it.

2.2 forms of energy

The liberal use of energy is characteristic of our present civilisation. Energy is used to build houses, to run factories, to light streets, to make cars go, to cultivate land. Electricity, gas, petrol, coal, are all readily available sources of energy; five hundred years ago man depended on wind, water and animal power to alter his environment.

To find out whether a body possesses energy, we may apply a test suggested by the definition above. *Is the body able to alter its surroundings in any way?*

Consider the case of something moving. If you are running down a street and collide with someone, you might come to a halt and send the other person sprawling. In scientific terms you possessed energy of motion and transferred it to the other person, changing his condition from stationary to moving. Your movement enabled you to change the condition of your surroundings.

An important source of energy is flowing water. From early times it has been used to turn water-wheels to grind corn. Nowadays it is used chiefly to rotate turbines, producing electricity in hydroelectric power stations. In short, moving bodies possess energy; this type of energy is called **kinetic energy.**

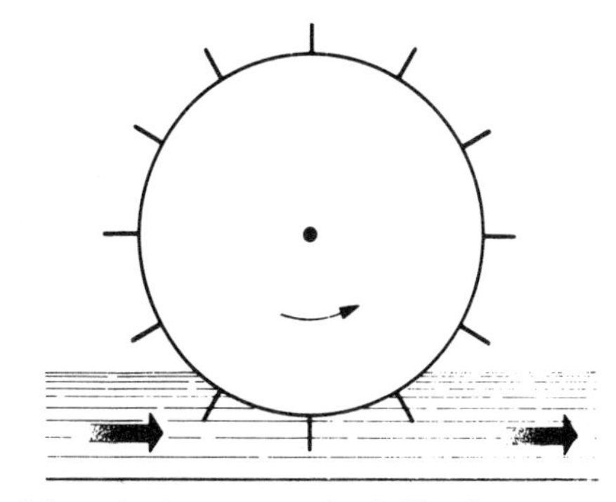

Fig. 2.1 undershot water-wheel: kinetic energy of water turns wheel

Think now of a body stored at a height. A brick placed on your head would have little effect on you. The same brick, dropped by a bricklayer from the top of a house, would have a very considerable effect on your head. The brick

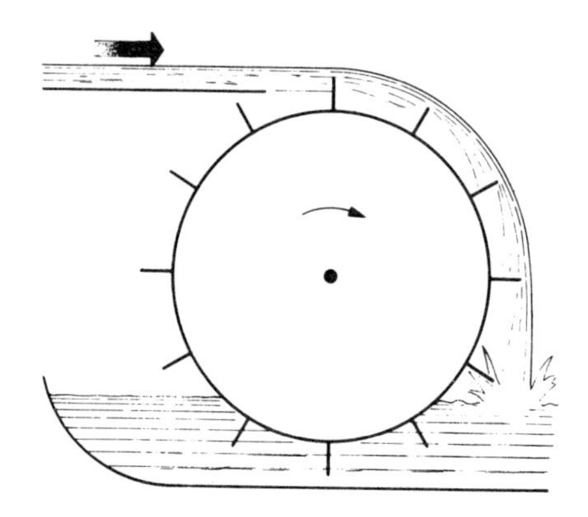

Fig. 2.2 overshot water-wheel: potential energy of water turns wheel

at the top of the house possessed energy not present at ground level. Some water-wheels work on this principle: the weight of water falling from top to bottom of the wheel overbalances it. The energy possessed by a body by virtue of its height is one kind of **potential energy.**

An object which is radiating light also has energy. The light rays reaching you cause a change in your eye and you say that you see the object. If the light intensity is sufficiently high, you may feel pain: the light has caused a change in your condition. A film in a camera exposed to light is altered so that the film may be developed to a negative. Before the exposure, the film would only develop to a uniform white. The light has caused a change in the condition of the film.

Heat and **electricity** are the two forms of energy of most interest to chemists, since they are often obtained from chemical reactions. They will also cause chemical reactions to occur and for this reason a Bunsen burner is one of the most useful pieces of apparatus in the chemistry laboratory. If we apply our original test for the presence of energy (can it cause a change in the surroundings?), we see that heat can certainly cause changes. Heat from hot water will hard-boil an egg. Further heat applied to the hot water will change the water into steam, which can be used to drive an engine. The heat of a flame from a painter's blowlamp will blister paint and make it more easily removed.

Electrical energy is a more recently discovered form which now has very widespread uses: electric fires are used to warm rooms, model trains are made to move by electric motors and this principle is being increasingly extended to power real trains.

The principal forms of energy are **kinetic energy, potential energy, light, heat and electricity.** Matter, sound and magnetism might also be considered as forms of energy: why? In which category should chemical energy (section 2.4) be placed?

2.3 transformation of energy

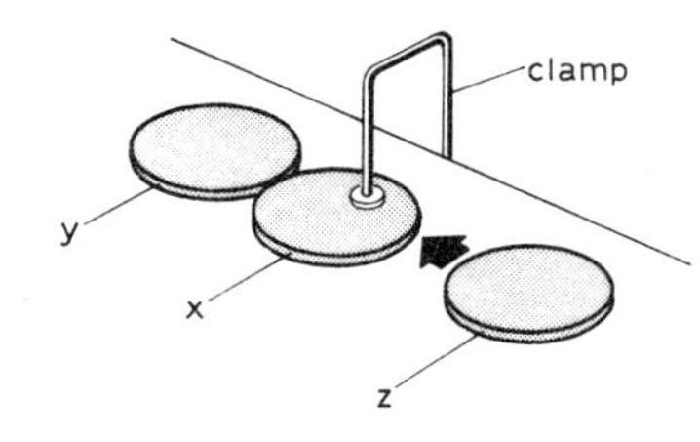

Fig. 2.3

A coin (X) is clamped firmly to a table and another coin (Y) is placed in contact with it. A third coin (Z) is fired at the clamped coin. Y moves away as the motion of Z is arrested. The energy of motion of Z is not lost; it is converted into energy of motion of coin Y.

Energy can be converted from one form to another. For instance, water at the top of a waterfall possesses more potential energy than water at the bottom. As the water falls down, its potential energy (due to height) becomes smaller and its kinetic energy (due to motion) becomes greater. Finally, as it hits the bottom of the fall, the kinetic energy is transformed into heat energy: the water is slightly warmer at the bottom of the waterfall than it was at the top.

Transformation from one form to another is an important characteristic of energy which has been universally recognised and is stated as the **Law of Conservation of Energy:**

Energy is neither created nor destroyed but is only changed from one form to another.

Using this principle, we can trace the passage of energy from its source through many different forms in an energy chain. The main source of energy for the earth is heat from the sun. This is used in many ways, including the evaporation of water from the sea. The water condenses in clouds: it falls as rain and runs in rivers, thus having kinetic energy which may be used to turn a turbine attached to a generator, producing electricity which can be used to heat your home.

2.4 energy in chemistry

The following experiment can be dangerous; it is advisable to use small quantities, to wear safety spectacles and to protect others by the use of a transparent screen.

Zinc powder and sulphur powder are sifted together to form a grey mixture and the mixture is placed on an asbestos sheet. It is ignited at arm's length, preferably using a fuse. A brilliant flash of light occurs and a mushroom-shaped cloud of smoke rises rapidly. The product of the reaction is off-white in colour and is too hot to touch.

The interaction between zinc and sulphur to produce zinc sulphide has evolved a large amount of energy: light seen in the flash, kinetic energy seen in the moving smoke cloud and heat energy observed in the products. Using the principle of conservation of energy, it follows that the zinc and sulphur powders must possess energy stored in them which is released as they react together.

Most chemical reactions release stored energy to the surroundings. A reaction which does not produce obviously different products, as does the zinc–sulphur reaction, may be detected by making use of a thermometer. If the thermometer shows a rise in temperature, a reaction is taking place (see Fig. 2.4).

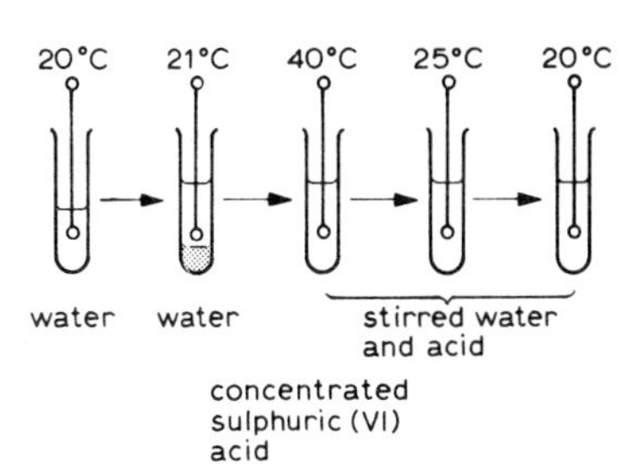

Fig. 2.4 At which stage does a reaction occur?

The energy contained in substances which can be released during a reaction is often called **chemical energy.** Zinc and sulphur powders contain individual particles which are rearranged during the reaction to form a lattice-work of alternating zinc and sulphur particles. Energy is released in the formation of the lattice-work; in general, energy is obtained from chemical reactions as a result of rearranging the particles involved. It is possible that the zinc sulphide produced as described above still contains some stored energy which is not released in its formation, but which may be released in other reactions.

In this country, chemical reactions constitute our principal sources of energy. Fuels such as coal, oil and gas contain carbon and hydrogen which burn in the oxygen of the air, forming two new substances, carbon dioxide and water. The reactions produce a great deal of heat, which may be used directly (as a coal, oil or gas fire heats a room) or indirectly, using the hot gases produced to push a piston in an engine; alternatively, the heat may be used to boil water, the steam produced being used to drive a turbine in an electricity power station.

Our own bodily energy is chemical in nature. The foods which we eat contain carbon and hydrogen which we convert in our muscles to carbon dioxide and water, simultaneously releasing energy for our use. This is the type of energy referred to on the cereal packet, which enables us to climb mountains or run about on a football pitch. We are part of an energy chain which starts with the sun. The chemical processes which release energy from fuels and foods are discussed in more detail in Chapter 19.

2.5 measurement of energy

The total energy possessed by a body cannot be measured; since energy is only detectable when it is transferred, causing a change, we can only measure transferred energy. The situation is similar to what you would see at the counter of a shop: you cannot see how much money is in the till, but you can see how much has been transferred between the till and a customer during a transaction.

The international unit for measuring energy changes is the joule, after James Prescot Joule (1818–1889) who performed the first accurate experiments to determine the equivalence between heat and mechanical energy. Some experiments which determine the amount of energy released in chemical reactions, measured in joules, are described in Chapter 20. It is useful at this stage to see approximately what a joule of energy can do: it is rather a small unit, as Fig. 2.5 shows.

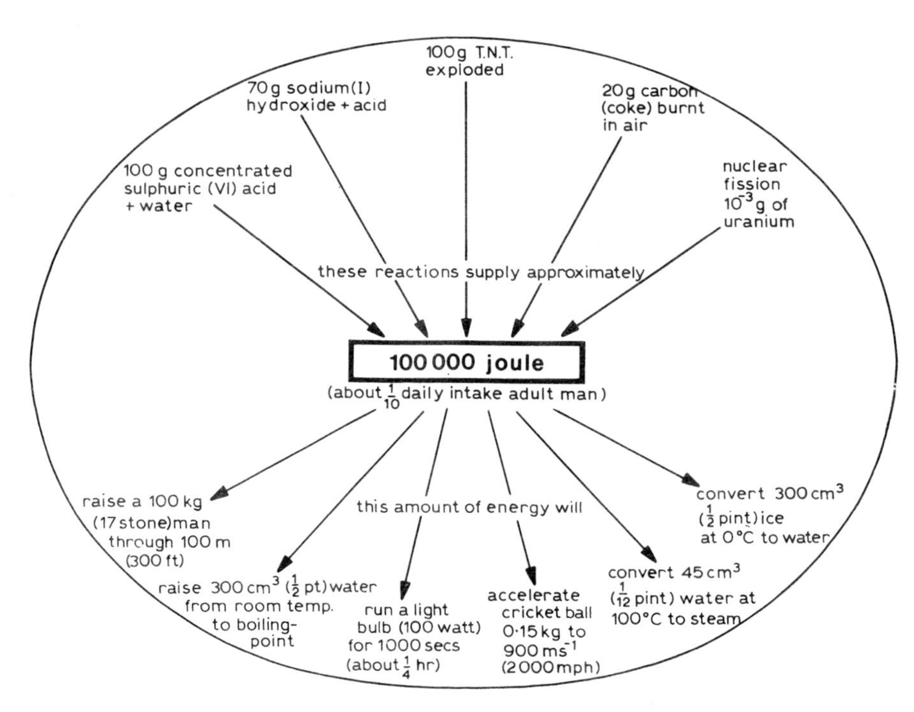

Fig. 2.5

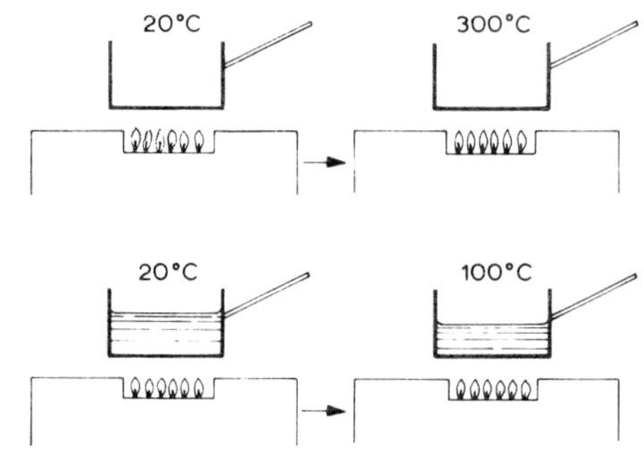

Fig. 2.6

2.6 the action of heat on a familiar substance–water

If an empty saucepan is placed on a stove, the heat energy supplied by the stove will make the saucepan red-hot. If the pan contains water, it does not become red-hot; instead, the water boils. Furthermore, no matter how rapidly the pan is heated, the temperature does not rise above 100 °C (the boiling point of water). This has long been known to cooks: an egg takes the same length of time to cook when placed in boiling water, whether the setting of the stove is high or low. The extra energy supplied by a high flame must be used to some purpose, but it is not being used to raise the temperature of the pan above 100 °C. The only change taking place while the water boils is that water is being turned into steam; it would therefore seem that in turning into steam, water absorbs energy. The various ways in which water can absorb energy are shown more extensively in a laboratory experiment using the apparatus shown in Fig. 2.7.

Crushed ice (30g) is placed in an insulated beaker which is fitted with a stirrer, thermometer and electric heater. Let us suppose that the starting temperature of the ice is observed to be −10 °C. The following changes are observed when the heater is switched on and the changes are summarised graphically in Fig. 2.8.

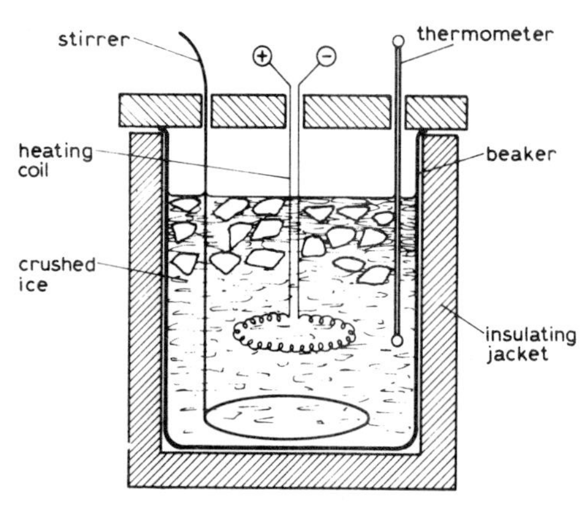

Fig. 2.7 apparatus to determine latent heat

1) The temperature rises steadily to 0 °C.
2) The temperature remains at 0 °C for several minutes. During this period the ice is seen to melt to water.
3) When no ice remains, the temperature begins to rise again and rises steadily to 100 °C.
4) The temperature remains at 100 °C for a very long time, during which steam issues from the top of the beaker.
5) When the beaker is virtually dry the temperature begins to rise sharply.

Since energy is supplied at a constant rate throughout the experiment, it is evident that the energy is absorbed in three different ways:

a) raising the temperature of the beaker and its contents (*stages 1, 3, 5*),
b) converting ice into water (*stage 2*),
c) converting water into steam (*stage 4*).

Some important conclusions may be drawn from this experiment:

i) When ice changes to water it does so at a fixed temperature called its melting point. Ice and water can exist together only at the melting point. Similarly, water changes to steam at a fixed temperature called the boiling point of water (N.B. the actual temperatures of the melting point and boiling point are affected by changes in pressure).

ii) Energy is required to change ice into water and to change water into steam. This energy is called the latent heat of fusion (for melting) or the latent heat of vaporisation (for boiling). The experiment shows that the heater takes a longer time to convert water into steam than to convert ice into water; since the heater supplies energy at a constant rate it follows that the former conversion requires more energy. The latent heat of vaporisation of water is greater than the latent heat of fusion of ice, for the same amount of material.

Experiments similar to that described above may be performed using substances other than water; they show that the conclusions drawn with reference to ice, water and steam may be applied to all solids, liquids and gases.

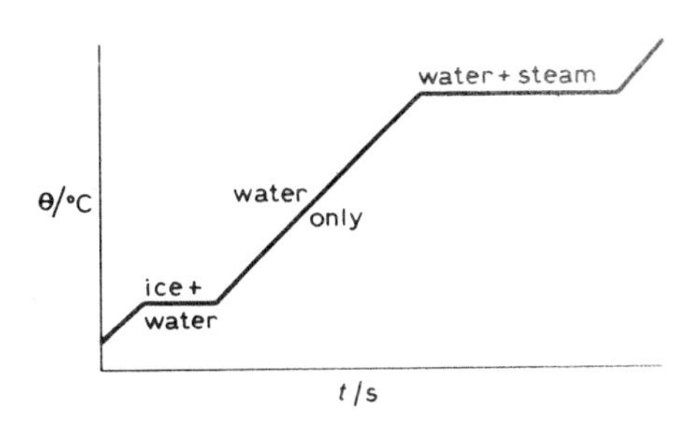

Fig. 2.8 temperature–time graph for energy supplied at constant rate to ice

2.7 measurement of latent heat

The experiment described in section 2.6 may be carried out quantitatively if the rate at which energy is supplied by the electric heating coil is known. Suppose that the power of the heater is 50 watts, i.e. it supplies energy at a rate of 50 joules every second. The beaker containing some water is first weighed, dry crushed ice is added and the beaker and contents are re-weighed, the apparatus is assembled and the heater is switched on. The time for which the temperature remains steady at 0 °C is measured, as is the time for which the temperature remains steady at 100 °C, while a certain amount of water is allowed to boil away as steam. Finally, the beaker and contents are weighed at the end of the experiment. A typical set of results is given below:

mass of beaker and water	= 120·0 g
mass of beaker, water and ice	= 150·0 g
mass of beaker after some water has boiled	= 145·0 g
time taken to convert ice into water	= 210 s
time taken to convert water into steam	= 225 s
mass of ice used	= 30·0 g
mass of water converted to steam	= 5·0 g

conversion of ice into water:

30 g of ice require 50 W for 210 s

30 g of ice require $50 \times 210\,\text{J} = 10\,500\,\text{J}$

$$1\,\text{g of ice requires } \frac{10\,500}{30} = 350\,\text{J}$$

The (specific) **latent heat of fusion of ice is 350 J g^{-1}**

conversion of water into steam:

5·0 g of water require 50 W for 225 s

5·0 g of water require $50 \times 225\,\text{J} = 11\,250\,\text{J}$

$$1\,\text{g of water requires } \frac{11\,250}{5} = 2250\,\text{J g}^{-1}$$

The (specific) **latent heat of vaporisation of water is 2250 J g^{-1}**

2.8 differences between solids, liquids and gases

Sections 2.6 and 2.7 give evidence that a given mass of the same substance contains more stored energy in the gaseous state than it does in the liquid state and that it contains more stored energy in the liquid state than it does in the solid state. It is useful at this stage to review the well-known differences between solids, liquids and gases.

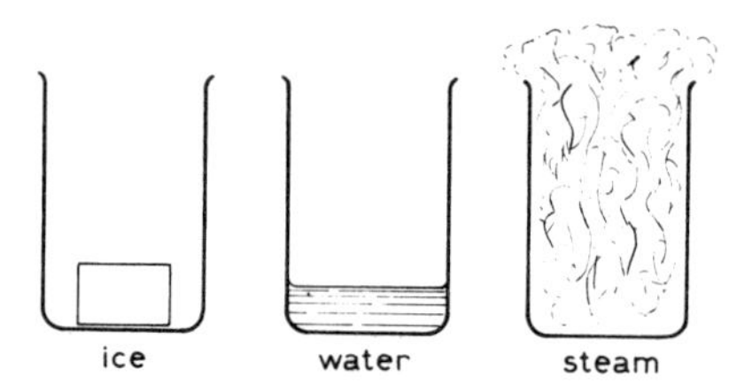

Fig. 2.9 physical properties of solid, liquid and gaseous water

Fig. 2.10 (a) solid tyres, incompressible–uncomfortable, (b) pneumatic tyres, compressible–comfortable

Fig. 2.11 (a) air-filled football–low mass, easy to kick, (b) water-filled football–high mass, bruised toe

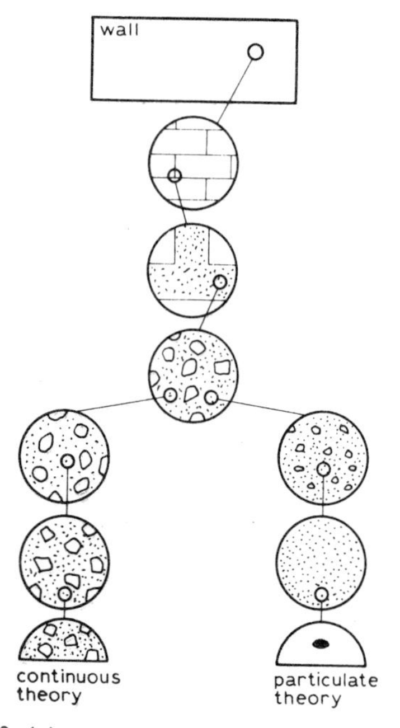

Fig. 2.12 (a) can we go on dividing for ever? (continuous theory), or (b) can there be a final limit? (particulate theory)

If a cube of ice, some water and some steam are each separately placed in beakers, certain differences are immediately apparent. The ice keeps its own shape; it does not expand or contract, but keeps its original volume. The water covers the bottom of the beaker and takes up the shape of the beaker. The shape of the water is altered by placing it in the beaker but its volume is not altered. The steam is seen to escape from the beaker; it has no shape of its own, neither has it a definite volume. It disperses so that it occupies any space which is available to it.

A solid has a fixed volume and a fixed shape. A liquid has a fixed volume but no fixed shape. A gas has neither a fixed volume nor a fixed shape.

If you put your finger over the end of a bicycle pump full of air, you will be able to push the handle in for some distance before your finger is forced off. If the pump is filled with water instead of air, you will not be able to push the handle in at all. **Gases can be compressed readily. Solids and liquids cannot be compressed.**

The other main difference between the three states of matter is in their **densities.** Gases have very low densities; a little mass occupies a lot of space. Liquids and solids have higher densities and more mass is required to occupy the same volume. The range of densities for gases (at s.t.p.) is from about 0·1 $kg\,m^{-3}$ for hydrogen to about 17 $kg\,m^{-3}$ for uranium hexafluoride. All known solids and liquids are more dense, from 100 $kg\,m^{-3}$ for hydrogen to 22000 $kg\,m^{-3}$ (22 tonnes m^{-3}) for osmium.

The division of matter into three states–solid, liquid and gas–is one of convenience; there are a number of borderline cases. Ice in glaciers takes up the shape of its valley and flows slowly downhill. Glass in old windows and lead on old roofs is found to be thicker at the bottom than at the top. Gases under high pressure become increasingly similar to liquids. You may be able to think of other substances which are difficult to classify as solid, liquid or gas.

2.9 composition of matter

To explain the differences between the three states of matter we must look carefully at the way in which matter is composed.

From a distance, a wall may appear to have a flat and uniform surface. As we approach more closely, we can see the bricks and mortar of which it is composed. Still closer examination reveals the individual grains in the mortar. If the grains were further magnified, would we always be able to see further divisions, or is there a fundamental particle which cannot be sub-divided? This question was argued for two thousand years, but not until John Dalton put forward his atomic theory in 1803 was it possible to design experiments to decide between the alternatives.

We now have more evidence than was available to Dalton, but we agree with his main conclusions. **Matter is composed of particles** which are not capable of sub-division. On magnification, it would look more like a sand-pie (which cannot, without difficulty, be divided further than into grains of sand) than like a jelly (which can, apparently, be divided without limit).

2.10 evidence for the existence of particles

1) *The laws of chemical combination* can be more easily explained if matter is assumed to be composed of particles (see Chapter 10).

2) *Brownian motion.* This can be demonstrated in an apparatus represented diagrammatically in Fig. 2.13. Smoke is suspended in air in a small cell fitted to the stage of a microscope. The cell is illuminated from the side and viewed from the top. The smoke particles appear as points of light against a dark background (think of dust in a ray of sunlight). These points of light are

seen to jump about in all directions, changing their motion rapidly and randomly. If the air in which the smoke is suspended is made up of particles, then the abrupt changes of direction of smoke particles can be visualised as resulting from collisions between the smoke particles and particles of gas in the air. This is the simplest direct evidence in favour of the particulate theory; the abrupt changes of direction of the smoke particles are difficult to explain using a continuous theory.

3) In this century, *X-rays* have been used to look at crystals. The patterns obtained are regular arrangements of spots on a photographic plate. The regular arrangement of the spots may be simply explained in terms of interaction between the X-rays and regular arrays of particles in the crystal.

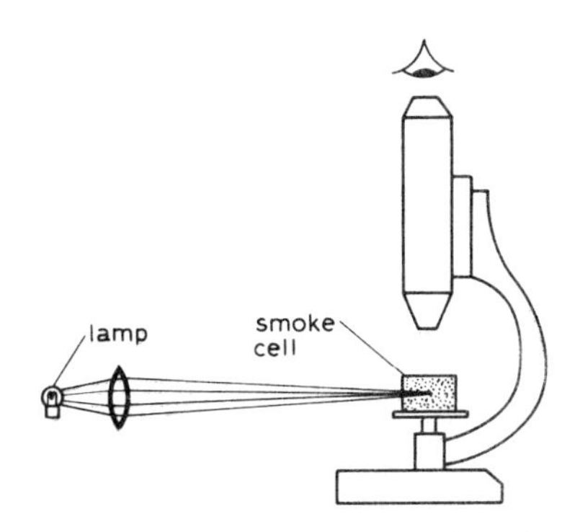

Fig. 2.13 Brownian motion: observation of illuminated smoke particles through microscope

2.11 the nature of the particles in matter

The particles with which we are concerned at this stage are called **molecules**: a molecule is defined as **the smallest unit of a substance which can exist independently.** Each molecule is made up of one or more **atoms,** an atom being defined as **the smallest particle of an element which can take part in a chemical change.** The atoms in a molecule are bound tightly to each other.

Solid ice, liquid water and gaseous steam all contain molecules of the chemical substance water; these molecules are composed of two hydrogen atoms and one oxygen atom and the formula for water is said to be H_2O (see Chapter 9). The molecule is the characteristic unit which is responsible for the chemical properties of the substance; ice, water and steam all have the chemical properties of water. Only if the molecule is broken into fragments do substances with different chemical properties appear.

If the energy needed to divide up water into smaller and smaller units is measured, certain distinct stages emerge and each stage requires successively more energy. The stages are summarised in the following table.

separation of	name of process	energy required (approximately)
1 g water into droplets	splashing	less than 1 joule
1 g water into molecules	boiling	2000 joules
1 g water molecules into individual atoms of hydrogen and oxygen	decomposition	50000 joules

Individual molecules are so small that they cannot be seen under a microscope. The following experiment shows how the approximate size of a large molecule can be determined in the laboratory.

2.12 assessment of the size of a molecule of oil

A layer of oil one molecule thick is formed on the surface of water; the thickness of the oil film is estimated and this gives an indication of the size of an oil molecule.

A clean water surface is produced by connecting a filter funnel with a tap as shown in Fig. 2.14. A little talcum powder is sprinkled on to the surface of the water. One drop of a solution of oil in petrol ether is added to the water surface from a dropping pipette; the petrol ether evaporates, leaving a film of oil on the surface. The diameter of the oil film is measured using a ruler. The number of drops in 1 cm^3 of oil solution is calculated by counting drops from the dropping pipette into a burette. The oil solution should contain 0·1 cm^3 of oil in 1 dm^3 of solution in petrol ether. The thickness of the oil

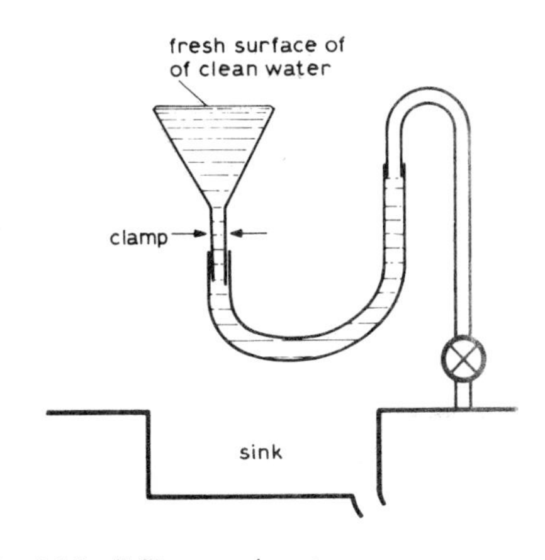

Fig. 2.14 oil-film experiment

film in cm (t) may be estimated as follows:

approximate diameter of oil film	$\doteq$ 6 cm
approximate area of oil film	$=$ 25 cm^2
approximate volume of oil film	$= 25 \times t\,\text{cm}^3$ (1)
number of drops of oil solution in 1 cm^3 of solution	$=$ 20 drops

1000 cm^3 of oil solution contains 0·1 cm^3 of oil

1 cm^3 of oil solution contains 1×10^{-4} cm^3 of oil

1 drop of oil solution contains $\dfrac{1 \times 10^{-4}}{20} = 5 \times 10^{-6}\,\text{cm}^3$ of oil (2)

From the two values (1) and (2) for the volume of a drop of oil, it follows that

$$25 \times t = 5 \times 10^{-6}$$

$$t = \frac{5}{25} \times 10^{-6}\,\text{cm}$$

$$t = 2 \times 10^{-7}\,\text{cm}$$

The molecule of oil cannot be larger than 2×10^{-7} cm (20 Ångstrom units).

2.13 three states of matter—a molecular picture

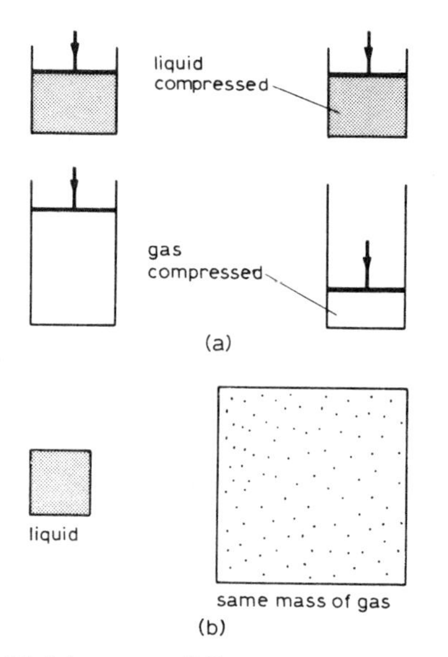

Fig. 2.15 (a) compressibility and (b) density of a gas compared with a liquid

Let us suppose that molecules are widely separated from each other in gases but are close together in solids and liquids. Let us further suppose that molecules attract each other and that the force of attraction between molecules becomes greater as the molecules approach each other more closely.

Solids and liquids have a fixed volume because there is no space between the molecules. Gases, having space between the molecules, can be made to reduce this space on compression. The weak attractive forces between the widely separated molecules in a gas cannot resist expansion.

Gases have low densities because of the considerable distances between molecules. A volume containing molecules widely separated from each other will possess less mass than the same volume containing molecules which are closely packed together.

The molecules in a solid occupy fixed positions about which they vibrate, resulting in solids having a characteristic rigidity. The molecules in a liquid are more randomly arranged and can 'tumble over' each other as the liquid is poured. The forces of attraction between the molecules in a gas are weak, those in a liquid are stronger and those in a solid stronger still; these differences are caused largely by the different distances of separation of molecules in the three states of matter.

When a solid is converted into a liquid, energy must be supplied to break down the close-packed, uniform arrangement of molecules characteristic of the solid state. When a liquid is converted into a gas, a large amount of energy must be supplied to overcome the attractive forces between the fairly close-packed molecules in the liquid state and separate them to the considerable distances apart characteristic of a gas.

revision summary: Chapter 2

energy:	capacity to cause a change in the condition of the surroundings
main forms:	kinetic, potential, heat, light, electrical, chemical
Law of Conservation of Energy:	energy is neither created nor destroyed but may be changed from one form to another

energy and chemistry:	substances can release energy as particles are rearranged in reactions
solids:	fixed volume and shape, incompressible, generally high densities
liquids:	fixed volume, no fixed shape, incompressible, moderately high densities
gases:	no fixed volume or shape, compressible, low densities
change of state:	solid → liquid and liquid → gas: require energy
molecule:	smallest unit of a substance which can exist independently
atom:	smallest particle of an element which can take part in a chemical change

questions 2

1. Describe the following events in terms of energy transformation and/or transfer, stating carefully how the energy was stored before and after the event.
(a) a bonfire burns,
(b) a stone falls from the top of a cliff,
(c) a game of 'conkers',
(d) a bomb explodes,
(e) a nuclear power station operates,
(f) a plant grows,
(g) a boy leaves his house and catches a bus to school.

2. Contrast the properties of solids, liquids and gases when examined for (a) rigidity, (b) density, (c) fluidity. Show how a stone, a glass of water and a balloon full of air can be used to illustrate these properties.

3. How do the following differ from each other?
(a) a bar of lead from a typical solid,
(b) 'silly putty' from a typical liquid,
(c) polystyrene foam from a typical solid.

4. Investigate the properties of a thick paste made by adding a large amount of cornflour, slowly, to a little water. Is it a solid or a liquid?

5. If a small quantity of ether is placed on one's skin a cold sensation is felt. Explain this effect. (S)

6. The evolution of gas in a chemical reaction is called 'effervescence'; the evolution of bubbles of vapour from the body of a liquid is called 'boiling'. How would you tell the difference between two solutions, both at about 100 °C, one of which is effervescing and the other of which is boiling?

7. Describe an experiment which could be used to find the specific latent heat of fusion of water, pointing out especially the errors which might arise in the experiment.

8. Two liquids, prop-1-yl ethanoate and water, have very nearly the same boiling-point. 1 g of each, in separate test-tubes, was placed in a brine bath at 120 °C. The prop-1-yl ethanoate boiled away completely in 17 seconds but the water took 110 seconds to boil away completely.
(a) Which has the larger specific latent heat of vaporisation? Justify your answer.
(b) If the specific latent heat of vaporisation of water is 2200 J g^{-1} what is the approximate value of this property for prop-1-yl ethanoate?

9. Calculate the specific latent heat of vaporisation of (a) oxygen; 4·0 g evaporated by 100 W supplied for 9·2 seconds, (b) hydrogen; 6·0 g evaporated by 30 W supplied for 10 seconds, (c) carbon dioxide; 3·0 g evaporated by 600 J.

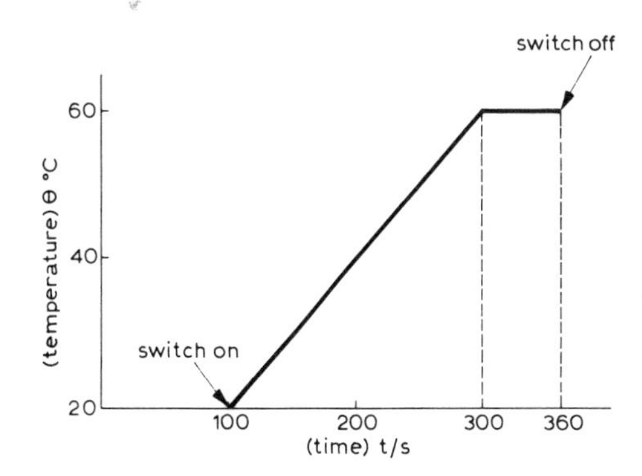

Fig. 2.16 see question 10

10. A substance X is heated in a beaker using a 50 W heating coil, with the results shown in Fig. 2.16. At the end of the experiment the total mass of beaker and contents had decreased by 7·5 g. What is (a) the boiling-point of X, (b) the specific latent heat of vaporisation of X? Do the results tell you anything about the melting-point of X? What is the significance of the horizontal section of the graph? Would this temperature be maintained indefinitely?

the behaviour of gases

Since the molecules in gases are widely separated, they have little effect on each other. It should be expected that different gases have certain properties in common.

3.1 the effect of heating a gas

cold

hot

Fig. 3.1 a simple dilatometer

The next time that you go out in a car, you can try an experiment. Before you start, measure the tyre pressure. Measure it again after the car has been going for some distance, especially after high-speed running, and you will notice that it has increased. You will also notice that the tyre is hot to the touch. It can be shown in the laboratory that these two effects are definitely linked. If a gas is heated in a fixed volume, such as the car tyre, its pressure increases.

What happens if the pressure is held fixed instead of the volume and the gas is again heated? The result can be demonstrated easily in the laboratory, using the apparatus shown in Fig. 3.1. A flask containing air is fitted with a stopper carrying a capillary tube which contains a drop of water. If the flask is placed in hot water, the drop is observed to move upwards. The volume of the gas increases under the constant pressure of the atmosphere, when the temperature is increased.

The volume of the gas can also be altered by simply increasing the pressure on the container. You can show this by putting your finger over the end of a bicycle pump and pushing on the handle. As the pressure applied increases, the volume of the air trapped in the pump decreases.

3.2 experimental investigation of pressure, volume and temperature relationships in gases

1. the effect of pressure on the volume of a gas at constant temperature

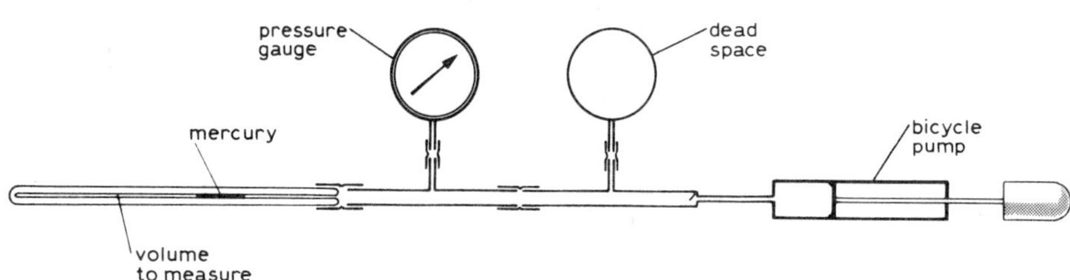

Fig. 3.2 variation in volume of a gas with changing pressure, at constant temperature

The apparatus used is shown in Fig. 3.2. The small volume of air to be investigated is trapped behind a bead of mercury in a capillary tube. The pressure is recorded on the pressure gauge, the dead space minimising small fluctuations in pressure as the bicycle pump is operated. As different pressures are obtained, the lengths of the air column trapped behind the mercury are measured. The volume of air can be calculated by multiplying this length by the cross-sectional area of the tube. All the apparatus is at a constant room temperature. The table below shows some readings which were obtained using the apparatus of Fig. 3.2.

p represents the pressure of air in the column, l the length of the air column and V the volume of air in the column.

cross-sectional area of capillary tube $= 1 \times 10^{-6} \mathrm{m}^2$ ($= 1 \mathrm{mm}^2$)

p/kPa	100	150	200	250	300	350	400	450
l/m	0·200	0·135	0·100	0·080	0·067	0·058	0·051	0·045
$10^6 V/\mathrm{m}^3$	0·200	0·135	0·100	0·080	0·067	0·058	0·051	0·045
$10^6 pV$/J	20·0	19·8	20·0	20·0	20·1	20·3	20·2	20·2

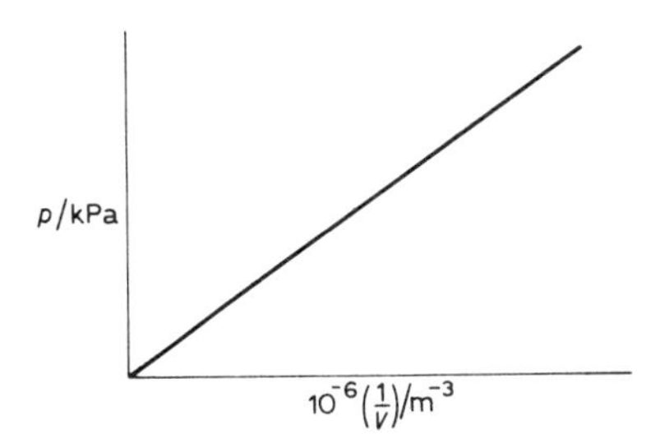

Fig. 3.3 since graph is of form $y = mx$, or $y/x =$ constant, it follows that $pV =$ constant

From these figures, we can conclude that **the pressure of a fixed mass of gas at constant temperature multiplied by its volume is a constant,** or $pV =$ constant (m, T). This is shown graphically in Fig. 3.3; the plot of p/kPa against $10^{-6}(1/V)/\mathrm{m}^{-3}$ gives a straight line passing through the origin (see section 1.4). The value for the constant in this experiment is about $20 \times 10^{-6} = 2 \times 10^{-5}$ J. This law, called **Boyle's law** after its discoverer, applies to any fixed mass of gas at constant temperature. These two conditions must be quoted as part of the law.

Pressures have been given here in kilopascals (kPa). These are the recently agreed scientific units for pressure, but in everyday life you may see other units used. The following list gives some approximate factors for conversion:

Atmospheric pressure = 100 kilopascals (kPa) (*science laboratory*)
(1 atmosphere)
= 760 mm of mercury (*barometer*)
= 15 lb/in² (*tyre gauge*)
= 1000 millibar (*weather chart*)
= 1 kg force/cm² (*continental tyre gauge*)
= 33 feet (11 m) of water (*diving, gas mains*)

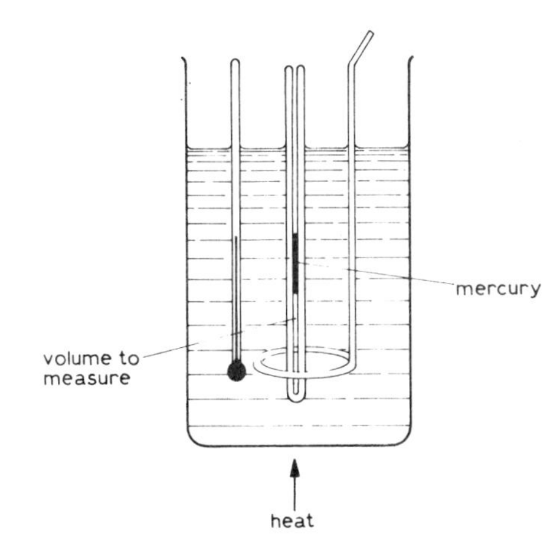

Fig. 3.4 variation in volume of a gas with changing temperature, at constant pressure

2. *the effect of temperature on the volume of a gas at constant pressure*

A volume of air is trapped behind mercury in a capillary tube as shown in Fig. 3.4. In this experiment it is mounted vertically in a water bath which can be varied in temperature from 0 °C to 100 °C. As the water bath is warmed, it is stirred vigorously. Readings of the lengths of the trapped column of air are taken, with readings of temperature, about every 10 °C. Low-temperature readings can also be taken by surrounding the capillary tube with an ice and salt mixture or other coolants giving even lower temperatures. Readings obtained using the apparatus of Fig. 3.4 are shown in the following table; t represents temperature, l length of air column, V the volume of the air column and T the temperature converted to the altered scale as subsequently explained.

If the temperature in degrees Celsius is plotted against volume as shown in Fig. 3.5(a), the graph predicts that at a certain temperature the volume of the gas will be zero. This is clearly a special temperature in the behaviour of a gas. The results at the top of the following page give this temperature as −273 °C. The altered scale of temperature given is calculated by adding 273 to the Celsius temperature.

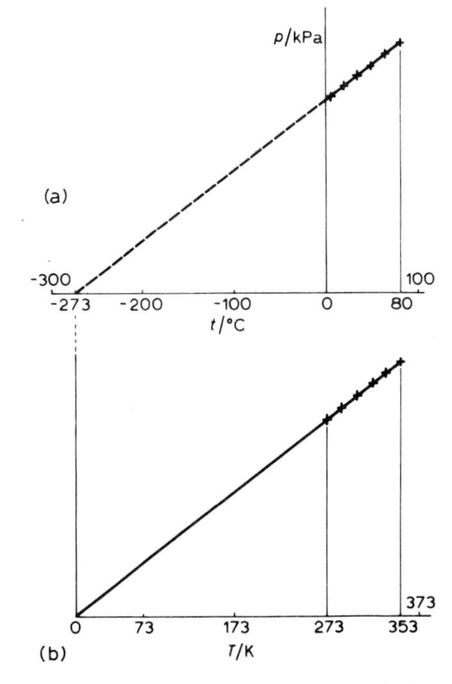

Fig. 3.5 graph (b) using the 'altered scale' of temperature, is of form $y = mx$ or $y/x =$ constant. It follows that $V/T =$ constant

cross-sectional area of capillary tube $= 1 \times 10^{-6} m^2$ ($= 1 mm^2$)

$t/°C$	0	10	20	30	40	50	60	70	80
l/m	0·182	0·189	0·196	0·202	0·208	0·215	0·222	0·229	0·235
$10^6 V/m^3$	0·182	0·189	0·196	0·202	0·208	0·215	0·222	0·229	0·235
T/K	273	283	293	303	313	323	333	343	353
$10^6 \left(\frac{V}{T}\right) / m^3 K^{-1}$	0·67	0·67	0·66	0·67	0·66	0·67	0·67	0·67	0·66

Now the results are easier to analyse. If the volume is divided by the altered temperature, a constant value is obtained. In this experiment, it is $6{\cdot}7 \times 10^{-7} m^3 K^{-1}$. Graphically, the plot of $10^6 V/m^3$ against T/K gives a straight line passing through the origin (Fig. 3.5(b)). The new scale of temperature is called the **absolute scale**, and its units are called **kelvins** (K) (sometimes, degrees Absolute). Celsius temperatures are denoted by t and absolute temperatures by T.

Thus, $$T/K = t/°C + 273$$

for example $-100°C = 173 K$, $0°C = 273 K$, $273°C = 546 K$

The conclusion drawn from this experiment, sometimes called **Charles's law,** is that **the volume occupied by a fixed mass of gas at constant pressure, divided by its absolute temperature, is constant,** or $V/T =$ constant (m, p). This law holds if the mass of gas is fixed and its pressure is constant; these conditions must be quoted when the law is stated.

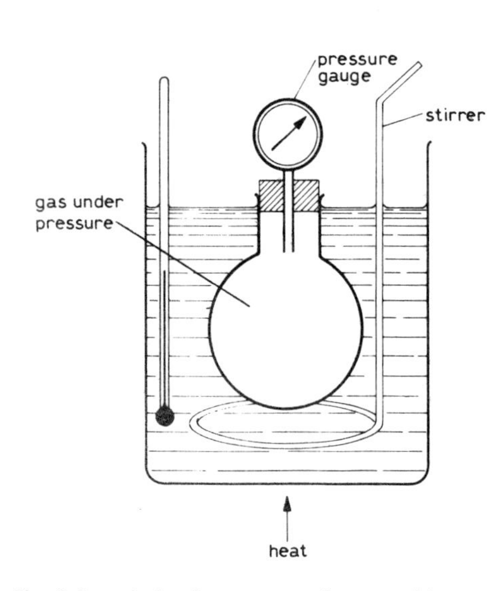

Fig. 3.6 variation in pressure of a gas with changing temperature, at constant volume

3. the effect of temperature on the pressure of a gas at constant volume

A large flask, fitted with a stopper and pressure gauge, is immersed in a water bath as shown in Fig. 3.6. The bath is heated slowly from 0 °C with constant stirring. Readings of pressure and temperature are taken at intervals of about 10 °C, or 10 K. The following results were obtained using the apparatus shown in Fig. 3.6; the meanings of the symbols should be clear from the earlier sections (1) and (2).

$t/°C$	0	10	20	30	40	50	60	70	80
p/kPa	100	104	108	111	115	118	122	126	130
T/K	273	283	293	303	313	323	333	343	353
$\left(\frac{p}{T}\right) / kPa K^{-1}$	0·37	0·37	0·37	0·37	0·37	0·37	0·37	0·37	0·37

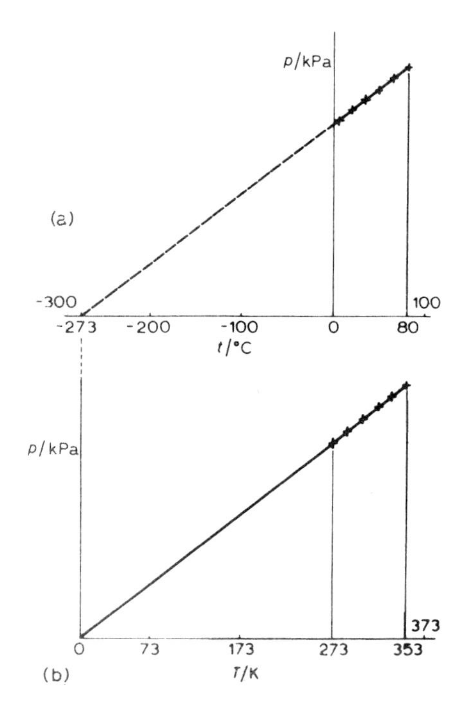

Fig. 3.7 graph (b) is of the form $y = mx$, or $y/x =$ constant. It follows that $p/T =$ constant

If a graph of pressure against temperature (Celsius) is plotted (Fig. 3.7(b)) it is exactly similar to Fig. 3.5(a), while Fig. 3.7(b) is exactly similar to Fig. 3.5(b). The conclusion is similar, that **the pressure of a fixed mass of gas at constant volume divided by its absolute temperature, is constant,** or $p/T =$ constant (m, V).

3.3 relationships between the pressure, volume and temperature of a gas

1. volume varying with pressure

As shown in the previous section, when the pressure of a gas increases, its volume decreases. If a mass of gas is at a pressure p_1 and occupies a volume V_1 and it is then compressed or expanded to a pressure p_2 and a volume V_2 while the temperature is kept constant, the exact relationship between the original and final pressures and volumes is

$$p_1V_1 = p_2V_2 \quad \text{because } pV = \text{constant } (m, T)$$

In words: **the volume of a fixed mass of gas at constant temperature is inversely proportional to its pressure.**

Example: 100 strokes of a bicycle pump of volume 0·12 dm³ are required to fill a bicycle tyre, the final volume of which is 1·8 dm³. The air drawn into the pump must originally be at atmospheric pressure (100 kPa). What is its final pressure? (Assume that time is allowed for the air to cool to its original temperature.)

original pressure $p_1 = 100\,\text{kPa}$
final pressure p_2 to be found
original volume $V_1 = 100 \times 0{\cdot}12 = 12\,\text{dm}^3$
final volume $V_2 = 1{\cdot}8\,\text{dm}^3$

$$p_1V_1 = p_2V_2$$
$$100 \times 12 = p_2 \times 1{\cdot}8$$
$$p_2 = \frac{100 \times 12}{1{\cdot}8} = 667\,\text{kPa}$$

The final pressure is 667 kPa

2. volume varying with temperature

It was also shown in the previous section that when the temperature of a gas increases its volume increases. The law connecting temperature and volume was originally stated by Gay-Lussac in the form 'for every °C rise or fall in temperature, the volume of a fixed mass of gas at constant pressure alters by 1/273 of its volume at 0 °C'. Thus if a gas has a volume of 273 cm³ at 0 °C, its volume will vary as follows:

V/cm^3	273	274	275	283	546	272	271	0
$t/°\text{C}$	0	+1	+2	+10	+273	−1	−2	−273

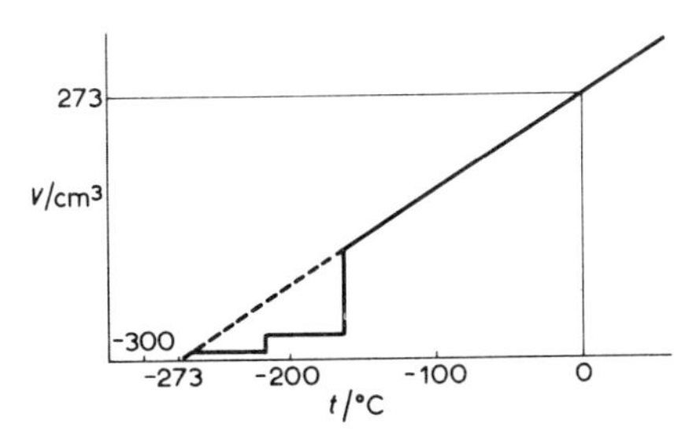

Fig. 3.8 variation of volume with temperature for carbon monoxide—the broken line shows the theoretical values, the continuous line the real values

The most noticeable thing about this table is the last value: at −273 °C a gas, theoretically, has no volume. (In practice it liquefies before being cooled to this temperature.) When the temperature is raised from −273 °C to 0 °C, the gas expands by a certain amount; if this rise in temperature is doubled, from −273 °C to +273 °C, the volume of gas doubles also. Therefore the law connecting volume and temperature is very simple, if the temperatures are measured from −273 °C. Such a scale is the absolute scale of temperature.

In exact terms, $V/T = \text{constant } (m, p)$ where V = volume of gas and T = absolute temperature. In words: **the volume of a fixed mass of gas at constant pressure is directly proportional to its absolute temperature.** Absolute temperatures are measured in kelvins, and are related to Celsius temperatures by the relationship $T/\text{K} = t/°\text{C} + 273$. Zero kelvin (0 K, −273 °C) is called the absolute zero, and is believed to be the lowest attainable temperature.

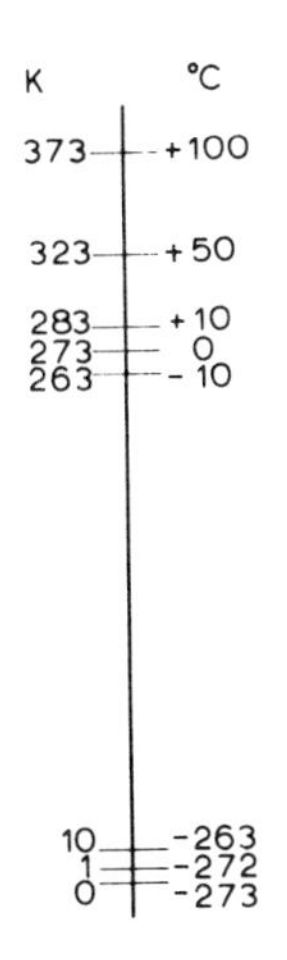

Fig. 3.9 Kelvin and Celsius temperature scales

3. pressure varying with temperature

Lastly, as temperature is increased, the pressure of a gas held in a fixed volume increases. In fact pressure varies with temperature in the same way as does volume, as described above. If p represents the pressure of a gas and

T its temperature then the exact law relating the pressure and temperature of a gas is $p/T =$ constant (m, V). In words: **the pressure of a fixed mass of gas at constant volume is directly proportional to its absolute temperature.**

An example should make use of absolute temperature clear. A road tanker is filled with ethene gas from cold store at $-23\,°C$ and 1000 kPa. As the tanker travels, the contents warm to $27\,°C$. Is it safe to send the tanker on the road if it has been tested as safe to 2000 kPa?

original pressure $p_1 = 1000\,kPa$
final pressure p_2 to be found
original temperature $t_1 = -23\,°C$
final temperature $t_2 = 27\,°C$

$$T_1 = (-23 + 273)\,K = 250\,K$$
$$T_2 = (27 + 273)\,K = 300\,K$$
$$\frac{p}{T} = \text{constant}; \quad \text{hence } \frac{p_1}{T_1} = \frac{p_2}{T_2}$$
$$\frac{1000}{250} = \frac{p_2}{300}$$
$$p_2 = 1200\,kPa$$

This is less than the maximum pressure at which the tanker can travel with safety.

3.4 the general gas law

All the three laws described in the preceding section can be combined into a single law called the general gas law. For a constant mass of gas, at pressure p, absolute temperature T and occupying a volume V,

$$\frac{pV}{T} = \text{constant}$$

This may be expressed in a more useful form for two sets of readings $p_1V_1T_1$ and $p_2V_2T_2$, as

$$\frac{p_1V_1}{T_1} = \frac{p_2V_2}{T_2}$$

Example: a gas at a pressure of 200 kPa and a temperature of $127\,°C$ is expanded and cooled to 100 kPa and $27\,°C$. If its original volume was $80\,dm^3$, what is the new volume?

$p_1 = 200\,kPa$ $\qquad p_2 = 100\,kPa$
$V_1 = 80\,dm^3$ $\qquad V_2$ to be found
$T_1 = 273 + 127 = 400\,K$ $\qquad T_2 = 273 + 27 = 300\,K$

$$\frac{200 \times 80}{400} = \frac{100 \times V_2}{300} \quad \text{or} \quad V_2 = 120\,dm^3$$

The new volume is $120\,dm^3$

All gases are very much alike in that the volume of a gas depends on its pressure and temperature, but not on the chemical constitution of the gas. This is because the molecules of the gas are widely separated from each other in comparison with their sizes. The molecules therefore have little effect on each other. Thus properties like pressure and volume, which are determined by the number of molecules present and the energy possessed by them and are independent of the identity of the molecules, are common to all gases.

3.5 a reference state

Scientist Smith announces that 10 g of substance A decomposes to give $2{\cdot}0\,dm^3$ of gas B. Scientist Jones repeats the experiment and finds only $1{\cdot}9\,dm^3$. These two results are not in conflict, since no temperature or pressure is quoted. Jones's 1·9 dm³ could, at a lower pressure, be Smith's 2·0 dm³.

For this reason, volumes of gases are always referred to the same state of pressure and temperature when an experiment is reported. The conditions chosen are 273·18 K (0 °C) and 101·325 kPa (1 atmosphere); these conditions are **standard** (or normal) **temperature and pressure (s.t.p.).**

Example: 10 dm³ of gas at 100 kPa and 100 °C are converted for reference to s.t.p. What volume does the gas occupy at s.t.p.?

$$p_1 = 100\,\text{kPa} \qquad p_2 = 101{\cdot}325\,\text{kPa}$$
$$V_1 = 10\,\text{dm}^3 \qquad V_2 \text{ to be found}$$
$$t_1 = 100\,°\text{C} \qquad t_2 = 0\,°\text{C}$$
$$T_1 = 373\,\text{K} \qquad T_2 = 273\,\text{K}$$
$$\frac{p_1V_1}{T_1} = \frac{p_2V_2}{T_2}$$
$$\frac{100 \times 10}{373} = \frac{101{\cdot}32 \times V_2}{273}$$
$$V_2 = 7{\cdot}41\,\text{dm}^3$$

The volume at s.t.p. is 7·41 dm³

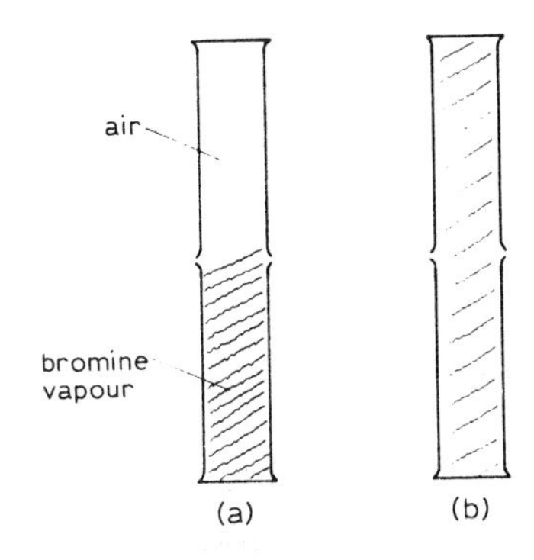

Fig. 3.10 diffusion of bromine vapour and air: the change from (a) to (b) takes about half an hour

Gases which obey the general gas law exactly are called ideal gases. Helium is an ideal gas at room temperature, as are other gases with low boiling-points. Easily liquefied gases like carbon dioxide are not ideal and at high pressures occupy smaller volumes than those predicted by the general gas law. The inaccuracy is small but significant; after reading section 18.18 it might be possible to explain non-ideal behaviour in gases.

3.6 diffusion

Spontaneous mixing on contact is another property common to all gases; it is also observed in liquids. The process in both cases is known as diffusion.

A jar of bromine vapour, mouth upwards, is placed mouth to mouth with a jar of air, mouth downwards (Fig. 3.10). Bromine is a dense vapour and should not 'flow uphill'. However, after half an hour, both jars are brown in colour, showing the presence of bromine in both. The bromine has diffused into the air and the air into the bromine.

A few potassium(I) manganate(VII) crystals in the bottom of a jar of water form a purple solution which diffuses throughout the rest of the water. The time taken, however, is about a month.

Different gases diffuse at different rates, as can be shown in the apparatus illustrated in Fig. 3.11.

The porous pot is an unglazed piece of pottery which possesses numerous tiny holes. These holes are too small to allow draughts to blow through, but allow diffusion to take plac

If a gas jar of hydrogen (which is less dense than air) is placed over the porous pot as shown in Fig. 3.11(a), bubbles are seen coming from the end of the tube in the water. If carbon dioxide (which is more dense than air) is used as shown in Fig. 3.11(b), the water rises in the tube.

These results can be explained by supposing that **the less dense a gas is, the more rapidly it diffuses.** In Fig. 3.11(a), the less dense hydrogen must diffuse into the porous pot more rapidly than the air can diffuse out. The extra pressure inside the pot causes gas to bubble out of the delivery tube. Conversely, the air in Fig. 3.11(b) must diffuse out of the pot more rapidly than the more dense carbon dioxide can diffuse in. The decreased pressure inside the pot causes water to rise up the tube.

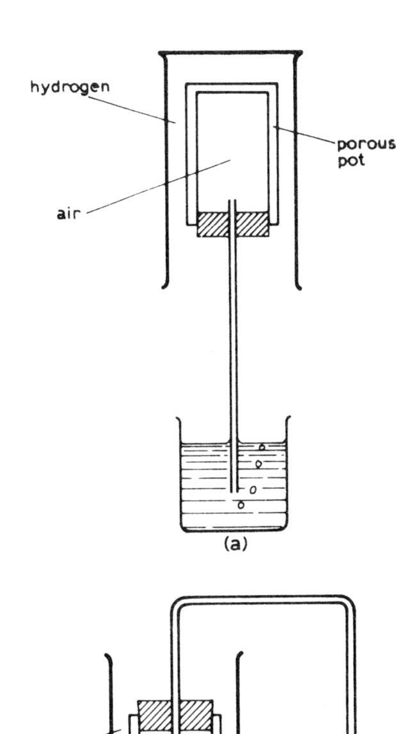

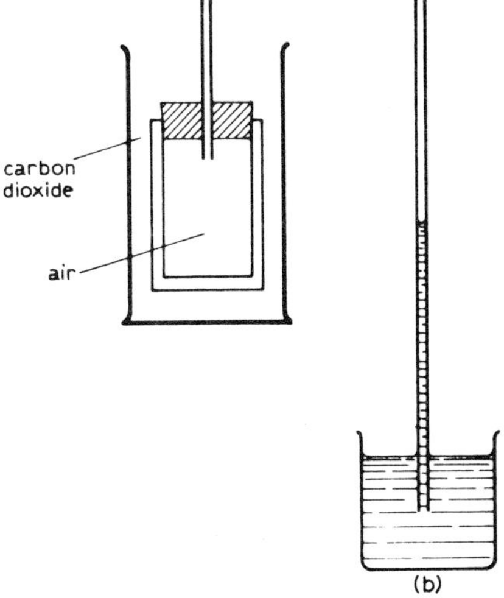

Fig. 3.11 diffusion experiments

3.7 the kinetic molecular theory of gases

The motion of molecules of gases is described by the kinetic molecular theory.

The theory is supported by evidence which has already been described and it can be used to explain the gas laws and other properties of gases. The main postulates of the theory are as follows:

i) the molecules of a gas are in continuous chaotic motion at high speeds (at 273 K, the hydrogen molecule moves at about $18\,km\,s^{-1}$). This can be deduced from observations on Brownian motion (see section 2.10). The smoke particles under observation are made up of millions of atoms; if one molecule of a gas, or even a few molecules, can make such a large particle jump about, the molecule must be moving at a very high speed.

ii) the molecules of a gas, when not actually in collision, are a long way apart compared with the size of a molecule. In support of this, it can be calculated that one molecule of water, as a liquid at 100 °C, occupies a volume of $3 \times 10^{-29}\,m^3$, while the same molecule as steam occupies a volume of $5000 \times 10^{-29}\,m^3$. We may calculate that, since $\frac{5000}{3} \approx 1700 \approx 12^3$, water molecules in steam are about twelve times further apart than they are in liquid water. Because of this separation, one molecule has very little effect on neighbouring molecules and all gases obey the same physical laws.

iii) the energy possessed by a given number of molecules in a gas is high compared with the energy possessed by the same number of molecules of the same substance in the liquid state. About 2000 joules must be supplied to convert 1 gram of liquid water into steam. This is more than six times as much energy as is required to heat the water from room temperature to boiling-point.

iv) no energy is lost by the molecules of a gas when they collide. If it were, then assuming the truth of (vi) below, a gas would cool down spontaneously and all gases would long ago have liquefied.

v) the pressure exerted by a gas on its container is caused by collisions of the molecules with the walls of the container. This can be likened to the pressure exerted by a sand-blast on any article being cleaned in its jet. The pressure experienced is caused by millions of grains of sand impinging on the article.

vi) the faster the molecules are moving, the higher the temperature of the gas; more precisely, the average kinetic energy of the molecules in a gas is proportional to the absolute temperature of the gas.

3.8 applications of the kinetic molecular theory

i) **Boyle's law, pV = constant** The pressure of a gas is caused by collisions of the molecules with the container. Reducing the size of the container will mean that molecules collide with the walls more often. More frequent collisions cause a greater pressure–therefore the pressure increases as the volume decreases.

ii) **p/T or V/T = constant** On increase of temperature the molecules move more rapidly. If the volume remains the same, the faster-moving molecules make more frequent and more energetic collisions with the walls of the container, so that the pressure increases. On the other hand, in order to keep the pressure constant when the molecules have more kinetic energy, the volume increases so that the molecules have further to travel and therefore collide with the container less frequently.

iii) **diffusion** Rapid random motion of widely separated particles ensures rapid mixing of gases. The rate of diffusion of a gas in air is much lower than the actual speed of the molecules would predict, because the diffusing gas makes so many collisions with the molecules in air. This is similar to the experience you would have if trying to run down a crowded street. Your running speed may be 15 m.p.h., but you would collide with so many people that your overall progress as a velocity in a given direction might be only 4 m.p.h. You would also be very unpopular.

revision summary: Chapter 3

General Gas Law: $\frac{p \times V}{T} = $ a constant, or $\frac{p_1V_1}{T_1} = \frac{p_2V_2}{T_2}$
(fixed mass of gas)

Boyle's Law: (m, T constant) $p \times V = $ a constant,
or $p_1V_1 = p_2V_2$

Charles' Law: (m, p constant) $\frac{V}{T} = $ a constant,
or $\frac{V_1}{T_1} = \frac{V_2}{T_2}$

Law of Pressures: (m, V constant) $\frac{p}{T} = $ a constant,
or $\frac{p_1}{T_1} = \frac{p_2}{T_2}$

standard temperature and pressure: (s.t.p.) 273 K or 0 °C, 101·325 kPa

diffusion: rate of diffusion is high if density of gas is low

kinetic molecular theory of gases: molecules in continual chaotic motion
widely spaced from each other
collisions (perfectly elastic) cause pressure
average kinetic energy proportional to absolute temperature

questions 3

1. The volume of a bicycle tyre is 1280 cm³. A boy pumps up the tyre using 64 strokes of a pump, the volume of which is 160 cm³, using air at 101 kPa pressure. What is the final pressure in the tyre?

2. Calculate the figures omitted from the following table, all being for a constant mass of gas.

	original values			*final values*		
	p/kPa	V/cm³	t/°C	p/kPa	V/cm³	t/°C
a)	100	290	17	100	?	37
b)	110	50	2	?	50	27
c)	95	43	29	?	19	29
d)	40	273	7	100	?	7
e)	80	160	47	90	?	−3
f)	111	222	−73	?	1369	127
g)	70	41	14	55	60	?
h)	2	2	2	128	?	277
i)	105	202	273	*find the volume at standard temperature and pressure*		
j)	140	55·5	21			

3. A room in a house obviously has a constant volume. Such a room was kept at a constant temperature, but it was observed that the air pressure varied with the weather. Is this contrary to Boyle's law? Justify your answer.

4. 250 cm³ of nitrogen is collected at 102 °C. What volume will it occupy if it is cooled to 27 °C, at the same pressure?

5. Describe an experiment to determine the connection between the volume and temperature of a constant mass of gas held at a constant volume. What sources of error are there? What precautions are taken to minimise these? What conclusion can be drawn from such an experiment?

6. What does the kinetic molecular theory of gases state about (a) the speed, (b) the direction, (c) the energy of movement, of molecules? How does it help to explain (i) the increase of pressure with increase of temperature for a constant mass of gas, (ii) the observation that gases diffuse in all directions, (iii) Brownian motion?

7. Do you think that all molecules of a given gas at a given temperature and pressure move with the same speed all the time? What is meant by saying that hydrogen molecules have a speed of 18·4 km s⁻¹? Why does hydrogen diffuse in air much more slowly than this quoted speed?

8. Draw a diagram to show how you would use two gas jars, one containing a colourless gas of vapour density 14 and the other a green gas of vapour density 35·5, to illustrate gaseous diffusion.

State what you would see after leaving your apparatus intact for two days. (JMB)

mixtures and compounds

4.1 formation of complex substances

Chemists are concerned with the effects of adding different substances to each other; the product may be interesting or useful, or energy, which can be put to good use, may be released by the reaction. If ammonia gas is added to sulphuric(VI) acid, a solid product (ammonium sulphate(VI)) which is an efficient fertiliser can be isolated. Natural gas and air interact when ignited, releasing considerable energy which can be used domestically.

In this Chapter we look at the various effects which are obtained when simple substances are mixed under a variety of conditions to form more complex substances.

4.2 sodium and chlorine

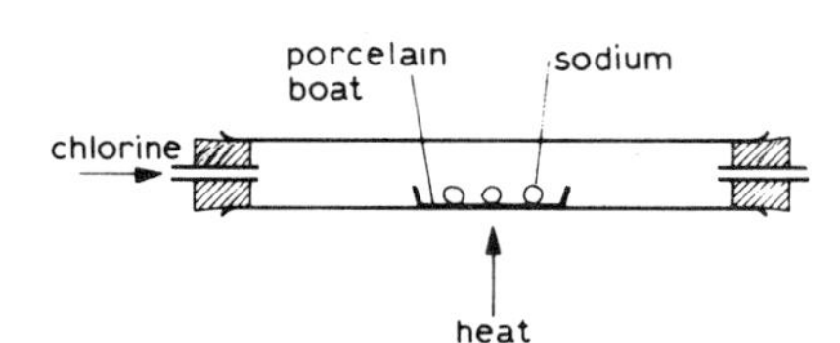

Fig. 4.1 action of chlorine on warm sodium

Chlorine gas is passed over warm sodium in a hard glass tube as shown in Fig. 4.1. The sodium burns with a bright yellow flame, which continues until no more sodium can be seen in the porcelain boat. The white solid product in the boat, when cooled and finely ground, is seen on examination with a lens to be quite different from the original metallic sodium and the green chlorine gas. The product also has a different action on water from that of either the original sodium or chlorine. The differences in properties of the two original substances and the product of the reaction are summarised in the following table.

property	sodium	chlorine	product
physical state	solid	gas	solid
colour	silver	green	white
action on water	fizzes, melts, moves round surface	dissolves giving green solution	dissolves giving colourless solution

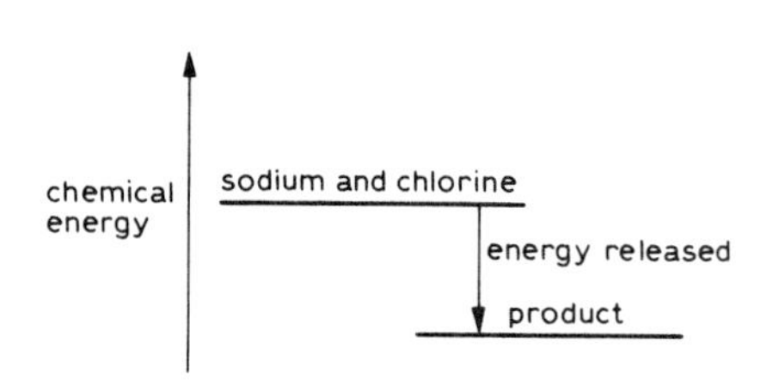

Fig. 4.2 when energy is released in a process the product contains less energy than the original substances

The simple substances sodium and chlorine interact to produce a product which has properties different from either of the two simple substances. Energy in the form of heat and light is given out and the product must therefore contain less chemical energy than the original simple substances (see Fig. 4.2).

4.3 instant coffee and desiccated coconut

If these two familiar substances are stirred thoroughly together, there is no obvious energy change. Examination of the product with a hand lens reveals separate particles of coffee and coconut. Observations on the nature of the separate substances and the product on mixing are summarised in the following table.

property	instant coffee	desiccated coconut	product
physical state	solid	solid	solid
colour	dark brown	white	light brown
taste	bitter	sweet	bitter and sweet
action of water	dissolves giving brown solution	insoluble: white solid floats on surface	brown solution and white solid floating on surface

The simple substances instant coffee and desiccated coconut stirred together form a light brown solid which has the properties of both of the original substances. No energy is released in the mixing process.

Fig. 4.3 some substances show no energy change on mixing

4.4 'quicklime' and water

Drops of water are added from a dropping pipette to a piece of 'quicklime' in an evaporating dish. The solid becomes hot and emits a hissing sound and clouds of steam appear: obviously energy is released when water is added to quicklime. If the addition of water is continued until no more steam is produced a crumbly white solid remains ('slaked lime'). If the product is touched with blotting paper it is seen to be dry–all of the added water has been absorbed. Observations on the nature of the quicklime, water and slaked lime, are summarised in the following table.

property	quicklime	water	slaked lime
physical state	solid	liquid	solid (powder)
colour	white	colourless	white
action of heat	none	boils at 100 °C giving steam	steam given at temperature much higher than 100 °C

The simple substances quicklime and water interact evolving energy, giving a product which has properties different from those of either of the original substances.

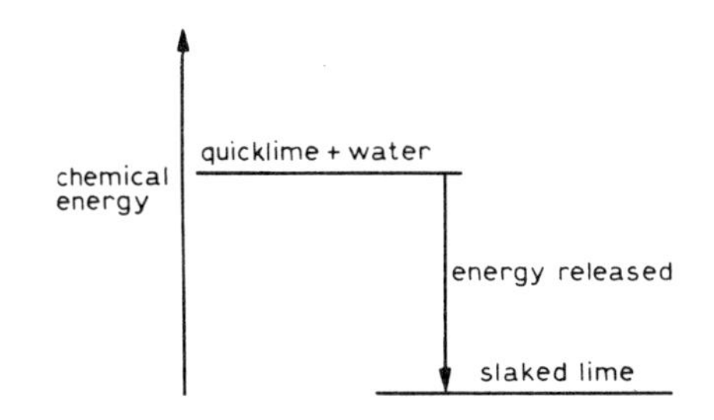

Fig. 4.4 slaked lime contains less chemical energy than the quicklime and water from which it is formed

4.5 nickel and sulphur

Equal volumes of nickel and sulphur powders are mixed gently but thoroughly in an evaporating basin using a plastic spoon. A yellowish-grey product is obtained with no accompanying energy change. The original nickel powder, the sulphur powder and this first product (which we will call product A) are each separately subjected to the tests outlined in the table on the following page.

A porcelain boat is filled with some of the product A and placed across the corner of a tripod. One end of the boat is heated with a Bunsen burner. The powder suddenly becomes red-hot and a glow spreads through the whole of

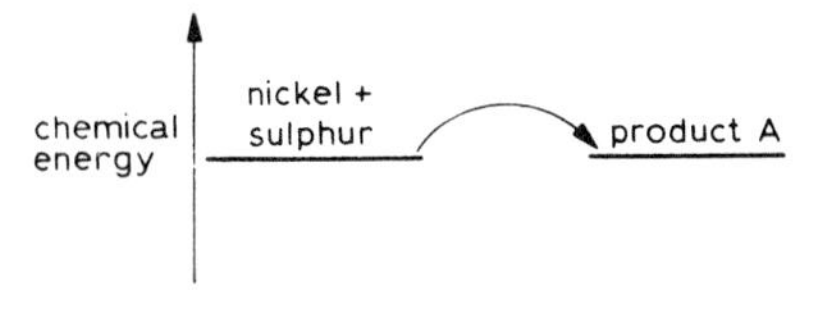

Fig. 4.5 A contains as much energy as nickel and sulphur

tests on nickel powder, sulphur powder, product A and product B

property or test	nickel	sulphur	product A	product B
physical state	powder	powder	powder	powder
colour	grey	yellow	yellow-grey (different particles apparent through lens)	grey (no different particles apparent through lens)
powder sprinkled on paper, magnet moved below paper	powder moves with magnet	no effect	powder separated by magnet	no effect
powder warmed with dilute hydrochloric acid	effervescence, green solution	no effect	effervescence, green solution, yellow residue	effervescence, green solution
gas tested by (a) smell (*care!*)	no strong smell		no strong smell	smell of bad eggs
(b) lighted splint	burns with squeak or pop		burns with squeak or pop	burns, no squeak, yellow deposit
(c) moist lead(II) ethanoate paper	no effect		no effect	paper turns black
(*in fume-cupboard, no exposed flames*) powder shaken with carbon disulphide (*care–poisonous fumes*)	liquid remains colourless	yellow solution	yellow solution	liquid remains colourless
resulting liquid decanted from any remaining solid on to watch-glass,	grey solid residue	no residue	grey solid residue	grey solid residue
liquid allowed to evaporate	no residue on watch-glass	yellow crystals formed	yellow crystals formed	no residue on watch-glass

(N.B. Similar results may be obtained using powdered cobalt in place of powdered nickel. Iron filings will also suffice in place of nickel, but the observations are less well defined) the powder, even after the Bunsen burner has been removed. The grey solid remaining in the boat (which we will call product B) is cooled, finely ground and subjected to the tests outlined in the table on the previous page.

Conclusions from the nickel–sulphur experiment:

i) Product A and product B are both more complex substances than either nickel or sulphur.
ii) Product A is formed from nickel and sulphur with no obvious energy change and has the composite properties of nickel and sulphur.
iii) Product B is formed from nickel and sulphur with a release of energy in the forms of heat and light and has properties different from those of either nickel or sulphur.

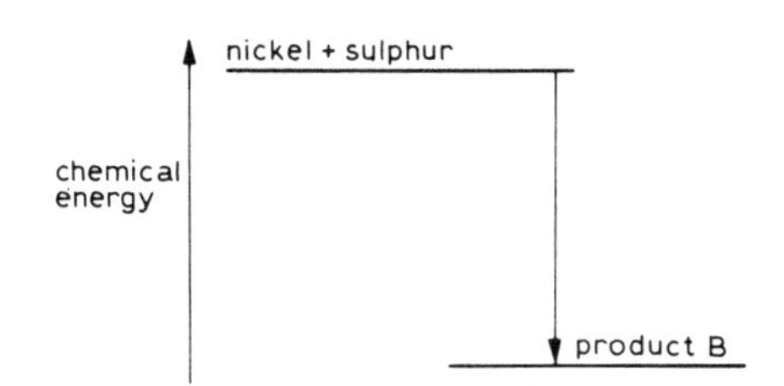

Fig. 4.6 B contains less energy than nickel and sulphur

A product which has the composite properties of the components and is formed without an energy change is called a mixture.

A product which has properties different from those of either of the component substances and which is formed with an accompanying energy change is called a compound.

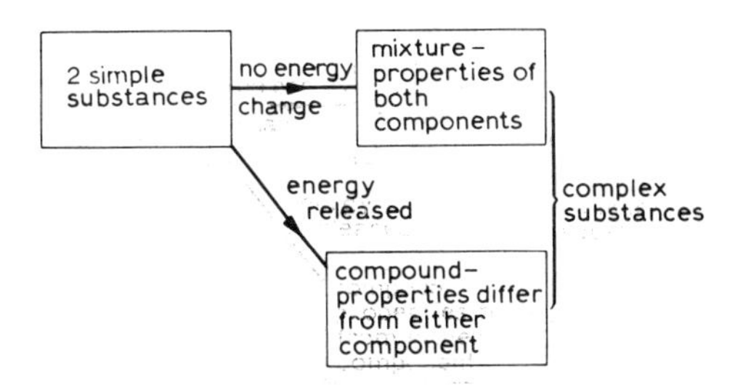

Fig. 4.7 characteristics of mixtures and compounds

Which of the products described in sections 4.2, 4.3, 4.4 and 4.5 are mixtures and which are compounds?

4.6 melting points of mixtures and compounds

When a solid is heated it usually turns to liquid at a well-defined temperature called its melting point. If the hot liquid produced is allowed to cool it becomes solid (i.e. freezes) at the same temperature as the original solid melted. The melting point of a solid and the freezing point of the corresponding liquid are synonymous terms. Melting points can be measured using the apparatus shown in Fig. 4.8. The capillary tube containing the finely powdered sample may be fixed to the thermometer using a rubber band, but it will usually stay in position without this. The beaker and its contents are heated slowly using a low flame, while the paraffin is stirred vigorously. The temperature at which the substance in the capillary tube is seen to melt is taken as the melting point of the sample.

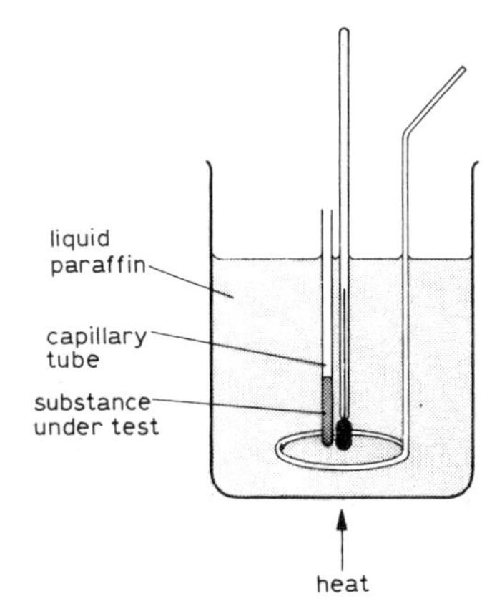

Fig. 4.8 melting point apparatus

A solid **compound** treated in this way is found to melt suddenly and completely when the paraffin bath reaches a certain temperature, i.e. it has a 'sharp melting point'. A solid **mixture** melts slowly over a wide range of temperature, i.e. it does not have a sharp melting point.

This fact is often used to check the purity of a compound which has been prepared in the laboratory. If the compound is pure it will have a sharp melting point which is characteristic of that compound. An impure compound (i.e. a *mixture* of the compound with some impurity) will not have a sharp melting point. **An impure compound always melts at a lower temperature than the corresponding pure compound,** even though the melting point of the impurity may be higher. Acetanilide as usually prepared in the laboratory will melt over the temperature range 109 °C–112 °C, whereas the melting point of pure acetanilide is 114 °C.

'Antifreeze' added to water in a motor-car radiator lowers the freezing point of the water below the temperature of the surrounding atmosphere so that the water in the radiator remains in the liquid state at a temperature at which pure water would freeze.

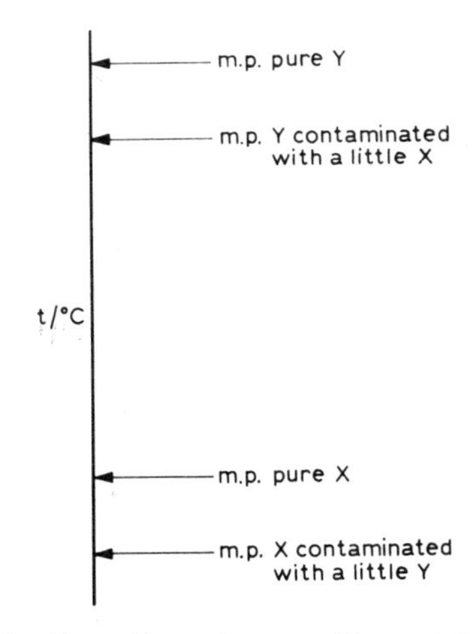

Fig. 4.9 effect of impurity on melting point

4.7 boiling points of mixtures and compounds

If water (a pure compound) and salt solution (a mixture of water and common salt) are heated in flasks as shown in Fig. 4.10 (on the next page), the water boils at 100 °C and the reading on the thermometer remains constant, while the salt solution boils at a temperature above 100 °C and the temperature rises as steam is given off. Three general points are illustrated by this experiment:

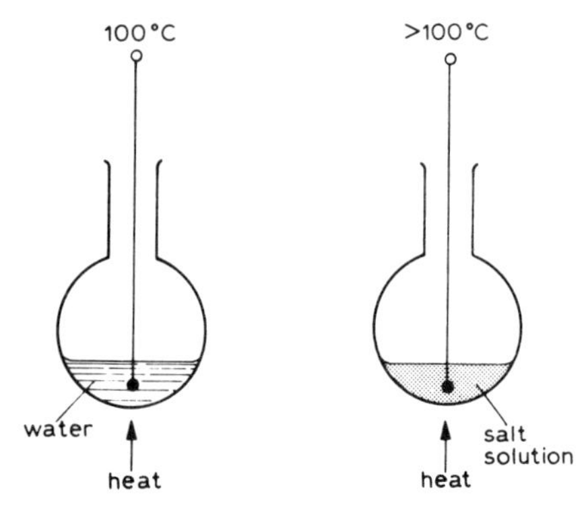

Fig. 4.10 boiling point of water and salt solution

i) A pure compound has a definite and constant boiling point (at constant pressure).
ii) An impure compound has a boiling point higher than the pure compound, provided that the impurity is non-volatile.
iii) The boiling point of a mixture is not constant. The actual value of the boiling point of a mixture depends upon the proportions in which the components are mixed.

4.8 further differences between mixtures and compounds

In general, mixtures are separated into their components much more easily than are compounds. This subject is dealt with fully in the next Chapter. Also, the components of a mixture can be present in any ratio, whereas the components of a compound are present in a fixed ratio by mass. Compounds are said to have a definite composition; this subject is discussed more fully in Chapter 10.

revision summary: Chapter 4

complex substances	
compounds	mixtures
formation accompanied by energy change	formation not accompanied by energy change
properties different from components	properties composite of those of components
sharp melting point	melt over temperature range
constant boiling point	boil over temperature range
difficult to separate into components	easy to separate into components
definite composition	variable composition

questions 4

1. Arrange the following complex substances into two lists, one labelled 'compounds' and the other labelled 'mixtures': water, tomato sauce, wood, salt, ice, whisky, paint, sugar.

2. List four differences in properties between mixtures and pure substances. Use water and ink as examples to illustrate the differences.

3. Four students, Black, Brown, Green and Dunn, were required to prepare samples of a certain compound; then to purify it; finally to measure its melting point. The melting points obtained were:

Black	168 °C	Green	172 °C
Brown	165–167 °C	Dunn	169–171 °C

Assuming that the melting points were accurately measured, which student prepared (a) the purest sample, (b) the least pure sample?

4. Devise two simple experiments in each case which you might perform to distinguish between (a) iron(II) sulphide and a mixture of iron and sulphur, (b) steam at 150 °C and a mixture of hydrogen and oxygen at 150 °C.

5. Benzoic acid is a white solid which melts at 122 °C. α-Naphthol is a white solid which melts at 95 °C. Which of the following do you think will be the melting point of a mixture of 100 g of α-naphthol with 1 g of benzoic acid: (a) 95 °C, (b) 120 °C, (c) 96 °C, (d) 94 °C or (e) 103 °C?

5.1 energy changes and separation

In Chapter 4 we considered the changes which can occur when complex substances are formed from simple substances. In this Chapter we consider the reverse process–the splitting-up of complex substances into simpler components.

The formation of a **compound** from simpler components involves an energy change, usually a release of energy. A consequence of the Law of Conservation of Energy (section 2.3) is that to separate a compound into simpler components, energy will need to be supplied. On the other hand, since **mixtures** are formed without an energy change, it should be possible to separate a mixture into simpler components without supplying energy.

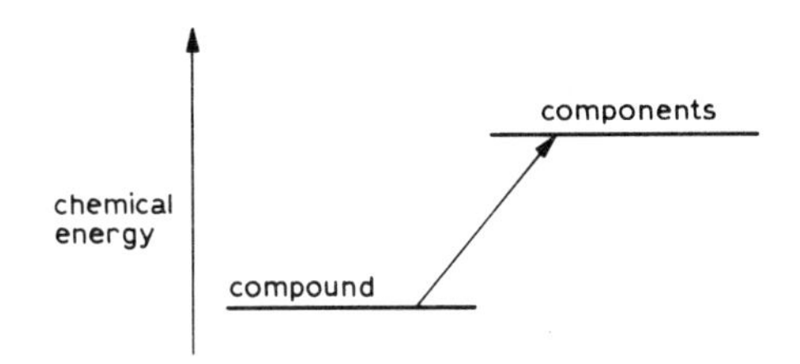

Fig. 5.1 energy is required to split up a compound

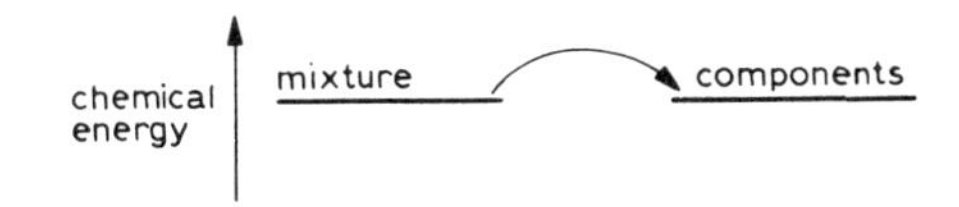

Fig. 5.2 no energy required to split up a mixture

5.2 separation of the components of mixtures

Since a mixture possesses the properties of all components, mixtures are usually separated by finding a property of one component not shared by the others and applying an operation based on the difference in properties.

1. red marbles and blue marbles

Since the eye can detect the difference in colour, a mixture of red and blue marbles can easily be separated by **hand-picking.** It is not necessary to supply a significant quantity of energy in this process. Although the method of hand-picking finds little use in a chemistry laboratory since the particles involved are usually very small, it was used in a famous experiment conducted by Pasteur, who separated a mixture of two forms of tartaric acid by picking out crystals of different shapes.

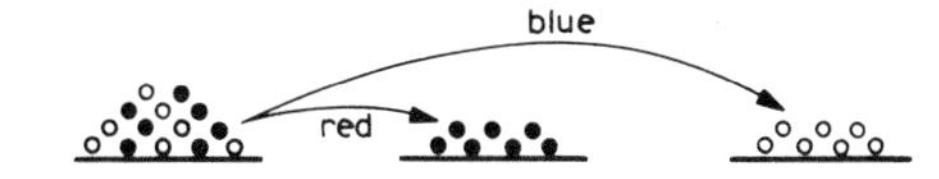

Fig. 5.3 hand-picking red and blue marbles

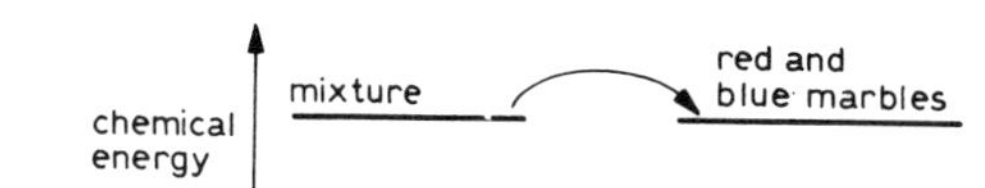

Fig. 5.4 no net energy required in separating mixture

2. nickel and sulphur

Nickel is attracted by a magnet, while sulphur is not. If a magnet is stirred in a mixture of nickel and sulphur powders and subsequently withdrawn the nickel adheres to the magnet and the sulphur does not; the small quantity of sulphur which sticks to the nickel may be removed by gentle tapping. The nickel may be removed from the magnet using a non-magnetic scraper. No significant quantity of energy is supplied to the nickel or sulphur during this process. Although this method of separation finds little use in the laboratory, relatively few substances being magnetic, it is widely used in scrap yards for separating mixtures of scrap metal into ferrous metals like iron and steel and non-magnetic materials such as glass, aluminium and plastics.

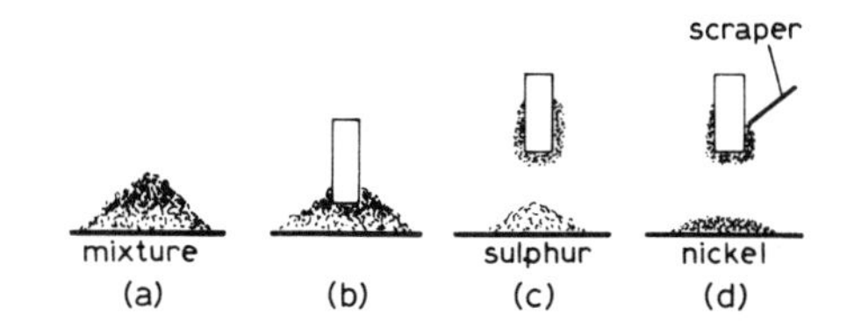

Fig. 5.5 a mixture of nickel and sulphur (a) separated magnetically (b), into sulphur (c) and nickel (d)

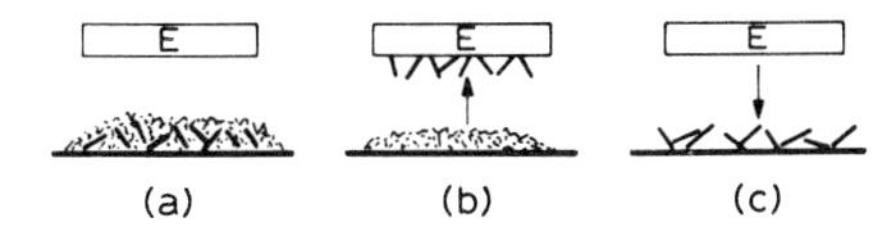

Fig. 5.6 'ground scrap' (a), separated by electromagnet E, leaving non-magnetic scrap (b) and depositing iron and steel scrap (c) when electromagnet switched off

3. water and petrol

If water and petrol are stirred together in a beaker, a fairly intimate mixture

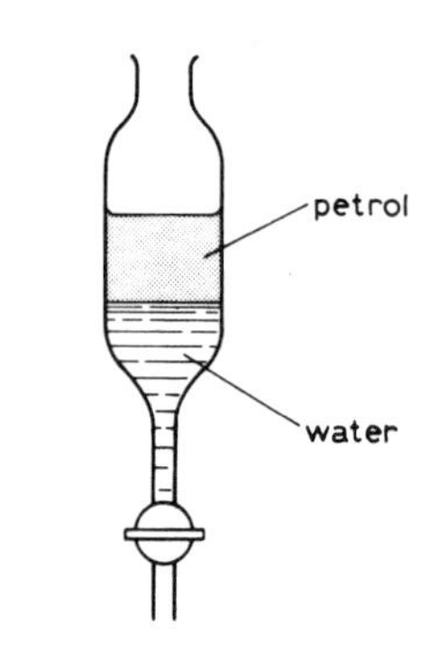

Fig. 5.7 separating funnel

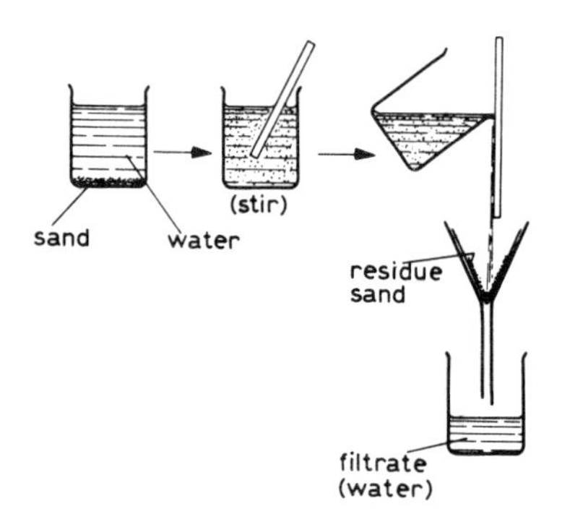

Fig. 5.8 separation by filtration

Fig. 5.9 sewage filtration beds

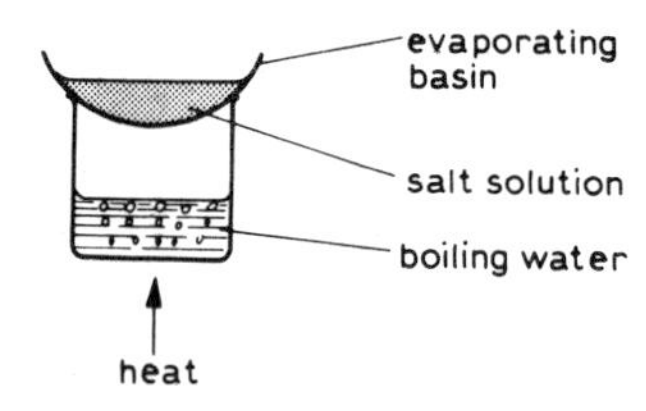

Fig. 5.10 separation by evaporation (recovery of solid)

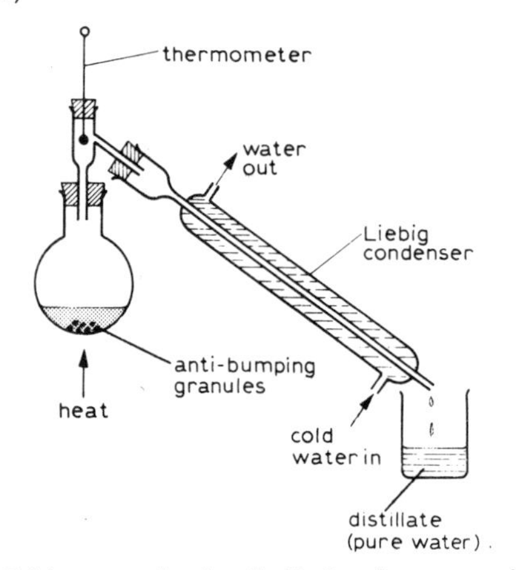

Fig. 5.11 separation by distillation (recovery of volatile liquid)

of the two liquids can be obtained. On standing, the liquid separates into an upper layer of petrol (which is less dense than water) and a lower layer of water. The two liquids may then be separated using a **separating funnel** as shown in Fig. 5.7. No energy is necessary in effecting this separation. This technique of separating immiscible liquids is widely used in the laboratory, especially as part of the process known as solvent extraction.

4. water and sand

Sand is insoluble in water, but when sand is stirred in water it forms a suspension; the liquid is not clear and particles of sand can be seen in the liquid. Water, being a liquid, will pass through the pores of filter paper, whereas sand, being a solid, will be retained in the filter paper. The process is called **filtration.** The suspension of sand in water is poured into a cone of filter paper supported in a filter funnel. The water passes through the funnel and the sand remains in the paper. The last traces of water may be removed from the sand by heating or by pressing the sand between sheets of dry filter paper. No energy is supplied in effecting this separation. The process of filtration is widely used in the laboratory, although it has been to a great extent superseded by the use of the centrifuge. In this latter process, small tubes containing the suspension to be separated are spun at high speeds in a horizontal plane so that the solid particles are packed at the bottom of the tubes. The clear liquid can then be removed from the upper part of the tubes using a dropping pipette. Large filtration plants are necessary to remove suspended solid matter from water supplies and from sewage; in these cases the filtering medium is a very tightly packed bed of shingle.

5. water and salt

Salt dissolves in water to form a clear solution which contains no solid particles; the components of such a mixture cannot therefore be separated by filtration. Separation can be effected by utilising the fact that water is volatile (boiling point 100 °C) and salt is not.

If it is desired to recover only the solid salt, the process of **evaporation** is used (Fig. 5.10). The solution is heated in a shallow dish in order to expose the maximum surface area of liquid to the atmosphere and when evaporation nears completion the last traces of water are driven off from the pasty mass by heating on a boiling water bath, thus reducing 'spitting' of solid from the basin.

The liquid component may be recovered from a salt solution by a process of evaporation and subsequent condensation of steam; the whole process is known as **distillation** and the apparatus required is shown in Fig. 5.11. Small pieces of broken pot are often added to the liquid in the flask to minimise 'bumping' and promote smooth boiling; it is not customary to heat the contents of the flask to dryness as this would risk cracking the flask. Distillation is a common laboratory technique, used especially when handling organic compounds, many of which are volatile liquids. It is the traditional method for producing pure (or 'distilled') water, though many laboratories now employ ion-exchange columns for this purpose. It is used in the wine and spirit industry for producing volatile spirits from dilute solutions containing ethanol.

Separations by the processes of evaporation and distillation require energy to be supplied to the solution, but the latent heat of vaporisation supplied to the water when it boils is subsequently regained on condensation and the separated salt and water will finally contain the same amount of energy as was contained in the original solution.

6. ethyl acetate and ethyl acetoacetate

Ethyl acetate and ethyl acetoacetate are liquids which mix completely with each other and they can therefore not be separated using a separating funnel. Since both liquids are volatile (ethyl acetate boils at 77 °C and ethyl acetoacetate at 181 °C) a simple process of distillation would yield a distillate containing both liquids; the liquids can, however, be separated by a process known as **fractional distillation,** the apparatus for which is shown in Fig. 5.12. When the liquid in the flask is boiled, the vapour which rises into the fractionating column contains *both* substances, though it is richer in the more volatile component (ethyl acetate) than is the liquid in the flask. The vapour condenses in the bulbs of the air-cooled fractionating column and the resulting liquid is subsequently re-boiled by hot vapours rising from the flask. This process is repeated many times over as the vapours rise up the column, while the vapour becomes increasingly rich in the more volatile ethyl acetate. The temperature recorded on the upper thermometer rises suddenly to 77 °C and pure acetate distils over. The temperature recorded by the upper thermometer remains steady at 77 °C while ethyl acetate continues to distil; the temperature recorded by the lower thermometer rises gradually throughout this period. The upper thermometer then registers a *sudden* change and its temperature rises to 181 °C; when this occurs, the receiver containing ethyl acetate is removed and replaced by a second receiver which collects the fresh distillate of pure ethyl acetoacetate.

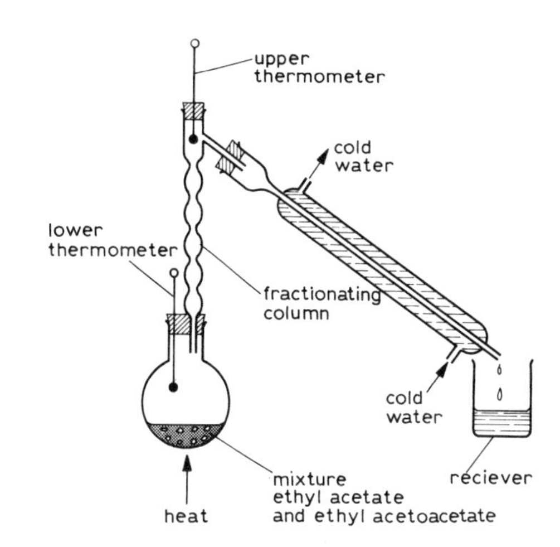

Fig. 5.12 separation by fractional distillation

The efficiency of this separating process can be shown clearly by using a mixture of 25 cm^3 of each component in the flask and collecting the distillate in separate tubes, 5 cm^3 at a time. Five such fractions are collected before the temperature shown on the upper thermometer rises above 77 °C and these fractions can be shown to be free from ethyl acetoacetate by adding to each a drop of aqueous iron(III) chloride; no colour change is seen. The ensuing fractions which distil at 181 °C all give a deep pink colour with aqueous iron(III) chloride.

Fractional distillation is widely used in the laboratory, in the wine and spirits industry and in separating the components of crude oil (see Chapter 19).

7. salt and sand

This is a mixture of two solids, one of which is water-soluble and the other of which is not. The solid mixture is stirred with excess water in a beaker and the resulting suspension is filtered. The filtrate (salt solution) is transferred to an evaporating dish and the solid salt is recovered by evaporation. The insoluble sand remains in the filter paper; it is washed free from salt solution by pouring successive small quantities of water through the sand, which may then be dried as previously described.

8. colour pigments in ink

Ink is a mixture of several differently coloured solids dissolved in water. The pigments can be separated by utilising the fact that they adhere to filter paper with different forces; the technique used is called **paper chromatography.**

A strip of filter paper is dipped into ink diluted with water so that a coloured band forms at the bottom of the paper. The paper is dried, then supported so that the coloured band dips just below the surface of water in a trough. As water rises up the paper, different coloured bands are seen to separate. The paper is removed and dried and the coloured portions are separated by cutting with scissors. Similarly coloured bands from several experiments may then be combined and stirred with water to form a coloured solution from which the solid pigment may be recovered by evaporation. One of the many applications of the technique of paper chromatography is described in section 19.14.

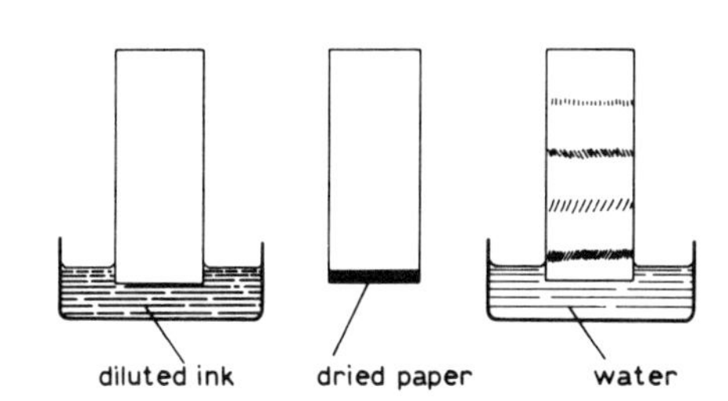

Fig. 5.13 formation of pigment bands from ink by paper chromatography

The techniques for separating the components of mixtures illustrated by (1)–(8) in this section may be summarised as follows:

a) *mixture of solids*
 i) visual difference and particles large: hand-picking
 ii) one component attracted by magnet: magnetic separation
 iii) one component soluble, the other insoluble: solution, filtration, evaporation
 iv) both components soluble: paper chromatography

b) *mixture of solid and liquid*
 i) solid insoluble in liquid: filtration
 ii) solid soluble in liquid: evaporation or distillation

c) *mixture of liquids*
 i) liquids do not mix: separating funnel
 ii) liquids mix: fractional distillation

The above list is not exhaustive and it should be remembered that the technique chosen for the separation of the components of a mixture must be matched with the property difference of the components. In all cases there is **little net energy change** in the process of separation.

5.3 separation of the components of compounds

Since energy must be supplied to separate the components of compounds, none of the methods outlined in section 5.2 are applicable. The two commonest forms of energy used in the decomposition of compounds are heat and electricity.

1. *decomposition using heat energy (thermal decomposition)*

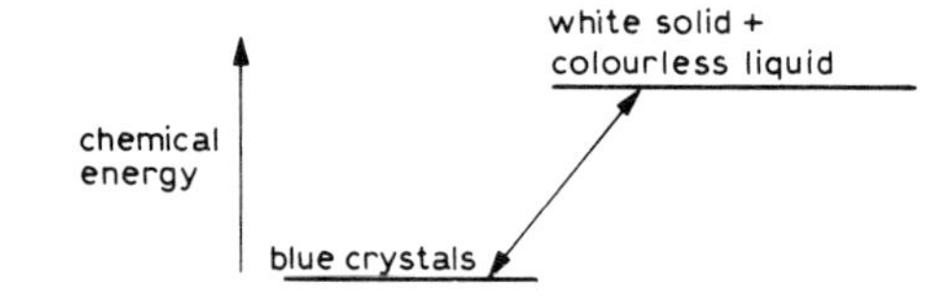

Fig. 5.14

a) If blue crystals of copper(II) sulphate(VI) are heated very gently in a test-tube, droplets of a colourless liquid form on the cool upper part of the tube and a white residue remains. The white solid and the colourless liquid when separated contain more chemical energy than the original blue crystals; this can be shown by placing some of the white solid on a watch-glass held in the palm of the hand and adding some of the colourless liquid to it. The heat energy given out in this process is the same as that which must be supplied in decomposing the blue crystals. The colourless liquid can be shown to be water and the white residue is anhydrous copper(II) sulphate(VI); the water is chemically combined in the blue crystals as 'water of crystallisation' (see section 8.16).
b) If lead(II) nitrate(V) crystals are heated in a test-tube fitted with a bung and a delivery tube, the crystals decompose with a crackling sound (*decrepitation*), leaving a yellow solid residue. Brown gas issues from the delivery tube and if this gas is collected in a test-tube over water it is seen to be colourless. Two gases are formed in the decomposition, one of which is brown and soluble in water and the other colourless and insoluble. Thermal decomposition of lead(II) nitrate(V) produces three components: a yellow solid, a brown gas and a colourless gas.

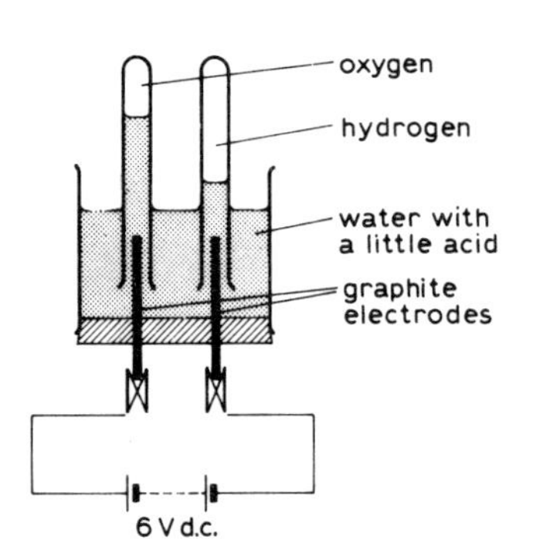

Fig. 5.15 electrolysis apparatus

2. *decomposition of compounds by electrical energy (electrolysis)*

Some compounds are more easily decomposed by electrical energy than they are by heat energy. One such compound is water; it will not conduct electricity in the pure state, but if a few drops of acid are added to it, a direct current readily produces two colourless gases. An appropriate apparatus which may be used to demonstrate this is shown in Fig. 5.15. The gases may be shown by appropriate tests to be hydrogen and oxygen and the fact that the gases separately contain more chemical energy than the water from which they are formed is amply illustrated by the experiments outlined in Chapter 8.

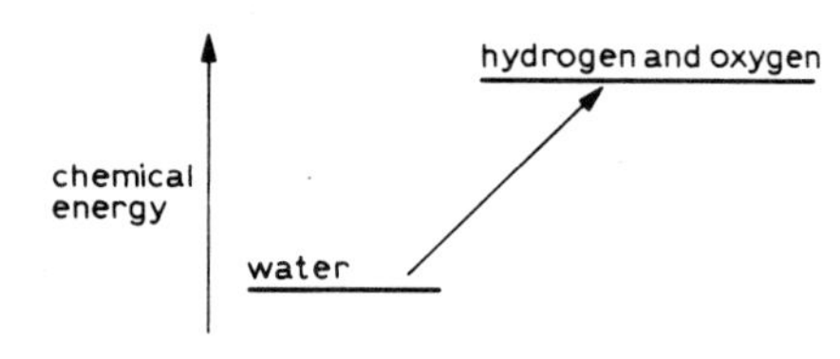

Fig. 5.16 energy required to split up water

5.4 elements

The action of heat on hydrated copper(II) sulphate(VI) produces water and anhydrous copper(II) sulphate(VI). The application of electrical energy to the water formed decomposes it into the gases hydrogen and oxygen. If anhydrous copper(II) sulphate(VI) is heated further, it produces a black solid, copper(II) oxide, and a gas which condenses to a white solid, sulphur(VI) oxide. Using different and more complicated techniques, these two products can be further decomposed, giving copper and oxygen from the black solid and sulphur and oxygen from the white solid. The complete breakdown of hydrated copper(II) sulphate(VI) is shown diagrammatically in Fig. 5.17.

The products sulphur, oxygen, copper and hydrogen resist further decomposition into simpler substances by the application of any form of energy; they are chemical substances in their simplest, or most elementary, form and are called elements. **An element is a substance which cannot be split up into simpler substances.**

To the present day, 92 elements have been discovered in nature. All other substances on our planet are complex substances made from this number of chemical building units. If elements join together with an accompanying energy change, usually a release of energy, compounds are formed; if there is no accompanying energy change, mixtures are formed. Mixtures of compounds occur widely but compounds of mixtures do not exist. Elements differ widely from each other in colour and physical state: chlorine is green, one form of phosphorus is red, sulphur is yellow and liquid oxygen is blue; hydrogen, oxygen and chlorine are gases; bromine and mercury are liquids; copper, lead and gold are solids. The property which they have in common is that no element can be split up into simpler substances.

The relative abundance of the elements in our planet is shown in the accompanying pie diagram (Fig. 5.18). It shows that about 50% by mass of our planet is composed of oxygen, either in its elementary form or as a component of complex substances, while many of the elements are extremely rare. Some of the more common elements are listed below:

argon	copper	mercury	silicon
arsenic	gold	nickel	silver
aluminium	helium	nitrogen	sodium
bromine	hydrogen	neon	sulphur
calcium	iodine	oxygen	tin
carbon	iron	phosphorus	titanium
chlorine	lead	platinum	uranium
cobalt	magnesium	potassium	zinc

Since the Ancient Greeks produced the first list of elements (air, earth, fire and water) the list has been constantly modified as new elements have been discovered and substances previously thought to be elements have been shown either by a process of analysis or synthesis to be composed of simpler substances. Like compounds, elements have sharp and constant melting and boiling points; elements and compounds are sometimes referred to as 'pure substances'. The classification of all substances into elements, mixtures and compounds is summarised below.

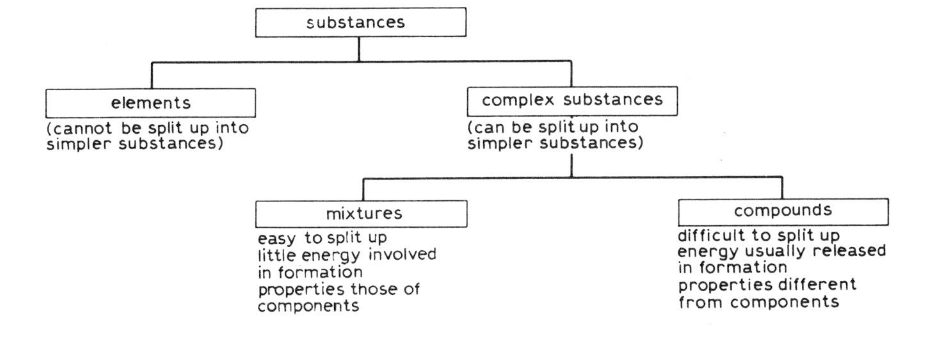

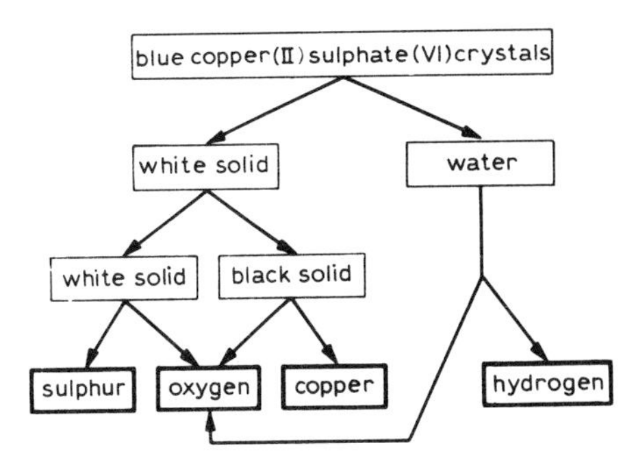

Fig. 5.17 successive decomposition of blue copper(II) sulphate(VI) crystals

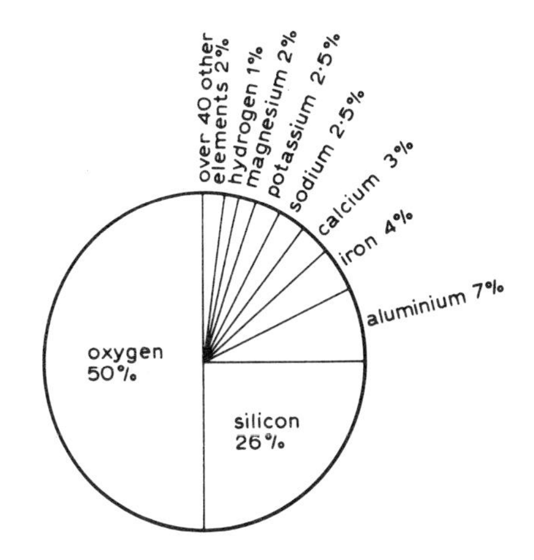

Fig. 5.18 abundance of elements

questions 5

1. From the list of elements in section 5.4 pick out:
(a) two elements which are solids at room temperature,
(b) one element which is liquid at room temperature,
(c) two elements which are gaseous at room temperature,
(d) one element which is often used inside the tubes of illuminated advertising signs,
(e) one element which may be employed to fill balloons,
(f) one element which is frequently added to the water of swimming pools,
(g) one element which is used for making wire to conduct electricity,
(h) one element which is very poisonous,
(i) two elements which are frequently used for making jewellery.

2. Which of the following is (are) possible?
(a) an element containing two compounds,
(b) a compound containing two elements,
(c) a mixture containing six elements and two compounds,
(d) a compound containing six elements,
(e) a compound containing seven elements and one mixture,
(f) an element containing three mixtures and four compounds.

3. The following is a list of techniques frequently employed to separate complex substances into their components:
(a) filtration, (b) electrolysis, (c) heating with carbon, (d) distillation, (e) sublimation, (f) heating in a stream of hydrogen, (g) chromatography.

For each member of the list state whether it is more appropriate to the separation of mixtures or of compounds.

4. Describe briefly how you would separate a pure sample of the first-named substance from the impurity in each of the following mixtures:
(a) iron filings from a sample contaminated with oil,
(b) sodium(I) chloride crystals from a sample contaminated with powdered glass,
(c) pure water from sea water,
(d) an orange dye from a blue dye which together form black ink,
(e) pure copper(II) sulphate(VI) from a sample contaminated with a small amount of sodium(I) sulphate(VI),
(f) almost pure alcohol from whisky. (JMB, modified)

5. A mixture containing benzene, boiling point 80 °C, and methylbenzene (toluene), boiling point 111 °C, is amongst the products of the thermal decomposition of coal.

Explain, with the aid of a diagram, how you would obtain samples of benzene and methylbenzene (toluene) by fractional distillation of the mixture. (C)

6. Below are five methods of separating mixtures.
A fractional distillation, B sublimation, C filtration, D chromatography, E evaporation.

Give the most suitable method for obtaining
(a) the red pigment from a purple ink,
(b) liquid nitrogen from liquid air,
(c) ammonium chloride from a mixture of potassium(I) chloride and ammonium chloride,
(d) calcium(II) carbonate from a suspension of calcium(II) carbonate in water,
(e) ethyl alcohol (ethanol) from a mixture of ethyl alcohol and water,
(f) sodium(I) chloride from a solution of sodium(I) chloride. (JMB)

7. It is required to separate a mixture of three solid dyes; one red, one yellow and one blue. The following facts are known about the dyes. The blue and yellow dyes are soluble in cold water, while the red dye is insoluble. When an excess of aluminium oxide is added to a stirred, green aqueous solution of the mixed blue and yellow dyes and the aluminium oxide is filtered off and washed with water, it is found that the solid residue is yellow and the filtrate blue. When the yellow solid is stirred with ethanol (ethyl alcohol) and the mixture filtered, the solid residue is white and the filtrate yellow.

Describe how you would obtain dry samples of the three dyes. (JMB, part question)

8. Describe and explain experiments by which you could obtain:
(a) hydrogen from a mixture of hydrogen with a little oxygen,
(b) carbon from a mixture of carbon and sulphur,
(c) pure water from a salt solution,
(d) iodine from a mixture of iodine with sodium(I) chloride. (AEB)

9. Describe briefly how you would separate pure samples of both components from the following mixtures:
(a) sugar and flour,
(b) sodium(I) chloride and ammonium chloride,
(c) water and copper(II) sulphate (VI),
(d) sand and sawdust,
(e) petrol and water.

substances in air

Our atmosphere extends to a height of about 500 km above the surface of the earth. It is composed of the substance which is called air; in this Chapter we consider the composition and properties of this familiar substance.

6.1 air: element, mixture or compound?

Air is drawn by means of a suction pump through a U-tube surrounded by a cooling mixture of crushed ice and salt, as shown in Fig. 6.1. The temperature of the ice–salt mixture is approximately −10 °C. A calcium(II) chloride tube is interposed between the U-tube and the suction pump in order to prevent any water or water vapour from the suction pump entering the U-tube. When air has passed through the apparatus for about half an hour, the U-tube is removed from the cooling mixture and is seen to contain a white solid; this solid has a melting point of 0 °C and the colourless liquid so produced has a boiling point of 100 °C–the white solid is ice. This experiment shows that air contains water vapour and since air is clearly not pure water there must be at least one other gaseous component present in air. Since water vapour can be separated from the other components of air without supplying energy, **air is a mixture of at least two components, water vapour being one of them.**

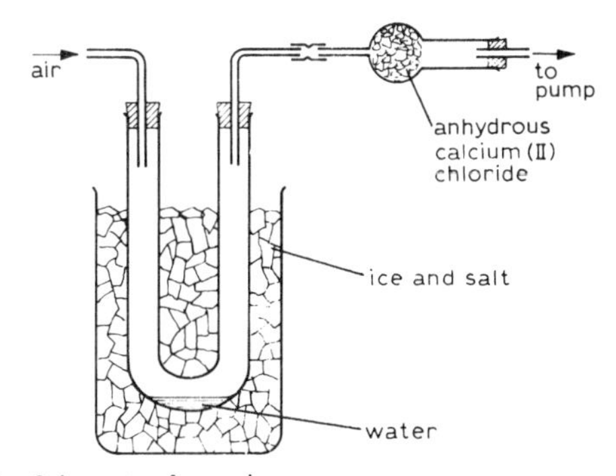

Fig. 6.1 water from air

The above experiment could be extended so that air is passed through a series of U-tubes which are cooled to successively lower temperatures, as shown in Fig. 6.2.

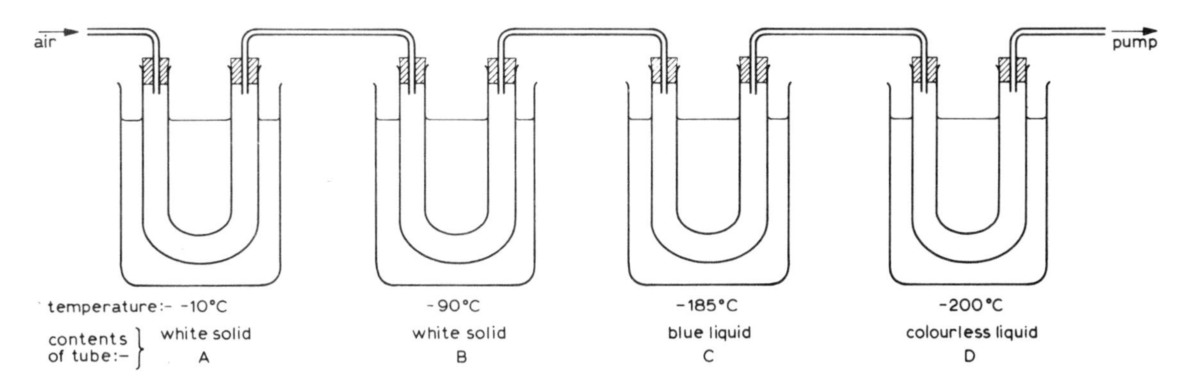

Fig. 6.2 a possible separation of the components of air

The substances condensing in the U-tubes could be warmed to room temperature and tested; the following results would be obtained:
A melts at 0 °C giving a colourless liquid which boils at 100 °C at atmospheric pressure.

B sublimes giving a colourless gas which when bubbled through aqueous calcium(II) hydroxide ('limewater') turns it milky.
C boils to give a colourless gas which rekindles a glowing splint.
D boils to give a colourless gas which extinguishes a burning splint.

These tests indicate that *A*, *B*, *C* and *D* are respectively water, carbon dioxide, oxygen and nitrogen and the experiment indicates that **air is a mixture containing at least these four components.**

component	b.p/°C (p = 100 kPa)
helium	-269
hydrogen	-253
neon	-246
nitrogen	-196
argon	-186
oxygen	-183
krypton	-152
xenon	-108
carbon dioxide	-78
water	+100

Fig. 6.3 boiling points of components of air

6.2 separation of the components of air

The separation of air into its components is very difficult to perform in the laboratory. The separation is achieved on the industrial scale by compressing air and subsequently allowing it to expand rapidly; the expansion causes the gas to cool. This is called the 'Joule–Thomson effect' and is the principle used in refrigeration: it is possible to explain the effect by applying the kinetic molecular theory (Chapter 3) to the expansion of a compressed gas. Successive treatments of compression and expansion cool the air to a temperature at which the gases liquefy. At this temperature air is a mixture of liquids and when the liquid air is allowed to warm up slowly below a fractionating column the components listed in Fig. 6.3 distil into cooled receivers at the temperatures shown. In certain localities air contains constituents not shown in the list: above active volcanoes sulphur dioxide is present, while in industrial areas air contains soot, sulphur dioxide and carbon monoxide.

6.3 percentage composition of air

To separate the components of air in the laboratory, we must utilise a property of each component in the mixture which is not possessed by any other component. Oxygen forms a solid compound when it comes into contact with hot copper; no other component of air has this property. Oxygen may therefore be removed from air by passing the air over heated copper and the proportion of oxygen in the air is indicated by the consequent diminution of volume.

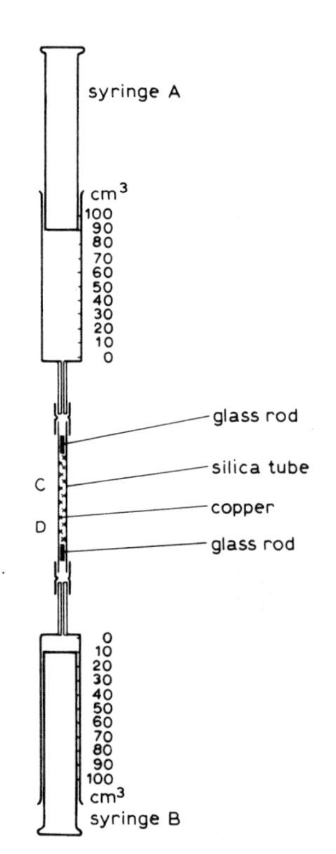

Fig. 6.4 determination of percentage of oxygen in air

1. percentage of oxygen in air

The apparatus used is shown in plan in Fig. 6.4. 100 cm³ of air is introduced into syringe A with syringe B empty. The copper in the silica tube is heated vigorously at C and air is passed slowly backwards and forwards over the heated copper for about two minutes. The silica tube is cooled using a damp cloth, the residual air in the apparatus being transferred to syringe A and the reading noted. The copper at C is found to have lost its bright red-brown colour and is dull black. Heating at C is resumed and the air is passed once more over the hot solid until, after cooling, the volume of gas in syringe A remains constant, showing that no more of the gas is affected by the hot copper. The tube is then heated at D and the air passed backwards and forwards through the hot tube; the copper at D remains bright red-brown, showing that all oxygen has been removed from the air.

$$\begin{aligned}
\text{original volume of air in syringe A} &= 100\,\text{cm}^3\\
\text{final volume of air in syringe A} &= 80\,\text{cm}^3\\
\text{volume of oxygen in } 100\,\text{cm}^3 \text{ of air} &= 20\,\text{cm}^3\\
\text{percentage by volume of oxygen in air} &= \frac{20}{100} \times 100\\
&= 20\%
\end{aligned}$$

2. percentages of other substances in air

Other experiments of a rather more complicated nature than that described above can be devised to determine the percentages by volume of the other components in air. The volume composition of air is shown in Fig. 6.5. Note that the values do not add up to 100%; this is because the percentage of water vapour varies from day to day and from place to place. The variable composition of air is further evidence that it is a mixture and not a compound and this variation is not confined to the water content.

6.4 properties of air

Being a mixture, air has the properties of its components. In this section some of the properties of air are examined to find out which component is responsible for each property. The specific property is then examined in more detail in the chapter devoted to the individual component. The density of air is found to be 14·4 times that of hydrogen; this is to be expected for a mixture of four parts of nitrogen (density 14 times that of hydrogen) with one part of oxygen (density 16 times that of hydrogen).

component	%
nitrogen	78
oxygen	21
argon	0·94
water vapour	0·5 - 3·0
carbon dioxide	0·03
hydrogen	0·01
neon	$1{\cdot}5 \times 10^{-3}$
helium	5×10^{-4}
krypton	$1{\cdot}1 \times 10^{-4}$
xenon	9×10^{-6}

Fig. 6.5 percentage by volume of components of air

1. burning substances in air and in its components

A wooden splint burns when heated in a Bunsen burner flame and continues to burn when it is withdrawn from the flame. We say that the splint ***burns*** in air, or alternatively that air ***supports the combustion*** of the splint.

Several gas jars are filled separately with oxygen and with nitrogen from cylinders by displacement of water as shown in Fig. 6.6. A burning wooden splint plunged into a jar of nitrogen is immediately extinguished: nitrogen will not support the combustion of a splint. A burning wooden splint plunged into a jar of oxygen burns much more brightly than it does in air. **Oxygen supports the combustion of a splint.** Similar experiments may be performed using other substances which burn in air, e.g. paper, sulphur and carbon. In each case it is found that nitrogen will not support combustion, whereas if a substance burns in air it will burn more brightly in oxygen. The experiment does not give any information about the role which the minor components of air play in burning. Similar experiments would show that the minor components will not support the combustion of most materials which burn in air, although magnesium will continue to burn with difficulty in carbon dioxide and in nitrogen. **Oxygen is the component of air which supports the combustion of substances which burn in air.** The chemistry of burning is considered in greater detail in Chapter 7.

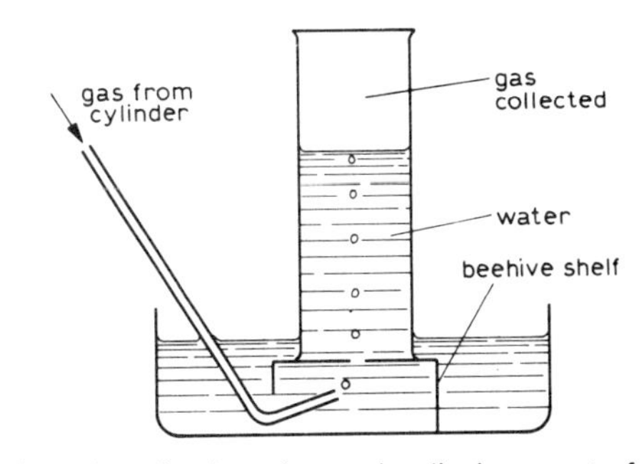

Fig. 6.6 collection of a gas by displacement of water

2. air necessary for respiration

Animals depend upon air for their existence and if the air supply is cut off they rapidly die. In a significant, if cruel, experiment, John Mayow (1645–1679) placed a mouse in a fixed volume of air trapped over water in a bell-jar. He showed that about one-fifth of the original volume of air was used up before the animal died, and this experiment would suggest to us now that oxygen is the component of air which is vitally important in sustaining the life of animals. **Oxygen is the component of air which is necessary for respiration**; the chemistry of the respiration process is considered in detail in Chapter 19.

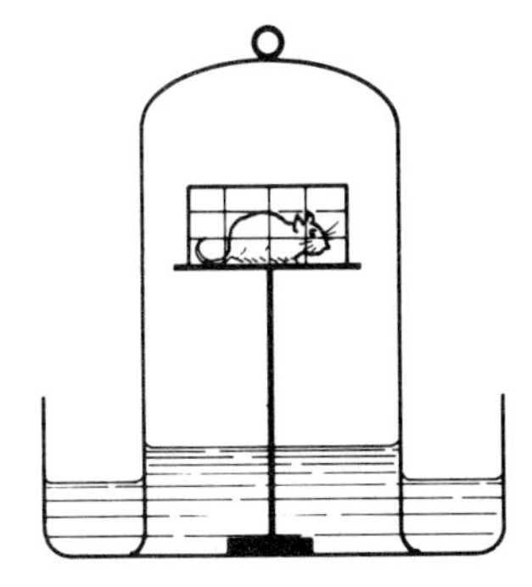

Fig. 6.7 Mayow's experiment

6.5 the noble gases

Argon, helium, krypton, neon and xenon comprise a 'family' or group of elements present in air known as the noble gases. They are all colourless gases at room temperature and pressure and are very unreactive; indeed, it was once thought that these gases had no chemical reactions at all and the group was named the 'inert gases'. Compounds containing xenon and krypton have now been prepared and although the gases are very unreactive, the term 'inert' is no longer strictly applicable. The group has been called the 'rare gases', but since there is approximately thirty times as much argon in the atmosphere as there is carbon dioxide, which is by no means rare, this name is not the most appropriate. The term **'noble gases'** is used in this book to attribute to the gases a distinctive lack of reactivity, just as the metals gold, silver and platinum which do not tarnish on prolonged contact with air, are called the 'noble metals'. This lack of chemical reactivity causes us to consider the gases to be a 'family' of similar elements; other 'families' of elements are considered in Chapters 15, 16 and 17. The densities of the gases, taken in conjunction with their relative atomic masses (Chapter 9) show them to be monatomic, i.e. each molecule consists of only one atom.

There is no convenient laboratory method for preparing samples of individual noble gases. They are obtained industrially by careful fractional distillation of liquid air and are packed under pressure in heavy metal cylinders. Helium is also obtained from the natural gas which is obtained from boreholes in south-western areas of the United States; concentrations of helium up to 7 or 8% by volume have been reported in this gas, but concentrations of 3 to 4% are more common. Helium is also found trapped in minerals containing radioactive substances.

6.6 uses of the noble gases

Extreme resistance to chemical attack under a wide range of conditions furnishes the main usefulness of the noble gases.

Argon is used for filling electric light bulbs. A light bulb filled with air causes the very hot tungsten filament to burn away rapidly, while if nitrogen is used in place of air the tungsten filament slowly forms a compound with the nitrogen which renders the filament brittle. An early solution to this problem was to evacuate the bulb, but this caused the hot metal to sublime from the filament and deposit on the inside of the cooler glass wall of the bulb, rendering it opaque. Evacuated bulbs are also dangerous because of the risk of implosion. Bulbs filled with argon suffer from none of the disadvantages enumerated above; any of the noble gases would serve this purpose, but argon is the cheapest as it is the most abundant. Argon is occasionally used in research work to provide an unreactive atmosphere in cases where the more commonly used nitrogen would be inappropriate, and also in welding.

Neon is extensively used for filling electric discharge tubes for illumination and for advertising signs. When an electric current is passed through neon at low pressures a red glow is observed and the colour of the glow can be modified by the introduction of other substances like mercury vapour or by the use of coloured glass. Neon is also used for filling electric light bulbs used for special purposes, e.g. in stroboscopes.

Helium, which is comparatively readily available, finds a variety of uses. It is the second least dense substance known and is useful for filling balloons; its non-inflammability renders it preferable to hydrogen. It has the lowest boiling point of all known substances (only four degrees above absolute zero). A cooling bath of boiling helium has proved of inestimable value in cryoscopics–the study of the effect of very low temperatures. Helium has also been used, in a mixture with oxygen, for providing the 'air' supplied to deep-sea divers. It is much less soluble in blood than nitrogen; nitrogen dissolves in blood at the high pressures experienced in the ocean depths and is subsequently released in small bubbles within the veins and arteries when the pressure is released on the diver's return to the surface. This is the major cause of the complaint known as 'the bends'.

Krypton and *xenon* are not used on a large scale, mainly because their rarity makes them very expensive. They have certain specialised uses, e.g. filling certain photographic flash bulbs.

revision summary: Chapter 6

air:	mixture of nitrogen, oxygen, carbon dioxide, water vapour, noble gases
separation:	fractional distillation
volume composition:	major components nitrogen 80%, oxygen 20%
main properties:	oxygen in air necessary for combustion, respiration

questions 6

1. Summarise the evidence which we have for believing that air is a mixture.

2. What names have been given to the family of elements called noble gases and why are these names now considered inappropriate?

3. The presence of oxygen in air is often demonstrated by burning phosphorus in a sample of air trapped over water (e.g. in a bell-jar). Could carbon be used as an alternative to phosphorus? Give reasons for your answer.

4. Solutions of substances called alkalis will absorb carbon dioxide, but not oxygen. Solutions containing both an alkali and a substance called pyrogallol will absorb oxygen and carbon dioxide, but none of the other gases present in air. Giving details of the apparatus you would use, describe how you would determine the percentage by volume of:

(a) carbon dioxide in air,

(b) oxygen in air,

(c) carbon dioxide in exhaled air,

(d) oxygen in exhaled air.

5. Nitrogen is slightly more soluble in water than is oxygen and the solubility of gases increases as the pressure increases. Blood is a mixture consisting mainly of water. What difficulties can these facts cause? How can these difficulties be overcome?

6. Describe an experiment for finding the percentage by volume of oxygen in air. How does the composition of dissolved air differ from that of ordinary air, and how do you account for the difference? (O)

see also Chapter 7

7. In what respects would you expect the nitrogen obtained by removing oxygen, carbon dioxide and water vapour from air to differ from the nitrogen obtained by decomposition of a nitrogen-containing compound? Give reasons for your answer.

oxygen and burning

Before investigating the properties of the gas which is probably the best known component of air, it is necessary to devise a convenient method for generating the gas in the laboratory. Many compounds yield oxygen on thermal decomposition, but a number of these decompositions are dangerous. One of the most satisfactory sources of oxygen is found to be the substance hydrogen peroxide.

7.1 decomposition of hydrogen peroxide

If a glowing splint is held at the mouth of the bottle in which aqueous hydrogen peroxide is stored, the splint glows more brightly and often bursts into flame. The rekindling of a glowing splint is the usual test for oxygen. In this case, the oxygen is produced by the slow decomposition of the hydrogen peroxide in the solution into water and oxygen. Since hydrogen peroxide is a compound, energy must be supplied to decompose it. What is the source of energy for the room-temperature decomposition of hydrogen peroxide?

The molecules of hydrogen peroxide are continually moving and colliding with each other and with water molecules in the solution. Each collision involves a transfer of energy and the molecules of hydrogen peroxide are therefore constantly gaining and losing energy as a result of collisions. A molecule which gains sufficient energy from collisions with other molecules may decompose forming water and oxygen. At room temperature this decomposition takes place much too slowly for it to act as a conveniently rapid source of oxygen gas; some method must be found to increase the rate of decomposition.

7.2 increasing the rate of decomposition of hydrogen peroxide

1. by heating

If a few drops of aqueous sodium(I) hydroxide are added to about 5 cm^3 of '20-volume' aqueous hydrogen peroxide in a test-tube and the resulting solution is heated, the rate of evolution of bubbles of gas is seen to increase as the tube gets hotter. Heating increases the kinetic energy of the molecules in the solution, which in turn increases the frequency of collisions and also the energy exchanged in any one collision; consequently, more hydrogen peroxide molecules have sufficient energy to decompose in a hot solution than is the case in a solution at room temperature.

2. using manganese(IV) oxide

If a small pinch of manganese(IV) oxide is added to about 5 cm^3 of '20-volume' aqueous hydrogen peroxide in a test-tube, an immediate vigorous effervesence occurs and oxygen is produced at a much higher rate than in (1) above. If a weighed quantity of manganese(IV) oxide is used in this experiment and is recovered and re-weighed at the end of the experiment, it can be shown that no manganese(IV) oxide has been used in the reaction. It speeds up the decomposition of hydrogen peroxide without itself being changed chemically or in mass during the process: such a substance is called a **catalyst.** One of the mechanisms by which catalysts perform their function is consider in detail in Chapter 9.

7.3 laboratory preparation of oxygen

A small measure of manganese(IV) oxide is placed in a Buchner flask and approximately 20 cm³ of '20-volume' aqueous hydrogen peroxide is placed in the tap-funnel of the apparatus shown in Fig. 7.1. As aqueous hydrogen peroxide is added to the solid, oxygen is generated and may be collected over water as shown; the rate of production of oxygen is easily controlled by varying the rate of addition of the solution. The first three test-tubes of gas collected should be rejected, as these will contain air which has been displaced from the apparatus by the oxygen produced. Test-tubes of gas collected subsequently may be used for investigation of the properties of oxygen, outlined in sections 7.4 and 7.5.

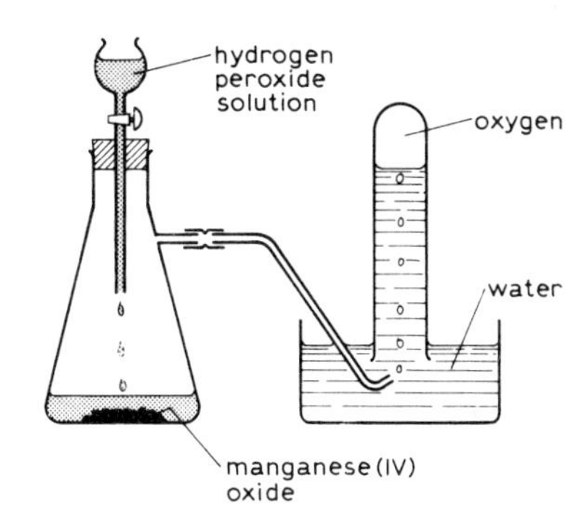

Fig. 7.1 preparation of oxygen

meaning of '20-volume' aqueous hydrogen peroxide

The concentration of a solution is usually expressed in moles per dm^3 or litre ($mol\,dm^{-3}$ or $mol\,l^{-1}$), the mass concentration in grams per dm^3 or litre ($g\,dm^{-3}$ or $g\,l^{-1}$)–see Chapter 11–but the concentration of solutions of hydrogen peroxide is often expressed as a 'volume concentration'. This expresses the concentration of the solution in terms of the volume of oxygen obtained from unit volume of the solution if all the hydrogen peroxide present decomposes to oxygen and water. Hence a '20-volume' solution of hydrogen peroxide has a concentration such that $1\,cm^3$ of it produces $20\,cm^3$ (measured at s.t.p.) of oxygen if all of the hydrogen peroxide in it decomposes.

7.4 physical properties of oxygen

1. colour

Observation of the samples prepared shows that oxygen is a colourless gas. Liquid and solid oxygen, however, are bright blue.

2. odour

Common experience tells us that oxygen is odourless.

3. taste

Oxygen has no taste.

4. liquefaction

Oxygen is not easy to condense to a liquid as it has a very low boiling point (−183 °C). High pressure and drastic cooling are necessary.

5. solubility in water

Oxygen is only slightly soluble in water; note that it is collected by displacement of water in the usual laboratory preparation. Even though its solubility is low, fish and other forms of marine life depend for their existence on oxygen dissolved in sea and river water (see the last part of section 7.8).

comparative solubilities in water of oxygen and nitrogen

The water in rivers and seas, being in continual contact with air, becomes saturated with dissolved air. Since air is a mixture, its components dissolve

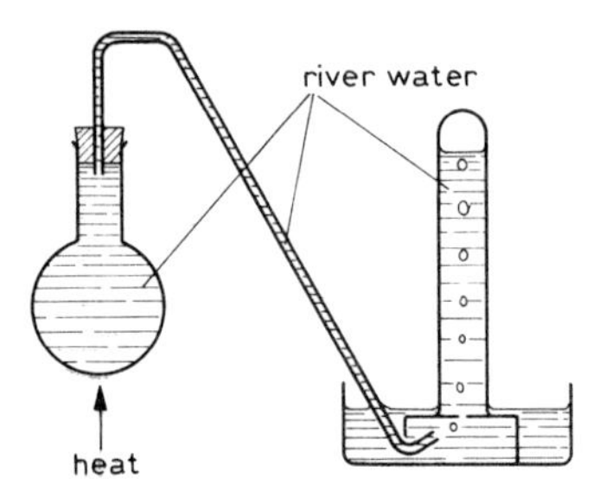

Fig. 7.2 recovery of dissolved oxygen and nitrogen from river water

to the limit of their individual solubilities. Analysis of the air dissolved in river water therefore affords a comparison of the solubilities in water of oxygen and nitrogen.

The apparatus shown in Fig. 7.2 is filled completely with river water; the delivery tube must be flush with the bottom of the bung in the flask in order to avoid trapping bubbles of gas. The flask is heated until the water boils for a short time and the expelled gas is collected as shown. The experiment should be repeated several times in order to obtain a quantity of gas sufficient to analyse.

The percentage of oxygen in the mixture is determined by the method described in section 6.3. Such an analysis shows that the percentage by volume of oxygen in the air dissolved in river water is approximately 36%, compared with about 20% in ordinary air. Since a certain volume of water will dissolve 36 cm³ of oxygen and 64 cm³ of nitrogen when the pressure of nitrogen is four times that of the oxygen, then under conditions of equal pressure of each gas the same volume of water would dissolve 36 cm³ of oxygen and 64/4 = 16 cm³ of nitrogen. Under conditions of equal pressure of each gas, it follows that oxygen is approximately twice as soluble in water as is nitrogen.

N.B. This argument assumes that the solubility of a gas is proportional to its pressure. Can you think of evidence to justify this assumption?

6. density

The relative vapour density (see Chapter 10) of oxygen is 16. This means that oxygen has a density 16 times as great as that of hydrogen; oxygen is 1·11 times as dense as air.

7.5 burning elements in oxygen

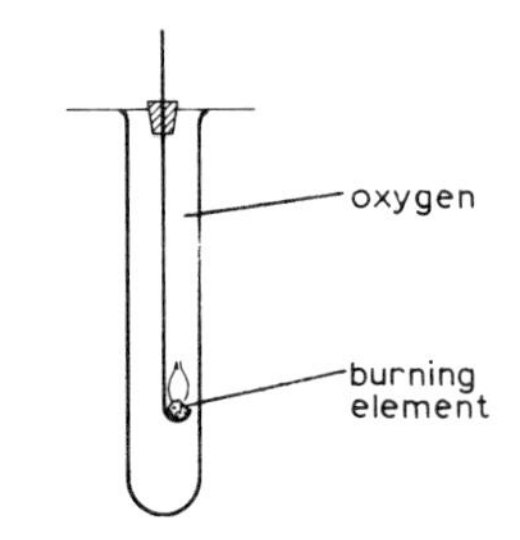

Fig. 7.3 burning an element in oxygen

We saw in section 6.4 (1) that oxygen is the component of air which supports the combustion of burning substances. The first step in understanding the chemistry of this process is to investigate the products formed when elements burn in oxygen. Elements can be conveniently burned in oxygen using boiling-tubes of the gas (prepared from hydrogen peroxide or obtained from a cylinder and collected over water) and small deflagrating spoons as shown in Fig. 7.3. A small quantity of each element is placed in the spoon and ignited in air by heating in a Bunsen flame; the spoon is then placed rapidly in a boiling-tube of the gas. When burning has ceased the spoon is removed and is placed on an asbestos board (*not* on a wooden bench) while the tube is rapidly corked. About 2 cm³ of water is added to the contents of the tube, which is rapidly re-corked and shaken. Any residue on the deflagrating spoon should be scraped into the water in the boiling-tube. After shaking the tube, a drop of purple litmus solution is added to the liquid in the tube and the resulting colour is noted. Typical observations for ten elements burned in oxygen are shown in Fig. 7.4.

7.6 oxides

When an element burns in oxygen, the element combines chemically with oxygen producing a compound known as an oxide, releasing energy as chemical bonds are formed.

e.g. magnesium + oxygen ⟶ magnesium(II) oxide

The suffix *-ide* to the name means that the compound contains only the two elements which form the name. Section 4.2 described the formation of sodium(I) chloride (a compound containing only sodium and chlorine) and section 4.5 the formation of nickel(II) sulphide (a compound containing only nickel and sulphur).

an oxide is a compound containing oxygen and one other element.

element	observations	litmus colour
*barium	burns with a green flame, white residue	blue
calcium	burns with a red flame, white residue	blue
carbon (charcoal)	burns with a yellow flame, no residue	claret-red
iron (wool)	crackles and sparks, dark brown residue	purple
magnesium (ribbon)	intense white flame, white residue	blue
*phosphorus (white)	burns with yellow flame, white residue	red
*potassium	burns with lilac flame, white residue	blue
*sodium	intense yellow flame, white residue	blue
sulphur	bright blue flame, no residue	red
zinc	burns, residue yellow when still hot but turns white on cooling	purple

Fig. 7.4 observations on burning certain elements in oxygen. The experiments marked * can be dangerous and should be performed only by an experienced demonstrator

Burning is an example of **oxidation**; *a substance is oxidised if oxygen is added to it.* The substance which provides the oxygen for this process is known as an **oxidising agent** or **oxidant.** When magnesium burns in oxygen, magnesium is oxidised to magnesium(II) oxide and oxygen gas is acting as the oxidant.

Litmus is a substance which changes colour according to whether the solution in which it is placed is acidic or alkaline: it is an **indicator.** Use of this indicator enables us to classify the oxides formed when elements burn in oxygen into three classes:

i) insoluble oxides — do not dissolve in water, litmus remains purple

ii) acidic oxides (soluble) — dissolve in water to give an acidic solution, litmus turns red

iii) alkaline oxides (soluble) — dissolve in water to give an alkaline solution, litmus turns blue

The oxides formed as described in Fig. 7.4 fall into the following groups:

acidic oxides	*alkaline oxides*	*insoluble oxides*
carbon dioxide	barium(II) oxide	iron(III) oxide
sulphur dioxide	calcium(II) oxide	zinc(II) oxide
phosphorus(V) oxide	sodium(I) oxide	
	potassium(I) oxide	
	magnesium(II) oxide	

Note that **metals** form alkaline or insoluble oxides and **non-metals** form acidic oxides; other differences between metals and non-metals are outlined in section 7.10.

The name oxygen means 'acid-producer' (*Greek–oxus = sour, gennao = produce*). The name was given because early chemists thought that all elements which burned in oxygen produced acidic oxides and that the presence of oxygen in a compound was an essential feature of acidity. This error is understandable if you recall that barium, calcium, sodium and potassium burn in oxygen with a considerable release of energy and this energy would have to be supplied to form the element from its oxide. Since early chemists had no technique for doing this, the elements which form alkaline oxides were unknown at the time when the name oxygen was given.

7.7 burning compounds in oxygen

It is well known that petrol burns in air. Petrol is a mixture of carbon hydrides (i.e. compounds containing carbon and hydrogen only); it combines with the oxygen in the air to form a mixture of carbon dioxide and hydrogen oxide (water). Hydrogen sulphide, the gas which smells of bad eggs (see section 4.5), is a compound of hydrogen and sulphur only and burns in oxygen forming a mixture of water and sulphur dioxide. In general, when a compound burns in oxygen a mixture of the oxides of its constituent elements is formed.

7.8 use of oxygen in the body: breathing

Man can live for several days without food or water but can exist for only a few minutes without oxygen. We are now in a position to understand the function of oxygen in the body.

Air is breathed in through our mouths and nostrils and passes into our lungs, where some of its oxygen combines with a component of blood called haemoglobin, forming the compound oxyhaemoglobin. Blood containing the bright red oxyhaemoglobin is pumped by the heart through the arteries to the muscles and other body tissues where a process of oxidation chemically very similar to combustion occurs. Oxyhaemoglobin acts as an oxidant, giving up its oxygen for the oxidation of various sugars, which contain carbon and hydrogen. The energy set free in this reaction is used by our bodies in performing the necessary functions, e.g. heart action, growth and movement of limbs. The carbon dioxide and haemoglobin produced in the reaction are returned in the blood stream through the veins to the heart and thence to the lungs, where the carbon dioxide passes into exhaled air and the haemoglobin receives a fresh supply of oxygen.

Fish use a similar method in obtaining oxygen to maintain their life, but they 'breathe' water instead of air and remove some of the oxygen dissolved in the water.

If exhaled air is collected over water and analysed by the technique described in section 6.3, it is found to contain about 16% (by volume) of oxygen. It is also possible to determine the percentage of the acidic oxide carbon dioxide in inhaled and exhaled air by absorbing the carbon dioxide in the alkali, aqueous sodium(I) hydroxide. The results of such an analysis are shown in Fig. 7.6.

If an athlete requires a rapid supply of extra energy, he can obtain it by supplying his body with either extra sugar (dextrose tablets are often used) or extra oxygen, depending on which his body lacks. An insufficient supply of oxygen, coupled with a high rate of energy output, produces lethargy and, in more serious cases, illness. This is particularly true at high altitudes where the air is rarefied. Himalayan climbers carry small oxygen cylinders and it was a familiar sight at the Olympic Games in Mexico in 1968 for exhausted athletes to be treated with oxygen at the end of a race.

Oxygen is continually removed from the atmosphere by the breathing of

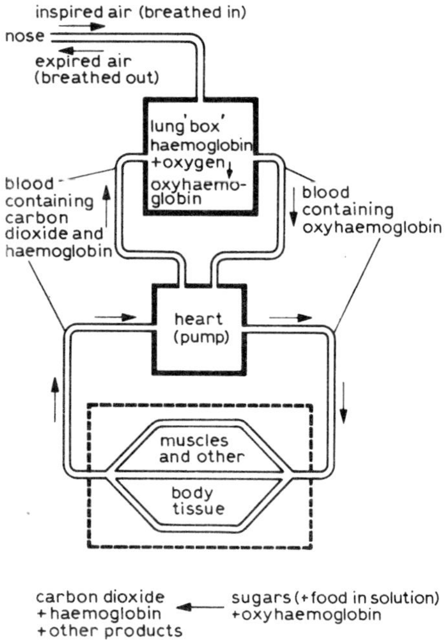

Fig. 7.5 use of oxygen in the body

gas	% by volume	
	inspired air	expired air
nitrogen	78·0	78·0
oxygen	21·0	16·4
carbon dioxide	0·03	4·0

Fig. 7.6 respiration

animals and by the many industrial and domestic processes which involve burning. Fortunately, the oxygen is being continually replaced by plants and trees in a process known as photosynthesis (see section 19.13).

7.9 uses of oxygen

1. in hospitals

Oxygen is frequently administered to patients suffering from difficulty in breathing, especially in cases of pneumonia, asthma, and heart failure. A 60% mixture of oxygen with air can safely be breathed indefinitely, but, surprisingly, pure oxygen, if breathed for several hours, will cause death.

2. to improve inadequate atmospheres

Men involved in activities in places where there is little air or where poisonous gases are present carry oxygen suitably diluted with another gas, usually nitrogen or air. Examples of such activities are diving, rescue work in mines, mountaineering above about 6000 metres and fire-fighting.

3. to provide high temperatures

Substances which burn in air burn with a brighter flame in oxygen; this flame is also hotter than that obtained in air. A mixture of hydrogen, ethyne (acetylene) or propane with oxygen, ignited at a jet, produces a very hot blow-pipe flame which has many uses including welding and the cutting of metals. An oxy-hydrogen torch flame reaches temperatures of about 2000 °C, while the oxygen–acetylene combination gives a temperature of about 3000 °C.

4. rockets

Spacecraft and some high-altitude and supersonic aircraft use rockets, which are forms of jet engine in which the jet is provided by a cone of burning gas. A rocket must carry a fuel (usually petrol, paraffin or hydrazine) and a supply of oxygen for the combustion of the fuel. To save space, liquid oxygen ('*lox*') under high pressure is usually used.

5. production of steel

A major use of oxygen is in the conversion of iron into steel (see Chapter 26). Many thousands of tons of oxygen ('tonnage oxygen') are used annually in this process.

7.10 classification of elements

Man is continually searching for patterns in order to help him understand his environment. The chemist is no exception to this; indeed the classification of substances into groups possessing common characteristics is one of the most important aspects of the study of chemistry.

There are many ways of classifying elements. They might be classified according to their colour: copper and bromine are red, sulphur and fluorine are yellow – but this classification is not found to be very useful. They might be classified according to their physical state at standard temperature and pressure; such a classification would give eleven gases, two liquids and about ninety solids. A much more useful classification emerges on examination of the nature of the compounds formed by those elements which burn in oxygen as described in section 7.5.

A clear pattern is evident in the lists on the following page; the elements in the left-hand column belong to a group having the characteristics: liquid chloride, gaseous hydride, acidic oxide and element a poor conductor of electricity. The elements in the right-hand column belong to a group having

physical state of chlorides at s.t.p.	
elements forming liquid chlorides	elements forming solid chlorides
carbon	barium
phosphorus	calcium
sulphur	iron
	magnesium
	potassium
	sodium
	zinc

nature of oxides	
elements forming acidic oxides	elements forming insoluble (i) or alkaline (a) oxides
carbon	barium (a)
phosphorus	calcium (a)
sulphur	potassium (a)
	sodium (a)
	iron (i)
	magnesium (a)
	zinc (i)

electrical conductivity	
insulators	conductors
carbon (diamond)	barium
phosphorus	calcium
sulphur	carbon (graphite)
	iron
	magnesium
	potassium
	sodium
	zinc

physical state of hydrides at s.t.p.	
elements forming gaseous hydrides	elements forming solid hydrides
carbon	barium
phosphorus	calcium
sulphur	magnesium
	potassium
	sodium

the characteristics: solid chloride, solid hydride, insoluble or alkaline oxide, element a good conductor of electricity. Those elements in the right-hand column are given the group name **metals** and those in the left-hand column are given the group name **non-metals.** The group characteristics can be extended as shown in the following table:

non-metals	metals
liquid (or gaseous) chlorides	solid chlorides
gaseous (or liquid) hydrides	solid hydrides
acidic oxides	insoluble or alkaline oxides
non-conductors	conductors of heat and electricity
not malleable or ductile	malleable and ductile
form negative ions (see section 13.9)	form positive ions

The division of properties between metals and non-metals is not clear-cut; there are certain exceptions to the broad statement in the above table. Graphite is a form of carbon which conducts electricity (see section 21.3) although carbon is classified as a non-metal because of its other properties. The majority of the properties of silicon lead to its classification as a non-metal, although it forms an insoluble oxide.

A more extensive classification of elements, which includes the above classification into metals and non-metals, is described in Chapter 17.

revision summary: Chapter 7

oxygen: *preparation:* decomposition of aqueous hydrogen peroxide
$H_2O_2(l) \rightarrow H_2O(l) + \frac{1}{2}O_2(g)$
collection: over water
test: rekindles glowing splint
reactions: good supporter of combustion; oxidant; necessary for respiration

catalyst: alters rate of chemical reaction without itself being changed chemically at end of process

metals and non-metals: operational classification based on similarities in nature of elements and their compounds

questions 7

1. Give a brief account of the industrial manufacture of oxygen. How is the oxygen transported to the user? Find out and write down as many uses of oxygen as you can.

2. Oxygen can be prepared using aqueous hydrogen peroxide and manganese(IV) oxide.
(a) What is the function of the manganese(IV) oxide?
(b) What is this type of reaction called?
(c) Oxygen is usually collected over water. This method of collection may be used because oxygen is not very soluble in water. How do we know that oxygen is soluble in water to some extent?
(d) Bottles of aqueous hydrogen peroxide are often labelled '20-volume'. What does this mean?
(e) Give one use of hydrogen peroxide solution other than in the laboratory preparation of oxygen.

3. Write an account of the work of Lavoisier on the constitution of air, explaining its importance in developing our understanding of the process of burning.

4. In (i) and (ii) below, choose the one best answer.

When copper is heated in a measured volume of air until no further change takes place, the copper becomes coated with a black substance and the air decreases to about 4/5 of its original volume. The same result is obtained each time the experiment is repeated, provided that sufficient copper is used to leave some unchanged at the end of the experiment.
(i) The evidence from this experiment indicates that air consists of:
A one substance only,
B two substances only,
C at least two substances,
D more than two substances.
(ii) If more copper was used in the experiment, which of the following would you expect?
A the volume of air would decrease by more than 1/5,
B all of the air would disappear,
C the volume of air would decrease by 1/5,
D the volume of air would decrease by less than 1/5.

5. Write a short account of the similarities and differences between the processes of burning and breathing.

6. State what class of oxide each of the following belongs to: (a) copper(II) oxide (CuO), (b) sulphur dioxide, (c) aluminium(III) oxide, (d) carbon dioxide.

In each case give **one** fact to support your statement.
(S, modified)

7. Copy the list below. Underline the name of any oxide having acidic properties. Put a tick (✓) opposite the name of any oxide having basic properties. (N.B. It is possible for an oxide to possess one, both or neither of these properties.)
carbon monoxide,
magnesium(II) oxide,
nitrogen dioxide,
phosphorus(V) oxide,
sulphur(VI) oxide,
zinc(II) oxide. (JMB)

8. Carbon at red heat will remove oxygen from the oxides of the metals *A*, *B* and *C*, but not from the oxide of metal *D*. Metal *C* will remove oxygen from the oxide of *A*, but not from the oxide of *B*.

List the metals *A*, *B*, *C* and *D* in order of decreasing activity. (JMB)

9. From the oxides of the following elements: calcium, carbon, copper, sodium, sulphur, zinc, name **one oxide** which
(a) is very soluble in water,
(b) is insoluble in water,
(c) dissolves in water forming a solution which has a pH value of less than 7,
(d) gives an alkaline solution in water,
(e) is gaseous at room temperature and pressure.
(JMB, part question)

10. Give four differences between the properties of a typical metal and those of a typical non-metal. Illustrate your answer by reference to iron and sulphur.

11. A white solid is used as a fuel in picnic stoves. On burning in oxygen it produces only carbon dioxide and water. Which of the following statements is (are) true?
(a) the white solid is an element,
(b) the white solid must be a mixture,
(c) the white solid contains carbon and hydrogen,
(d) the white solid may contain oxygen.

12. By considering the evidence below, classify the elements A, B, C, D and E as either metal or non-metal, giving reasons.
A has a melting point of 1535 °C and conducts electricity.
B is a gas at room temperature and pressure.
C burns in air to produce an oxide which dissolves in water giving a solution which turns blue litmus red.
D is a solid which burns in oxygen to produce a solid which is insoluble in water but which will neutralise an acid.
E has a melting point of −38·9 °C and a boiling point of 357 °C and conducts electricity when in the liquid state.

water and hydrogen

Water is perhaps the most familiar of all chemical substances. It is essential to all forms of life and constitutes over 70% of the surface of the earth. Throughout the ages it has acted as nature's sculptor in forming rivers, carving out valleys and eroding mountains. Without water, plant and animal life would rapidly expire–as, unhappily, is the case in arid regions during prolonged drought. It is of such importance that it merits a close and detailed study.

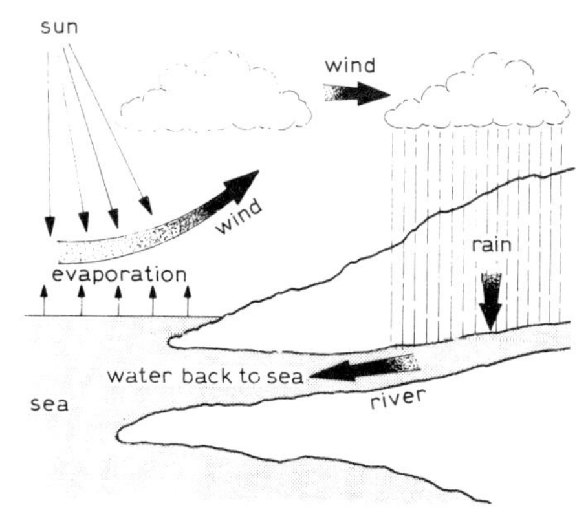

Fig. 8.1 the water cycle

8.1 the water cycle

Water in our lakes and rivers comes from the sea by means of rainfall or snow deposits. Heat energy from the sun causes constant evaporation from the surface of the sea, a process much assisted by wind; the water rises as vapour and condenses into clouds in the cooler upper atmosphere. The water falls to earth as rain, or snow when the atmosphere is sufficiently cold. About 30% of the rain returns directly to the sea in rivers, thus completing the natural water cycle which is represented diagrammatically in Fig. 8.1. This water cycle furnishes a means of harnessing solar energy, as explained in section 2.3.

8.2 some well-known facts about water

1) Pure water is a colourless liquid which freezes at 0°C (273 K) and boils at 100°C (373 K) under a pressure of 1 atmosphere (101 325 Pa). At higher pressures the boiling-point of water is raised and at lower pressures it is decreased; food cooks more rapidly in a 'pressure cooker' than in an open pan, while it is impossible to brew tea satisfactorily on a high mountain. The effect of pressure on the boiling-point of water may be explained by application of the kinetic molecular theory (see Chapter 3).

2) Addition of water to white anhydrous copper(II) sulphate(VI) turns it blue due to the formation of the blue hydrate, copper(II) sulphate(VI)-5-water:

$$\underset{\text{(white)}}{CuSO_4(s)} + 5H_2O(l) \rightleftharpoons \underset{\text{(blue)}}{CuSO_4.5H_2O(s)}$$

An alternative test for water is that it turns blue anhydrous cobalt(II) chloride pink due to the formation of the hydrate, cobalt(II) chloride-6-water:

$$\underset{\text{(blue)}}{CoCl_2(s)} + 6H_2O(l) \rightleftharpoons \underset{\text{(pink)}}{CoCl_2.6H_2O(s)}$$

Positive results for either of these tests do not indicate the presence of *pure* water; any aqueous solution would produce the colour changes described.

3) Pure water has a pH value of 7 at 25°C (298 K); any neutral aqueous solution also has a pH of 7 (see Chapter 11).

4) Water contracts as it is cooled from room temperature, but on cooling below 4°C it expands slightly and at 0°C it expands considerably as it solidifies. It has a maximum density of 1 g cm^{-3} at 4°C. Burst pipes and leaking motor-car radiators in winter are familiar reminders that water expands as it freezes.
5) Water has a higher specific heat capacity than most other common substances, hence its use in 'hot-water bottles'.
6) Water is a very good and widely used solvent. Most substances dissolve in it, even if only to a limited extent. When you drink a cup of tea, you also drink some of the spoon which you use to stir it, although the quantity dissolved is so minute as to be harmless.
7) Water obtained from lakes and reservoirs often contains dissolved salts which render it 'hard'. The hardness of water is discussed in detail in Chapter 22.

8.3 purification of water supplies

Cholera, dysentery and typhoid are diseases which result directly from the presence of harmful bacteria in drinking water. Water for human consumption must have such bacteria removed; the main stages involved in purification of water supplies are as follows:

a) *sedimentation*
Most of the insoluble matter carried by rivers settles on the bed of the reservoir.

b) *filtration* (see section 5.2 (4))
Solid particles which do not settle on sedimentation are removed by filtration. The water is continually spread over a filter bed consisting of a deep layer of fine sand on a gravel base (see Fig. 8.2). This process removes much bacterial content as well as suspended solid impurities; it is effective but slow. A more rapid filtering system uses pressure to force the water through the sand bed.

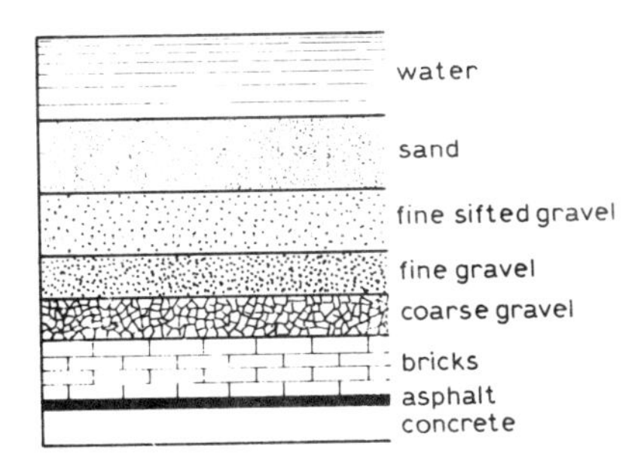

Fig. 8.2 gravity filtration

c) *removal of dissolved salts*
Dissolved salts are removed by precipitation methods or by ion-exchange (see Chapter 22). Large-scale de-ionisation plants purify up to 60000 gallons of water every minute.

d) *removal of dissolved gases*
Water from underground sources often contains dissolved carbon dioxide and even hydrogen sulphide. These gases are removed by aerating the water; it is allowed to fall over a series of shelves, thus saturating the water with air. During this process the other gases escape.

e) *removal of harmful bacteria*
The concentrations of harmful bacteria in samples of water may be estimated using sensitive biochemical tests. The World Health Organisation of the United Nations has set rigid standards of safety regarding the permitted level of disease bacteria in water for human consumption: the concentration of disease bacteria must not exceed one bacterium per 100 cm^3 of water. If a sample of water does not conform to this standard, the bacterial content is reduced by chemical treatment. One common method employs chlorine gas, a very powerful disinfectant (see Chapter 15).

8.4 water pollution

Modern technology brings problems as well as benefits. Although soapless detergents are a great domestic and industrial asset, their introduction into rivers via industrial effluents constitutes a serious threat to water supplies. Some types of detergent molecules are not broken down by bacteria in sewage waste and this makes water containing them very difficult to purify.

Fig. 8.3 foam pollution of a river

Research is presently in progress aimed at developing detergents which are both cheap and susceptible to bacterial decomposition. Plastic detergent containers also cause problems of disposal, for such a container left undisturbed in the countryside would still be present, totally unaltered, in one hundred years' time. The practice of 'dumping' radioactive waste and drums containing toxic chemicals on the bed of the ocean presents a further very serious threat of water pollution and alternative means of disposal of such substances must be found if future generations are to survive.

8.5 chemical reactivity of water

Since water is so widely used as a solvent it is often wrongly assumed that it takes little part in chemical reactions. In fact it is a very reactive substance, as is shown by its action on metals (sections 8.12 and 8.13), hydrogen chloride (sections 15.6 and 18.23), ammonia (section 24.8) and anhydrous salts (section 20.8).

8.6 synthesis of water

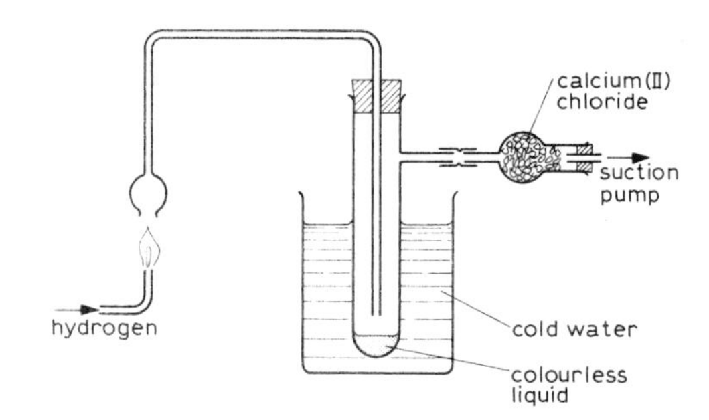

Fig. 8.4 product from hydrogen burning in air

You may have noticed the mist which forms on the outside of a beaker of cold water when you begin to heat it with a Bunsen burner. The mist consists of fine droplets of water formed by oxidation of hydrogen in the flame.

Water may be synthesised (i.e. built up from its constituent elements) by burning hydrogen in air as shown in Fig. 8.4. The colourless liquid which collects in the cooled receiver may be shown to be water by measuring its freezing point and boiling point. The purpose of the calcium(II) chloride tube is to prevent possible entry of water vapour into the receiver from the suction pump. The water obtained in the receiver is therefore a product of the burning of hydrogen in air. Since other elements burn in air to form oxides, it is reasonable to assume that **water is an oxide of hydrogen.**

N.B. Since air contains water vapour which may be condensed on cooling (see section 6.1), it is possible that the water formed in the above experiment comes from the air drawn through the apparatus. How would you modify the apparatus shown in Fig. 8.4 to eliminate this possibility?

8.7 reciprocal combustion

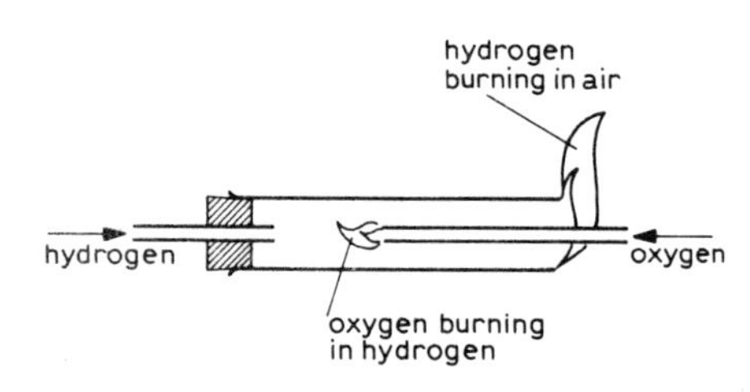

Fig. 8.5 reciprocal combustion

If hydrogen is passed through a wide glass tube, it will burn in the oxygen of the air if ignited at the mouth of the tube. If a glass tube attached to an oxygen supply is introduced through the flame of burning hydrogen as shown in Fig. 8.5, the oxygen is seen to burn in the atmosphere of hydrogen in the tube. Thus oxygen will burn in an atmosphere of hydrogen, just as hydrogen will burn in an atmosphere of oxygen. In both cases the product is water. The formation of water is the result of chemical combination between the elements hydrogen and oxygen, during which process energy is released.

8.8 the effect of supplying energy to water

1. heat energy

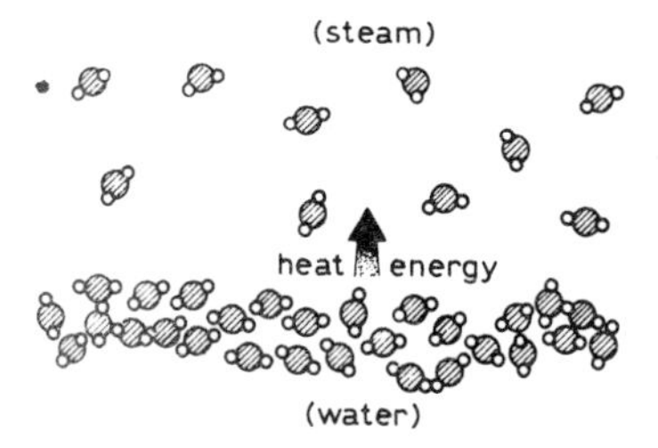

Fig. 8.6 heat energy separates the molecules in water, forming steam, but does not break them down

If water is heated to 100 °C at atmospheric pressure it boils, forming steam. The latent heat of vaporisation supplied (see section 2.7) does not split up the water into hydrogen and oxygen, for steam will neither burn nor rekindle a glowing splint. The energy absorbed by the water at its boiling-point is used to separate the *molecules* of water from their close-packed condition in the condensed liquid phase to the considerable distances apart which are characteristic of the gaseous phase, as shown in Fig. 8.6. Thus the application of heat energy to liquid water merely *separates the molecules from each other* and does not break down the molecules into their constituent elements.

2. chemical energy from a reactive element

An absorbent plug is placed at the bottom of a hard-glass test-tube and is soaked with water using a dropping pipette. A spiral of magnesium ribbon is placed in the tube, which is then fitted with a bung and a delivery tube as shown in Fig. 8.7. The tube is clamped horizontally near the bung and the magnesium is heated strongly until it appears to melt, whereupon the heat is transferred to the wet absorbent plug. Steam passes over the hot magnesium, which becomes incandescent as an inflammable gas issues from the delivery tube; this gas is obviously hydrogen. The white solid remaining in the tube can be shown to be identical with that formed when magnesium burns in oxygen: it is magnesium(II) oxide. Thus red-hot magnesium has sufficient energy to break down the water molecules in steam, absorbing the oxygen and liberating hydrogen gas.

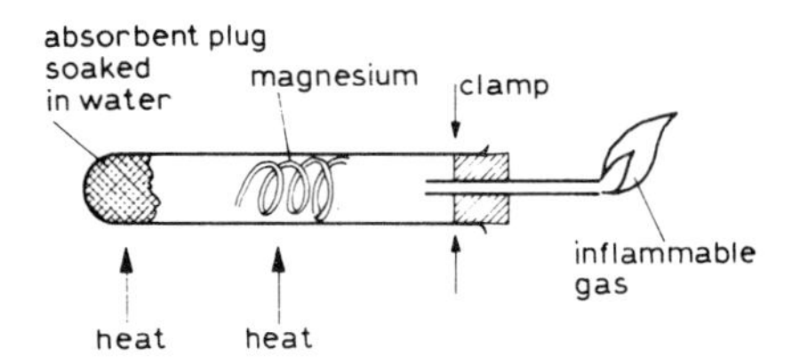

Fig. 8.7 action of steam on magnesium

3. electrical energy

An apparatus which permits electrical energy to be supplied to water is shown in Fig. 8.8. If pure water is used in the cell, no change is observed when the 6-volt d.c. supply is switched on. When a little sodium(I) chloride is added to the water in the cell, bubbles of gas stream from the electrodes and may be collected in tubes as shown. The electrode connected to the positive terminal of the electricity supply is called the **anode** and that connected to the negative terminal of the supply is called the **cathode.** The volume of gas collected at the cathode is twice that of the gas collected at the anode.

Simple tests on the gases collected reveal that hydrogen is produced at the cathode and oxygen at the anode; neither gas can have been produced by the sodium(I) chloride added to the water, as this substance contains sodium and chlorine only (see section 4.2). The application of electrical energy to water (electrolysis) decomposes water into the gases **hydrogen and oxygen in the ratio by volume of 2 : 1.**

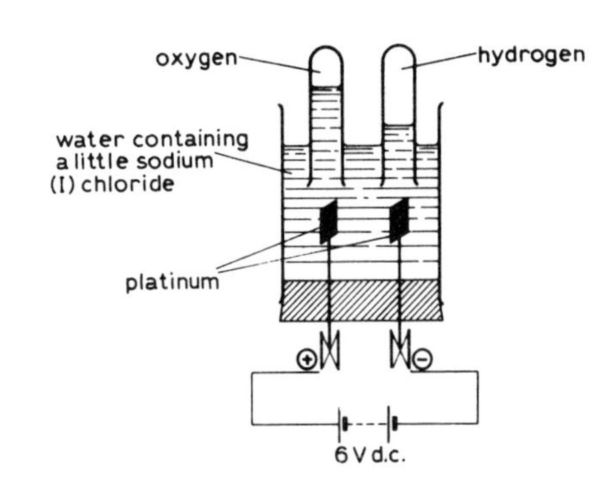

Fig. 8.8 electrolysis of water

8.9 ignition of a 2 : 1 volume mixture of hydrogen and oxygen

Since the electrical decomposition of water produces hydrogen and oxygen in the ratio by volume of 2 : 1, it ought to be possible to synthesise water from a 2 : 1 volume mixture of hydrogen and oxygen.

An empty plastic detergent container is one-third filled with oxygen and two-thirds filled with hydrogen by displacement of water, and the container is securely corked and shaken. The container is placed upright on a bench, the cork is removed and a light is applied to the open mouth of the container by means of a spill fixed to the end of a metre rule held at arm's length. Combination occurs with a deafening report which leaves witnesses in no doubt that a considerable release of energy accompanies the formation of water from a 2 : 1 volume mixture of hydrogen and oxygen. This particular ratio is obviously a highly significant feature of the composition of water.

8.10 determination of the ratio by mass of hydrogen to oxygen in water

Hydrogen combines so readily with oxygen that it will remove oxygen from certain oxides; this property may be employed to determine the composition of water by mass.

If dry hydrogen gas is passed over hot copper(II) oxide, copper and water are formed. The increase in mass of the whole apparatus during the experiment is equivalent to the mass of hydrogen which is absorbed in the formation of the water, while the decrease in mass of the copper(II) oxide is equivalent to the mass of oxygen absorbed in the formation of the same quantity of water. The ratio by mass of hydrogen to oxygen in water can therefore be calculated.

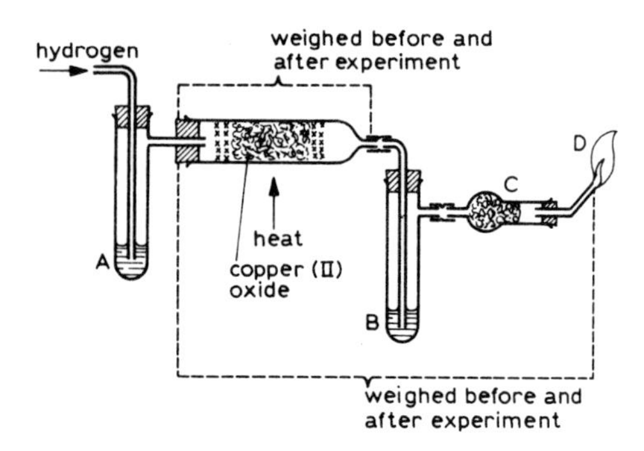

Fig. 8.9 determination of ratio by mass of hydrogen to oxygen in water

The appropriate parts of the apparatus shown in Fig. 8.9 are weighed, and hydrogen gas, preferably from a cylinder, is passed through the apparatus. When all the air has been displaced from the apparatus the excess gas is ignited at D. Concentrated sulphuric(VI) acid in tube A serves both to dry the incoming hydrogen and to indicate the rate of flow of gas. The copper(II) oxide is heated at one end and the flame is moved gradually along the tube as hydrogen is passed through the hot oxide for about 15 minutes. The water formed in the reaction is absorbed in concentrated sulphuric(VI) acid and calcium(II) chloride in tubes B and C respectively. When the apparatus is quite cool, the supply of hydrogen is turned off, the flame at D is extinguished, and dry air is drawn through it to displace the hydrogen. The apparatus is then disconnected and the appropriate sections re-weighed. A typical set of readings is given below:

mass of combustion tube and copper(II) oxide	=	37·253 g
mass of combustion tube and copper	=	37·095 g
mass of whole apparatus before experiment	=	106·800 g
mass of whole apparatus after experiment	=	106·820 g
mass of hydrogen in the water formed	=	(106·820 g − 106·800 g)
	=	0·020 g
mass of oxygen in the water formed	=	(37·253 g − 37·095 g)
	=	0·158 g

0·020 g of hydrogen combines with 0·158 g of oxygen

1 g of hydrogen combines with $\frac{0{\cdot}158}{0{\cdot}020} \times 1$ g of oxygen

$\approx$ 8 g of oxygen

The ratio by mass of hydrogen to oxygen in water is 1: 8

8.11 classification of water as element, compound or mixture

Water can be decomposed into simpler substances; it is therefore a complex substance and not an element. If the criteria of section 4.8 are applied to the case of water, we note that much energy is released on its formation, the properties of water are quite different from its components hydrogen and oxygen, it has a sharp melting point (0 °C) and a constant boiling point (100 °C), it is separated into its constituent elements hydrogen and oxygen only when energy is supplied to it and it has a definite composition both by volume and mass. Clearly, **water is a compound** of hydrogen and oxygen.

Evidence from the mass spectrometer (Chapter 9) reveals that the mass of an atom of oxygen is sixteen times that of the mass of an atom of hydrogen. The ratio by mass of oxygen to hydrogen in water is 8 : 1. It follows that there are twice as many hydrogen atoms as oxygen atoms in the water molecule and the formula for a water molecule can be written H_2O or a multiple of this (H_4O_2 or H_6O_3, etc.). Further evidence from the mass spectrometer (Fig. 9.9) reveals that the formula for a water molecule is indeed H_2O and further, that the two hydrogen atoms in the molecule are separately joined to the oxygen atom; the water molecule has the structure H—O—H rather than H—H—O. Water should therefore be considered to be hydrogen hydroxide rather than dihydrogen oxide.

8.12 the action of metals on water

1. sodium

If a piece of freshly cut sodium, about the size of a small pea, is dropped into a trough of cold water, it becomes spherical in shape and floats and skims over the surface of the water with a violent fizz; it may even inflame. In the trough, a clear solution remains which turns red litmus paper blue, showing that it is alkaline. The gas produced during this process may be identified using the apparatus of Fig. 8.10. When arranging the apparatus, it is important not to wet the inside walls of the glass tube, which should be of about 1 cm internal diameter. If very small pieces of sodium (about the size of finger-nail clippings) are used, the escaping gas may be ignited at the mouth of the tube and shown to be hydrogen.

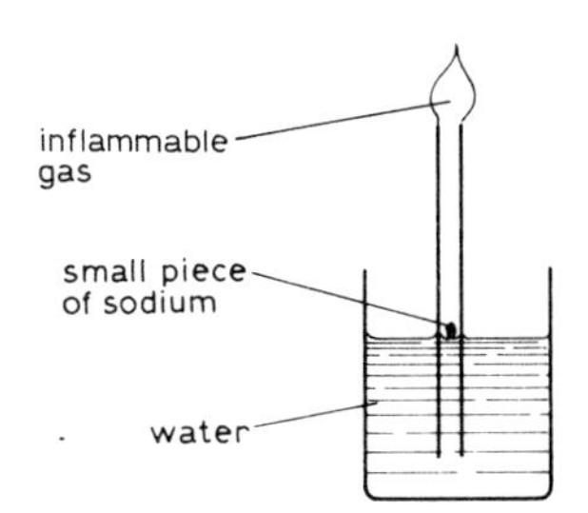

Fig. 8.10 action of sodium on water

Evaporation of the alkaline solution remaining in the beaker reveals a white solid; this solid clearly contains sodium and oxygen, since the only substances lost during the experiment are hydrogen gas and water during evaporation–it may or may not also contain some hydrogen, depending on whether the sodium has replaced one or both of the atoms in the water molecule. If the white solid is mixed with dry powdered aluminium and the mixture is heated strongly, an inflammable gas is evolved and since this gas cannot be sodium, oxygen or aluminium it is clearly hydrogen. It therefore appears that when sodium reacts with water, it replaces only one of the hydrogen atoms of the water molecule, liberating hydrogen gas and leaving in solution a compound containing sodium, oxygen and hydrogen–sodium(I) hydroxide:

$$\text{sodium} + \text{water} \longrightarrow \text{sodium(I) hydroxide} + \text{hydrogen}$$

2. potassium

A small piece of freshly cut potassium dropped into a trough of cold water produces a reaction similar to that of sodium but even more violent: the potassium burns with a lilac-coloured flame. Experiments similar to those described in the preceding section establish that the gas evolved is hydrogen and the substance remaining in solution is potassium(I) hydroxide:

$$\text{potassium} + \text{water} \longrightarrow \text{potassium(I) hydroxide} + \text{hydrogen}$$

3. calcium

An ignition tube full of water is inverted in a beaker of water and a large piece of calcium metal is dropped into the beaker. The calcium sinks to the bottom of the beaker and bubbles of gas rise rapidly from it; these bubbles may be collected in the ignition tube and the gas ignited when the tube is full–it proves to be hydrogen. The remaining solution is alkaline to litmus but is cloudy in nature, suggesting that the calcium(II) hydroxide formed is not very soluble in water.

$$\text{calcium} + \text{water} \longrightarrow \text{calcium(II) hydroxide} + \text{hydrogen}$$

4. other metals

Similar experiments using aluminium, copper, iron, magnesium and zinc show that these metals have no action with cold water, although magnesium powder reacts slowly with hot water. Apparently these metals possess less chemical energy than potassium, sodium and calcium and are insufficiently energetic to decompose cold water. The extra energy present in steam may be sufficient to cause decomposition using these metals.

8.13 action of metals on steam

Steam is generated by heating a water-soaked absorbent plug at the bottom of a hard-glass tube and passed over a heated metal which is held in position half-way up the tube by loose plugs of inert wadding (see Fig. 8.11). Any gas issuing from the delivery tube may be tested with a lighted splint; if it burns it may be assumed to be hydrogen. Hydrogen is released from steam by magnesium, zinc, aluminium and iron (the latter with difficulty) but not by copper, and the vigour of the reaction decreases from magnesium to iron. By comparison with the products formed when the metal burns in oxygen, the product of reaction in this case may be shown to be the oxide of the metal

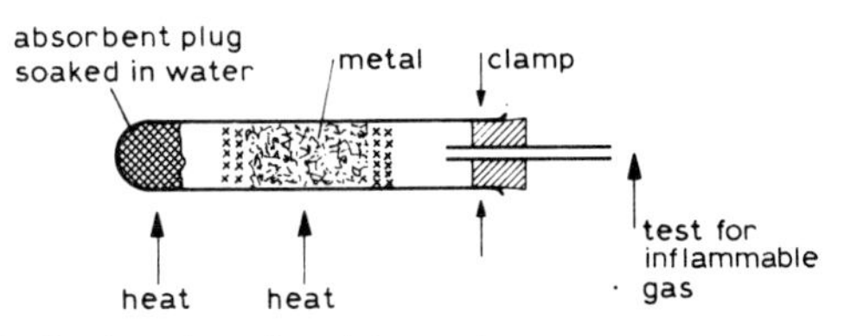

Fig. 8.11 action of metals on steam

and not the hydroxide. This is consistent with the fact that most hydroxides decompose on heating leaving the oxide.

magnesium + steam ⟶ magnesium(II) oxide + hydrogen

8.14 relative reactivity of metals with water

potassium
sodium
calcium
magnesium
zinc
aluminium
iron
(hydrogen)
copper

Fig. 8.12

Fig. 8.12 shows the order of reactivity of the metals used towards water, either in the cold or as steam. Hydrogen is included for reference, showing that the metals above it will displace it from water, whereas the metal below it will not. Potassium, at the top of the table, is the most reactive metal of those used and copper is the least reactive.

It was stated in section 7.6 that a substance is said to be oxidised if oxygen is added to it and that a substance which gives oxygen to another substance is called an oxidising agent or oxidant. *If oxygen is removed from a substance, the substance is said to be reduced*; a substance which removes oxygen from another substance is said to be a **reducing agent** or **reductant.** A full treatment of oxidation and reduction processes is given in Chapter 14.

The experiments described in section 8.10 and 8.22 show that hydrogen is a reducing agent, as it will remove oxygen from copper(II) oxide. Note the relative positions of hydrogen and copper in Fig. 8.12. The metals above hydrogen in the table are reducing agents, as they will remove oxygen from water or steam. The table gives an order of the reducing ability of the metals, with the most powerful reductant (potassium) at the top.

8.15 action of metals on hydrochloric acid

In section 5.3(2) we saw that pure water, which was not decomposed by electrical energy, could be so decomposed if a little acid was added to it. It may be that a dilute aqueous acid is more readily decomposed by a metal than is pure water. In the following experiment, only those metals are used which will not release hydrogen from cold water; those which do would produce an explosive reaction with acids.

A little of the powdered metal under investigation is placed in a test-tube and sufficient dilute hydrochloric acid is added to one-third fill the tube. If reaction is not rapid, the tube may be warmed. Any gas evolved is tested for inflammability and the vigour of the reaction is noted in each case. Magnesium, zinc, aluminium and iron are found to liberate hydrogen from dilute hydrochloric acid, the vigour of the reaction decreasing in this order. Copper does not react. The order of reactivity of the metals used in this experiment is seen to be identical with that shown in Fig. 8.12. This simple reactivity table is closely related to the electrochemical series described in section 14.24.

8.16 water of crystallisation

The energy changes associated with the interconversion of copper(II) sulphate(VI) between its white anhydrous form and its blue hydrated form were discussed in section 5.3 (1)(a). The water present in the dry blue hydrated crystals was referred to as water of crystallisation; it is chemically bonded in the crystal and energy is released when such bonds are formed. To remove water from the blue crystals, such links must be broken and energy is necessary to do this.

When copper(II) sulphate(VI) crystallises from aqueous solution, blue crystals are produced which can be shown on analysis to contain 5 moles of water for each mole of copper(II) sulphate(VI); the meaning of the term mole is explained fully in Chapter 10. The water present in the crystals has lost its characteristic property of 'wetness', for it is chemically united in the crystal to form a compound. This is represented in the full name for the blue crystals, which is copper(II) sulphate(VI)-5-water. The correct method of writing the formula for the crystals is indicated in the following equation:

$$CuSO_4.5H_2O(s) \rightleftharpoons CuSO_4(s) + 5H_2O(g)$$

(used to indicate chemical attachment of the water in the compound

Some substances crystallise from solution without water of crystallisation, e.g. lead(II) nitrate(V), sodium(I) chloride, sodium(I) nitrate(V) and all potassium salts. Other compounds crystallise with varying amounts of water of crystallisation, e.g. cobalt(II) chloride-6-water, magnesium(II) chloride-6-water, iron(II) sulphate(VI)-7-water, barium(II) chloride-2-water.

8.17 water in air

The early morning dew on grass and on the windscreens of motor-cars is a familiar reminder of the water content of the atmosphere which was demonstrated in the experiment described in section 6.1. The water vapour in air can react with certain chemical substances.

If some pellets of sodium(I) hydroxide are exposed to the air for a short time they become shiny and sticky and after prolonged exposure to the air a colourless liquid is formed. The liquid is concentrated aqueous sodium(I) hydroxide, formed by the solid dissolving in water absorbed from the atmosphere. Anhydrous calcium(II) chloride behaves in a similar way. Substances which absorb water from the atmosphere in sufficient quantities to dissolve in it are termed **deliquescent,** the process being known as **deliquescence.** Deliquescent substances are widely used in the laboratory as drying agents; anhydrous calcium(II) chloride is a well-known example of such a substance. Many other substances absorb water from the atmosphere without dissolving in it: newspaper stored for a long period will feel damp even though it has been kept in a relatively dry place, and glass heated in a Bunsen flame is initially seen to 'sweat'. Many fine powders absorb water in this way, apparently without becoming wet. Copper(II) oxide does this and when it is to be used in an analysis involving water (as in section 8.10) it must be heated strongly and cooled in a desiccator before use. Substances which absorb water from the atmosphere without dissolving in it are termed **hygroscopic.** A well-known example is concentrated sulphuric(VI) acid, which is widely used as a drying agent.

Some hydrated crystals crumble to a powder if exposed to the atmosphere; this is caused by the breakdown of the crystal lattice consequent on loss of water of crystallisation from the lattice to the atmosphere. Such substances are said to **effloresce** and the process is known as **efflorescence.** A familiar example is the conversion of colourless crystals of 'washing-soda', sodium(I) carbonate-10-water, to the white powdery sodium(I) carbonate-1-water on exposure to air.

It is instructive to trace the chemistry involved in the changes which occur to a freshly cut piece of sodium which is left exposed to the atmosphere for a few days on a watch-glass. The main features in the chain of changes are as follows:

i) bubbles of gas appear on the surface of the metal
ii) a colourless solution is formed
iii) colourless crystals appear from the solution
iv) the colourless crystals slowly change to a white powder which fizzes when an acid is added.

8.18 uses of water

Water is such an important and widely used substance that an exhaustive list of its uses is impossible to make. The following list constitutes only a selection.

i) *hydro-electric power* In addition to the familiar examples of the conversion into electrical energy of the potential energy of the water in lakes and reservoirs, schemes are presently being devised to utilise the ebb and flow of tides in driving turbines.

ii) *water-cooling in industry*

iii) *as a solvent*

iv) *domestic consumption* Most of the solids which we eat, as well as virtually all liquids which we drink, have a high water content.

v) *irrigation* In parts of the world where the regularity of rainfall is inadequate to sustain crops, expensive and complicated systems of irrigation are necessary to supply water to the land.

hydrogen

8.19 reactions generating hydrogen

Although hydrogen may be obtained by the electrolysis of water and by the action of reactive metals on cold water or steam, the most convenient method for generating a supply of hydrogen in the laboratory is by the action of a moderately reactive metal on a dilute aqueous acid. Nitric(v) acid cannot be used for this purpose, as the nitrate(v) ion in the aqueous nitric(v) acid has oxidising properties which interfere with the production of hydrogen. The combination usually used is zinc with dilute sulphuric(vi) acid.

If a piece of pure zinc is placed in a test-tube and dilute sulphuric(vi) acid is added to it, very little reaction occurs. The addition of a few drops of aqueous copper(ii) sulphate(vi) to the contents of the tube produces a smooth and rapid evolution of gas. The explanation of the action of the aqueous copper(ii) sulphate(vi) is by no means simple, but it may be partly understood by placing a separate piece of pure zinc in a second test-tube and covering it with aqueous copper(ii) sulphate(vi); on pouring off the blue solution and washing the remaining solid with water, the zinc is seen to be coated with a red-brown deposit which is obviously copper. The zinc has displaced copper from the aqueous copper(ii) sulphate(vi): this process is explained in section 14.5. The zinc and copper together form a metallic 'couple' having more powerful reducing properties than zinc itself.

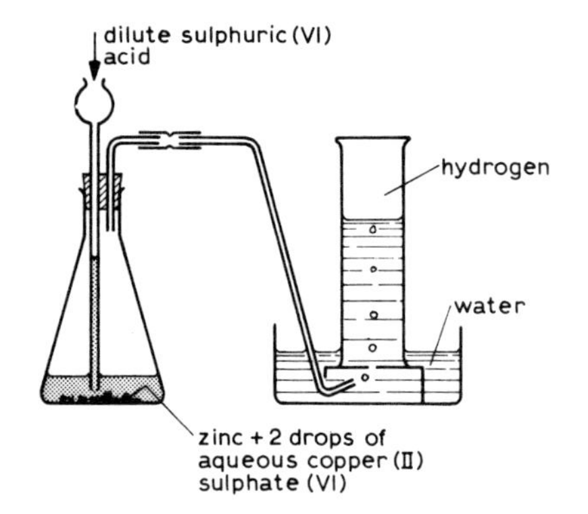

Fig. 8.13 preparation and collection of hydrogen

8.20 laboratory preparation of hydrogen

Several jars of hydrogen may be obtained using the apparatus shown in Fig. 8.13; the lower end of the thistle-funnel must dip below the surface of the liquid in the flask and the first jar of gas must be rejected as it will be contaminated with air displaced from the flask. Remembering the danger of igniting a mixture of hydrogen and oxygen (section 8.9) it is advisable to reject the first two jars of gas collected before carrying out any tests on hydrogen prepared in this way. If the gas is required dry, it may be bubbled through concentrated sulphuric(vi) acid and collected by downward displacement of air as explained in section 15.7.

Hydrogen is one of three gases (the other two are carbon dioxide and hydrogen sulphide) which are conveniently obtained in the laboratory using **Kipp's apparatus,** shown in Fig. 8.14. Large pieces of impure zinc are placed in the central compartment and acid is added to the top compartment; when the acid reaches the zinc, reaction occurs and the hydrogen escapes through the open tap. When the tap is closed, the pressure developed by the hydrogen trapped in the central compartment forces the acid away from the zinc into the upper compartment. The apparatus may be stored in this state until a supply of hydrogen is required once more.

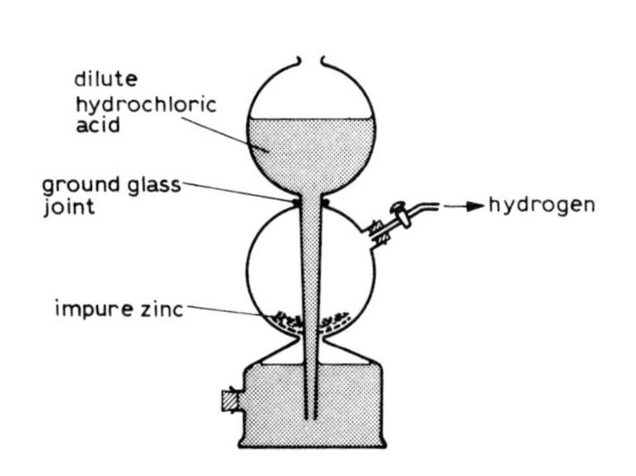

Fig. 8.14 Kipp's apparatus for generating hydrogen

8.21 physical properties of hydrogen

Hydrogen is colourless and in the pure state is odourless and tasteless, though the gas prepared in the laboratory usually possesses a slightly fishy smell because of impurities in the metal and acid from which it is prepared. It requires a high pressure and a very low temperature to liquefy it. It is insoluble in water and may be collected by displacement of water as described in the preceding section.

Hydrogen is the least dense gas known; one litre of the gas at s.t.p. has a mass of 0·09 g. Its low density is strikingly demonstrated by the following experiments.

a) If the wide mouth of a thistle-funnel is dipped into a bowl of diluted detergent or 'bubble mixture' while a slow stream of hydrogen (free from acid spray) is passed into the stem of the funnel, bubbles full of hydrogen are formed which rise rapidly in the air. A member of the group standing on the bench next to this experiment holding a lighted taper will give his speed of reaction a fair test if asked to ignite the bubbles before they reach the ceiling.

b) If two jars of hydrogen are clamped side by side, A mouth upwards and B mouth downwards (Fig. 8.15), the application of a lighted splint to the mouth of each jar seven or eight seconds after the cover-slips have been removed gives no result with jar A and a loud squeak or pop with jar B. The difference in behaviour is explained by the low density of hydrogen. If the experiment is repeated, allowing 30 seconds to elapse before application of the splint, no explosion occurs in either case, showing that hydrogen will diffuse even out of jar B in spite of its low density.

c) It is quite a commonplace operation to pour water from one gas jar into a second empty gas jar without spilling any. It is also possible to pour hydrogen from one jar into a second empty jar using exactly the same technique, but holding the jars 'upside-down'. The efficiency of the process may be tested by applying a lighted splint to the second jar.

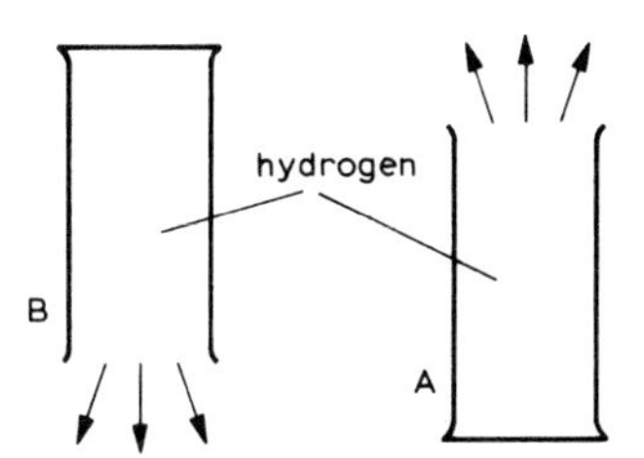

Fig. 8.15 hydrogen will diffuse downwards in spite of its low density

8.22 reactions of hydrogen

1. *with oxygen*

If a long splint is ignited and plunged well inside a jar of hydrogen which is clamped mouth downwards, the splint inside the jar is extinguished while the gas burns at the mouth of the jar. Hence **hydrogen burns in air but does not support combustion** (see reciprocal combustion, section 8.7). The explosive reaction of a 2 : 1 volume mixture of hydrogen with oxygen is described in section 8.9.

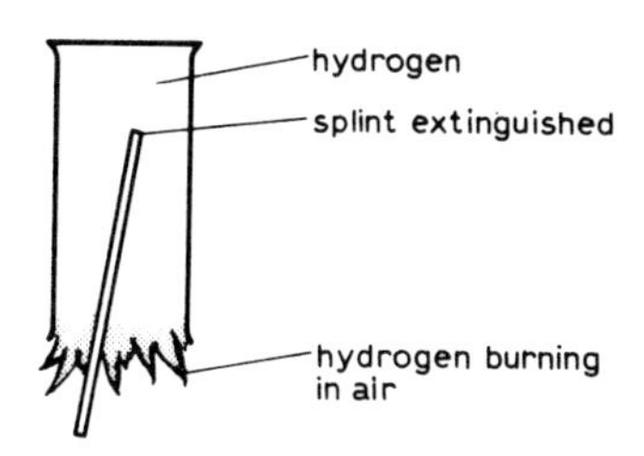

Fig. 8.16 hydrogen burns but does not support combustion

2. *with halogens (see Chapter 15)*

A mixture of equal volumes of hydrogen and chlorine explodes violently on exposure to direct sunlight or on ignition, yielding hydrogen chloride. Hydrogen will also burn in an atmosphere of chlorine:

$$\text{hydrogen} + \text{chlorine} \longrightarrow \text{hydrogen chloride}$$

A similar but less violent reaction occurs between hydrogen and bromine vapour, whereas the reaction between hydrogen and iodine vapour is feeble and reversible.

3. *as a reductant*

The energy released when hydrogen reacts with oxygen is sufficient to decompose many metal oxides; hence hydrogen will reduce copper(II) oxide and lead(II) oxide to copper and lead respectively. This may be demonstrated in the apparatus shown in Fig. 8.17. The supply of hydrogen should be allowed to pass through the tube for 20 seconds before igniting the excess gas at the small hole A. Gentle heating of the oxide is usually sufficient to initiate the reaction, which then proceeds without further heating, showing that the process liberates energy. At the end of the experiment, the flame at A should

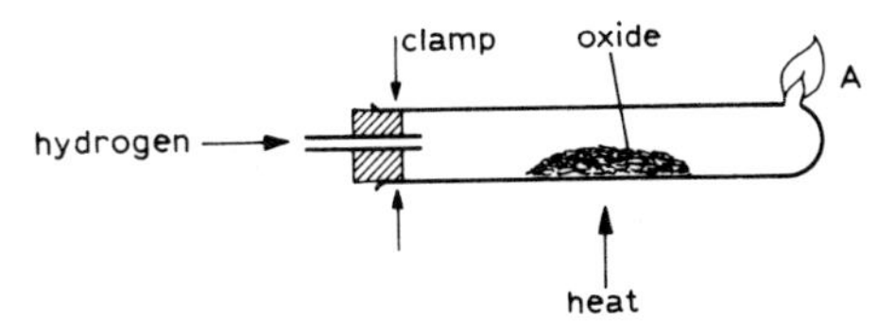

Fig. 8.17 reduction of oxides using hydrogen

be extinguished carefully *before* the bung is removed from the test-tube. Why?

hydrogen + copper(II) oxide ⟶ copper + water (steam)
hydrogen + lead(II) oxide ⟶ lead + water (steam)

4. with metals

Metals in Groups I and II of the Periodic Table combine directly with hydrogen on heating forming solid crystalline hydrides:

e.g. sodium + hydrogen ⟶ sodium(I) hydride
calcium + hydrogen ⟶ calcium(II) hydride

These hydrides are rapidly decomposed on contact with water.

5. with nitrogen

In the presence of a catalyst, hydrogen can be made to combine directly with nitrogen to form ammonia. This important process, known as the Haber process, is described in Chapter 24.

8.23 occurrence of hydrogen in nature

The atmosphere of the earth contains only about one part of hydrogen in 20000 parts of air, but the sun is believed to consist largely of a mass of very hot hydrogen. 'Solar energy' is thought to be the result of the thermonuclear conversion of hydrogen to helium. Hydrogen occurs widely on our planet in combination with oxygen as water, in combination with carbon as petroleum and natural gas, and in combination with carbon, oxygen and other elements as a constituent of all plant and animal tissue.

8.24 uses of hydrogen

1. in the Haber process

Large quantities of ammonia gas are needed for the production of fertilisers; this is produced by the direct combination of nitrogen with hydrogen (see Chapter 24).

2. for hydrogenation

Hydrogenation is the addition of hydrogen to 'unsaturated' organic compounds (see section 19.9). Hydrogenation of certain liquid animal and vegetable oils hardens them into solid fats (margarine), while unsaturated hydrocarbons from the 'cracking' of petroleum are hydrogenated in the production of petrol.

3. as a fuel

Hydrogen is a major constituent of coal-gas and town-gas which until recently were our main domestic fuels. A mixture of hydrogen and carbon monoxide (water-gas) is widely used as an industrial fuel. Some rockets are powered by the combustion of hydrogen in oxygen, this familiar reaction producing the immense energy required to thrust such projectiles into outer space.

4. the oxy-hydrogen flame

The very high temperature of this flame (2000 K) is employed in welding and cutting metals.

5. liquid hydrogen as a coolant

Liquid hydrogen boils at 20 K and is consequently used as a coolant for liquefying other gases and for the investigation of the behaviour of substances at very low temperatures.

6. for levitation

Being the least dense of all gases, hydrogen was commonly used for filling balloons and airships until a series of terrible disasters brought the era of the airship to a close. It has now been largely replaced by the non-inflammable gas helium.

8.26 large-scale manufacture of hydrogen

Hydrogen is readily produced industrially in a mixture with carbon monoxide, either by the action of steam on white-hot coke or by the action of steam under pressure on hydrocarbons from petroleum, in the presence of a catalyst. The carbon monoxide is eliminated from such mixtures by passing the gases under pressure, together with excess steam, over a heated iron(III) oxide catalyst. Much of the carbon monoxide is oxidised to carbon dioxide and more hydrogen is produced. The carbon dioxide is removed from the gas mixture by dissolving it in water under pressure and the hydrogen is subsequently dried.

revision summary: Chapter 8

water:	*formation:*	exothermically from hydrogen and oxygen
	properties:	liquid, pH 7, m.p. 273 K (0 °C), b.p. 373 K (100 °C)
	tests:	physical constants, turns anhydrous $CuSO_4$ blue
	reactions:	electrolyses to hydrogen and oxygen if impure
		reactive metals liberate hydrogen
		exists in salts as water of crystallisation
hydrogen:	*preparation:*	zinc (impure) on dilute sulphuric(VI) acid
		$Zn(s) + H_2SO_4(aq) \longrightarrow ZnSO_4(aq) + H_2(g)$
	collection:	over water, or by downward displacement of air
	reactions:	burns in oxygen and in chlorine
		reduces oxides of unreactive metals
	test:	burns when ignited in air, often with squeak or pop

questions 8

1. Find out all you can about, and write notes on, the following topics:
(a) pollution of river water,
(b) pollution of sea water,
(c) the use of water for the production of energy.

2. Ordinary tap-water always contains some air in solution. Describe in detail how you would collect a quantity of this air from tap-water. How could you find the proportion by volume of oxygen in such air? Explain why the composition of this dissolved air will be different from that of 'ordinary' air. (JMB)

3. Imagine that you wish to prove experimentally to someone that 18 g of water contain 2 g hydrogen and 16 g oxygen. Give a labelled sketch of the apparatus you would use; indicate the two chief precautions you would adopt to ensure an accurate result; show how you would use the data you obtain to prove the above statement. (JMB)

4. Describe how you would obtain hydrogen from water by a chemical reaction. How would you use this gas to synthesise a small sample of water from its elements?
(b) Describe briefly (i) a test for the presence of water in a liquid, (ii) a test to show that a sample of water contains no dissolved solid impurities, (iii) **two** physical measurements to prove that a given liquid is pure water. (C)

5. Draw a labelled diagram of the apparatus you would use to prepare and collect gas jars of **either** oxygen **or** hydrogen (**not** by electrolysis).

Describe briefly **three** experiments you have seen that demonstrate physical or chemical properties of the gas that you have chosen above. In any chemical reaction mentioned, name and describe the product(s). Give **two** important different uses of **each** gas (excluding balloons). (C)

6. How would you prepare dry hydrogen and show that it forms a liquid when burned in air? Draw the apparatus you would use. By what experiments could you show that the liquid (a) contained water, (b) contained nothing but water? Give full details. (JMB)

7. Describe a laboratory method for the preparation of hydrogen. Give two uses of hydrogen. Describe experiments to show (a) that the gas has a low density, (b) that when it burns, water is formed, (c) that it is a reducing agent.

Explain why a beaker full of cold water 'mists' on the outside when it is first placed over a Bunsen flame.

8. State the reactions (if any) of sodium, calcium, iron and copper with (a) oxygen, and (b) water, giving the necessary conditions. From these reactions deduce the relative positions of the four metals in the 'activity series'.

State **two** reactions by which the relative positions of iron and copper could be confirmed. (AEB)

9. Describe the action (if any) of (a) water, (b) dilute hydrochloric acid and (c) oxygen, on the metals calcium, copper, iron and sodium. In what order (most active first) do you deduce that these metals should be placed in the electrochemical series? Explain your deductions. (S)

10. Dry hydrogen was passed over 3·18 g of copper(II) oxide heated in a hard-glass test-tube. When the reduction was complete, 2·54 g of copper remained and 0·72 g of water was collected. Calculate the composition of water by mass. Draw a labelled diagram of the apparatus necessary to perform the above experiment.

11. What is water of crystallisation? Give the formulae for two substances containing water of crystallisation. Describe and explain what happens when an efflorescent substance is exposed to air. (O)

12. What is meant by 'deliquescence' and 'efflorescence'? Samples of each of the following substances were weighed on watch-glasses and left in the air for some time before being re-weighed. State in each case whether there would be any gain or loss in mass, giving any necessary explanation: (i) anhydrous calcium(II) chloride, (ii) sodium(I) carbonate-10-water, (iii) concentrated sulphuric(VI) acid.

13. 'Water is so often regarded solely as a solvent that its capacity to take part in a chemical change is sometimes overlooked.'

Classify the behaviour of water into the following types of reaction. Summarise one experiment to illustrate each type of reaction.

(i) The formation of hydrates.
(ii) The action of water on elements.
(iii) The action of water with substances to form acidic solutions.
(iv) Hydrolysis.
(v) The formation of hydrogen and oxygen from water.

L (NUFFIELD)

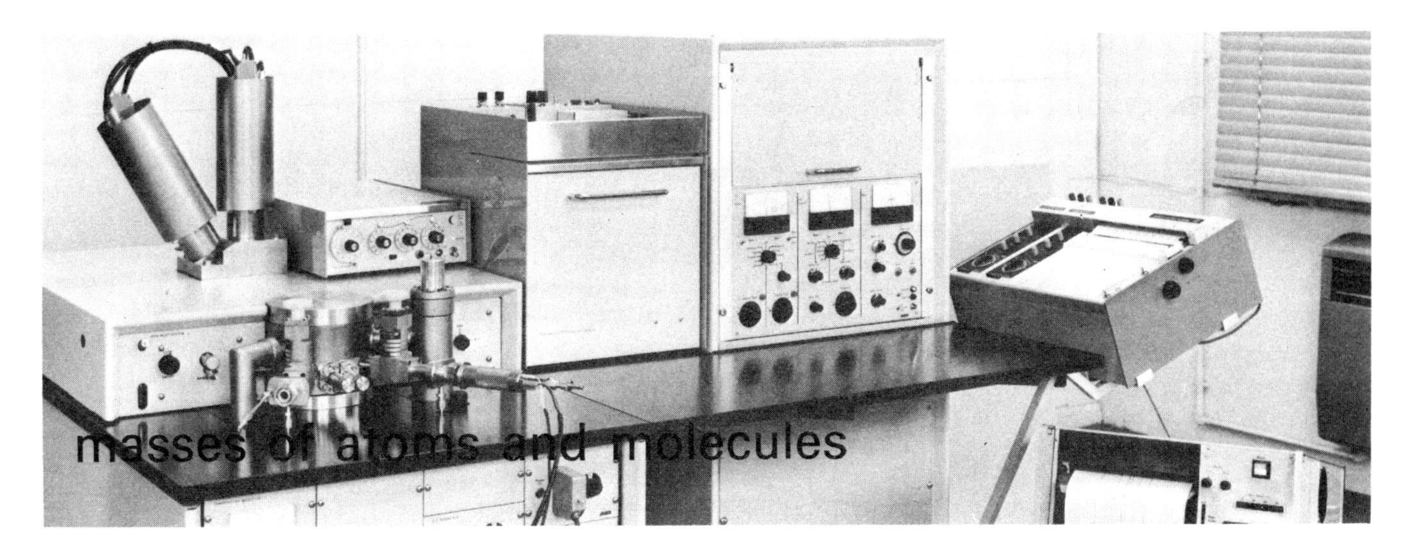

9.1 splitting molecules into atoms

We learned in Chapter 2 that energy must be supplied to convert liquid water into steam, the energy being necessary to overcome the forces of attraction between the closely packed water molecules and separate them to the greater distances characteristic of the gaseous state. In this process each molecule is not split apart; it is merely separated from its neighbouring molecules.

Molecules are assemblages of smaller particles called atoms. In this Chapter we will seek evidence to justify certain important assumptions concerning atoms and molecules. The assumptions are:

a) Atoms have characteristic masses. Those of the same element have the same mass, those of different elements have different masses (see section 1.3 for the distinction between mass and weight).

b) Atoms are held together in molecules by *strong* forces of attraction, often called **bonds.** These attractive forces are usually much stronger than the forces holding neighbouring molecules to each other in the solid or liquid states; the energy required to overcome such forces between atoms is correspondingly greater than that required to separate molecules from each other (Fig. 9.1).

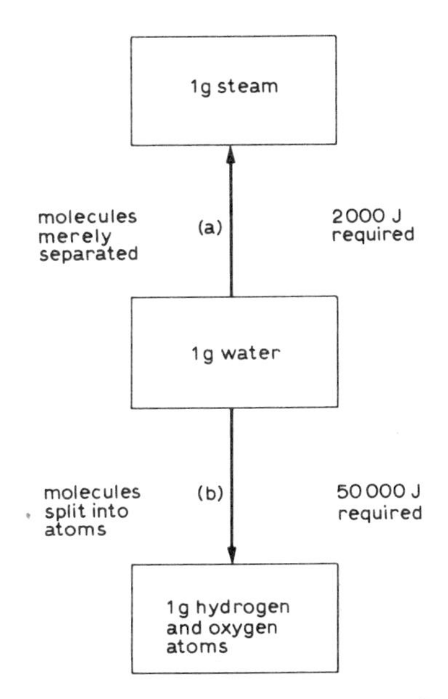

Fig. 9.1 energies required to separate water into (a) molecules, (b) atoms

9.2 masses of molecules and atoms

Atoms and simple molecules are much too small to be seen (see section 2.12). Fortunately, much can be learned about them by examining the behaviour which results from their differences in mass.

Imagine that you closed your eyes and that you were struck on the head by three objects in turn travelling at the same speed: a light plastic ball, a tennis ball and a cricket ball. You would not need to see them to tell which was which. It would be possible to identify them by the different effects on you caused by their differences in mass. Now imagine throwing each of these objects in a strong cross-wind: which would be deflected from its original path most? Which would be deflected least?

If a large box were placed in position to catch the plastic ball (Fig. 9.3(a)), it would also catch all other plastic balls, thrown similarly. In Fig. 9.3, P represents plastic balls, T tennis balls and C cricket balls. If the strength of the wind were gradually increased, then at a certain wind strength the tennis balls would be collected in the box and the others would miss (Fig. 9.3(b)). In a very strong wind the cricket balls would be collected and the others would miss (Fig. 9.3(c)).

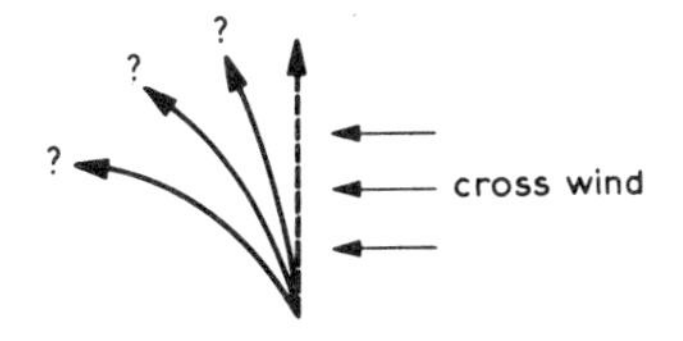

Fig. 9.2 which object will be deflected least in a cross-wind ?

Objects of great mass require more energy to make them move than do objects of small mass. Once they are moving, objects of great mass are more difficult to stop or deflect than are objects of small mass.

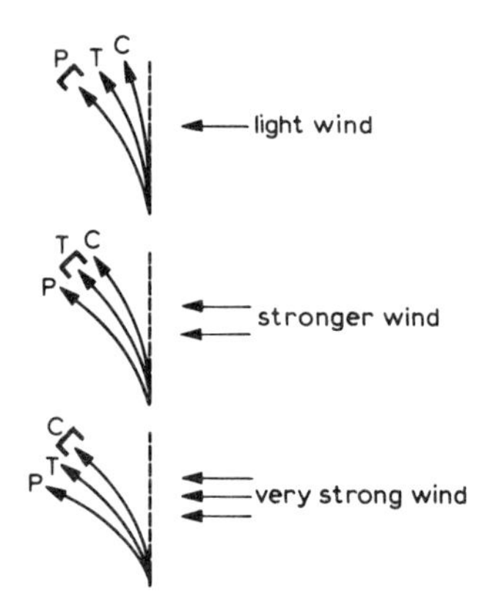

Fig. 9.3 a box in a fixed position relative to the source catches each object in turn as the wind strength increases

This simple principle, applied to molecules and atoms in an instrument called the mass spectrometer, enables us to learn much about these invisible particles.

9.3 the mass spectrometer

The substance to be examined is vaporised under the influence of a very high vacuum. This has the effect of separating the molecules. The molecules in the sample are then bombarded by a stream of tiny, negatively charged particles (electrons) which are emitted from their source at high speed. The bombarding electrons have the surprising effect of imparting a *positive* charge to each molecule; they are also sufficiently energetic to split many (but not all) of the molecules into smaller fragments. All possible fragments may be produced, down to individual atoms; each fragment has a positive charge imparted to it.

The positively charged molecules and fragments are attracted to a negatively charged plate containing a slit (see Fig. 9.4), through which they pass at high speed. They then enter a magnetic field (which has the effect on the charged particles of the 'cross-wind' of the previous section) and are deflected according to their mass. The unfragmented molecules, having the greatest mass, are deflected least and the individual atoms, being fragments of low mass, are deflected most. The atoms of smallest mass are deflected most of all.

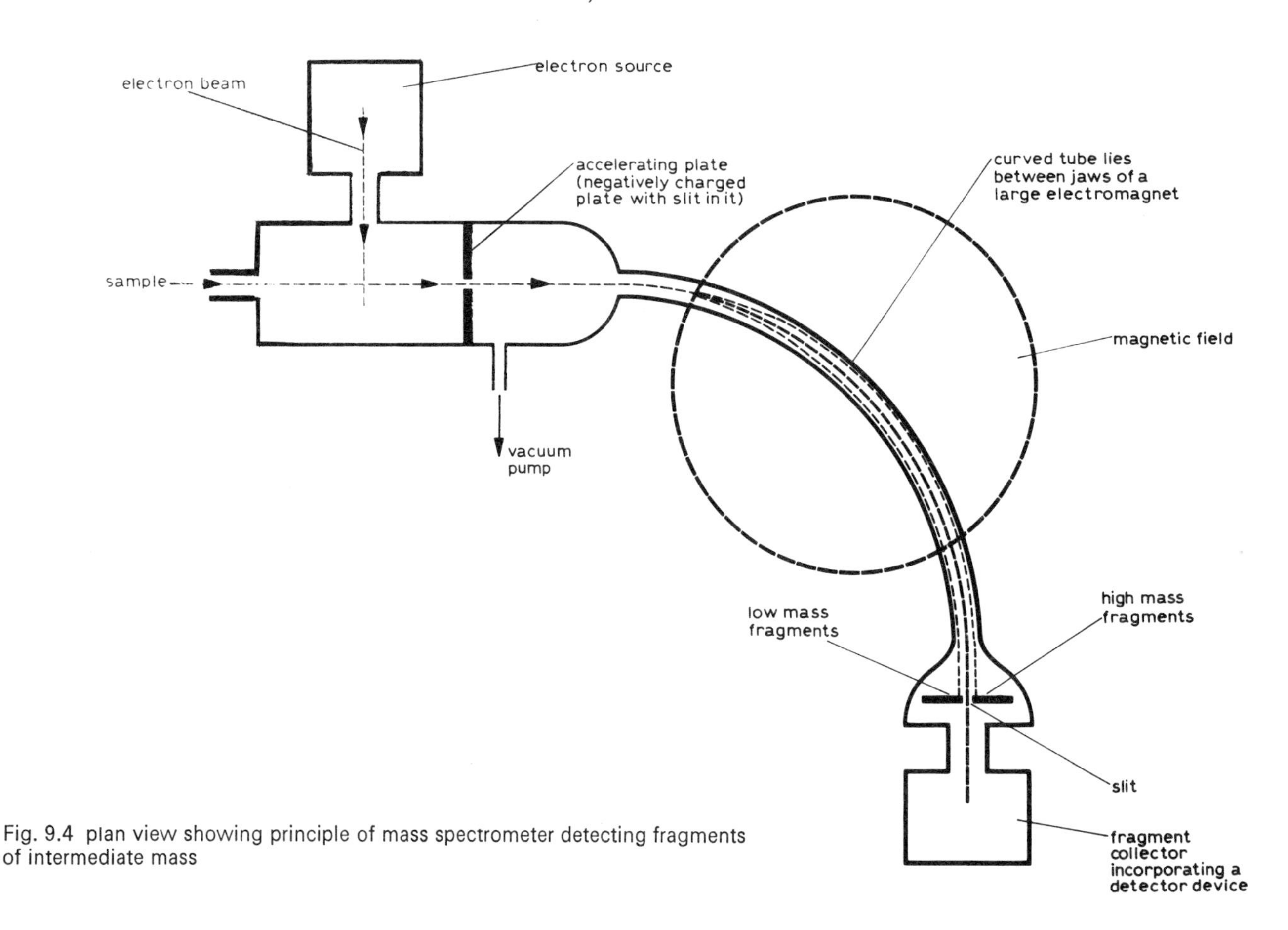

Fig. 9.4 plan view showing principle of mass spectrometer detecting fragments of intermediate mass

The streams of particles, now separated according to their masses, can be made to fall in turn on to a detecting device by varying the intensity of the magnetic field (this corresponds with varying the 'wind strength' of the previous section). The stream of charged particles falling on the detector constitutes an electric current, the magnitude of the current being a measure of the abundance of that type of particle. The current produced is recorded as a peak on a chart. The horizontal position of the peak is associated with the strength of the applied magnetic field, which in turn is associated with the mass of the particle:

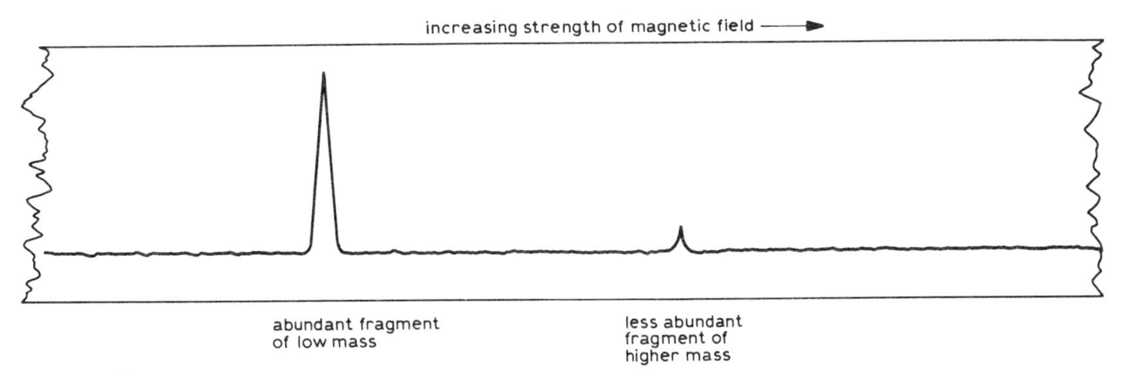

Fig. 9.5 the form and meaning of a mass spectrograph

Since the relative abundance of fragments is of no great significance for the purposes of this discussion, subsequent examples will show the peaks of equal heights. The vitally important feature is the horizontal spacing of the peaks.

9.4 interpretation of mass spectrographs

The chemistry of hydrogen, oxygen and water contains features which can be explained only if we can answer questions like 'how many atoms are there in the molecules of hydrogen, oxygen and water?'; 'what are the masses of hydrogen and oxygen atoms?'; 'how are the atoms arranged in the molecules?' The mass spectrometer provides the answers to these questions and to many others.

1. hydrogen

If hydrogen is studied in the mass spectrometer, a chart showing only two peaks is obtained. The more massive particle must be the unfragmented hydrogen molecule and the less massive must be a single hydrogen atom. **The hydrogen molecule thus contains two hydrogen atoms**: it is said to be **diatomic.** Since the hydrogen atom is represented symbolically as H, the hydrogen molecule is written H_2.

It is possible to calculate that the mass of a hydrogen atom is $1{\cdot}66 \times 10^{-27}$ kg, but the kilogram is not an appropriate unit in which to express masses as small as this. Since no atom has a mass smaller than that of a hydrogen atom, the masses of other atoms can be expressed as the number of times greater they are than the mass of a hydrogen atom. This number is called the **relative atomic mass** (A_r) of the element (a fuller discussion of this topic is given in section 18.7, *3*).

Molecular masses can be expressed on the same scale of H = 1 unit; the hydrogen molecule has a mass of $3{\cdot}32 \times 10^{-27}$ kg and therefore has a **relative molecular mass** (M_r) of 2 units. This information is summarised in Fig. 9.6.

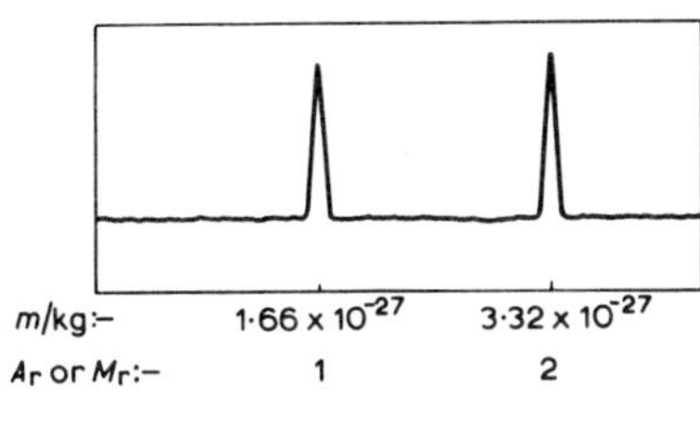

Fig. 9.6 hydrogen

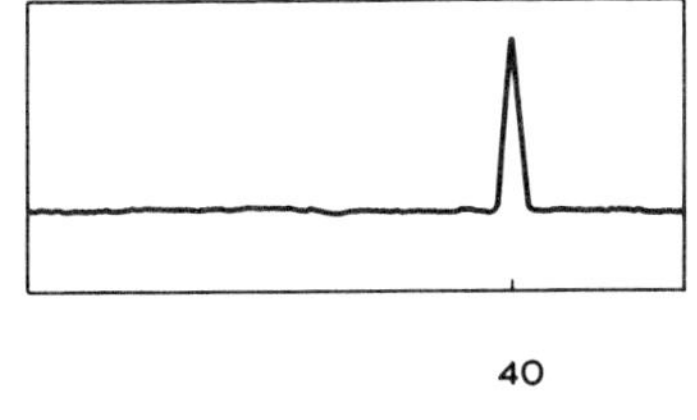

Fig. 9.7 argon

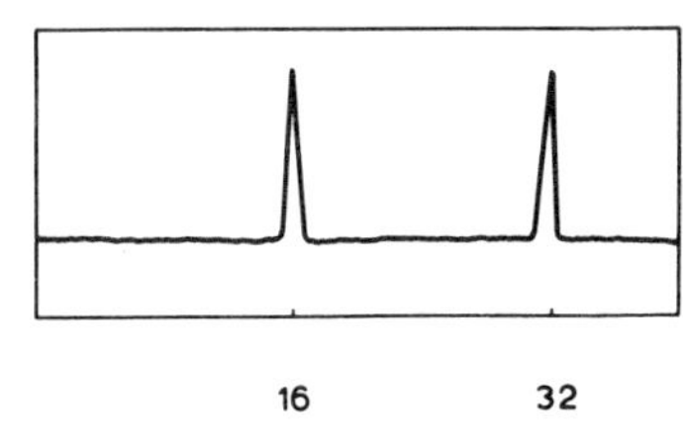

Fig. 9.8 oxygen

2. argon

The noble gas argon gives a mass spectrograph with only one peak at a relative atomic mass of 40 (Fig. 9.7). Argon molecules consist of single atoms and the element is said to be **monatomic.** All the noble gases have monatomic molecules (see section 6.5).

3. oxygen

The mass spectrograph for oxygen (Fig. 9.8) tells us that

a) its relative molecular mass is 32,
b) its relative atomic mass is 16,
c) the oxygen molecule is diatomic: it is written O_2.

4. water

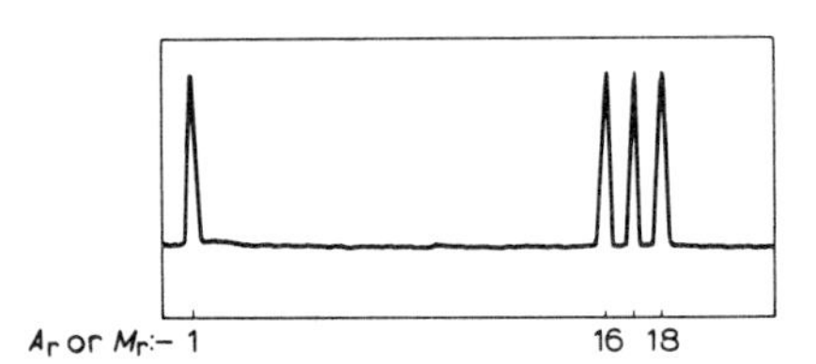

Fig. 9.9 mass spectrograph for water

The mass spectrograph for water (Fig. 9.9) reveals much valuable information:

a) water contains hydrogen and oxygen only,
b) the water molecule has a relative molecular mass of 18,
c) the water molecule contains two hydrogen atoms and one oxygen atom,
d) the two hydrogen atoms are joined separately to the oxygen atom and are not joined to each other. (*How is this deduced from the chart?*)

A table of approximate relative atomic masses, which you will find useful in exercises and calculations, is given at the beginning of this book.

The 'shorthand' representations of the constitution of molecules, e.g. H_2, O_2 and H_2O for hydrogen, oxygen and water molecules respectively, are called **formulae** (or formulas–both forms of the plural are used). Note that there is another meaning associated with these formulae, which is discussed in section 10.2. The mass spectrometer is very valuable in determining formulae; the following examples show how the mass spectrographs of some simple compounds can be interpreted.

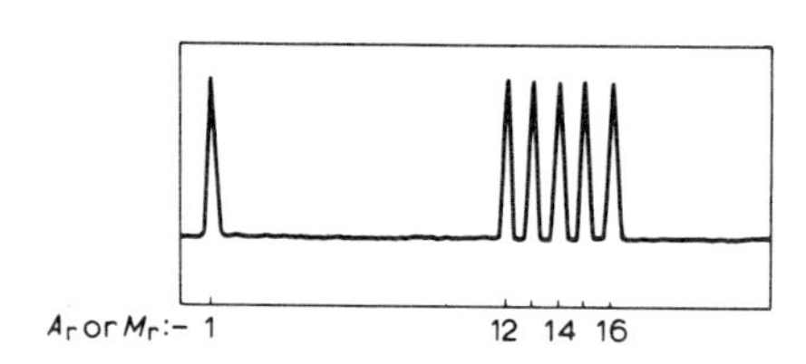

Fig. 9.10 mass spectrograph, example 1

Example 1

i) the relative molecular mass of the compound is 16
ii) the compound contains hydrogen (relative atomic mass 1)
iii) the compound contains carbon (relative atomic mass 12)
iv) the formula for the compound is CH_4
v) the four hydrogen atoms are joined separately to the carbon atom (no peaks at 2, 3 or 4)

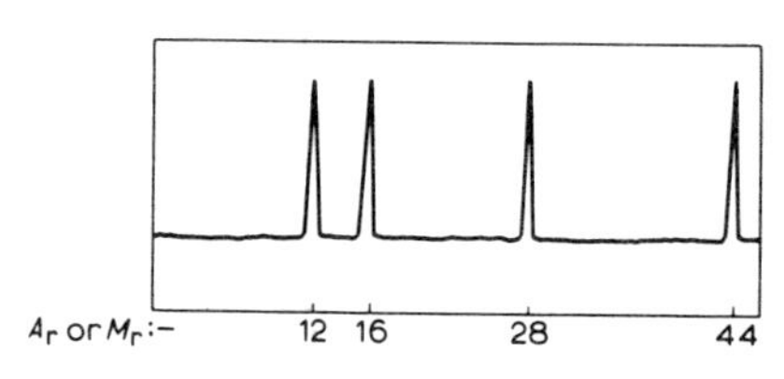

Fig. 9.11 mass spectrograph, example 2

Example 2

i) the relative molecular mass is 44
ii) the compound contains carbon (relative atomic mass 12) and oxygen (relative atomic mass 16)
iii) the peak at 28 cannot be due to another element (*why not?*); it must be due to the fragment CO
iv) the formula for the compound is CO_2
v) the oxygen atoms are joined separately to the carbon atom

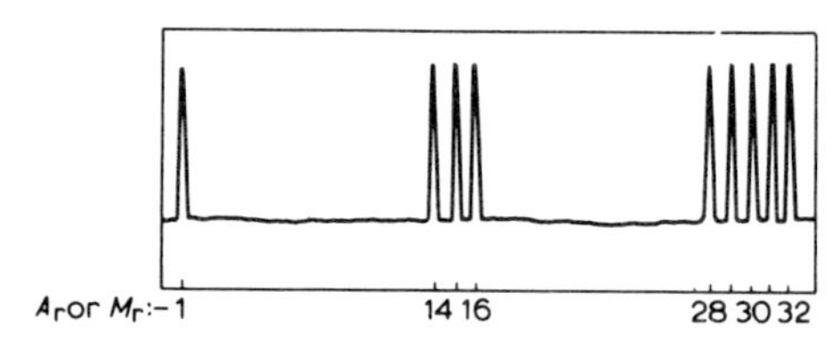

Fig. 9.12 mass spectrograph, example 3

Example 3

This compound, relative molecular mass 32, contains hydrogen ($A_r = 1$) and nitrogen ($A_r = 14$). It can *not* contain oxygen ($A_r = 16$), for the formula H_2NO does not explain the peak at 29. The peak at 16 must be due to NH_2 and the compound must contain hydrogen and nitrogen and no other element. Its formula is thus N_2H_4 and the molecule contains two nitrogen atoms joined to each other (peak 28), each bearing two hydrogen atoms (no peak at 17).

9.5 energy and chemical bonds

If molecules contain atoms held together by strong attractive forces, then to decompose the molecule the atoms must be separated and energy (heat, light, electrical or chemical energy) must be supplied to do this. Conversely, when atoms come together and bonds are formed, energy must be released; this is usually in the form of heat or light. This is fundamental to the understanding of chemical processes and the following statements should be learned thoroughly:

energy must be supplied to break bonds
energy is released when bonds are formed

The nature of the chemical bonds involved is considered in greater detail in Chapter 18, but it is interesting to see how the above principle enables us to explain a number of problems involved in the chemistry of hydrogen, oxygen and water described in Chapter 8.

9.6 hydrogen, oxygen and water: some facts explained

1) *At normal temperatures hydrogen and oxygen are gases, water is a liquid.* Hydrogen and oxygen molecules contain pairs of atoms. Water molecules contain hydrogen atoms joined separately to single oxygen atoms. It is not unreasonable to suppose that this different arrangement of atoms confers different properties on the compound from those of the elements (see also differences between mixtures and compounds, Chapter 4).
2) *When hydrogen and oxygen form water, much energy is released, but some energy must be put in to start the process* (see sections 8.7, 8.9).

The energy is released in the formation of bonds between separate hydrogen and oxygen atoms when water molecules are formed. Before this can happen, some energy must be supplied to break the bonds holding the hydrogen atoms in pairs in diatomic molecules. The same is true for oxygen. The thermal energy available at room temperature is insufficient to break these bonds, but that from a lighted match or splint is sufficient.

Some bonds are stronger than others: hydrogen–oxygen bonds are stronger than hydrogen–hydrogen or oxygen–oxygen bonds. In the formation of two water molecules, three relatively weak bonds are broken and four relatively strong bonds are formed. Overall, the formation of water molecules from hydrogen and oxygen molecules releases energy and the reaction is said to be **exothermic.**

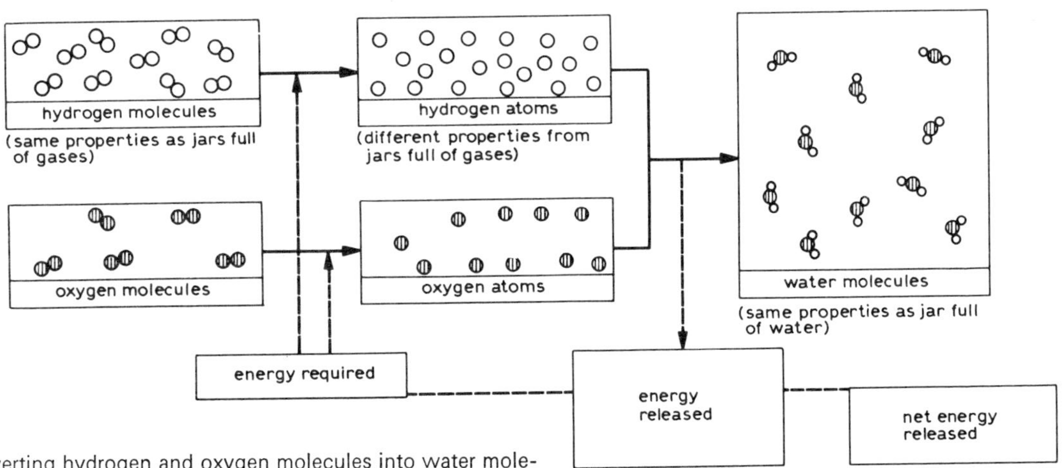

Fig. 9.13 converting hydrogen and oxygen molecules into water molecules: energy required to break bonds and released when bonds formed

3) *The frayed edge of a piece of torn filter paper is dipped in aqueous platinum*(IV) *chloride and heated in a jet of hydrogen burning in air. If the hydrogen supply is momentarily turned off so that the flame is extinguished then turned on again, the edge of the frayed paper glows and eventually re-lights the jet of gas.*

On heating, platinum(IV) chloride decomposes forming finely divided platinum metal: it is this platinum metal which catalyses the room-temperature combination between hydrogen and oxygen which generates sufficient energy to ignite the hydrogen.

This is the principle on which 'automatic' gas-lighters depend. The energy required to decompose the diatomic hydrogen and oxygen molecules is the 'barrier' preventing the formation of water molecules at room temperature: it is called the **activation energy** of the reaction, and this is usually supplied either thermally or electrically. If the diatomic molecules approach a platinum surface, they form bonds with the surface which weaken the bonds between the atoms. The energy required to separate the atoms is consequently reduced and thermal energy at room temperature is sufficient to overcome the activation energy 'barrier' (see Fig. 9.14).

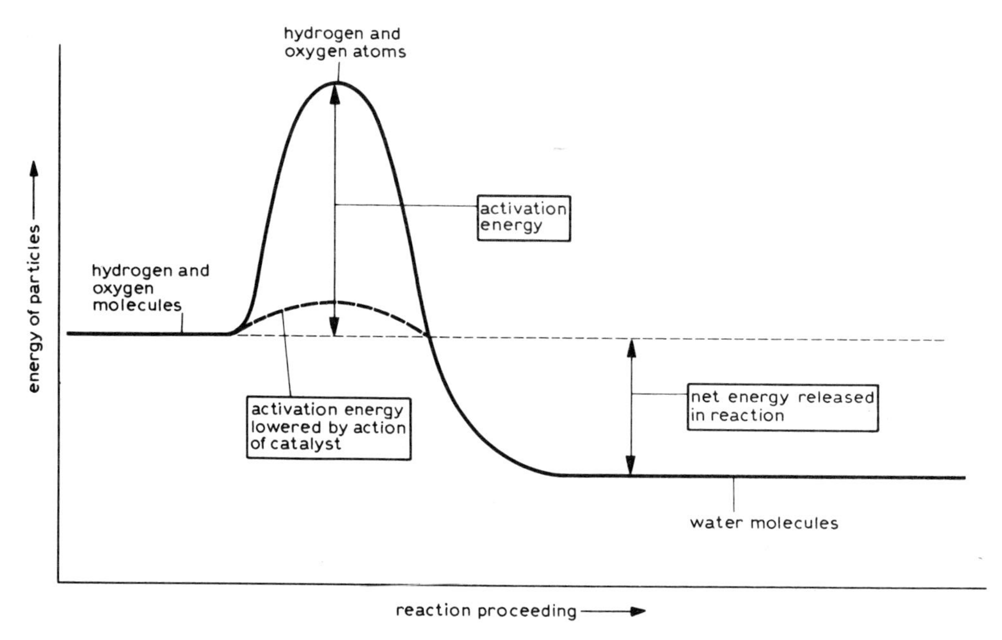

Fig. 9.14 energy diagram for formation of water molecules from hydrogen and oxygen molecules

4) *The ratio* BY MASS *of hydrogen to oxygen in water is always 1 : 8* (see section 8.10).

All water molecules contain two hydrogen atoms and one oxygen atom. Since the oxygen atom has a mass sixteen times that of a hydrogen atom, it will have a mass eight times that of two hydrogen atoms.

5) *The ratio* BY VOLUME *of hydrogen gas to oxygen gas produced when water is decomposed is always 2 : 1* (see section 8.8).

The decomposition of water molecules produces twice as many hydrogen molecules as oxygen molecules. This may be represented symbolically by an equation:

$$2H_2O(l) \longrightarrow 2H_2(g) + O_2(g)$$

It is not unreasonable that the hydrogen molecules will occupy twice the volume occupied by the oxygen molecules: the capacity of a football stadium is governed by the number of spectators which it will hold (and not by how heavy the individual spectators are). The connection between the volume occupied by a gas and the number of molecules present is considered in detail in sections 10.24 and 10.27.

revision summary: Chapter 9

molecules:	contain atoms joined by forces much stronger than forces between molecules
formulae:	can be used to represent individual molecules
mass spectrometer:	provides evidence about relative atomic and molecular masses and about formulae
energy:	must be supplied to break bonds is released when bonds are formed

questions 9

1. Distinguish carefully between the meanings of the terms 'mass' and 'weight'. Why is the former term more useful when describing atoms and molecules?

2. Summarise briefly the evidence for the existence of (a) atoms, (b) molecules.

3. What is meant by the term 'activation energy'? For a reaction between the diatomic molecules X_2 and Y_2, state why it is necessary to supply 'activation energy' before reaction can occur. What factors determine whether the formation of molecules of XY from X_2 and Y_2 will be exothermic or endothermic?

4. Explain why a mixture of hydrogen and oxygen gases, which can be stored indefinitely at room temperature without reaction, will ignite at room temperature in the presence of platinum.

For the following questions, first sketch the mass spectrometer trace, then deduce as much information as you can about the structures of the molecules concerned.

5. Compound A shows peaks at 1; 14, 15, 16, 17 only.

6. Compound B shows peaks at 1; 32, 33, 34 only.

7. Compound C shows peaks at 14; 16; 30 only.

8. Compound D shows peaks at 16; 48; 64 only.

9. Compound E shows peaks at 1; 12, 13, 14, 15; 24, 25, 26, 27, 28, 29, 30 only.

10. Compound F shows peaks at 14; 16; 28; 30; 44 only.

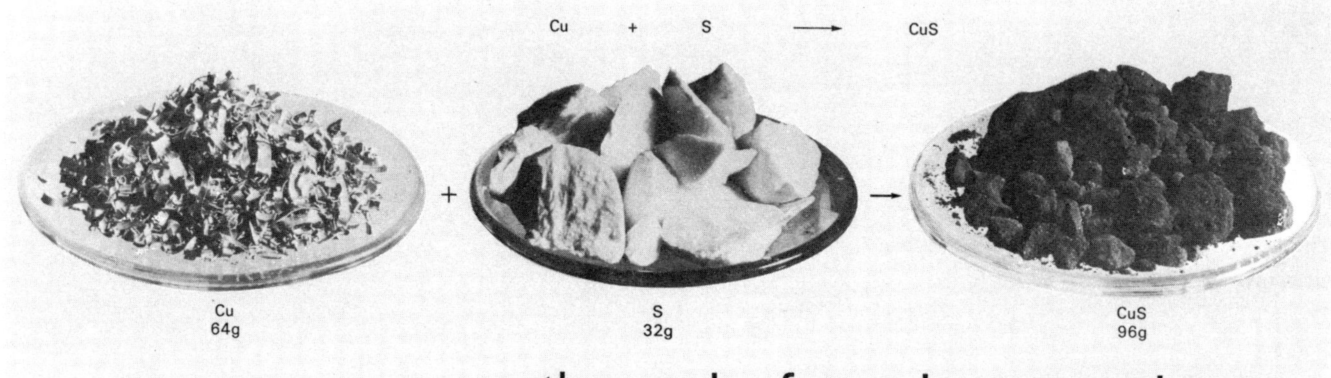

the mole, formulae, equations

10.1 the gram-atom

One of the interesting features of the study of water in the previous two Chapters is the idea that two hydrogen atoms combine with one oxygen atom to make one water molecule. In many other cases it is found that combination between atoms involves definite numbers of atoms, and to synthesise (build up) molecules we therefore need to be able to 'count out' the correct number of atoms of each element in the molecule. Since atoms are far too small to be seen, how is this counting-out process to be achieved?

Let us think for a moment on a different scale. A steel ball-bearing has five times the mass of a glass 'marble'. You have a bag containing glass marbles which weighs one kilogram. Your friend has a bag of steel bearings which weighs five kilograms. You would not need to open the bags to know that your friend has the same number of steel bearings in his bag as you have marbles in yours. If a cricket ball has twenty times the mass of a glass marble, then a bag of cricket balls equivalent in number to the marbles and steel bearings will have a mass of twenty kilograms. Thus, if we know the relative masses of a cricket ball, a steel bearing and a marble, we can effectively count numbers of these objects by weighing quantities of them: we do not need to see the individual objects.

We cannot see atoms, but we do know their relative masses (section 9.4 and end-paper). We can therefore count out numbers of atoms relative to each other by weighing large quantities of them.

A carbon atom has twelve times the mass of a hydrogen atom; a sulphur atom has thirty-two times the mass of a hydrogen atom. Twelve grams of carbon and thirty-two grams of sulphur are alike in one very important respect: they contain the *same number of atoms*. **The relative atomic mass of an element expressed in grams is called one gram-atom of the element.** One gram-atom of carbon weighs 12g; one gram-atom of sulphur weighs 32g.

One gram-atom of any element contains the same number of atoms as one gram-atom of any other element.

The actual number of atoms in one gram-atom is very large indeed: it is 6×10^{23} and is called the **Avogadro constant** (symbol L). The work of the nineteenth-century scientist Avogadro is discussed later in this Chapter.

If all the land on the earth was a lawn with 100000 blades of grass on each square metre, the earth would be covered by 2×10^{19} blades of grass. It would take 30000 worlds like this to make room for the number of blades of grass equal to the Avogadro constant.

10.2 symbols

In Chapter 9 we used the symbol H to represent one atom of hydrogen and O to represent one atom of oxygen. When we use these symbols in equations (see section 10.20) the symbols do not represent atoms, but gram-atoms. The symbol C may represent one atom of carbon, or it may represent 12g of carbon (one gram-atom) depending upon the context in which it is used. In either case, however, the symbol represents a definite quantity of the element (one atom or one gram-atom); it should *not* be used as lazy shorthand meaning 'some carbon', except in certain forms of reporting in which lack of space necessitates abbreviation.

In this Chapter we will concentrate on the symbol as representing the gram-atom. Fig. 10.1 shows the symbols for one gram-atom of some elements to which we will refer later in the text: you should learn these symbols. It is a useful test for the memory to construct your own list of symbols, arranged in alphabetical order of the symbols, and fill in the names of the elements without further reference to Fig. 10.1.

aluminium	Al	cobalt	Co	nitrogen	N
antimony	Sb	copper	Cu	oxygen	O
argon	Ar	fluorine	F	phosphorus	P
arsenic	As	gold	Au	platinum	Pt
barium	Ba	helium	He	potassium	K
beryllium	Be	hydrogen	H	silicon	Si
bismuth	Bi	iodine	I	silver	Ag
boron	B	iron	Fe	sodium	Na
bromine	Br	krypton	Kr	strontium	Sr
cadmium	Cd	lead	Pb	sulphur	S
caesium	Cs	lithium	Li	tin	Sn
calcium	Ca	magnesium	Mg	titanium	Ti
carbon	C	manganese	Mn	uranium	U
chlorine	Cl	mercury	Hg	xenon	Xe
chromium	Cr	neon	Ne	zinc	Zn
		nickel	Ni		

Fig. 10.1 symbols for gram-atoms in alphabetical order of elements

10.3 determination of the formula for an oxide of copper

The following experiment illustrates how the formula for a compound can be discovered by finding the mass of the elements present in it and using the gram-atom principle.

A stream of hydrogen is passed over a known mass of dry copper(II) oxide. The hydrogen removes oxygen from the oxide, forming steam, and the mass of copper remaining is determined. The mass of oxygen in the oxide is calculated.

A small metal boat is weighed and two-thirds filled with the oxide, which must be dried by strong heating and cooled in a desiccator before the experiment. The boat is re-weighed and is placed in a test-tube with a small hole in the end, as shown in Fig. 10.2. The gas supply is turned on and after a pause of ten seconds the excess gas is lit at the hole (*what is the reason for the pause?*). The boat is heated strongly for about three minutes, when all the oxide will have assumed a pink colour. The tube is cooled with the gas supply still running (*why?*); the gas is then switched off and, after the small flame has been completely extinguished with a damp cloth, the boat and contents are removed from the tube and weighed. The boat is then returned to the tube and the whole procedure is repeated until the mass of the boat and contents shows no further change. This last precaution, called 'heating to constant weight', ensures that the oxide has been converted completely to copper.

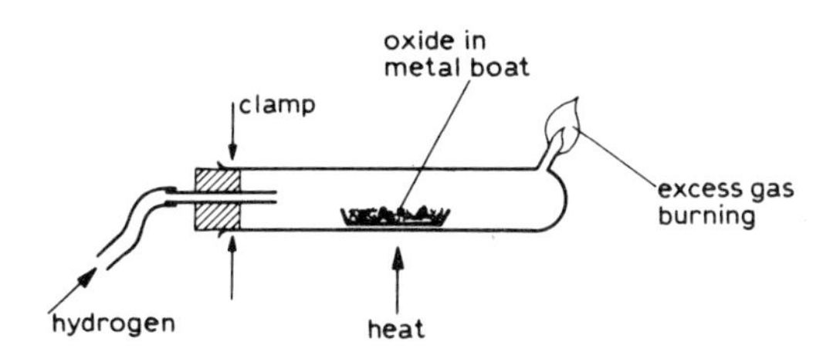

Fig. 10.2 determination of formula for an oxide by reduction

1) mass of empty boat = 1·34 g
2) mass of boat + oxide = 7·46 g
mass of boat + copper (first reading) = 6·31 g
(second reading) = 6·24 g
(third reading) = 6·24 g
3) mass of oxide used (2 — 1) = 6·12 g
mass of copper formed (3 — 1) = 4·90 g
mass of oxygen in oxide = 1·22 g

1·22 g of oxygen combines with 4·90 g of copper

1 g of oxygen combines with $\frac{4\cdot90 \times 1}{1\cdot22}$ g of copper

= 4·01 g of copper

16 g of oxygen combines with 64·16 g of copper

1 g-atom of oxygen combines with $\frac{64\cdot16}{63\cdot5}$ g-atom of copper

≈ 1 g-atom of copper

the formula for the oxide of copper is CuO

(N.B. This formula represents 79·5 g of the oxide of copper)

This reaction, in which copper(II) oxide is converted into copper using hydrogen, can be expressed in the form of an **equation,** as is explained in section 10.20:

$$CuO(s) + H_2(g) \rightarrow Cu(s) + H_2O(g)$$

The technique of obtaining information about the constitution of a compound by splitting it up into simple, recognisable parts is called **analysis.** Many substances containing oxygen can be analysed by the method of the above experiment; for compounds containing other elements, different techniques of analysis must be used. If the masses of the elements in a compound can be found by analyses, the formula for the compound can be determined. The following examples should make this clear.

10.4 calculation of formulae

Example 1
An oxide of chromium contains 48% by mass of oxygen. What is the simplest formula for the oxide?

100 g of oxide contains 48 g of oxygen
52 g of chromium combines with 48 g of oxygen

$\frac{52}{52}$ g-atoms of chromium combine with $\frac{48}{16}$ g-atoms of oxygen

1 g-atom of chromium combines with 3 g-atoms of oxygen

the simplest formula for the oxide is CrO_3

Example 2
Calcium(II) bromide contains exactly 20% by mass of the metal. What is its simplest formula?

20 g of calcium combine with 80 g of bromine

$\frac{20}{40}$ g-atom of calcium combines with $\frac{80}{80}$ g-atoms of bromine

0·5 g-atom of calcium combines with 1 g-atom of bromine
1 g-atom of calcium combines with 2 g-atoms of bromine

the simplest formula for calcium(II) bromide is $CaBr_2$

10.5 extension of the gram-atom principle to compounds

The gram-atom principle is invaluable in calculating chemical quantities, but it applies only to atoms of elements. It is equally important that we should be able to count atoms which are combined in compounds: sometimes these atoms are combined to form molecules, but sometimes they are not.

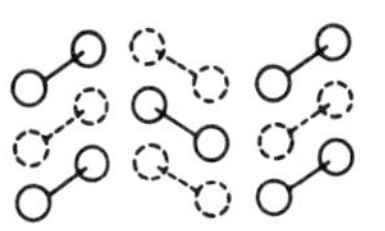

Fig. 10.3 iodine (an easily sublimed solid) is molecular

10.6 molecular and non-molecular substances

Molecular substances are composed of separate molecules; the forces holding the atoms in the molecule are strong, but the forces holding the molecules to each other are weak. Because of these weak forces, molecular substances are **volatile** (gases, liquids or easily sublimed solids).

Non-molecular substances are not composed of molecules; the forces holding the particles together are uniformly strong throughout the structure. Because of this, non-molecular substances are **involatile** (solids with high melting-points).

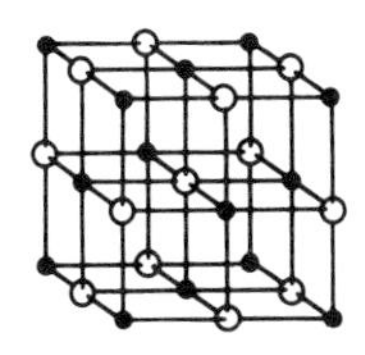

Fig. 10.4 sodium(I) chloride (solid, high melting point) is non-molecular

10.7 the gram-molecule

One gram of hydrogen atoms contains 6×10^{23} (the Avogadro constant) atoms; this is symbolised by H. Two grams of hydrogen atoms contain 12×10^{23} (twice the Avogadro constant) atoms; this is symbolised by 2H. If the atoms in 2g of hydrogen are combined in pairs to form molecules, there will be 6×10^{23} (the Avogadro constant) molecules; this is symbolised by H_2.

H: hydrogen atoms(1 g)
2H: hydrogen atoms(2 g)
H_2: hydrogen molecules(2 g)
$\frac{1}{2}H_2$: hydrogen molecules(1 g)

The relative molecular mass (formerly called molecular weight) **of an element or compound expressed in grams is called one gram-molecule of the element or compound.** One gram-molecule of a substance contains the same number of molecules as one gram-molecule of any other substance; the number is 6×10^{23}, the Avogadro constant.

(*Compare:* one gram-atom of any element contains 6×10^{23} atoms.)

The term 'gram-molecule' applies only to molecular substances. If the formula for a compound is known, the mass of one gram-molecule of it may be calculated by adding the masses of the elements in it. For example:

The formula for carbon disulphide is CS_2.

One molecule of carbon disulphide contains one atom of carbon and two atoms of sulphur.

One gram-molecule (6×10^{23} molecules) of carbon disulphide contains 1g-atom (6×10^{23} atoms) of carbon and 2g-atoms (12×10^{23} atoms) of sulphur. It therefore contains 12g of carbon and $2 \times 32 = 64$g of sulphur, and the mass of one gram-molecule of carbon disulphide is $12 + 64 = 76$g.

Further examples are given in Fig. 10.5.

physical state	name	formula	mass of 1 gram-molecule
gas	hydrogen	H_2	$1 \times 2 = 2$g
	oxygen	O_2	$16 \times 2 = 32$g
	carbon dioxide	CO_2	$12 + (16 \times 2) = 44$g
	phosphine	PH_3	$31 + (1 \times 3) = 34$g
liquid (easily vaporised)	bromine	Br_2	$80 \times 2 = 160$g
	water	H_2O	$(1 \times 2) + 16 = 18$g
	ethanol	C_2H_6O	$(12 \times 2) + (1 \times 6) + 16 = 46$g
solid (easily sublimed)	iodine	I_2	$127 \times 2 = 254$g
	naphthalene	$C_{10}H_8$	$(12 \times 10) + (1 \times 8) = 128$g

Fig. 10.5 gram-molecules of some molecular compounds

Using this principle, it is possible to calculate rapidly the masses of compounds which we may need to use in order to obtain equivalent numbers of molecules.

Example: What mass of carbon dioxide contains the same number of molecules as i) 9g of water, ii) 8g of oxygen?

i) 9g of water contains $\frac{9}{18} = \frac{1}{2}$g-molecule of water
$\frac{1}{2}$g-molecule of carbon dioxide has a mass of $\frac{1}{2} \times 44 = $ **22g**

ii) 8g of oxygen contains $\frac{8}{32}$ (*care!*) $= \frac{1}{4}$g-molecule of oxygen
$\frac{1}{4}$g-molecule of carbon dioxide has a mass of $\frac{44}{4} = $ **11g**

10.8 the 'gram-formula-weight'

This term can be applied to any compound, but it is normally used to describe non-molecular compounds, to which the term 'gram-molecule' obviously cannot be applied. The 'gram-formula-weight' means the total mass of all the gram-atoms represented by the formula for a compound. Some examples are given in Fig. 10.6.

name	formula	'gram-formula-weight'
nickel(II) oxide	NiO	$(59 \times 1) + (16 \times 1) =$ **75g**
copper(II) chloride	$CuCl_2$	$(63 \cdot 5 \times 1) + (35 \cdot 5 \times 2) =$ **134·5g**
calcium(II) carbonate	$CaCO_3$	$(40 \times 1) + (12 \times 1) + (16 \times 3) =$ **100g**
tin(IV) oxide	SnO_2	$(119 \times 1) + (16 \times 2) =$ **151g**
potassium(I) nitrate(V)	KNO_3	$(39 \times 1) + (14 \times 1) + (16 \times 3) =$ **101g**
sodium(I) sulphate(VI)	Na_2SO_4	$(23 \times 2) + (32 \times 1) + (16 \times 4) =$ **142g**

Fig. 10.6 gram-formula-weights of some non-molecular compounds

NiO means one 'gram-formula-weight' of nickel(II) oxide (75g); this contains 6×10^{23} atoms of nickel and 6×10^{23} atoms of oxygen, held together by uniformly strong forces in a lattice.

$CaCO_3$ means one 'gram-formula-weight' of calcium(II) carbonate (100g); this contains 6×10^{23} atoms of calcium, 6×10^{23} atoms of carbon and $3 \times (6 \times 10^{23}) = 1 \cdot 8 \times 10^{24}$ atoms of oxygen, held together in a lattice.

10.9 the mole

The Avogadro constant, 6×10^{23}, has obviously become a very important collective unit and it is convenient to give it a name. One Avogadro constant of anything is called a **mole**; the name can be applied to atoms, to molecules or to collections of atoms held in the lattice of a non-molecular compound. The chemist expresses all his quantities most conveniently in moles and it is most important that you should strive always to think in terms of moles rather than in terms of grams when considering chemical quantities. This will save you much unnecessary calculation–and be much more meaningful. For example, the instruction 'add 98g of sulphuric(VI) acid to 85g of sodium(I) nitrate(V)' does not tell us much; if the same instruction reads 'add 1 mole of sulphuric(VI) acid to 1 mole of sodium(I) nitrate(V)', it tells us rather more. Thinking in terms of moles can prevent errors: If 4g of hydrogen is passed over 239 g of lead(IV) oxide, which substance is 'in excess'? There are in fact twice as many moles of hydrogen as there are moles of lead(IV) oxide!

A mole is a collective number, just as is 'a dozen'. One can have 'a dozen' of many different things and one can have 'a mole' of many different chemical things. This idea is summarised in Fig. 10.7.

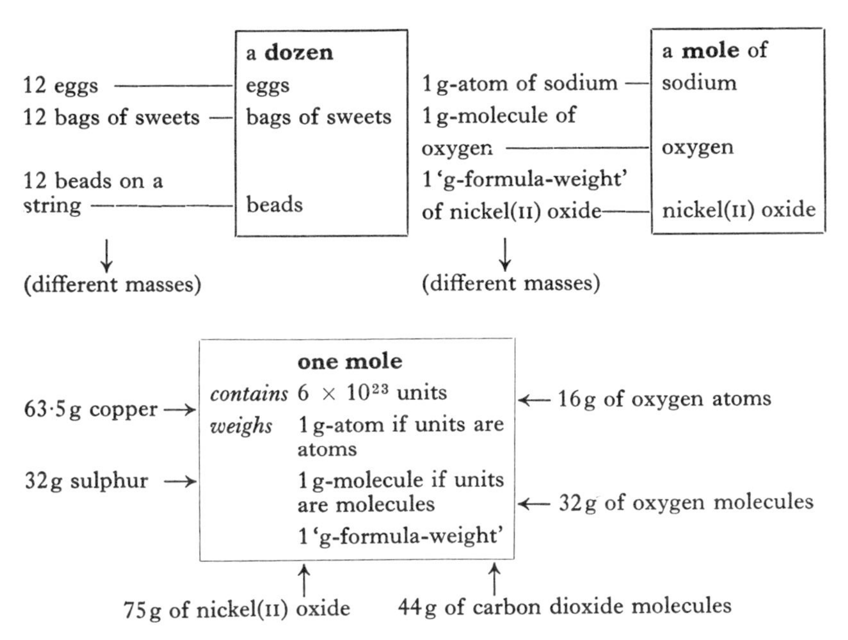

Fig. 10.7 some examples of one mole of substances

Once the meaning of the mole has been thoroughly understood, this term should be used in preference to the terms gram-atom, gram-molecule and gram-formula-weight, which it renders unnecessary. In subsequent sections the mole will be used throughout.

The mole has recently been uniformly accepted as defining an **amount of substance.** Such a definition must be expressed very carefully and the convention used is explained fully in section 18.7(3).

10.10 the law of constant composition

If the analysis of copper(II) oxide described in section 10.3 is carried out using specimens of oxide obtained from different sources (e.g. by heating the hydroxide, the carbonate, or the nitrate(V)), the same results are obtained: 4 g of copper combines with 1 g of oxygen in each specimen of oxide. This is true of a large number of compounds and leads to the statement of the **Law of Constant Composition, or Definite Proportions:**

A pure compound contains the same elements, combined in the same proportion by mass, no matter how it is made.

This law was readily explained by John Dalton in the early nineteenth century, by his assumption that atoms of elements have characteristic masses and that compounds are formed by the combination of definite small whole numbers of atoms. Thus if all specimens of copper(II) oxide contain exactly one atom of copper for each atom of oxygen, then since the relative atomic masses of copper and oxygen are 63·5 and 16, the mass ratio of copper to oxygen in all specimens of copper(II) oxide will be 4 to 1.

Another chemist, Berthollet, indicated the existence of compounds which apparently were not in agreement with this law; specimens of iron(II) sulphide, for example, are found to have a variable composition. Such compounds become known as *Berthollide compounds* as opposed to *Daltonide compounds* of fixed composition. Although many Berthollide compounds exist, it is more fruitful for us to accept constancy of composition as characterising the ordered state of true compounds and to explain the existence of Berthollide compounds in terms of imperfections in the lattice arrangement; thus in a lattice like that depicted in Fig. 10.4, if a variable number of the sites are unoccupied, variable composition will result.

10.11 three important characteristics of a compound

We can now extend the distinction between mixtures and compounds made in Chapter 4 by attributing to compounds the following characteristics:

1) the properties of a compound differ from the properties of its constituent elements,
2) a compound has a constant composition,
3) a considerable energy change is associated with the formation and decomposition of a compound–this energy change has a fixed value for a particular compound.

The energy value in item (3) can be very useful in explaining the chemical behaviour of compounds. For example:

When 1 mole (18 g) of liquid water is formed from hydrogen gas and oxygen gas, the energy given out is always 286 kJ.

When 1 mole of hydrogen sulphide gas is formed from hydrogen gas and solid (rhombic) sulphur, the energy given out is always 11·3 kJ.

Now consider the decomposition of water and hydrogen sulphide. One mole of water requires almost thirty times as much energy to decompose it as does one mole of hydrogen sulphide: water is the more 'chemically stable' compound in this respect. Further, since hydrogen sulphide can readily act as a source of hydrogen, which can be used to remove oxygen from other compounds, hydrogen sulphide is a good reductant. Water does not share this property.

10.12 the law of multiple proportions

Dalton knew of the existence of pairs of compounds, possessing different properties, which contained the same elements. He explained this by assuming that the two compounds in the pair contained different numbers of atoms; thus the black oxide of copper (the oxide referred to in section 10.3) has the formula CuO, whereas another oxide of copper, red in colour, has the formula Cu_2O. He was able to predict a further law of chemistry on this basis, called the **Law of Multiple Proportions:**

When two elements A and B combine to form more than one compound, the different masses of A which combine with a fixed mass of B are in the ratio of small whole numbers.

When this law was found by experiment to hold for a large number of pairs of compounds, the belief in Dalton's Atomic Theory was greatly strengthened. Perhaps this success blurred his judgment, as we will see later in this Chapter.

10.13 analysis of the oxides of lead

Lead forms two oxides, a yellow one and a brown one, which can both be analysed using the technique described in section 10.3. The results of the analysis illustrate the Law of Multiple Proportions.

1. analysis of the yellow oxide

mass of empty boat	= 1·25 g
mass of boat + oxide	= 8·30 g
mass of boat + lead (first reading)	= 7·84 g
(second reading)	= 7·81 g
(third reading)	= 7·81 g

mass of oxide used	= 7·05 g
mass of lead formed	= 6·56 g
mass of oxygen in oxide	= 0·49 g

0·49 g of oxygen combine with 6·56 g of lead
1 g of oxygen combines with 13·1 g of lead

2. analysis of the brown oxide

mass of empty boat	= 1·30 g
mass of boat + oxide	= 8·82 g
mass of boat + lead (first reading)	= 7·85 g
(second reading)	= 7·83 g
(third reading)	= 7·83 g
mass of oxide used	= 7·52 g
mass of lead formed	= 6·53 g
mass of oxygen in oxide	= 0·99 g

0·99 g of oxygen combine with 6·53 g of lead
1 g of oxygen combines with 6·5 g of lead

Thus the masses of lead which combine with a fixed mass of oxygen (1 g) are 13·1 g and 6·5 g, which figures are in the ratio of small whole numbers, 2 : 1.

If the calculations are taken further, we can obtain the formulae for the two oxides, as we did for the oxide of copper in the experiment described earlier.

a) *formula for the yellow oxide*
1 g of oxygen combines with 13·1 g of lead
16 g of oxygen combines with 13·1 × 16 g of lead = 209·6 g of lead

this mass of lead is $\frac{209 \cdot 6}{207} \approx 1$ mole of lead

ratio of moles, lead : oxygen is 1 : 1

the formula for the yellow oxide of lead is PbO

b) *formula for the brown oxide*
1 g of oxygen combines with 6·5 g of lead
16 g of oxygen combines with 6·5 × 16 g of lead = 104 g of lead

this mass of lead is $\frac{104}{207} \approx 0 \cdot 5$ moles of lead

ratio of moles, lead : oxygen is 0·5 : 1
or 1 : 2

the formula for the brown oxide of lead is PbO_2

10.14 a meaningful pattern in the formulae for compounds

The formulae for many compounds can be determined by laboratory analysis followed by calculations in terms of moles. Examples of formulae so determined are given in Fig. 10.8, in which the metals and non-metals in the compounds are arranged alphabetically. Hydrogen is placed in both lists, in brackets, because it can take the place of either a metal or a non-metal in compounds.

Scrutiny of Fig. 10.8 reveals the following pattern:
i) hydrogen, sodium and potassium never have more than one mole of the other element in one mole of the compound – the same is true for the non-metals bromine, chlorine and iodine.
ii) magnesium and zinc have a maximum of two moles of the other element in one mole of each compound – the same is true for oxygen and sulphur.
iii) aluminium has a maximum of three moles of the other element in one mole of each compound – the same is true for nitrogen.

The same formulae are rearranged in Fig. 10.9 so that the elements with similar patterns of formulae are placed next to each other.

metals	aluminium	(hydrogen)	magnesium	sodium	potassium	zinc
non-metals						
bromine	$AlBr_3$	HBr	$MgBr_2$	$NaBr$	KBr	$ZnBr_2$
chlorine	$AlCl_3$	HCl	$MgCl_2$	$NaCl$	KCl	$ZnCl_2$
(hydrogen)	AlH_3	H_2	MgH_2	NaH	KH	ZnH_2
iodine	AlI_3	HI	MgI_2	NaI	KI	ZnI_2
nitrogen	AlN	H_3N	Mg_3N_2	Na_3N	K_3N	Zn_3N_2
oxygen	Al_2O_3	H_2O	MgO	Na_2O	K_2O	ZnO
sulphur	Al_2S_3	H_2S	MgS	Na_2S	K_2S	ZnS

Fig. 10.8 formulae determined by analysis

If we assign 'code numbers', I, II and III, to the appropriate elements as shown in Fig. 10.9, we can reduce the pattern shown by the formulae to one rule:

The 'code number' represents the number of moles of the other element in one mole of each compound, except when an element combines with another element of the same code number: in this case the formula is the simplest possible.

These 'code numbers' obviously have something to do with the capacity of elements to combine with other elements; they are called **valency numbers** and are very useful in helping to remember formulae – a necessary task which would otherwise be very burdensome for the chemist.

	I			II		III
metals	(hydrogen)	potassium	sodium	magnesium	zinc	aluminium
non-metals						
(hydrogen)	H_2	KH	NaH	MgH_2	ZnH_2	AlH_3
chlorine	HCl	KCl	$NaCl$	$MgCl_2$	$ZnCl_2$	$AlCl_3$
iodine	HI	KI	NaI	MgI_2	ZnI_2	AlI_3
oxygen	H_2O	K_2O	Na_2O	MgO	ZnO	Al_2O_3
sulphur	H_2S	K_2S	Na_2S	MgS	ZnS	Al_2S_3
nitrogen	H_3N	K_3N	Na_3N	Mg_3N_2	Zn_3N_2	AlN

Fig. 10.9 formulae rearranged to show meaningful pattern

10.15 valency number; formulae for compounds

The valency of an element is a number which represents its capacity to combine with other elements. It is equal to the number of hydrogen atoms which one atom of the element will combine with or replace.

Examples:

a) The formula for water is H_2O. One oxygen atom combines with two atoms of hydrogen. The valency of oxygen is two.

b) The formula for nickel(II) oxide is NiO. Comparing this with the formula for water, we see that each nickel atom is taking the place of two hydrogen atoms. The valency of nickel is two.

working out formulae from valencies

The formula for aluminium oxide is Al_2O_3. We notice that the valency number of aluminium appears after the symbol for oxygen, and the valency number of oxygen after the symbol for aluminium. Scrutiny of many formulae tells us that this is generally the case and we can consequently formulate a general rule to save us the need to remember the formula for every compound:

If the valency of element A is x and the valency of element B is y, the formula for the compound of A with B is:

$$A_yB_x$$

Note: (i) The number one is never written.

(ii) When the two valency numbers have a common factor, they are divided by that factor.

e.g. carbon(IV) oxygen(II): not C_2O_4, but CO_2

nickel(II) oxygen(II): not Ni_2O_2, but NiO

Fig. 10.10 lists the valencies of some commonly encountered elements and radicals (groups of atoms–see the next section). Some elements have more than one valency; examples already encountered are copper (I and II) and lead (II and IV). Other elements have even higher valencies, e.g. phosphorus (V), sulphur (VI); at this stage it should be sufficient to learn the commoner valencies listed in the figure.

valencies of metals

I	II	III	IV
sodium	calcium	aluminium	tin
potassium	magnesium	iron	lead
(ammonium)	zinc	chromium	manganese
silver	copper		
copper	iron		
mercury	lead		
	mercury		
	tin		
	barium		
	strontium		
	manganese		

Fig. 10.10(a) valencies of some commonly encountered elements and radicals

valencies of non-metals and radicals			
I	II	III	IV
chlorine	oxygen	nitrogen	carbon
bromine	sulphur	phosphorus	silicon
iodine	sulphate(VI)	phosphate(V)	
nitrate(V)	sulphate(IV)		
nitrate(III)	carbonate		
hydroxide			
chlorate(V)			
hydrogencarbonate			
hydrogensulphate(VI)			

Fig. 10.10(b) valencies of some commonly encountered elements and radicals

10.16 radicals; formulae for more complex compounds

Sodium reacts with water liberating hydrogen gas, but only one hydrogen atom is liberated from each water molecule (see section 8.12). The substance which remains in solution, sodium(I) hydroxide, contains three elements: sodium, hydrogen and oxygen, but the experiments described in section 12.29(4)(i) show that the oxygen and hydrogen behave in most reactions as though they were one unit.

A group of atoms which behaves as one unit is called a radical. The name for the radical described above is the hydroxide radical. There is similar evidence for regarding a group of one carbon atom with three oxygen atoms as one unit, called the carbonate radical; one sulphur atom with four oxygen atoms makes up the sulphate(VI) radical. Fig. 10.11 shows a list of the common radicals, with the symbols representing one mole of each radical and their valencies.

name	symbol for 1 mole	valency
hydroxide	OH	I
nitrate(V)	NO_3	I
nitrate(III)	NO_2	I
chlorate(V)	ClO_3	I
carbonate	CO_3	II
sulphate(VI)	SO_4	II
sulphate(IV)	SO_3	II
phosphate(V)	PO_4	III

Fig. 10.11 valencies of radicals

10.17 some notes on naming compounds

i) The appropriate valency of a metal is shown in the name as a Roman numeral in brackets after the name for the metal, thus:

copper(II) oxide, lead(IV) oxide

This notation replaces the older system, in which the lower valency was denoted by attaching the suffix '*ous*', and the higher valency by '*ic*', to the Latin name for the metal:

formula	old name	new name
Cu_2O	cuprous oxide	copper(I) oxide
$FeCl_3$	ferric chloride	iron(III) chloride

ii) The name for a compound ends in 'ide' when it contains only those elements mentioned in the name:

NaCl	sodium(I) chloride
MgO	magnesium(II) oxide
KOH	potassium(I) hydroxide

iii) The name for a compound ends in 'ate' when the compound contains oxygen as well as the elements mentioned in the name:

Na_2CO_3	sodium(I) carbonate
K_2SO_4	potassium(I) sulphate(VI)
KNO_3	potassium(I) nitrate(V)

iv) Some elements combine with oxygen to form more than one type of radical, e.g. NO_3 and NO_2, SO_4 and SO_3. The names for all such radicals end in 'ate', the difference being denoted by representing the valency state of the element in the radical in Roman numerals at the end of the name.

nitrogen	valency state III	nitrate(III)	$-NO_2$	(old name: nitrite)
	valency state V	nitrate(V)	$-NO_3$	(old name: nitrate)
sulphur	valency state IV	sulphate(IV)	$-SO_3$	(old name: sulphite)
	valency state VI	sulphate(VI)	$-SO_4$	(old name: sulphate)

10.18 writing formulae for compounds containing radicals

Analysis shows that calcium(II) hydroxide has the formula CaO_2H_2, molar mass 74g. The formula is never written like this; since calcium(II) hydroxide contains hydroxide radicals its formula is written $Ca(OH)_2$. Note the importance of the bracket: the molar mass is $40 + (16 + 1) \times 2 = 74$g.

The formula $CaOH_2$ would be incorrect, for this would represent $40 + 16 + 2 = 58$g. The bracket must be used when it is followed by a subscript other than 1. When the subscript is 1, both subscript and bracket are usually omitted, though it would not be wrong to include them.

name	formula
sodium(I) hydroxide	NaOH
barium(II) nitrate(V)	$Ba(NO_3)_2$
ammonium sulphate(VI)	$(NH_4)_2SO_4$
aluminium(III) sulphate(VI)	$Al_2(SO_4)_3$

Fig. 10.12 formulae including radicals

10.19 the law of conservation of mass

We have now learned sufficient about the meaning of formulae to be able to represent the changes which occur during the course of a reaction in a precise but conveniently brief way, using an equation. It is worthwhile to realise first, however, that there is one property which does not change during a reaction: that is the total mass of all of the elements present.

The Law of Conservation of Mass states:
No change in mass has ever been observed during the course of a chemical reaction.

Some of the most accurate experiments illustrating this law were devised and carried out by Landolt. An example is shown in Fig. 10.13. Two solutions A and B, which react when mixed, are placed in separate limbs of the tube, which is then sealed. The mass of the whole is determined as accurately as possible. The solutions are mixed by rocking the tube so that reaction can occur; after leaving the apparatus for a long period, the mass of the whole is again determined. In such experiments, no change in mass has ever been observed.

(Why is the apparatus left for a period before the final weighing? In view of the relationship between mass and energy, why is the Law stated in the above form?)

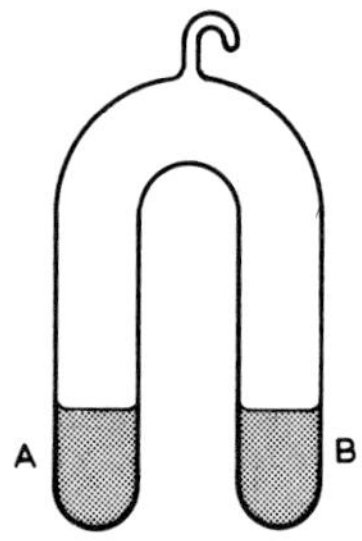

A B

Fig. 10.13 Landolt tube

how to state chemical quantities precisely using equations

The Law of Conservation of Mass means that the total mass of the products of a reaction is the same as the total mass of the reacting substances. One example of this would be the reduction of copper(II) oxide to copper using hydrogen. We could say:

'When 79·5 g of copper(II) oxide is heated with 2 g of hydrogen, a reaction takes place; 63·5 g of copper and 18 g of water are formed, the water being driven off as a gas.'

Nearly all of this information can be stated on one line using an equation; remember that each formula represents one mole of the substance.

$$CuO + H_2 \rightarrow Cu + H_2O$$

$$(63{\cdot}5 + 16)\,g \qquad\qquad (2 + 16)\,g$$

$$\underbrace{79{\cdot}5\,g \quad 2\,g} \qquad\qquad \underbrace{63{\cdot}5\,g \quad 18\,g}$$

$$81{\cdot}5\,g \text{ reactants} \rightarrow 81{\cdot}5\,g \text{ products}$$

If we start with 63·5 g of copper in the copper(II) oxide, we must end up with 63·5 g of copper metal in the products. In the terms of the equation, there must be the same number of symbols for each element on the right-hand side as there are on the left: **the equation must balance.**

10.20 building up an equation

1) *Make sure that the reaction works*
   $Cu + H_2O \rightarrow CuO + H_2$ is correctly balanced, but the reaction does not work!
   Example: iron(III) hydroxide gives on heating iron(III) oxide and steam–this is a correct statement which can be checked by experiment.
2) *You must know the identity of the reactants and products*
   $KNO_3 \rightarrow KNO + O_2$ balances, but KNO does not exist.
3) *You must know the correct formula for each compound*

These formulae must be consistent with the valencies of the elements and radicals in the compounds.
Example: $Fe(OH)_3 \rightarrow Fe_2O_3 + H_2O$
this is not yet balanced, but the formulae are correct.

4) *You can now attempt to balance the equation*, but in doing so you must not change the subscripts (valency numbers).

Example: To form one mole of iron(III) oxide you need $2 \times 56 = 112$g of iron; you may be tempted to write the formula for iron(III) hydroxide as $Fe_2(OH)_3$ in order to get it. This would mean that 112g of iron combines with 48g of oxygen and 3g of hydrogen; this is not so. 56g of iron combines with 48g of oxygen and 3g of hydrogen in iron(III) hydroxide and its formula must always be written $Fe(OH)_3$.

Once you have written the correct formulae, never alter the valency subscripts.

We need 112g of iron for one mole of iron(III) oxide, Fe_2O_3. There is only 56g of iron in one mole of iron(III) hydroxide, $Fe(OH)_3$. Therefore we must use TWO moles of iron(III) hydroxide, and this is written as $2Fe(OH)_3$. The large figure 2 in front of the formula for a compound doubles everything in the formula, for it represents 2 moles of the whole compound.

How many grams of hydrogen are there in $2Fe(OH)_3$?
$(1 \times 3) \times 2 = 6$g
How many grams of oxygen are there in $5Al_2(SO_4)_3$?
$(16 \times 4) \times 3 \times 5 = 960$g

The equation can now be balanced as follows:

$$Fe(OH)_3 \rightarrow Fe_2O_3 + H_2O$$

then: $$2Fe(OH)_3 \rightarrow Fe_2O_3 + H_2O$$

finally: $$2Fe(OH)_3 \rightarrow Fe_2O_3 + 3H_2O$$

At the end of the process, **check** to see that you have the same number of moles of each element on both sides of the equation–but beware that the subscript numbers and the large numbers can both 'act at a distance'.
Example: Aqueous lead(II) nitrate(V) mixed with aqueous potassium(I) iodide gives a yellow precipitate which analysis shows to be lead(II) iodide; aqueous potassium(I) nitrate(V) remains in solution. Here is the order of building up the equation:

formulae: $Pb(NO_3)_2$, KI, PbI_2, KNO_3

unbalanced equation: $$Pb(NO_3)_2 + KI \rightarrow PbI_2 + KNO_3$$

then: $$Pb(NO_3)_2 + 2KI \rightarrow PbI_2 + KNO_3$$

finally: $$Pb(NO_3)_2 + 2KI \rightarrow PbI_2 + 2KNO_3$$

10.21 using fractions in equations

The reaction between hydrogen and oxygen forming water may be represented by either

i) $$2H_2 + O_2 \rightarrow 2H_2O$$

or ii) $$H_2 + \tfrac{1}{2}O_2 \rightarrow H_2O$$

There is nothing wrong with equation ii); $\frac{1}{2}O_2$ means '16g of oxygen molecules'. Equation ii) represents the formation of 18g of water, whereas equation i) refers to the formation of 36g of water. When a specific quantity of a reagent or product is being referred to, fractions are often used in equations. In other cases it is more usual to use whole numbers.

10.22 calculating quantities using equations

We may well need to know the correct quantity of a reagent to use in a reaction, or the quantity of product formed. Since an equation represents quantities in terms of moles, it readily supplies the answers to such questions.

Example 1 What mass of hydrogen is required to reduce 7·95 g of copper(II) oxide to copper?

$$\underbrace{\underset{63{\cdot}5\ +\ 16}{CuO}}_{79{\cdot}5\,g} + H_2 \rightarrow Cu + H_2O$$

79·5 g requires 2 g
7·95 g requires 0·2 g

0·2 g of hydrogen is required

Example 2 What mass of magnesium is obtained by the electrical decomposition of 2 tonnes of magnesium(II) chloride into its elements?

$$Mg + Cl_2 \rightarrow Mg + Cl_2$$
$$24 + \underbrace{(35{\cdot}5 \times 2)}_{71}$$
$$\underbrace{24 + 71}$$

95 tonnes → 24 tonnes

2 tonnes → $\frac{24}{95} \times 2$ tonnes

= 0·5 tonne

0·5 tonne of magnesium is produced

10.23 the mole principle applied to gas volumes

The electrolysis of water (section 8.8) shows that the decomposition of water produces hydrogen and oxygen in the ratio 2 : 1 *by volume.* Having subsequently established the formula for water as H_2O, we could write the equation for its decomposition:

$$2H_2O \rightarrow 2H_2 + O_2$$

Thus the decomposition of water produces hydrogen and oxygen in the ratio 2 : 1 *by moles.* There is apparently a relationship between the volume occupied by a gas and the number of moles of gas present: the following experiment examines that relationship.

determination of the volumes occupied by one mole of various gases

A fixed volume of gas is weighed in air; after correcting for buoyancy the weight of the gas is found. The volume occupied by one mole of the gas under standard conditions of temperature and pressure may then be calculated.

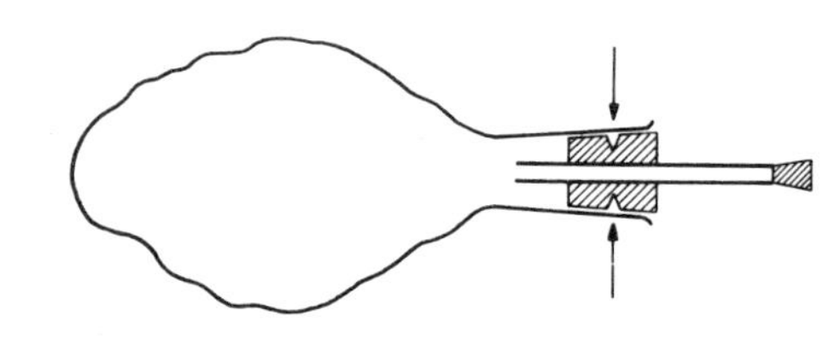

Fig. 10.14

A plastic bag is folded at the mouth in small pleats around a grooved rubber bung bearing a short delivery tube and stopper and is secured in place using a rubber band as indicated by the arrows in Fig. 10.14. All air is pressed out of the bag and the small bung is inserted into the delivery tube; the bag is then weighed. The bag is filled with air from a cylinder or bellows (*not* by blowing into it) until the plastic is fully extended but not under pressure. The small bung is replaced and the bag re-weighed. Perhaps surprisingly, the bag full of air weighs the same as the empty bag: can you explain this? The bag is emptied and filled with the gas under test, preferably from a cylinder. The bag full of gas is weighed. This procedure is repeated using different gases: hydrogen, oxygen, nitrogen, dinitrogen oxide, carbon dioxide and sulphur dioxide are suitable.

To find the volume of the bag it is refilled with air, which is then squeezed out gently through a rubber tube, displacing water from a large bottle inverted in a trough of water as shown in Fig. 10.15. The large bottle is stoppered, placed upright on the bench and refilled with water from a measuring cylinder, the volume of water needed being equal to the volume of air in the bottle. Finally, the laboratory temperature and pressure is measured. A list of specimen results for the gases oxygen and hydrogen is given on the next page. Note that the observations are expressed as weights, not masses (*why?*).

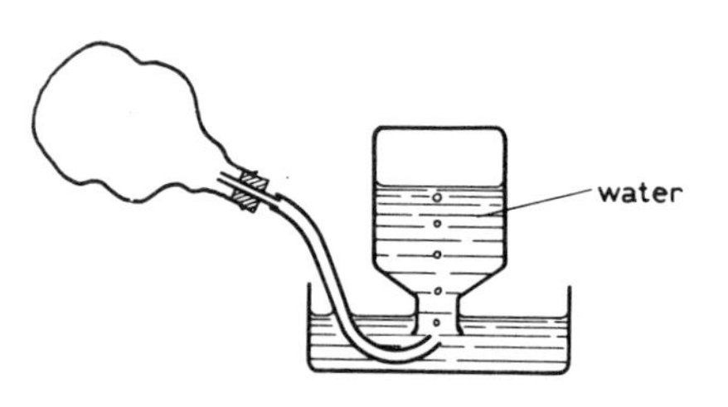

Fig. 10.15 finding the volume of the bag

weight of empty bag = 65·12 g-force
weight of bag full of air = 65·12 g-force
weight of bag full of oxygen = 65·32 g-force
weight of bag full of hydrogen = 62·23 g-force
initial volume of water in cylinder = 3·00 dm^3
final volume of water in cylinder = 0·44 dm^3
laboratory temperature = 15 °C
laboratory pressure = 100 kPa

t/°C p/kPa	10	15	20	25
96	1·18	1·16	1·14	1·12
97	1·20	1·18	1·16	1·14
98	1·21	1·19	1·17	1·15
99	1·22	1·20	1·18	1·16
100	1·23	<u>1·21</u>	1·19	1·17
101	1·25	1·23	1·21	1·19
102	1·26	1·24	1·22	1·20
103	1·27	1·25	1·23	1·21

Fig. 10.16 masses (in g) of 1 dm^3 of dry air at various pressures (p) and temperatures (t) – the value used in the specimen results is underlined

calculation using oxygen

apparent weight of oxygen = 0·20 g-force
volume of bag = 2·56 dm^3
mass of air displaced by oxygen = (2·56 × 1·21) g
= 3·10 g
real mass of oxygen = (0·20 + 3·10) g
= 3·30 g
volume of 1 mole of oxygen (at laboratory temperature and pressure) = $2{\cdot}56 \times \frac{32}{3{\cdot}30}$ dm^3
= 24·8 dm^3

volume of 1 mole of oxygen at s.t.p. = 22·5 dm^3

calculation using hydrogen

apparent weight of hydrogen = −2·89 g-force
volume of bag = 2·56 dm^3
mass of air displaced by hydrogen = (2·56 × 1·21) g
= 3·10 g
real mass of hydrogen = (−2·89 + 3·10) g
= +0·21 g
volume of 1 mole of hydrogen (at laboratory temperature and pressure) = $\left(2{\cdot}56 \times \frac{2}{0{\cdot}21}\right)$ dm^3
= 24·3 dm^3

volume of 1 mole of hydrogen at s.t.p. = 22·1 dm^3

All gases used in this experiment give approximately the same result: **one mole of the gas occupies a volume of about 22·4 dm^3 at s.t.p.** This means that the volume occupied by a gas at a fixed temperature and pressure depends only on the number of molecules of gas present and not on the identity of the gas.

10.24 the gram-molecular volume (GMV) or molar volume (V_m)

The general conclusion to be drawn from the experiments described in the preceding section is that one mole of any gas at 273 K and a pressure of 101 325 Pa occupies a volume of $2{\cdot}24 \times 10^{-2}\,m^3$. This volume is called the **molar volume** and is symbolised by V_m (until recently the value was represented as 22·4 litres and was called the gram-molecular volume, GMV). Being the volume occupied by a fixed amount of substance, the molar volume has the units $m^3\,mol^{-1}$.

In Chapter 3, the relationship between the pressure, temperature and volume of a fixed mass of gas was investigated and it was seen that if any one of the three properties is held at a fixed value, variation of the second caused a corresponding variation in the third so that the value of $\frac{p \times V}{T}$ remains constant for a fixed mass of one gas. For example:

$$\frac{p \times V}{T} = \text{a constant, for 1 kg of oxygen}.$$

$$\frac{p \times V}{T} = \text{a different constant, for 1 kg of nitrogen}$$

The numerical values of the two constants are *different* for the two gases because there is no fundamental similarity between 1 kg of oxygen and 1 kg of nitrogen. If we consider the relationship for a fixed amount of substance in each gas (1 mole), rather than a fixed mass, the following result is obtained:

for 1 mole of oxygen $V = 2{\cdot}24 \times 10^{-2}\,m^3$ at 273 K and 101 325 Pa

$$\frac{p \times V}{T} = \frac{101\,325 \times 2{\cdot}24 \times 10^{-2}}{273} = \mathbf{8{\cdot}3\ J\,mol^{-1}\,K^{-1}}$$

for 1 mole of nitrogen, $V = 2{\cdot}24 \times 10^{-2}\,m^3$ at 273 K and 101 325 Pa

$$\frac{p \times V}{T} = \frac{101\,325 \times 2{\cdot}24 \times 10^{-2}}{273} = \mathbf{8{\cdot}3\ J\,mol^{-1}\,K^{-1}}$$

The constant $8{\cdot}3\,J\,mol^{-1}\,K^{-1}$ is the same for all gases; it is called the **molar gas constant** and is symbolised by R.

For 1 mole of any gas, $\frac{p \times V}{T} = R$, or $pV = RT$;

for an amount of substance n, $\frac{p \times V}{T} = nR$, or $pV = nRT$

(where n is expressed in moles).

At this point it might be illuminating to think of the constancy of the expression $\frac{p \times V}{T}$ in a slightly different way. It means that one mole of any gas in a volume of $2{\cdot}24 \times 10^{-2}\,m^3$ at 273 K will exert a pressure of 101 325 Pa. The pressure exerted will be constant if the number of particles present is fixed and will remain constant even if the identity of the fixed number of particles is changed. Thus there are certain properties of all gases (e.g. pressure at constant volume and temperature), measurement of which will reveal the number of moles of gas present. If the relative molecular mass of the gas is unknown, it can be calculated by measuring the properties p, V, and T and the mass of gas present:

$$\frac{p \times V}{T} \quad = \quad n \times R$$

(all measurable) (R accepted, n can be calculated)

If the mass of gas used is m, the ratio m/n is known and this is equal to M_r, the relative molecular mass. By expressing m in grams, m/n has the units $g\,mol^{-1}$ and is numerically equal to M_r.

The use of the molar volume V_m and the molar gas constant R is very valuable in quantitative work in chemistry.

10.25 chemical calculations involving the molar volume

Since equations represent quantities in terms of moles, the volumes of gases (measured at s.t.p.) involved in a reaction may easily be found from an equation once the constancy of the molar volume is understood. Some examples should make this clear. Non-standard units have been used in some cases in order to illustrate the necessary conversions.

Example 1 Exactly 1 g of calcium(II) carbonate is dissolved in hydrochloric acid; what volume of carbon dioxide, measured at s.t.p., is liberated? (A_r: C = 12, O = 16, Ca = 40)

$$\underbrace{\underset{40 + 12 + 48}{CaCO_3}}_{} + 2HCl \rightarrow CaCl_2 + H_2O + CO_2$$

$$100\,g \rightarrow 2{\cdot}24 \times 10^{-2}\,m^3 \text{ (s.t.p.)}$$

$$1\,g \rightarrow 2{\cdot}24 \times 10^{-4}\,m^3$$

$$= 224\,cm^3$$

The volume of carbon dioxide liberated is 224 cm^3

Example 2 What volume of chlorine, measured at 91 °C and 101 325 Pa, is required to react with 4·6 g of sodium? (A_r: Na = 23.)

$$\underbrace{\underset{2 \times 23}{2Na}}_{} + Cl_2 \rightarrow 2NaCl$$

46 g require $2{\cdot}24 \times 10^{-2}\,m^3$ (s.t.p.)

4·6 g require $2{\cdot}24 \times 10^{-3}\,m^3$ (s.t.p.)

At 91 °C, this chlorine will occupy

$$2{\cdot}24 \times 10^{-3} \times \frac{364}{273}\,m^3$$

$$= 2{\cdot}24 \times 10^{-3} \times \frac{4}{3}\,m^3$$

$$= 2{\cdot}99 \times 10^{-3}\,m^3$$

The volume of chlorine required is 2·99 dm^3

Example 3 5 litres of hydrogen measured at s.t.p. is passed slowly over hot copper(II) oxide and the excess hydrogen is collected. If 12·7 g of copper is formed, how many cm^3 of hydrogen, measured at s.t.p., are collected? (A_r: Cu = 63·5)

$$CuO + H_2 \rightarrow Cu + H_2O$$

$$2{\cdot}24 \times 10^{-2}\,m^3 \rightarrow 63{\cdot}5\,g$$

$$2{\cdot}24 \times 10^{-2} \times \frac{12{\cdot}7}{63{\cdot}5}\,m^3 \rightarrow 12{\cdot}7\,g$$

$$= 4{\cdot}48 \times 10^{-3}\,m^3$$

Since there were originally 5 litres of hydrogen

$$= 5 \times 10^{-3}\,m^3 \text{ of hydrogen}$$

the volume of hydrogen remaining is

$$(5 \times 10^{-3}) - (4{\cdot}48 \times 10^{-3})\,m^3$$

$$= 0{\cdot}52 \times 10^{-3}\,m^3$$

$$= 520 \times 10^{-6}\,m^3$$

The volume of hydrogen collected is 520 cm^3

10.26 deduction of the equation for a reaction between gases by volume measurement and the molar volume

The previous section showed how volumes of gases involved in a reaction can be calculated using the molar volume if the equation for the reaction is known. If the equation is not known, it can be deduced by measuring the

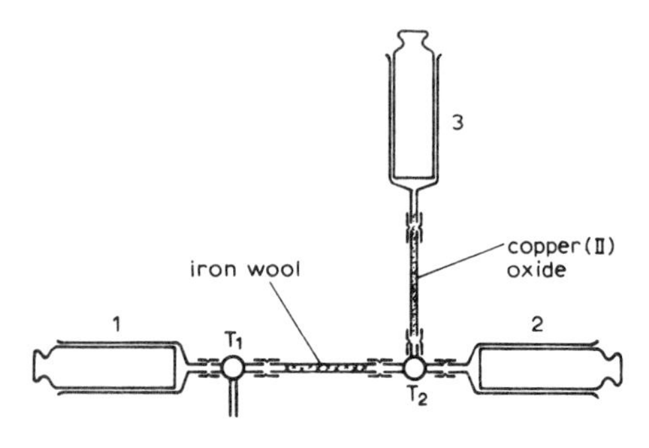

Fig. 10.17 determination of the formula for ammonia and the equation for its decomposition

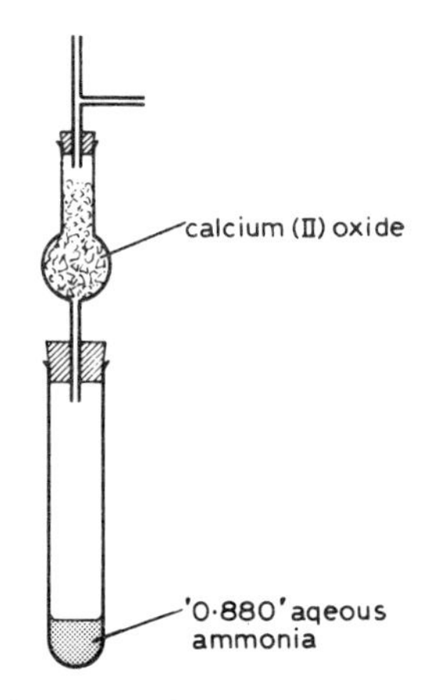

Fig. 10.18 dry ammonia generator

volumes of the gases involved in a reaction and using the same principle in reverse.

The following experiment measures the volumes of nitrogen and hydrogen produced by the thermal decomposition of ammonia, the composition of which you may not know at this stage in your studies. The results will tell us the equation for the decomposition and the formula for ammonia as well. Gas syringes are used and are arranged as shown in Fig. 10.17.

A known volume of ammonia is decomposed by heat into nitrogen and hydrogen; the volume of the products is measured. The hydrogen is removed by passing the products over hot copper(II) oxide; the volume of nitrogen remaining is measured.

The whole apparatus, which must be clean and dry, is flushed out with nitrogen from a cylinder and all syringes are pushed in. During this process the apparatus is tested for air-tightness. Dry ammonia is generated by warming concentrated aqueous ammonia in the apparatus shown in Fig. 10.18 and $40\,cm^3$ of gas is introduced into syringe 1. Taps T_1 and T_2 are turned to connect syringes 1 and 2 and the ammonia is passed over the hot iron wool (free from oxide) until the volume of the gas is constant. The apparatus is cooled and the volume measured. The gas is transferred to syringe 2 and tap T_2 turned to connect syringes 2 and 3; the gas is then passed over hot copper(II) oxide until the volume of gas is constant. The apparatus is cooled and the volume measured.

volume of ammonia used	$= 40\,cm^3$
volume of nitrogen and hydrogen formed	$= 80\,cm^3$
volume of nitrogen remaining	$= 20\,cm^3$
volume of hydrogen formed	$= 60\,cm^3$

$40\,cm^3$ of ammonia produces $20\,cm^3$ of nitrogen and $60\,cm^3$ of hydrogen
2 volumes of ammonia produce 1 volume of nitrogen and 3 volumes of hydrogen
$2 \times V_m$ of ammonia produce $1 \times V_m$ of nitrogen and $3 \times V_m$ of hydrogen
2 moles of ammonia produce 1 mole of nitrogen and 3 moles of hydrogen

or 2 moles of ammonia $\longrightarrow N_2 + 3H_2$

therefore **the formula for ammonia is NH_3** and the equation for its decomposition is

$$2NH_3 \longrightarrow N_2 + 3H_2$$

10.27 Dalton, Gay-Lussac, Avogadro and Cannizzaro

In 1804 John Dalton proposed his atomic theory. The two new and significant proposals which he made were:
i) All atoms of any one element are exactly alike and have identical masses. Atoms of different elements are different and have different masses.
ii) Chemical combination takes place between small whole numbers of atoms.

Using these postulates, he was able to explain the Law of Conservation of Mass (section 10.19), the Law of Constant Composition (section 10.10) and the Law of Multiple Proportions (section 10.12). Can you? He actually predicted the last law from his theory and the law was subsequently verified by experiment.

The atomic theory was next brought to bear on the results of a number of experiments carried out by the French scientist Gay-Lussac, which were summarised in the following statement:

Gay-Lussac's law of combining volumes

The volumes of gases which react, and the volumes of products (if gaseous), when measured under the same conditions of temperature and pressure, are in the ratio of small whole numbers.

Remember that nothing was known at this time about molecules; the existence of atoms was only beginning to gain acceptance and the particles of which compounds were composed were referred to as 'compound atoms'. Note, however, that Gay-Lussac did not seek to explain his findings according to any theory; he was content to state the factual result of a large number of experimental observations carried out with great skill and patience. These facts were ripe for explanation and the atomic theory was invoked to do so. The argument may be summarised as follows:

Gay-Lussac: Gases combine in small whole numbers of **volumes** (*fact*)
Dalton: All elements (including gases) combine in small whole numbers of **atoms** (*theory*)

To the Swedish chemist Berzelius the connection was apparently quite clear: he assumed that equal volumes of gases (measured under the same conditions of temperature and pressure) contained equal numbers of atoms. He examined one of Gay-Lussac's experimental results and set about explaining it in detail:

hydrogen +	chlorine	$\rightarrow$ hydrogen chloride	
1 volume	1 volume	2 volumes	(*Gay-Lussac*)
n atoms	n atoms	2n 'compound atoms'	(*Berzelius*)
1 atom	1 atom	2 'compound atoms'	

This result cannot be true. Can you see why?

Dalton's solution to the problem was simple: he claimed that Gay-Lussac's results were in error. Perhaps we can now forgive such dogmatism and intolerance in view of the nature of his profession (he was a schoolmaster in Manchester).

A more satisfactory solution was put forward in 1811 by the Italian chemist Avogadro. He amended Berzelius's assumption to read:

Avogadro's hypothesis (law)

Equal volumes of gases, under the same conditions of temperature and pressure, contain equal numbers of molecules.

He further assumed that the molecules of hydrogen and chlorine were diatomic and explained the above results as follows:

hydrogen	+ chlorine	$\rightarrow$ hydrogen chloride	
1 volume	1 volume	2 volumes	(*Gay-Lussac*)
n molecules	n molecules	2n molecules	(*Avogadro*)
1 molecule	1 molecule	2 molecules	

It need now only be assumed that the *molecules* of hydrogen and chlorine each split into two parts (*atoms*), one for each of the hydrogen chloride molecules, to overcome the difficulty in Berzelius's argument involving the splitting of hydrogen and chlorine atoms. You may already have noticed that Avogadro's hypothesis is an alternative statement of the principle inherent in the molar volume concept (section 10.24).

This admirable solution was ignored for over thirty years, for lack of evidence to justify the assumptions. We now have evidence to support the assumptions

a) that gases are composed of particles (molecules) which can be split into smaller particles if sufficient energy is supplied (mass spectrometer, see Chapter 9),

b) that the hydrogen molecule is diatomic (mass spectrometer),

c) that equal volumes of gases (same T, p) contain equal numbers of molecules (determination of molar volume, sections 10.23 and 10.24).

In 1848, without the aid of the evidence of *a*) and *b*) above, the Italian chemist Cannizzaro produced a brilliant but complicated argument to justify the idea that the molecule of hydrogen is diatomic. He then assumed the truth of Avogadro's hypothesis and used it to predict a result which could be checked against other experimental evidence. His argument is summarised on the next page.

density of a gas = mass of a given volume of gas

$$\text{relative vapour density of a gas} = \frac{\text{mass of a given volume of gas}}{\text{mass of an equal volume of hydrogen}} \quad (T, p \text{ const.})$$

$$(\textit{assuming Avogadro's hypothesis}) = \frac{\text{mass of n molecules of gas}}{\text{mass of n molecules of hydrogen}}$$

$$= \frac{\text{mass of 1 molecule of gas}}{\text{mass of 1 molecule of hydrogen}}$$

$$(\textit{assuming hydrogen diatomic}) = \frac{\text{mass of 1 molecule of gas}}{\text{mass of 2 atoms of hydrogen}}$$

$$= \frac{1}{2} \times \frac{\text{mass of 1 molecule of gas}}{\text{mass of 1 atom of hydrogen}}$$

$$= \tfrac{1}{2} \times \text{relative molecular mass}$$

relative molecular mass = 2 × relative vapour density

By measuring densities of gases and vapours, Cannizzaro obtained values for their relative molecular masses. He then used this information to deduce values for the relative atomic masses of elements. Some of these values could be checked against results obtained by independent means: they were found to correspond. Many relative atomic masses, hitherto unknown, were determined by his method, and for the first time a comprehensive set of reliable values for the relative atomic masses of elements was made available. Cannizzaro's brilliant researches did not end at this point, for amongst the audience at Cannizzaro's meeting was a Russian chemist who pursued the significance of the new information for over ten years, then to emerge with an idea which revolutionised the study and understanding of chemistry. His name was Mendeleeff and his monumental work is outlined in Chapter 17.

revision summary: Chapter 10

gram-atom: one gram-atom of any element contains 6×10^{23} atoms

symbols: can be used to represent gram-atoms as well as atoms

mole: one mole of substance contains 6×10^{23} units–the mole is a base unit in SI defining an amount of substance

valency: a number which represents the capacity of an element to combine with other elements–the number of hydrogen atoms which one atom of the element will combine with or replace. The idea is also applicable to radicals

molar volume: volume occupied by 1 mole of any gas at s.t.p., value $2{\cdot}24 \times 10^{-2}\,\text{m}^3$

molar gas constant R: $R = 8{\cdot}3\,\text{J}\,\text{mol}^{-1}\,\text{K}^{-1}$
$pV = RT$ for 1 mole of gas

Avogadro's hypothesis: equal volumes of gases, under the same conditions of temperature and pressure, contain equal numbers of molecules

relative molecular mass = 2 × relative vapour density

questions 10

1. When 6 g of magnesium burns in oxygen, 10 g of its oxide is formed. Find the simplest (empirical) formula for the oxide.

2. 254 g of iodine was obtained from 334 g of an oxide of iodine. What is the simplest formula for the oxide?

3. How many grams of hydrogen would be required to reduce the oxide in question 2?

4. An oxide of chromium contains 48% by mass of oxygen. What is its simplest formula?

5. A gas of formula SX_2 contains exactly 50% by mass of sulphur. Identify the element X.

6. Calculate the percentage by mass of titanium in rutile, TiO_2.

7. A crystalline salt was found to contain 20·8% chromium, 19·2% sulphur and 2·4% hydrogen by mass, the rest being oxygen. Find the empirical formula for the salt. Can you write this formula in any more meaningful way?

8. At s.t.p., a gas has a density of 15·7 g dm^{-3}. Assuming the value for the molar volume, calculate the relative molecular mass of the gas. If the gas is a compound of uranium and fluorine only, what is its formula?

9. The oxide of an element X, of relative atomic mass 39, is found to contain 15·6 g of X combined with 0·2 gram-atom of oxygen. What is the simplest formula of the oxide? (W)

10. What volume of hydrogen (measured at s.t.p.) would be liberated when 3 g of magnesium are treated with excess dilute hydrochloric acid?

What volume of oxygen (measured at s.t.p.) would completely combust the liberated hydrogen? (S)

11. (a) Complete the following sentence: The Avogadro constant, $6{\cdot}02 \times 10^{23}$, is the number of . . .

(b) A packet of sugar contains crystals of average mass $3{\cdot}42 \times 10^{-3}$ g. How many sugar molecules will an average crystal contain?

(Formula of sugar: $C_{12}H_{22}O_{11}$. Relative atomic masses: H = 1·0, C = 12·0, O = 16·0) (JMB)

12. Given that 32·0 g of oxygen contain N molecules, state in terms of N the number of

(a) atoms in 32·0 g of sulphur,

(b) molecules in 32·0 g of hydrogen,

(c) molecules in 32·0 g of sulphur dioxide,

(d) ions in 32·0 g of sulphate(VI) ions.

(Relative atomic masses: H = 1·0, O = 16·0, S = 32·0) (JMB)

13. (a) Find the ratio of the number of atoms in 16 g of sulphur to the number of atoms in 46 g of sodium. (Na = 23; S = 32)

(b) Find the ratio of the number of molecules in 32 g of methane, CH_4, to the number of molecules in 5·6 litres of hydrogen at s.t.p. (H = 1; C = 12)

(c) One oxide of manganese, Mn, contains 2·4 g of oxygen combined with 0·1 gram-atom of manganese. What is the simplest formula of this oxide? (O = 16) (W)

14. Calculate the volume of carbon dioxide produced at s.t.p. when 4·2 g of sodium(I) hydrogencarbonate are (a) heated until no further loss in mass occurs, (b) allowed to react with excess dilute hydrochloric acid. (S)

15. (a) State the *law of multiple proportions.*

Two oxides of a metal contain 90·7% and 92·8% of the metal respectively. Show that the two values support the law.

(b) An ore of copper has the formula $CuFeS_2$. What maximum mass of copper to the nearest kilogram could be extracted from 100 kg of this ore?

(c) A compound of carbon and sulphur contains 64% of sulphur. Find the simplest formula of the compound. (C)

16. State (a) the law of definite (constant) proportions, (b) the law of conservation of mass, (c) Graham's law of diffusion of gases.

Describe experiments by which the law of definite proportions can be illustrated.

What evidence do these laws provide for the atomic nature of matter? (AEB)

17. State (a) Gay-Lussac's law of gaseous volumes, (b) Avogadro's law.

Describe any one experiment you choose which fully illustrates Gay-Lussac's law.

Four grammes of sulphur on heating in oxygen give 2·8 litres of sulphur dioxide. Calculate the relative molecular mass (molecular weight) of sulphur dioxide. (AEB)

18. When 12·0 g of the anhydrous sulphate(VI) of a divalent metal (M) were strongly heated a residue of 4·0 g of the metal oxide was obtained.

(a) Assuming that only one volatile product was formed, write a possible equation for the reaction.

(b) Name the volatile product.

(c) What mass of volatile product was formed?

(d) Calculate the relative atomic mass of the metal.

(Relative atomic masses: O = 16·0, S = 32·0) (JMB)

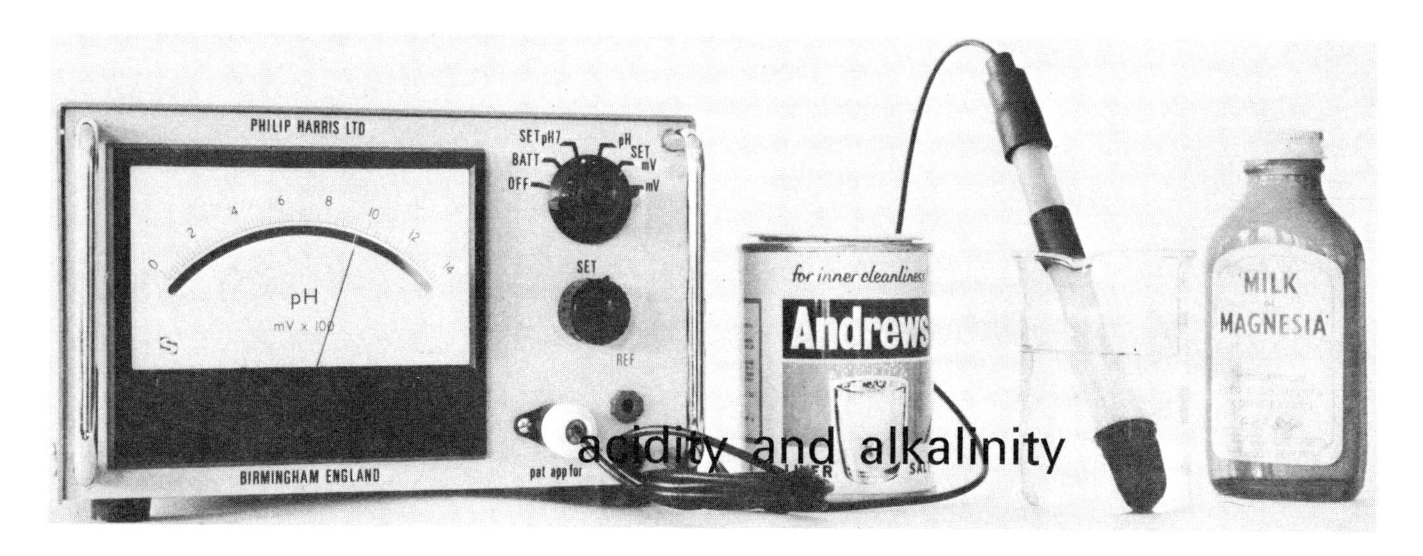

acidity and alkalinity

11.1 characteristics of acids

Everyone knows about acids–or do they? Certainly ideas of acidity and alkalinity are among the oldest in the history of science. When a citizen of ancient Rome sucked a lemon and pronounced it '*acidus*' (sour) he had noticed a characteristic property of lemon juice which it shares with a large number of compounds. Perhaps acids may be defined as substances with a sharp sour taste? This is an **operational definition** (a definition which relies only on what is observed or measured)–a very practical definition. However it would be a foolhardy chemist who relied upon taste; there must be other reasons for classifying substances as acids. You may consider looking for characteristics of appearance. Inspect any acids available in the laboratory and you will probably agree that physical appearance or state is hardly characteristic. Some acids are labelled *dilute*, some *concentrated*; some are solids, others are liquids.

Acids have a reputation for corrosiveness. We hear of acid 'burns' and think of acids eating through metals, cloth and flesh. In fact, while some acids well deserve this reputation, it is not essentially a characteristic of their *acidity* but of other properties shared by numerous non-acids–*oxidising* or *dehydrating* properties. Fortunately a great deal of information has been collected about the compounds called acids and alkalis and several characteristics are apparent. In this Chapter only operational definitions, derived from laboratory experience, are mentioned. Later, in section 18.24, acidity and alkalinity are considered from a structural point of view and the nature of acids and alkalis re-interpreted in terms of the Ionic Theory.

A number of investigations into the characteristic reactions of acids may be attempted. Suitable acids for study are: hydrochloric, sulphuric(VI), nitric(V), ethanoic (acetic) and citric acids. Great care must be taken when pouring acids, particularly if they are concentrated. If spills occur, treat the place with aqueous sodium(I) carbonate, then wash it thoroughly with water.

11.2 action of acids with indicators

About 2 cm^3 of the purest or most concentrated form of each of the acids is placed in a set of test-tubes–one acid to each tube. Another set of test-tubes contains 2 cm^3 of the dilute form of each acid. A piece of dry, blue litmus paper is added to each test-tube and all changes are noted.

Only the dilute acids show any constancy of behaviour; blue litmus turns red in each case. With the concentrated acids blue litmus turns red quickly only in the cases of hydrochloric and nitric(V) acids. There is a slow change in concentrated sulphuric(VI) acid and no change in pure ethanoic (acetic) or solid citric acid.

Aqueous acids turn blue litmus red. There are several other common indicators; screened methyl orange and phenolphthalein are found in most laboratories and their action with acids may be investigated.

11.3 action of acids with metals

Rather than try all acids with all metals, it is preferable to choose one typical metal. Magnesium is a metal sufficiently reactive to give useful observations. Sets of test-tubes containing 2 cm^3 of each acid are prepared as in section 11.2. A small piece (1 cm long) of magnesium ribbon is added to each.

hydrochloric
sulphuric(VI)
nitric(V)
ethanoic (acetic)
citric

As in 11.2, the concentrated or pure acids do not show consistent behaviour. Ethanoic (acetic) and citric acids do not react; sulphuric(VI) acid reacts slowly; brown fumes appear with nitric(V) acid; hydrochloric acid effervesces, liberating a colourless gas which 'pops' when ignited–hydrogen. Only when plenty of water is present do acids behave similarly. With all the **dilute acids** there is effervescence and **hydrogen is liberated.** There are some differences in the rates of reaction and the quantity of heat produced even with dilute acids. These differences are due to the strength or weakness of the acids (which should not be confused with dilution or concentration–this subject is discussed in section 18.24).

11.4 salts

When magnesium reacts with dilute sulphuric(VI) acid the metal disappears and the gas hydrogen appears. Any product other than hydrogen must remain dissolved in the solution. What is this product?

Small pieces of magnesium are added to 2 cm^3 of dilute sulphuric(VI) acid in a test-tube until excess metal remains, even when the solution is warmed. When a little of the solution is removed with a dropping pipette and a few drops are allowed to evaporate on a warm watch-glass or microscope slide, colourless crystals remain. This substance is magnesium(II) sulphate(VI) and it belongs to the class of compounds known as **salts.**

Salts are always named after the 'parent' metal and acid, though, as shown in section 12.16, salts are not always formed directly from their 'parents'.

acid	salt	example	formula
sulphuric(VI), H_2SO_4	sulphate(VI)	magnesium(II) sulphate(VI)	$MgSO_4$
hydrochloric, HCl	chloride	potassium(I) chloride	KCl
nitric(V), HNO_3	nitrate(V)	lead(II) nitrate(V)	$Pb(NO_3)_2$
ethanoic, CH_3CO_2H (acetic)	ethanoate (acetate)	sodium(I) ethanoate (acetate)	CH_3CO_2Na

Fig. 11.1

The investigation in section 11.3 shows that dilute aqueous acidic solutions react with magnesium to give hydrogen and a salt solution.

$$Mg(s) + H_2SO_4(aq) \longrightarrow MgSO_4(aq) + H_2(g)$$

Why is sodium or copper not chosen for this investigation? Which other metals might have been used? (see section 8.15).

11.5 action of acids with carbonates and hydrogencarbonates

Using the apparatus shown in Fig. 11.2, 2 cm^3 of the dilute aqueous acidic solutions are added to small measures (about 0·5 g) of a variety of carbonates and

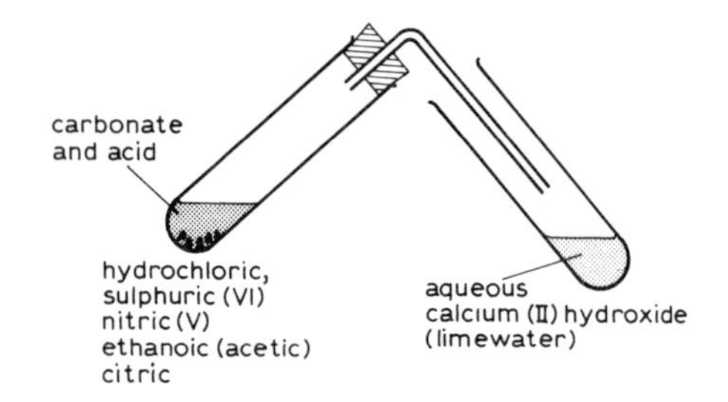

Fig. 11.2 action of acids on carbonates

hydrogencarbonates; copper(II) carbonate, calcium(II) carbonate, sodium(I) carbonate and sodium(I) hydrogencarbonate are good examples.

In most cases the gas carbon dioxide is liberated and is detected by the appearance of a white precipitate of calcium(II) carbonate in aqueous calcium(II) hydroxide (limewater). If more carbonate is added, evolution of carbon dioxide eventually ceases. Testing the solution with blue litmus paper shows that it is no longer acidic and if a few drops of the solution are removed and allowed to evaporate on a warm microscope slide, crystals of a salt are found in every case.

A conclusion from this investigation is that **acid solutions react with a carbonate or a hydrogencarbonate to yield carbon dioxide, a salt and water.**

$$H_2SO_4(aq) + MgCO_3(s) \longrightarrow MgSO_4(aq) + CO_2(g) + H_2O(l)$$
$$HCl(aq) + NaHCO_3(s) \longrightarrow NaCl(aq) + CO_2(g) + H_2O(l)$$

11.6 action of acids with bases

The investigations described in sections 11.3 and 11.5 show that acidity may be reduced or even removed by the action of a metal or carbonate; in both cases a gas appears as one of the products. Substances which counteract acidity without gas evolution are also known. A good example is magnesium(II) hydroxide, the active ingredient in several famous remedies for the indigestion pangs caused by excess stomach acidity. When magnesium(II) hydroxide and an acid react, the only products are the magnesium salt of the acid and water. Compounds which react with acids to produce salts and water only are called **bases.** They include all hydroxides of metals and many oxides of metals.

(N.B. another definition of bases is discussed in section 18.25.)

$$H_2SO_4(aq) + Mg(OH)_2(s) \longrightarrow MgSO_4(aq) + 2H_2O(l)$$
$$2HCl(aq) + CuO(s) \longrightarrow CuCl_2(aq) + H_2O(l)$$

11.7 alkalis

Many bases are solids which do not dissolve readily in water; those which are soluble are called **alkalis.** Their solutions are said to be alkaline (derived from the Arabic word for 'ashes'–it was from ashes of wood that the Arabs first extracted substances with alkaline properties).

The most familiar alkalis are the hydroxides of the 'alkali' metals potassium and sodium and the 'alkaline earth' metals calcium and barium; aqueous ammonia is the unusual fifth member of the group. The properties of alkalis may be investigated by finding answers to the following questions. 2 cm^3 volumes of each alkaline solution are used in each case and the techniques are identical to those for investigating acids. Care must be taken in pouring alkalis, particularly the 'caustic' alkalis, sodium(I) and potassium(I) hydroxides. If any is spilled, the alkali should be neutralised by pouring dilute ethanoic (acetic) acid on the place, then washing the area with water.

a) How do alkalis affect indicators, compared with acids?
b) How do alkalis react with metals? Aluminium, zinc and iron should be tried as well as magnesium; gentle warming may be required.
c) Do alkalis have any reaction with carbonates or hydrogencarbonates?
d) How does any one alkali react with a particular acid? Predictions may be justified using an indicator, followed by evaporation of the solution.
e) How does an aqueous solution of a typical alkali such as sodium(I) hydroxide affect aqueous solutions of metal salts? (This is more fully discussed in section 12.29.)

These investigations show that in some reactions alkalis reverse the action of acids. In (*a*), indicators have one colour in acidic solutions and another in

indicator	in acid	in alkali
litmus	red	blue
screened methyl orange	red	green
phenolphthalein	colourless	red

Fig. 11.3 colour changes of indicators

alkaline solutions (see Fig. 11.3). In (*d*), addition of excess alkali completely removes all acidity; in 'neutral' solutions salts and water are the only products.

e.g. $$NaOH(aq) + HCl(aq) \rightarrow NaCl(aq) + H_2O(l)$$

Alkalis have little reaction with many metals but potassium(I) and sodium(I) hydroxides do react with aluminium and zinc yielding hydrogen. Thus liberation of hydrogen by a metal is not an infallible test for acids. However (*c*) shows that alkalis have no effect on carbonates or hydrogencarbonates and this remains a good test for acids.

Addition of a hydroxide solution causes precipitation of the insoluble hydroxides of other metals from solutions of their salts. For example, when aqueous sodium(I) hydroxide is added to aqueous copper(II) sulphate(VI), a pale-blue precipitate of copper(II) hydroxide forms (see section 12.29 for more examples).

$$CuSO_4(aq) + 2NaOH(aq) \rightarrow Cu(OH)_2(s) + Na_2SO_4(aq)$$

A particularly important property of sodium(I) hydroxide is its reaction with oils, fats and greases, forming soaps. This is why aqueous sodium(I) hydroxide feels soapy on your hands and therefore must be washed off quickly.

11.8 degrees of acidity and alkalinity: pH scale

Simple indicators like litmus or screened methyl orange show only whether a solution is acidic, alkaline or perhaps neutral. Universal Indicator will distinguish different degrees of acidity or alkalinity by displaying a *range* of colours.

The colour scale shown in Fig. 11.4 is too imprecise for accurate work; instead the numerical **pH scale** is used. The pH scale has numbers which are mathematically related to the degree of acidity or alkalinity of a solution. The most commonly used numbers on the scale lie between 1 and 14; the number 7 represents neutrality, numbers below 7 represent acidity and those above 7 alkalinity.

The pH of pure water is 7.
All acids, when dissolved in water, change its pH to less than 7.
All alkalis, when dissolved in water, change its pH to more than 7.

Universal Indicator is useful for determining approximate pH values, but for really accurate work an electrical pH meter must be used (see Fig. 11.6).

A knowledge of the pH of solutions is immensely important in chemistry and particularly in the chemistry of biological systems. Mammals suffer ill-effects if the pH of their blood is not maintained at about 7·4. However, human gastric juices may have a pH as low as 1·5 and this acidic solution provides the most suitable conditions for enzymes to aid the digestion of food (see section 19.14).

Plant growth is affected by the pH of the environment. The pH of soil ranges from 10 or 11 in alkali deserts to 3·5 at the most acidic. Only very few plants can tolerate these extremes.

In this country the pH of soil ranges from approximately 8·0 in the chalk and limestone regions to about 4·5 in peaty areas. You should now understand why certain crops are grown in particular areas. Of course it is possible to change the pH of the soil; liming is the oldest and most common method of reducing soil acidity. Farmers and serious gardeners who wish to know the pH of soil use pH soil-testing kits.

11.9 concentrations of solutions of acids and alkalis

Acids and alkalis are most often used in diluted form and frequently chemists need to know how much *pure* acid or alkali is present in a given volume of solution.

	(*most acid*)
acid colours	red orange yellow
neutral	green
alkaline colours	blue violet
	(*most alkaline*)

Fig. 11.4 colours of Universal Indicator

	pH	Univ. Ind.
	1	
strongly	2	red
acidic	3	
	4	orange
weakly	5	yellow
acidic	6	
neutral	7	green
weakly	8	
alkaline	9	blue
	10	
	11	
	12	
strongly	13	violet
alkaline	14	

Fig. 11.5 colours of Universal Indicator in terms of pH values

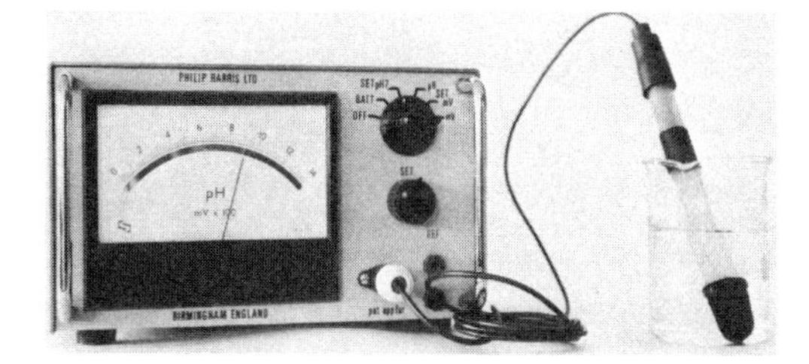

Fig. 11.6

crop	optimum pH
sugar beet	7·5–7
wheat	7·5–6
turnips	7·5–5·5
rye	6–4·6
swedes	5·5–4·6

Fig. 11.7 optimum pH for growth of crop

Fig. 11.8

a) Concentration measures the amount of solute per unit volume of solution. The unit of concentration should be $mol\,dm^{-3}$ but occasionally the same concentration is shown as $mol\,l^{-1}$ or mol/l. Chemists often abbreviate $mol\,dm^{-3}$ to M.

e.g. A solution having a concentration of $0{\cdot}1\,mol\,dm^{-3}$ is 0·1 M.

b) Mass concentration measures the mass of solute per unit volume of solution. The unit of mass concentration used in this book is $g\,dm^{-3}$, though elsewhere the same mass concentration may be shown as grams per litre, $g\,l^{-1}$ or g/l.

Example 1: What is the concentration of a solution containing 80 g of pure sodium(I) hydroxide in $1\,dm^3$ of solution?

First find the number of moles of sodium(I) hydroxide in 80 g:

1 mole of sodium(I) hydroxide Na O H has a mass of 40 g

$$(A_r)\ 23 + 16 + 1, \quad M_r = 40$$

$$80\text{ g of sodium(I) hydroxide} = \frac{80}{40}\text{ mol} = 2\text{ mol}$$

the solution is 2 M

Example 2: What is the concentration of a solution containing 4 g of sodium(I) hydroxide in $500\,cm^3$?

$$4\text{ g of sodium(I) hydroxide} = \frac{4}{40}\text{ mol} = 0{\cdot}1\text{ mol}$$

The solution contains 0·1 mol of sodium(I) hydroxide in $500\,cm^3$

The solution contains 0·2 mol of sodium(I) hydroxide in $1000\,cm^3$

the solution is 0·2 M

Example 3: What is the mass concentration of a 0·3 M solution of hydrochloric acid (hydrogen chloride solution)?

1 mol of hydrogen chloride has a mass of 36·5 g

0·3 mol of hydrogen chloride has a mass of $36{\cdot}5 \times 0{\cdot}3\,g = 10{\cdot}95\,g$

the mass concentration is $10{\cdot}95\,g\,dm^{-3}$

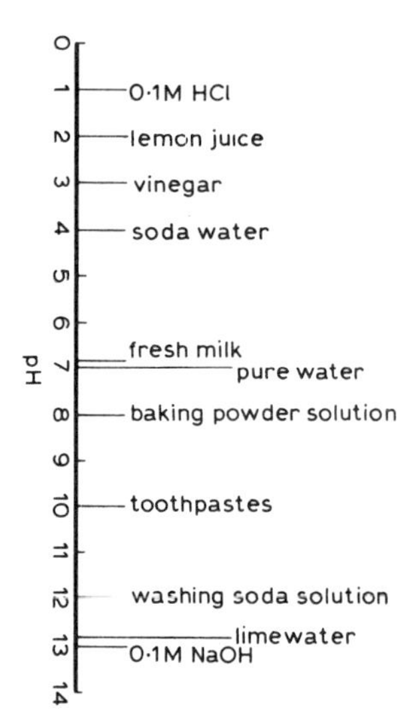

Fig. 11.9 pH values of some familiar substances

11.10 a more detailed study of an acid–alkali reaction

It is interesting to study some of the changes which occur when an alkali is added to an acid. In the experiment detailed here, changes in pH and temperature are investigated.

a) Four drops of Universal Indicator are added to $10\,cm^3$ of 4 M hydrochloric acid in a boiling-tube. The colour and temperature are noted in a table as shown in Fig 11.11. $1\,cm^3$ of 4 M aqueous sodium(I) hydroxide is added *quickly* from a syringe or pipette, the solution is gently stirred with the stirring rod (not the thermometer) and the colour and maximum temperature are noted *as soon as possible. Immediately* another $1\,cm^3$ of the alkaline solution is added and the changes are again observed. This is continued until about $15\,cm^3$ of alkaline solution have been added. pH values are found from the pH indicator colour table in Fig. 11.5.

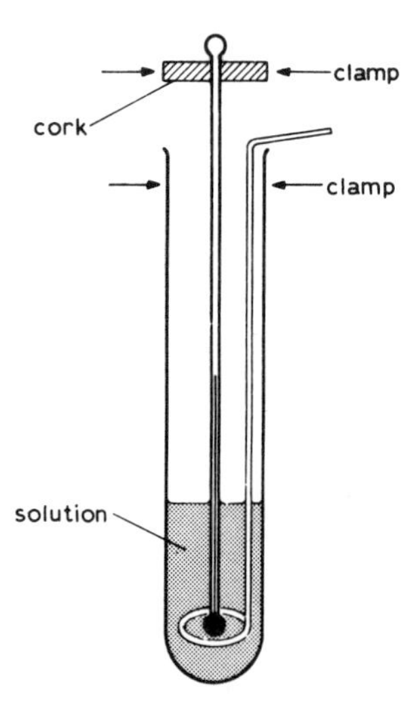

Fig. 11.10 thermometric titration

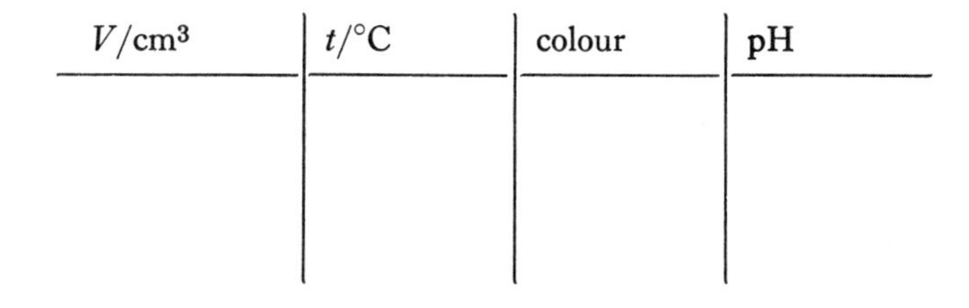

V/cm^3	$t/°C$	colour	pH

Fig. 11.11 table of results for volume of alkali added (V), temperature (t), colour and pH

b) The results are best expressed graphically as shown in Fig. 11.12. The continuous line shows the variation of pH with the volume of alkali added and the dotted line shows the variation in temperature with the volume of alkali added.

c) i) Graph (1) is in three parts. What does each part represent?
ii) Graph (2) shows that the temperature reaches a maximum, then falls. Why?
iii) What significant feature is common to both graphs?
iv) How many moles of hydrogen chloride are there in $10 cm^3$ of 4M hydrochloric acid?
v) With what volume of 4M sodium(I) hydroxide solution does this amount of hydrogen chloride react?
vi) How many moles of sodium(I) hydroxide are there in this volume?
vii) How many moles of sodium(I) hydroxide react with one mole of hydrogen chloride?

The answer to question (vii) justifies the 'balancing' of the equation for the reaction:

$$HCl(aq) + NaOH(aq) \longrightarrow NaCl(aq) + H_2O(l)$$

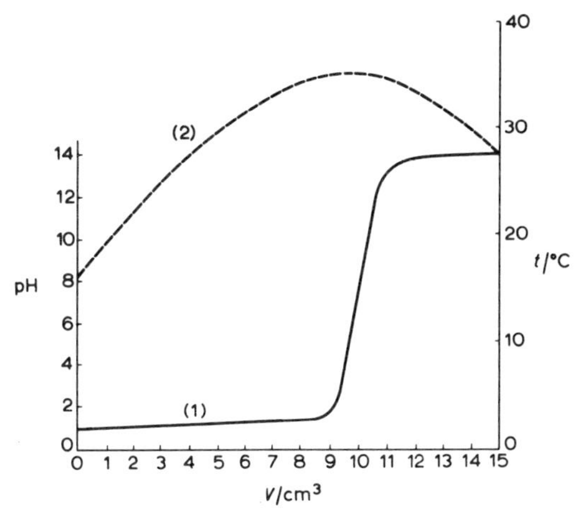

Fig. 11.12 variation in (1) pH and (2) temperature on addition of a volume of alkali to the acid

11.11 titrations

The process of adding a measured volume of one solution, in small quantities, to a measured volume of another solution, until reaction is complete, is called a **titration.** The stage of a titration when the reaction is just complete, and there is no excess of either reactant, is called the **equivalence point.** In the type of titration described in section 11.10 the equivalence point is detected in two ways:

Graph (1): the point through which the pH changes rapidly,
Graph (2): the point at which the temperature reaches a maximum.

11.12 volumetric analysis

Accurate titrations are often used to analyse solutions of unknown concentration. This is known as volumetric analysis. At least one of the solutions used in a volumetric analysis must be of known concentration and such a solution is called a **standard solution.**

The main volume-measuring devices are:

i) the **burette**, which delivers from 0 to $50 cm^3$ of solution accurately to $0{\cdot}1 cm^3$;
ii) the **pipette**, which delivers a single volume very accurately, if used properly.

Other apparatus and reagents required include two conical flasks, a small funnel for filling the burette, pipette-filler, burette-stand, white tile, wash-bottle of distilled water, some beakers, an indicator and the two solutions.

A very careful and methodical procedure is necessary to ensure accuracy.

11.13 standardisation of a solution of sodium(I) hydroxide given standard sulphuric(VI) acid

a) *steps in the procedure*
i) All glass vessels are rinsed with distilled water; they must be thoroughly clean.
ii) The burette is rinsed twice with a little of the acid, some of which is allowed to run through the tap. The rinsings are then rejected.
iii) The burette is filled with the acid and the funnel removed. The liquid must fill the space below the tap.
iv) The pipette is rinsed twice with a little alkali. If a pipette-filler is not

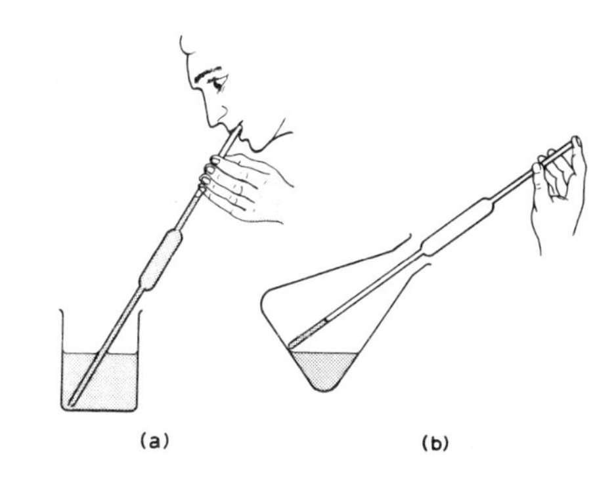

Fig. 11.13 (a) filling a pipette
(b) delivering from a pipette

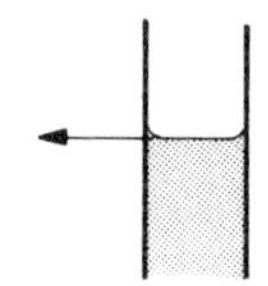

Fig. 11.14 always read from the bottom of the meniscus

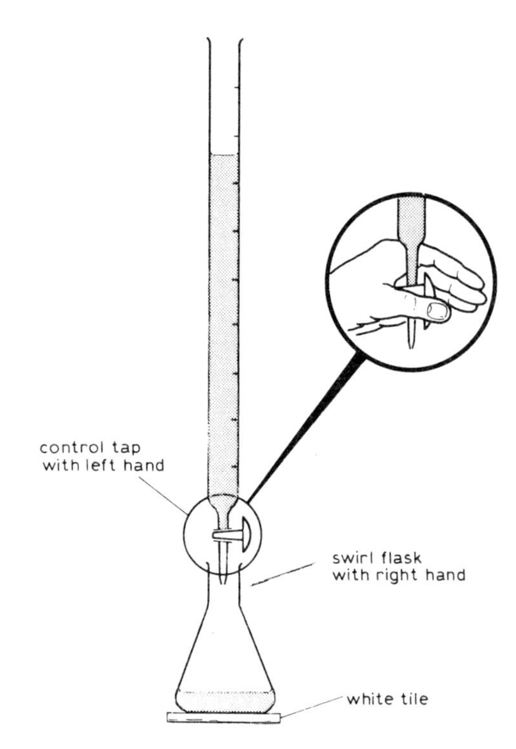

Fig. 11.15 operation of burette during titration

available, mouth-suction may be used. Great care must be taken in doing this; during the process, the tip of the pipette must not be allowed to rise above the level of liquid in the beaker or flask. In either case the pipette is filled until the meniscus just rests on the mark on the stem above the bulb (Fig. 11.14).

v) Using the pipette, $25\,cm^3$ of aqueous sodium(I) hydroxide is transferred to a $250\,cm^3$ conical flask. (Do not shake or blow into the pipette; just touch the tip against the bottom of the flask, held at an angle as shown in Fig. 11.13(b).)

vi) Two drops of screened methyl orange are added, as indicator, to the alkali in the flask.

vii) The burette reading is noted, reading from the bottom of the meniscus.

viii) The acid is added to the alkali approximately $1\,cm^3$ at a time with swirling, not vigorous shaking, until the indicator just changes colour. A good technique is to control the tap of the burette with the left hand, while swirling the flask with the right, as shown in Fig. 11.15. The change of colour is called the **end-point** of the titration; the end-point should indicate the equivalence point of the reaction (see section 11.10). In practice it may be difficult to stop the first titration just at the end-point and this is usually regarded as a trial.

ix) The burette reading at the end-point is noted, reading again from the bottom of the meniscus.

x) After washing the conical flask well with distilled water, the titration is repeated. This time the acid is added quickly until the end-point is approached, then drop by drop to the end-point. Before the end-point is reached, the inside walls of the conical flask should be washed down with distilled water using the washbottle. The indicator should change colour on the addition of one drop of acid.

xi) The burette reading is noted and the titration repeated until at least two readings for the volume of acid added agree within $0{\cdot}1\,cm^3$. The readings should be displayed in a table, as shown in Fig. 11.16.

b) *specimen readings*

concentration of sulphuric(VI) acid $= 0{\cdot}10\,M$

volume of alkali used $= 25{\cdot}0\,cm^3$

burette readings	titration number		
	1	2	3
final reading, V/cm^3	29·6	30·5	28·2
initial reading, V/cm^3	1·5	3·0	0·7
volume of acid added, V/cm^3	28·1	27·5	27·5
average volume of acid added, V/cm^3	*trial*	27·5	

Fig. 11.16

c) The reaction equation gives the relative number of moles of acid and alkali; the number of moles of acid in the measured volume of standard solution at the equivalence point enables the number of moles of alkali to be deduced from the equation. It is then a simple matter to find the concentration of the alkaline solution:

$$H_2SO_4(aq) + 2NaOH(aq) \longrightarrow Na_2SO_4(aq) + H_2O(l)$$

moles: 1 2

at equivalence point:

$\frac{27{\cdot}5 \times 0{\cdot}1}{1000}$ mol of acid has been used

$$\frac{2 \times 27{\cdot}5 \times 0{\cdot}1}{1000} \text{ mol of alkali was present in } 25\,\text{cm}^3 \text{ of solution}$$

$$\frac{2 \times 27{\cdot}5 \times 0{\cdot}1}{1000} \times \frac{1000}{25} \text{ mol of alkali was present in } 1000\,\text{cm}^3 \text{ of solution}$$

$$= 0{\cdot}22 \text{ mol}$$

the sodium(I) hydroxide solution is 0·22M

11.14 questions concerning technique in titrations

i) Why is the burette rinsed with acid before filling, while the conical flasks are not rinsed with alkali?
ii) Why must the last drop of liquid not be blown or shaken from the pipette?
iii) After delivering the alkali by pipette into a conical flask, why is it reasonable to add water to the flask but not to the burette?
iv) Why is the inside of the conical flask washed down with distilled water near the end-point?

11.15 indicators for different titrations

Any acid may be titrated with any alkali or soluble carbonate by the technique outlined in section 11.13. The only difference is that screened methyl orange is not effective when titrating weak acids such as ethanoic (acetic) acid, and phenolphthalein should not be used with weak alkalis such as aqueous ammonia or with carbonates. Litmus is not used for precise titrations because it does not give a sharp colour change at the end-point.

11.16 use of titration results

1. determining equations for reactions

If the concentrations of both solutions are known, the ratio of reacting moles may be calculated and the equation for the reaction may thus be determined.

Remember that reaction equations should really only be written after experiments have been performed to determine the number of moles of reactants and products. You will write many reaction equations which you do not determine personally, but somebody had to complete an experiment before the equation could be written honestly.

2. standardising solutions

Assuming that the reaction equation is known and given one standard solution, the concentration of the other solution may be found as in section 11.13.

revision summary: Chapter 11

acids:	havc a sharp taste change the colour of indicators reactive metals liberate hydrogen liberate carbon dioxide from carbonates and hydrogen-carbonates show common properties only in aqueous solution
salts:	formed when the hydrogen in an acid is replaced, directly or indirectly
bases:	react with acids to form salt and water only
alkalis:	water-soluble bases, have soapy feel

pH:	numerical scale of degree of acidity 7 for water or neutral aqueous solutions 1–6 for acid solutions in water 8–14 for alkaline solutions in water
concentration:	$mol\,dm^{-3}$ or $mol\,l^{-1}$ ($mol\,dm^{-3}$ abbreviated to M)
mass concentration:	$g\,dm^{-3}$ or $g\,l^{-1}$
standard solution:	a solution of known concentration
indicators:	do not use methyl orange with weak acids do not use phenolphthalein with weak alkalis

questions 11

1. Give details of a number of tests you might use to find whether (a) a fruit juice, (b) a soap powder, is acidic, alkaline or neutral.

2. The great French scientist Antoine Lavoisier (1743–1794) suggested that all acids contain the element oxygen. Why did he think so and was he justified according to the knowledge available at that time? What further discoveries may have changed his ideas?

3. You are given a number of solutions of known pH.

Solution:	A	B	C	D	E	F	G
pH:	2	9	4	7	8	5	6

a) Which solution is most acidic?
b) Which solution is most alkaline?
c) Are you likely to arrive at acidic, alkaline or neutral solutions if you mix equal volumes of (i) A and C, (ii) D and F, (iii) C and G, (iv) B and D, (v) A and E?

4. State, in each case giving a *brief* reason for your answer, whether you would expect an increase, a decrease or no change in pH of
a) limewater when carbon dioxide is bubbled into it.
b) an aqueous solution of ammonia when it is boiled.
c) water when a piece of sodium is added. (W)

5. Select from the pH values, 1, 5·5, 7, 8, 11, the one you consider most applicable to each of the following solutions:
limewater
household soap
hydrochloric acid
lemon-juice
sodium(I) chloride (W)

6. What is the concentration ($mol\,dm^{-3}$) of the following aqueous solutions?
a) sodium(I) hydroxide containing $120\,g\,dm^{-3}$
b) sodium(I) hydroxide containing 0·1 g in $50\,cm^3$
c) sulphuric(VI) acid containing $49\,g\,dm^{-3}$
d) sulphuric(VI) acid containing 19·6 g in $500\,cm^3$
e) nitric(V) acid containing 6·3 g in $250\,cm^3$

7. How many (a) moles, (b) grams of pure solute are there in
i) 500 cm^3 of 0·1 M calcium(II) hydroxide solution;
ii) 250 cm^3 of 0·5 M hydrochloric acid;
iii) 10 cm^3 of 0·2 M copper(II) sulphate(VI) solution;
iv) 25 cm^3 of 0·25 M nitric(V) acid;
v) 19·6 cm^3 of 0·3 M potassium(I) hydroxide solution;
vi) 22·5 cm^3 of 0·15 M aluminium(III) chloride solution.

8. You are provided with a solution of nitric(V) acid containing $9{\cdot}0\,g\,dm^{-3}$ of pure acid and a solution of sodium(I) carbonate, the mass concentration of which you are to determine. Explain how you carry out the necessary measurements naming the indicator and stating the colour change. Also show how you calculate the answer, given that 15·0 cm^3 of the acid solution neutralises 20·0 cm^3 of the carbonate solution.

9. a) 24·0 cm^3 of 0·1 M sulphuric(VI) acid required 32·0 cm^3 of aqueous potassium(I) hydroxide for neutralisation. Calculate the concentration of the alkali.
b) What volume of 0·5 M hydrochloric acid will neutralise 50 cm^3 of a solution containing 5·3 g of anhydrous sodium(I) carbonate in 250 cm^3?

10. 30 cm^3 of M sulphuric(VI) acid is mixed with 50 cm^3 of 2 M sodium(I) hydroxide solution. What volume of 0·1 M hydrochloric acid is required to neutralise the excess alkali?

11. An excess of powdered calcium(II) carbonate was added to 20 cm^3 of 2 M hydrochloric acid.
a) give the equation for the reaction;
b) find the mass of powder which reacts;
c) find the volume at s.t.p. of the gas evolved. (JMB)

12. (a) A solution of acid X was prepared containing 48·5 g of X per dm^3. 20·0 cm^3 of this solution exactly neutralised 25·0 cm^3 of a solution of sodium(I) carbonate, the concentration of which was 21·2 g of Na_2CO_3 per dm^3. Using this information calculate:
i) the volume of acid which neutralises one dm^3 of the alkali;
ii) the mass of X which reacts with 106·0 g of sodium(I) carbonate Na_2CO_3.
If the formula of X is H_3NSO_3, deduce
iii) the number of moles of the acid which react with one mole of the alkali, and hence
iv) the equation for the reaction.
(b) In the reaction described above an indicator is used to show when the alkali has neutralised the acid; explain why no indicator is used when preparing a salt by neutralising dilute sulphuric(VI) acid with copper(II) carbonate. (O)

solubility, salts, crystals

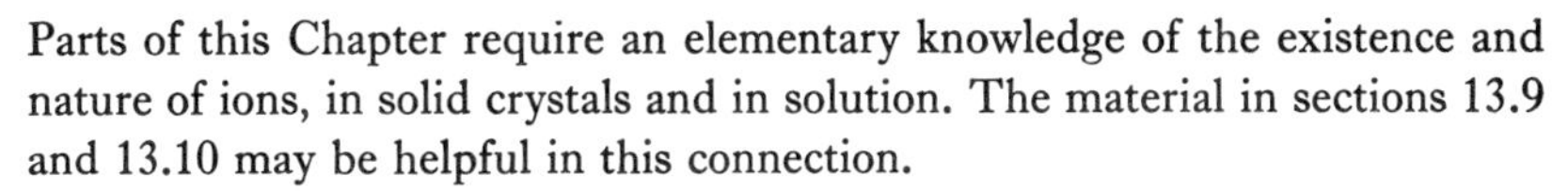

Parts of this Chapter require an elementary knowledge of the existence and nature of ions, in solid crystals and in solution. The material in sections 13.9 and 13.10 may be helpful in this connection.

12.1 solutions and solubility

How much glass did you drink today? Everything dissolves to some extent in water (see section 8.2) which may be a nuisance at times, but without this property life would be extraordinarily difficult, if not impossible. When water acts like this it is called a **solvent**; the substance which dissolves is called a **solute**; solute and solvent together form a **solution.**

Both solvents and solutes may be solids, liquids or gases, but liquid solvents are the most important.

In true liquid solutions the solute is so finely dispersed that no particles are visible even under a powerful microscope. Liquid solutions may be coloured but are always clear. Substances which dissolve in each other at all concentrations are said to be **miscible**: a well-known example is given by ethanol and water. Substances are partially miscible when there is a limit to the amount of mixing or dissolving. **Immiscible** substances do not mix and there is a well-defined boundary where they meet; examples are oil and water or benzene and water. A system of very small but visible particles dispersed in a liquid is called a **suspension.** Dispersed small droplets of a liquid, which is immiscible with the main liquid, form an **emulsion**, as in emulsion paints.

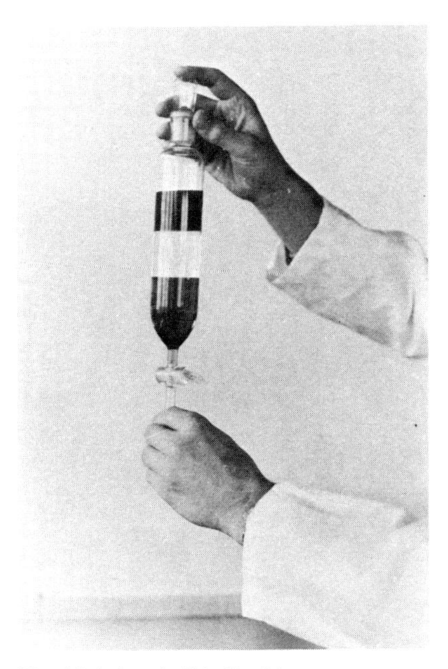

Fig. 12.1 immiscible liquids

12.2 saturation

A solid stirred into a solvent may all dissolve at first but, as more is added, at some stage undissolved solid remains. The solution is then said to be **saturated.** With many solutes and solvents the excess solute dissolves when the solution is heated, but if more solute is added the solution eventually becomes saturated at the higher temperature.

A saturated solution at a particular temperature is one which contains as much solute as can dissolve at that temperature, in the presence of undissolved solute.

12.3 solubility, crystallisation and equilibrium in solutions

The **solubility** of a solute at a particular temperature is the **mass of solute which saturates a known mass of solvent at that temperature.**

Strictly, solubility has no units; it is a dimensionless quantity, since the ratio $\frac{\text{mass of solute}}{\text{mass of solvent}}$ is a pure number. However, in practice it is found convenient to give solubility a unit and usually this is **grams of solute per 100 grams of solvent** (g/100g solvent).

The solubility of most solids increases with rise in temperature and decreases as the solution cools. Thus when most hot solutions are allowed to cool they eventually become saturated, at a lower temperature. On further cooling, crystals of the solute are deposited. This is the process known as **crystallisation.**

Dissolving and crystallisation are opposing processes and in a saturated solution there exists a situation of dynamic equilibrium (see section 23.8). Particles of solute continuously enter and leave the surface of a crystal, in contact with a saturated solution, in equal numbers; crystallisation and dissolving occur at equal rates:

$$\text{solid solute} \rightleftharpoons \text{dissolved solute}$$

This is shown pictorially in Fig. 12.2.

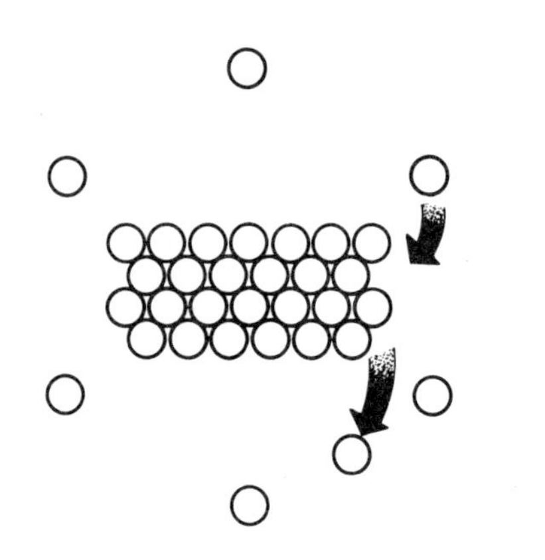

Fig. 12.2 particles entering and leaving solid solute

12.4 supersaturation

A test-tube one-third full of hydrated sodium(I) thiosulphate(VI) crystals, $Na_2S_2O_3.5H_2O$, is gently warmed and shaken until the crystals all dissolve. The tube is clamped firmly and the solution allowed to cool (under running water if time is short). If a thermometer is placed in the solution it is observed that the temperature can fall to room temperature without crystals appearing. Solutions in which crystals do not form, even though the temperature is well below saturation temperature, are said to be **supersaturated.** If a very small 'seed' crystal of the solute is dropped into the supersaturated solution, crystallisation begins immediately; crystals grow out rapidly from the 'seed' and the temperature of the system rises as this happens. Crystallisation may also be stimulated by 'scratching' the sides of the test-tube with a glass rod; even particles of dust or vigorous shaking may be effective.

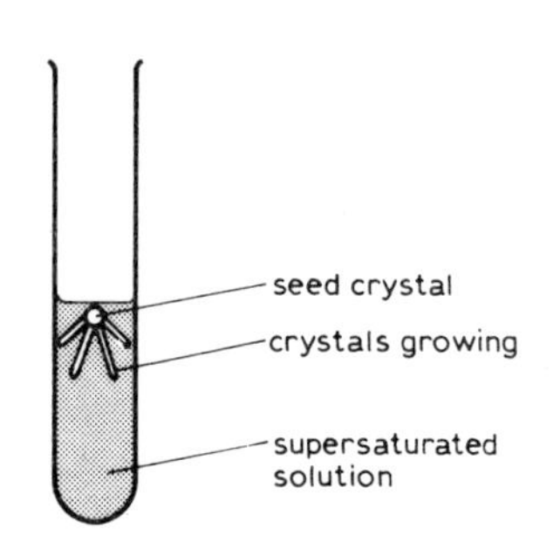

Fig. 12.3 seeding a supersaturated solution

All these observations lend support to the theory that crystals form easily only when they have something to form upon–a *nucleus*. Almost any solid particle will serve as a nucleus and it is the absence of such nuclei which causes supersaturation.

This is similar to the 'superheating' of liquids when, in the absence of nuclei for bubbles to form upon, boiling does not begin until the temperature is above the 'normal' boiling-point.

12.5 measurement of solubility of a solid: solubility curves

A variety of techniques is available for measuring solubilities, all of which involve making saturated solutions at different temperatures.

a) A saturated solution of the solute is made and some of the liquid is decanted carefully, at a known temperature, into a weighed evaporating basin. The basin and solution are weighed, the solvent is evaporated over a water or steam bath, and the dry solid and basin are weighed. The weighings give the mass of solute in a known mass of solvent. The mass of solute forming a saturated solution in 100g of solvent may then be calculated. The measurements are repeated at various known temperatures. A graph of solubility against temperature gives a **solubility curve** for that solute and using this, solubilities at any temperature may be found. Some solubility curves are shown in Fig. 12.4.

b) A simpler method requires only one weighing but is not very effective for substances which readily form supersaturated solutions.

About 5g of potassium(I) chlorate(V) is weighed into a boiling tube, then 10 cm^3 (10g) of distilled water is added from a burette. The solution is warmed until all the crystals dissolve, the heating is stopped and the cooling solution is stirred gently with a thermometer. The temperature at which crystals first appear is noted. The solubility at this temperature is 50g

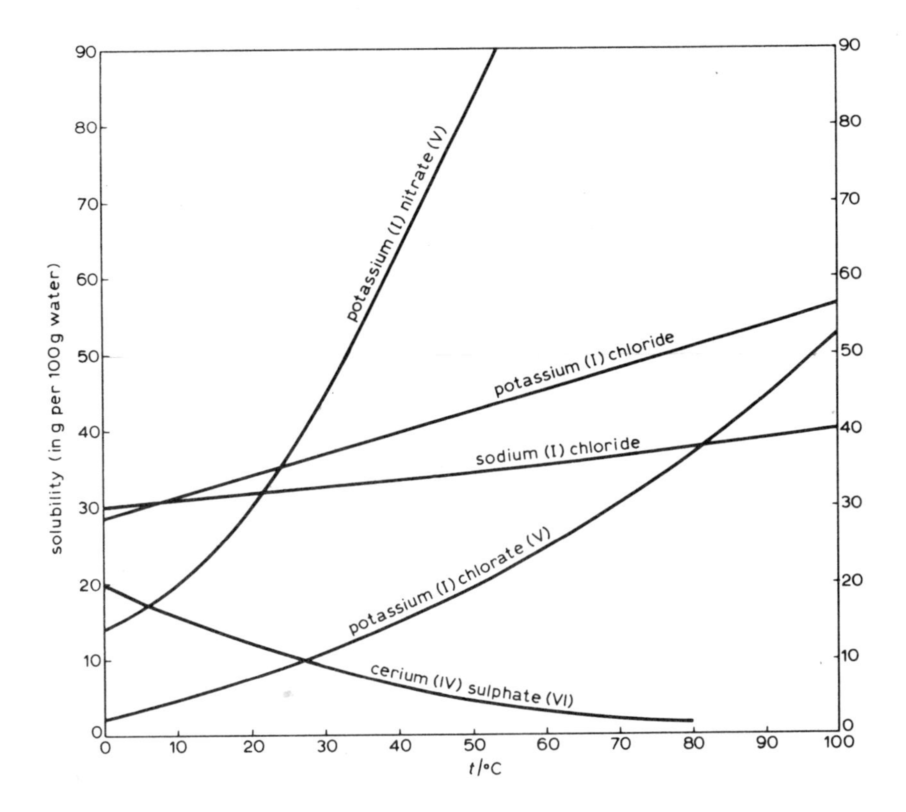

Fig. 12.4 solubility curves

potassium(I) chlorate(V) per 100g water. A further 5 cm^3 of water is added, and the mixture is warmed until the crystals dissolve; this liquid is again cooled and the temperature is noted at which crystals first appear. At this temperature the solubility is $\frac{5}{15} \times 100 = 33$ g/100g water. Repetition of the experiment using successive further 5 cm^3 additions of water enables a solubility curve to be plotted; several examples of such curves are shown in Fig. 12.4.

Consideration of the following questions will encourage an understanding of the meaning and use of solubility curves.

i) Which substance has the highest solubility at 70°C?
ii) For which substance does the solubility change most markedly with temperature?
iii) For which substance does the solubility decrease with rise in temperature?
iv) Saturated solutions of each salt are prepared at 40°C. One gram of each solution is evaporated to dryness. For which salt is the greatest mass of solid obtained?
v) What is the effect of cooling a solution of potassium(I) chlorate(V) containing 10g of salt in 60g of solution, from 80°C to 20°C?
vi) How would you attempt to obtain pure samples of potassium(I) chlorate(V) and potassium(I) chloride from a mixture of equal masses of the two salts?

12.6 types of solute and solvent

Approximately 0·5g of the powdered solids mentioned below is shaken in a test-tube with 3 cm^3 of distilled water. A further 0·5g sample of each is shaken with 3 cm^3 of tetrachloromethane. If the solid 'disappears' it is very soluble. Lower solubilities may be judged by allowing two or three drops of the resulting liquid to evaporate on a watch-glass or microscope slide over a beaker of boiling water or under a hot lamp. *Warning:* tetrachloromethane has a poisonous vapour and experiments using it should be carried out in a fume cupboard.

Suitable materials for study are: potassium(I) nitrate(V), iodine, naphthalene, glucose, sulphur, sodium(I) chloride, paraffin wax, potassium(I) chloride.

Certain patterns may be distinguished in these investigations: for example, that water is a good solvent for the *ionic* compounds and glucose, but a poor solvent for the rest.

Tetrachloromethane is a good solvent for most of the *molecular* substances but poor for the ionic substances.

12.7 dissolving ionic crystals in water

An ionic crystal consists of a **lattice** of positively and negatively charged ions so arranged in space that the electrostatic forces of attraction are greater than those of repulsion. If a crystal is placed in water, forces of attraction develop between water molecules and the ions in the crystal. This is due to the **polarity** of water molecules (see section 18.20). For many ionic substances, the forces of attraction between water molecules and ions overcome the attractive forces between oppositely charged ions in the lattice. The ions are 'pulled out' of the lattice by water molecules and the crystal dissolves. In the solution each ion is surrounded by a number of water molecules called **water of hydration.**

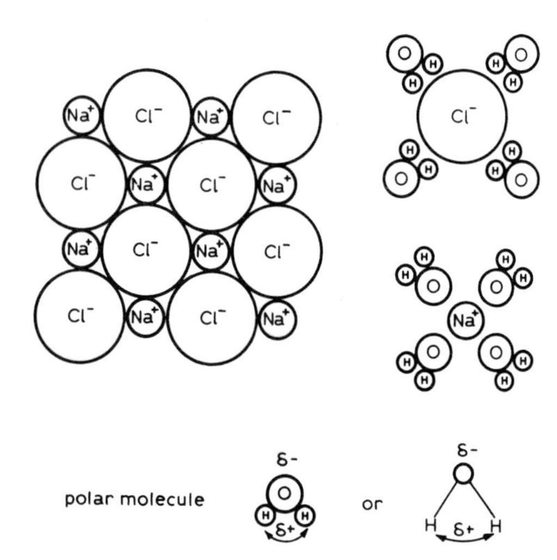

Fig. 12.5 hydration of ions by polar water molecules

The dissolving of sodium(I) chloride shown in Fig. 12.5 is a good example. The positive 'poles' of the water molecules of hydration are orientated around the negative (Cl^-) ions and the negative 'poles' around the positive (Na^+) ions.

In other ionic substances the forces holding the crystal together are stronger than the forces of attraction exerted by the water molecules. These substances have a very low solubility in water. Good examples are silver(I) chloride and aluminium(III) oxide.

12.8 non-polar solvents

Ionic substances do not dissolve in liquids like tetrachloromethane or benzene because the molecules of these solvents have little interaction with ions. Only solvents with polar molecules (like water) interact strongly with ions; tetrachloromethane and benzene have non-polar molecules.

Non-polar liquids do interact strongly with molecular substances, such as iodine, and are often excellent 'dry-cleaning' solvents for removing substances like oils and paints.

In brief, a solute will be only slightly soluble in a particular solvent if:

i) the solute particles (ions, atoms or molecules) are held together by strong bonding forces;
ii) the solvent particles have little attraction for the solute particles.

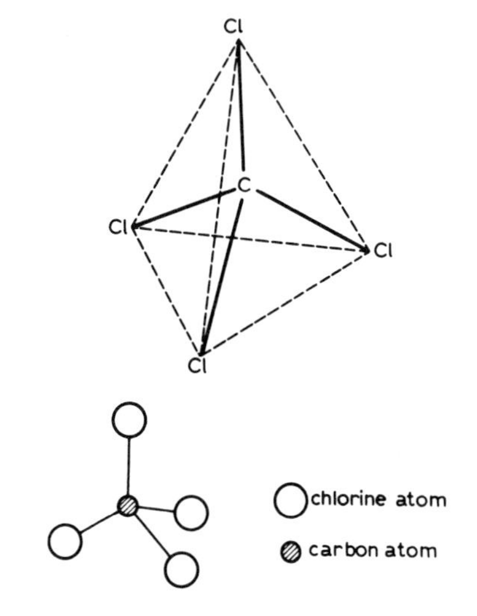

Fig. 12.6 structure of tetrachloromethane

12.9 energy changes as ionic crystals dissolve

1) The temperature of about 100 cm³ of water is measured, then 10 g of solid ammonium nitrate(V) or ammonium chloride is stirred in. The temperature drops quickly; the process of solution is **endothermic** for these solutes.
2) To another 100 cm³ of water, 10 g of anhydrous copper(II) sulphate(VI) is added and stirred. The temperature rises rapidly; the process of solution is **exothermic** for this solute.

Ionic solids may dissolve endothermically or exothermically. In an endothermic process the energy required to break bonds is greater than that released when bonds are formed; the reverse is true of an exothermic process. During dissolving of ionic solids

i) bond breaking takes place, as the lattice separates into individual ions, absorbing energy;
ii) bond formation takes place during hydration of the ions, liberating energy as heat of hydration.

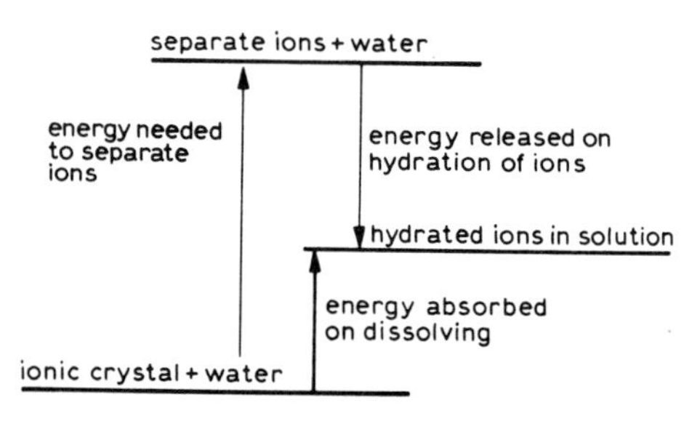

Fig. 12.7 endothermic dissolving (e.g. ammonium nitrate(V))

The process of solution is endothermic or exothermic depending on which of these two processes involves the larger amount of energy. This subject is considered in more detail in section 20.8.

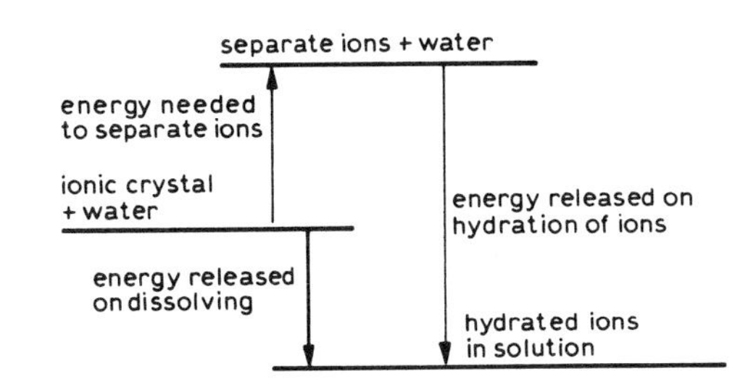

Fig. 12.8 exothermic dissolving (e.g. anhydrous copper(II) sulphate(VI))

12.10 energy changes and crystallisation

The factors causing energy changes in dissolving also apply to crystallisation. A substance which dissolves endothermically, crystallises exothermically. This explains the considerable rise in temperature when crystals form rapidly from a supersaturated solution of sodium(I) thiosulphate(VI). *Left to themselves*, substances which dissolve exothermically do not crystallise endothermically; external heating is needed to drive away the solvent.

12.11 effect of temperature change upon solubility equilibria

The solubility curves in Fig. 12.4 show that the solubilities of many substances rise with increase in temperature, but some fall. This is a good example of the effect of change in temperature on a chemical system in equilibrium (see section 23.9).

If a substance dissolves endothermically *in its near-saturated solution*, raising the temperature increases solubility; if a substance dissolves exothermically in its near-saturated solution, raising the temperature decreases solubility.

Of the compounds shown in Fig. 12.4 only cerium(IV) sulphate(VI) dissolves exothermically *in its near-saturated solution.*

12.12 how to increase the rate of solution of solids in liquids

1) *by crushing the solid to powder:* Powders have a much larger surface area for the solvent to act upon than the same mass of larger particles.
2) *by stirring:* If the solution is not stirred, the solvent immediately around the solid quickly becomes saturated. Agitation exposes the solid surface to fresh solvent.
3) *by heating:* This helps to agitate the bulk of the solution, as in (2). In addition, the increase in thermal energy of the particles of solute and solvent increases the rate of the dissolving process.

However, fast rate of dissolving must not be confused with high solubility. As in all reversible reactions, there is a distinction between 'how fast' and 'how far'. An important fact to remember in connection with this and the previous section is that many substances which dissolve exothermically to give *unsaturated* solutions dissolve endothermically in their *nearly saturated* solutions. This should be borne in mind when applying Le Chatelier's Principle (section 23.9) to the variation of solubility with temperature.

12.13 water of crystallisation

The effects of heat on the blue crystals of hydrated copper(II) sulphate(VI) and of water on anhydrous copper(II) sulphate(VI) are described in sections 5.3, 8.16 and 20.8.

When water is added to anhydrous copper(II) sulphate(VI), liberation of heat results from a reaction between water and the ions of the salt. The heat liberated by this interaction is called **heat of hydration** (see Figs. 12.7, 12.8). Hydrated copper(II) sulphate(VI) contains molecules of water chemically bound to ions within the crystal structure; this is called **water of crystallisation.** It may also be known as water of hydration – a term often applied to the water around ions in aqueous solution (see Fig. 12.5).

Many ionic substances crystallise with water of crystallisation. Examples of these **hydrates** are given in the list in section 12.15. The water often influences the colour of the hydrated crystals and is responsible for giving them a structure which is different from that of the anhydrous crystals.

The number of moles of water of crystallisation per mole of hydrate is shown in the formula and in the name for the compound, e.g. copper(II) sulphate(VI)-5-water, $CuSO_4.5H_2O$.

12.14 determination of the number of moles of water of crystallisation in one mole of salt

The experiment described uses hydrated copper(II) sulphate(VI); hydrated barium(II) chloride also gives good results.

i) A clean dry crucible and lid are weighed empty.
ii) The crucible is half-filled with powdered, hydrated copper(II) sulphate(VI) crystals and re-weighed.
iii) The crucible and contents are heated gently in a pipeclay triangle over a low Bunsen flame. At first the lid should be half-off; later it may be removed with tongs.
iv) When the blue crystals appear to have changed completely to white powder, heating is stopped and the crucible is transferred to the dry atmosphere of a desiccator to cool. The cool crucible, contents and lid are weighed.
v) The crucible and contents are re-heated. *At no time* may a 'roaring' Bunsen flame be used as this may decompose the salt to copper(II) oxide. Again the crucible is allowed to cool in a desiccator and is weighed. If the mass is the same as before, all the water has been driven off; if not, the process is repeated until 'constant weight' is achieved.

Sample results and calculation are set out below.

mass of crucible = 20·218 g
mass of crucible + hydrated salt = 28·313 g
final mass of crucible + anhydrous salt = 25·368 g
mass of anhydrous salt = 5·150 g
mass of water of crystallisation = 2·945 g

mass of one mole $CuSO_4$ = 159·5 g

number of moles $CuSO_4 = \frac{5·15}{159·5}$ = 0·03228

mass of one mole H_2O = 18 g

number of moles $H_2O = \frac{2·945}{18}$ = 0·1636

ratio of moles: $CuSO_4$: H_2O
0·03228 : 0·1636
1 : 5·066

hydrated copper(II) sulphate(VI) is $CuSO_4.5H_2O$

12.15 examples of hydrated and anhydrous crystals

copper(II) sulphate(VI)-5-water	$CuSO_4.5H_2O$
sodium(I) carbonate-10-water	$Na_2CO_3.10H_2O$
zinc(II) sulphate(VI)-7-water	$ZnSO_4.7H_2O$
sodium(I) thiosulphate(VI)-5-water	$Na_2S_2O_3.5H_2O$
sodium(I) chloride	NaCl
potassium(I) nitrate(V)	KNO_3
ammonium chloride	NH_4Cl
potassium(I) manganate(VII) (permanganate)	$KMnO_4$

12.16 methods of preparing salts

The compounds called **salts** are regarded as **crystalline solids containing any cation except the hydrogen ion and any anion except oxide or**

hydroxide ions. They may also contain water of crystallisation. (N.B. cations are positively charged and anions negatively charged, ions.)

Salts are prepared using a variety of reactions and techniques. All try to place cations and anions of the desired salt in a suitable environment for reaction and crystallisation. The methods are summarised in four groups:

a) direct combination of elements;
b) acid with an insoluble reactant (metal, carbonate or base), giving a soluble salt;
c) acid with a soluble reactant (carbonate or alkali), giving a soluble salt (volumetric titration used);
d) two soluble reactants giving an insoluble salt (precipitation used).

Solubility of both reactants and products is important in choosing which method to use, but the choice of method also depends on other factors, e.g. availability and expense of reactants, speed of reaction and ease of extraction of the salt.

12.17 solubility in water of oxides, hydroxides and salts

The following information is essential for an intelligent choice of method for the preparation of a salt.

nitrate(V)	all soluble
sodium(I), potassium(I), ammonium	all compounds soluble
carbonate	insoluble, except Na^+, K^+, NH_4^+
chloride, bromide	soluble, except Pb^{2+}, Ag^+, Hg^+
sulphate(VI)	soluble, except Pb^{2+}, Ba^{2+} (Ca^{2+} slightly)
sulphate(IV) (sulphite)	insoluble, except Na^+, K^+, NH_4^+
hydrogencarbonate, hydrogensulphate(VI), hydrogensulphate(IV)	all soluble
oxide, hydroxide	insoluble, except Na^+, K^+, NH_4^+, Ba^{2+} (Ca^{2+} slightly)

12.18 examples of each method for preparation of soluble salts

The experiments in this section (e.g. (2) and (3) below) are well suited to small-scale techniques and apparatus: a 10 cm^3 beaker for reaction, centrifuge tubes for centrifuging instead of filtering suspensions, a small crucible held in a wire ring-holder for evaporation and a dropping pipette for transferring small volumes of solutions drop by drop. Counting drops from a dropping pipette can replace the more accurate but more lengthy titration technique using pipettes, burettes, etc. and in this way crystals of potassium(I) sulphate(VI) and hydrogensulphate(VI) can be prepared and examined easily in a one-hour laboratory class. Microscopic examination of a single drop of hot saturated solution on a microscope slide enables crystal formation and growth to be studied in a fascinating way. Such small-scale techniques economise in time as well as in materials: two salt preparations can be completed in a one-hour class and the hazardous experience of preserving solutions, crystals, etc. from one class to the next can be avoided. The experiments are described in a general way, as the principles are the same whatever apparatus is used.

1. direct combination of elements

Direct combination is used only when the reactants are of high reactivity; it is not easy to obtain a pure product.

Direct reaction of chlorine with metals, e.g. sodium, aluminium, magnesium, iron and zinc, give reasonable yields of the chlorides. The preparation of chlorides is more fully discussed in section 15.13. Salts prepared in this way are always anhydrous and often covalent in their bonding, unless recrystallised from water.

2. aqueous acid with insoluble reactant

The insoluble reactant (a metal, carbonate or base) is added to warm aqueous acid until reaction ceases and excess solid is seen. Filtration gives the aqueous salt; the salt is obtained by crystallisation as outlined in section 12.21.

i) *aqueous acid + metal*

Salts prepared by this method may include the sulphates(VI) and chlorides of magnesium, aluminium, zinc and iron, nitrates(V) except aluminium, iron and alkali metals. Small pieces of the metal are added to dilute (2M) acid in a beaker and the solution is warmed and stirred. When evolution of hydrogen ceases and excess metal remains, the solution is filtered into a clean basin or crucible. Salt crystals are obtained by crystallisation. For magnesium and sulphuric(VI) acid the reaction equation may be written:

$$Mg(s) + 2H^+(aq) + SO_4^{2-}(aq) \rightarrow \underbrace{Mg^{2+}(aq) + SO_4^{2-}(aq)}_{\text{aqueous magnesium(II) sulphate(VI)}} + H_2(g)$$

It is instructive to weigh the metal used, calculate a theoretical yield of salt from the reaction equation and compare it with the actual yield. Which steps in the procedure lead to losses?

ii) *aqueous acid + insoluble carbonate*

The technique is exactly similar to that for acid + metal; addition of the carbonate should be stopped when evolution of carbon dioxide ceases and excess carbonate remains.

The reaction equation for the preparation of lead(II) nitrate(V) crystals using lead(II) carbonate and nitric(V) acid is:

$$PbCO_3(s) + 2H^+(aq) + 2NO_3^-(aq) \rightarrow \underbrace{Pb^{2+}(aq) + 2NO_3^-(aq)}_{\text{aqueous lead(II) nitrate(V)}} + CO_2(g) + H_2O(l)$$

iii) *aqueous acid + insoluble base (oxide or hydroxide)*

No gas is evolved and the reaction is assumed to be complete when an excess of solid oxide or hydroxide is present, even in very hot solutions. The reaction equation for the preparation of copper(II) chloride from copper(II) oxide and hydrochloric acid is:

$$CuO(s) + 2H^+(aq) + 2Cl^-(aq) \rightarrow \underbrace{Cu^{2+}(aq) + 2Cl^-(aq)}_{\text{aqueous copper(II) chloride}} + H_2O(l)$$

3. aqueous acid + aqueous alkali by titration

The technique of volumetric titration is outlined in section 11.13. With 2M solutions it is best to use burettes for both volume measurements. Alternatively, counting drops from dropping pipettes gives acceptable results in a shorter time.

20 cm^3 of 2M aqueous acid is measured into a conical flask and two drops of screened methyl orange indicator are added. Aqueous (2M) carbonate or alkali is added slowly from the second burette until the indicator just changes colour. The volume of carbonate or alkali added is noted and the titration repeated with fresh acid but *without* the indicator, using the same volumes. A salt solution only remains for crystallising.

Potassium(I) nitrate(V) may be prepared from potassium(I) hydroxide and nitric(V) acid:

$$K^+(aq) + OH^-(aq) + H^+(aq) + NO_3^-(aq) \rightarrow \underbrace{K^+(aq) + NO_3^-(aq)}_{\text{aqueous potassium(I) nitrate(V)}} + H_2O(l)$$

Sodium(I) sulphate(VI) crystals may be prepared from aqueous sodium(I) hydroxide and sulphuric(VI) acid:

$$2Na^+(aq) + 2OH^-(aq) + 2H^+(aq) + SO_4^{2-}(aq) \rightarrow 2Na^+(aq) + SO_4^{2-}(aq) + 2H_2O(l)$$

aqueous sodium(I) sulphate(VI)

12.19 normal and acid salts

One way of regarding the formation of salts is that metal or ammonium ions 'replace' the hydrogen ions in an acid. If all of the hydrogen ions are replaced, the salt formed is said to be the **normal** salt of the acid. For example: sodium(I) sulphate(VI), (Na_2SO_4) is the normal salt of sulphuric(VI) acid, H_2SO_4. The salt formed when only part of the 'acidic' hydrogen ions are replaced is called the **acid** salt. The acid salt of sulphuric(VI) acid is sodium(I) hydrogensulphate(VI), $NaHSO_4$. It is made by adding the same volume of sodium(I) hydroxide used in the titration which formed the normal salt, to *twice* the volume of sulphuric(VI) acid. The hydrogen sulphate(VI) ion, $HSO_4^-(aq)$ is formed:

$$Na^+(aq) + OH^-(aq) + 2H^+(aq) + SO_4^{2-}(aq) \rightarrow Na^+(aq) + HSO_4^-(aq) + H_2O(l)$$

aqueous sodium(I) hydrogensulphate(VI)

12.20 basicity of acids

The basicity of an acid measures the number of moles of hydrogen which are replaceable in one mole of an acid. It may also be regarded as the number of moles of 'acidic' hydrogen per mole of acid.

Monobasic acids, e.g. HCl, HNO_3, have only one mole of replaceable hydrogen in one mole of acid. They produce only normal salts. Ethanoic (acetic) acid, CH_3CO_2H, has four moles of hydrogen per mole of acid but only one mole of replaceable ('acidic') hydrogen; it is monobasic.

Dibasic acids, with two moles of replaceable hydrogen per mole of acid, give a normal and an acid salt, e.g. sulphuric(VI) acid, H_2SO_4, produces two sodium salts: $NaHSO_4$ and Na_2SO_4.

Tribasic acids, with three moles of replaceable hydrogen per mole of acid, give a normal and two acid salts, e.g. phosphoric(V) acid, H_3PO_4, produces three sodium salts, NaH_2PO_4, Na_2HPO_4 and Na_3PO_4.

12.21 technique for obtaining crystals from a solution

The preparation of salt solutions is described in sections 12.18(2) and (3). From these solutions it is possible to extract a high yield of good quality crystals.

The fundamental process in crystallisation from solution is **to cool a hot, saturated solution.**

a) First, the solution should be made hot and saturated, by heating it in an evaporating basin, so that some solvent evaporates until either
  i) small crystals are seen just above the liquid level *or*
  ii) on taking out a small sample with a dropping pipette, crystals form when the sample is cooled in a test-tube.

b) The saturated solution is allowed to cool as *slowly* as possible. This may take several hours, but the waiting is worthwhile, as only very small crystals appear on rapid cooling.

c) The cold, saturated solution (mother liquor) is decanted from the crystals into a clean evaporating basin.

d) Any remaining solution is removed from the crystals by filtering through a 'fluted' filter paper, adding the filtrate to that decanted.

e) If the crystals are not very soluble in cold water they may be given a quick rinse, while on the filter paper, with distilled water. Sometimes ethanol or propanone is used as a washing liquid. All washings are rejected.

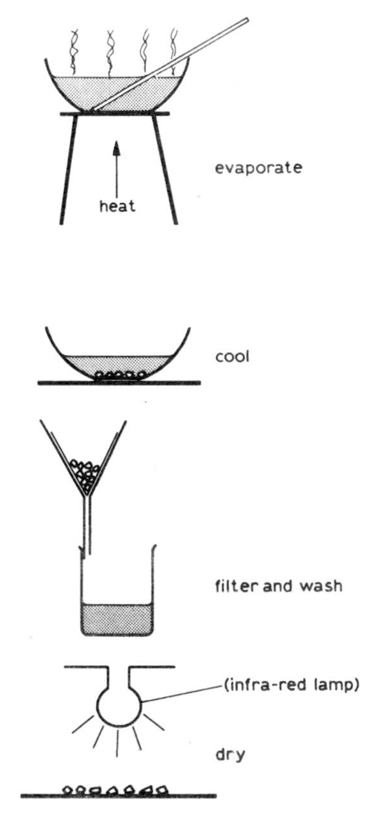

Fig. 12.9 obtaining dry crystals from a solution

f) The crystals are dried on filter paper or under an infra-red lamp or over a steam bath, and stored in a stoppered container.

This is not the maximum possible yield of crystals, as it is possible to obtain a further crop by reheating and evaporating, again forming a hot, saturated solution and proceeding as before.

Simply boiling off all solvent gives very poor results; many salt crystals contain water of crystallisation and boiling to dryness leaves only anhydrous salts; there is considerable loss of material by 'spitting' when concentrated solutions are heated; slow crystallisation from solution gives crystals in good size and shape.

12.22 recrystallising as a method of purifying crystals

All too often, crystals contain small quantities of impurities, particularly when too much solute is extracted from the solution. Recrystallising involves the following steps.

i) The crystals are dissolved in the minimum quantity of hot solvent.
ii) The solution is filtered while still hot–as rapidly as possible, to prevent crystallisation in the funnel. This removes the insoluble impurities.
iii) The solution is allowed to cool. The soluble impurities should be present only in small concentrations and do not form a saturated solution except at very low temperatures. The main solute, however, does form a saturated solution on cooling, and pure crystals are deposited.

12.23 growing large crystals

Crystal growing can be enjoyed wherever methods of mixing and heating solutions are available. Jam pots for mixing and old pans or tins for heating on kitchen cookers are as effective as beakers and Bunsens. Brief instructions are given here, but the best conditions are found by personal experiments. A good substance to practise with is 'alum', aluminium(III) potassium(I) sulphate(VI)-12-water, $KAl(SO_4)_2.12H_2O$.

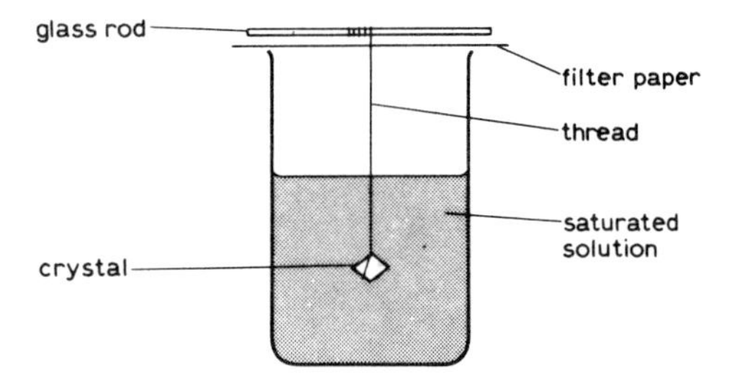

Fig. 12.10 growing a large crystal

A well-shaped small crystal should be obtained by the process outlined in sections 12.21 and 12.22. This 'seed' crystal is attached to a length of cotton using a very small quantity of adhesive or by securing with a knot.

About 500 cm³ of a cold saturated solution–supersaturated if possible–of the same solute, is made in a fairly tall container. Several holes are pierced in a piece of strong cardboard; using this as a lid for the container, the seed crystal is suspended in the solution. The container is placed where the temperature remains constant and cool and left until the crystal stops growing. This may take several days; even larger crystals are obtained if fresh saturated solution is added.

Tiny, secondary crystals may grow on the faces of the seed crystal and should be removed before they become large. When the main crystal is of a satisfactory size it should be removed, wiped with damp filter paper and stored in a stoppered container.

Fig. 12.11 a large crystal

12.24 natural crystals

Many natural substances crystallise from fused liquids rather than solutions and some very large crystals have been found in nature. Mica crystals up to four metres in diameter have been found; these must have taken centuries to grow. Diamonds, rubies and sapphires are all famous natural crystals which may now be grown in laboratories. Splendid examples of natural crystals may be seen in many museums throughout the country.

12.25 precipitation reactions

In chemistry, **precipitation** is the name given to a chemical reaction in which tiny crystals appear, usually very rapidly, when solutions of the reactants are mixed. The reactants must both be soluble in a solvent and one of the products, the **precipitate**, insoluble in the same solvent.

Any insoluble ionic compound may be obtained by first making aqueous solutions of two soluble substances; one of these must contain the appropriate cation, the other the anion. When the solutions are mixed, the ions of the insoluble product come together and form a precipitate. This is the only satisfactory method of preparing insoluble salts. **Any attempt to produce an insoluble salt directly from an insoluble reactant will fail,** as the insoluble product forms a coating on the particle of reactant and prevents further reaction.

The solubility data given in section 12.17 are essential when deciding which substances may be prepared by ionic precipitation. Any substance mentioned as insoluble or even slightly soluble may be prepared if suitable reactants are chosen. The following points are important:

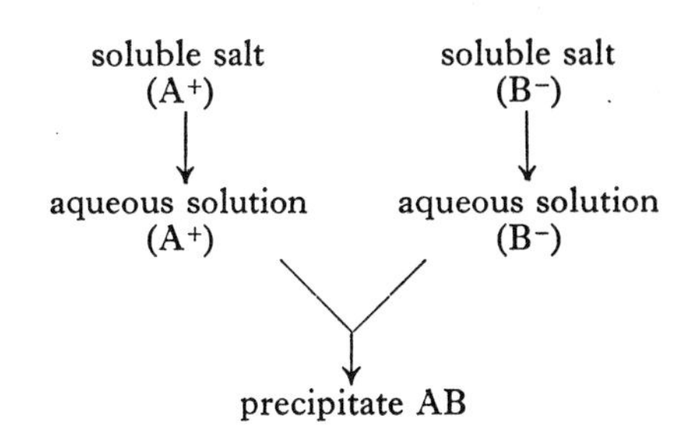

a) *Both reactants must be soluble in water.* If only an insoluble compound containing the cation or anion is available (e.g. an insoluble carbonate or base), it must first be made into an aqueous solution by reaction with a suitable acid, as in section 12.18(2). For example, if only lead(II) carbonate is available for the preparation of lead(II) iodide, aqueous lead(II) nitrate(V) is first obtained by reaction with dilute nitric(V) acid (section 12.18(2)(ii)):

$$PbCO_3(s) + 2H^+(aq) + 2NO_3^-(aq) \rightarrow Pb^{2+}(aq) + NO_3^-(aq) + H_2O(l) + CO_2(g)$$

aqueous lead(II) nitrate(V)

b) *The other product of reaction must be soluble in water.* The use of the nitrate(V) of the required cation or the sodium(I) salt of the required anion will ensure this.

12.26 preparation of lead(II) iodide

Approximately 2M solutions of a soluble lead(II) salt, e.g. lead(II) nitrate(V), and potassium(I) iodide are prepared in distilled water. The lead(II) nitrate(V) solution contains $Pb^{2+}(aq)$ and $NO_3^-(aq)$ ions; potassium(I) iodide contains $K^+(aq)$ and $I^-(aq)$ ions.

Aqueous potassium(I) iodide is added to an equal volume of aqueous lead(II) nitrate(V) with stirring. A heavy yellow precipitate appears and settles to the bottom of the beaker. A little more aqueous potassium(I) iodide is added to test for completion of reaction. If more precipitate appears, the reaction is not complete and further aqueous potassium(I) iodide is added until precipitation stops.

The precipitate is **filtered** off, **washed** well while still on the paper with several separate volumes of distilled water and allowed to drain. It may be **dried** beneath a hot lamp or over a boiling water bath.

The reaction involves only the lead(II) and iodide ions. Nitrate(V) and potassium(I) ions remain in solution as 'spectator' ions since potassium(I) nitrate(V) is very soluble in water. It is essential that in any preparation by precipitation the other products are highly soluble; only one compound must be precipitated.

12.27 precipitation titrations

The ratio of the reacting number of moles of lead(II) and iodide ions may be confirmed by a precipitation titration.

5·0 cm³ of M aqueous potassium(I) iodide is placed in each of six equal-sized test-tubes in a rack. From a burette, different volumes of M aqueous lead(II) nitrate(V) are added to each tube: to the first 1·0 cm³, to the second 1·5 cm³, to the third 2·0 cm³, to the fourth 2·5 cm³, to the fifth 3·0 cm³ and to the sixth 3·5 cm³. The contents of each tube are mixed by shaking, then the precipitate is left to settle for a fixed length of time. The test-tubes are re-

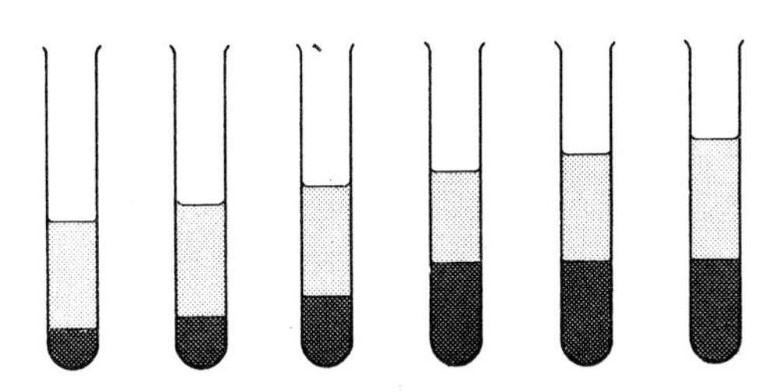

Fig. 12.12 depths of precipitates formed from $Pb^{2+}(aq)$ and $I^-(aq)$

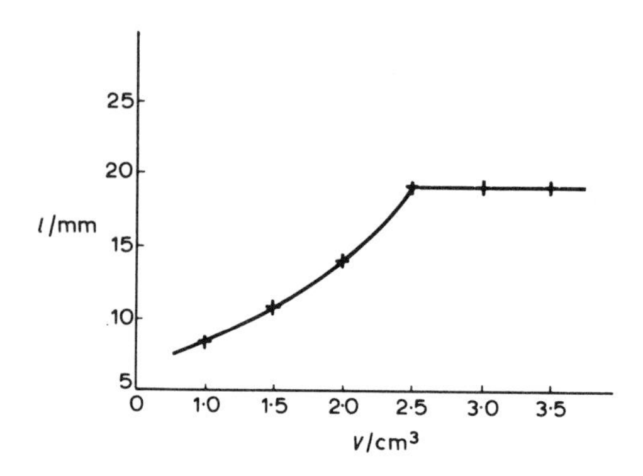

Fig. 12.13 variation in depth of precipitate with volume of $Pb^{2+}(aq)$ added

gas	from
$H_2O(g)$	water of crystallisation; most OH^-
O_2	$NaNO_3$, KNO_3
$NO_2 + O_2$	most NO_3^-
CO_2	HCO_3^-, most CO_3^-
SO_2	some SO_3^{2-} some SO_4^{2-}

Fig. 12.14 heat on solids

gas	from
CO_2	CO_3^{2-}, HCO_3^-
SO_2	SO_3^{2-}, HSO_3^-
H_2S	S^{2-}
NO_2	NO_2^-

Fig. 12.15 dil. HCl on solid

gas	from
NO_2	NO_3^-
HCl	Cl^-
HBr, Br_2 (*also* SO_2)	Br^-
I_2 (*also* H_2S)	I^-

Fig. 12.16 conc. H_2SO_4 on solid

placed in the rack and the height of each precipitate is measured as accurately as possible. A graph is plotted of the measured height of precipitate against the volume of aqueous lead(II) nitrate(V) added (see Figs. 12.12 and 12.13).

From the graph it is clear that no more precipitate appears after a certain volume of aqueous lead(II) nitrate(V) has been added.

In this experiment the volume is $2{\cdot}5\,cm^3$. $2{\cdot}5\,cm^3$ of M lead(II) nitrate(V) contains 1/400 mol $Pb^{2+}(aq)$; $5{\cdot}0\,cm^3$ of M potassium(I) iodide contains 1/200 mol $I^-(aq)$. The ratio of reacting moles of ions is:

$$Pb^{2+}(aq) : I^-(aq)$$
$$\frac{1}{400} : \frac{1}{200}$$
$$1 : 2$$

The equation for the reaction is:

$$Pb^{2+}(aq) + 2I^-(aq) \rightarrow PbI_2(s)$$

Any precipitation reaction may be investigated in this way, though some precipitates may not settle so quickly as does lead(II) iodide.

12.28 precipitation accompanied by other reactions

a) The reaction between barium(II) hydroxide and sulphuric(VI) acid involves both a **precipitation** reaction and an **acid–base** reaction.

The ions in aqueous barium(II) hydroxide are $Ba^{2+}(aq)$ and $OH^-(aq)$; the ions in aqueous sulphuric(VI) acid are $H^+(aq)$ and $SO_4^{2-}(aq)$.

The reactions are:

precipitation: $Ba^{2+}(aq) + SO_4^{2-}(aq) \rightarrow BaSO_4(s)$

acid–base: $H^+(aq) + OH^-(aq) \rightarrow H_2O(l)$

b) The 'softening' of temporarily hard water by boiling (section 22.9), involves **decomposition** of hydrogencarbonate ions. This releases carbonate ions and is followed by precipitation of insoluble calcium(II) carbonate. Calcium(II) ions, which cause hardness in water, are thus removed from solution.

$$2HCO_3^-(aq) \xrightarrow{heat} CO_3^{2-}(aq) + CO_2(g) + H_2O(l)$$

then: $Ca^{2+}(aq) + CO_3^{2-}(aq) \rightarrow CaCO_3(s)$

12.29 detection of particular ions—qualitative analysis

The presence of a particular ion in a compound is detected by a variety of simple tests in which observable changes occur. For example:

a gas may be liberated and identified;
a precipitate may be formed;
there may be a distinct colour change.

Tests for anions and cations are carried out separately; the order of tests shown is usually the most successful. The procedure for identifying gases is shown in section 12.30.

1. *tests for anions in solids*

i) *action of heat on dry solid*

About $0{\cdot}5\,g$ of the dry powdered solid is heated, gently at first then more strongly, in a small hard-glass test-tube (ignition tube). Many substances decompose under these conditions and a gas may be liberated and identified. Possible sources of some gases are listed in Fig. 12.14.

ii) *action of solid with dilute acid*

$2\,cm^3$ of M hydrochloric acid is added to $0{\cdot}5\,g$ of powdered solid. Bubbles indicate liberation of a gas which may be identified. If no gas appears immediately the reaction mixture may be warmed gently. Possible inferences are shown in Fig. 12.15.

iii) *action of solid with concentrated sulphuric*(VI) *acid*
This is only used if there is no result in (ii). Great care must be observed, as 2 cm³ of concentrated sulphuric(VI) acid is added to 0·5 g of powdered solid, in a test-tube. If warming is necessary, the tube should not be held in the hand.

colour	ion
white	Cl^-
cream	Br^-
yellow	I^-
black	S^{2-}

Fig. 12.17 colours of silver(I) salts

2. *tests for anions in solution (confirmatory tests)*

About 1 g of the solid is shaken with 10 cm³ of distilled water. If it does not all dissolve, the solution is separated from excess solid using a dropping pipette or, if necessary, by filtration. 2 cm³ volumes of the solution are used in the following tests for particular anions.

i) *tests for anions forming insoluble silver*(I) *salts*
1 cm³ of 2M nitric(V) acid is added to 2 cm³ of the test solution, followed by a few drops of aqueous silver(I) nitrate(V). The colour of the precipitate and the inference is shown in Fig. 12.17.

ii) *test for sulphate*(VI)
1 cm³ of 2M nitric(V) acid is added to 2 cm³ of the test solution, followed by a few drops of aqueous barium(II) nitrate(V). A white precipitate indicates the presence of sulphate(VI) ion.

iii) *the 'brown ring' test*
1 cm³ of fresh aqueous iron(II) sulphate(VI) is added to 2 cm³ of the test solution, followed by concentrated sulphuric(VI) acid poured slowly into the tube held at an angle, so that the acid forms a lower layer. A brown ring at the junction of the two liquids indicates the presence of nitrate(V).

flame colour	cation
yellow	Na^+
lilac	K^+
crimson	Sr^{2+}
brick-red	Ca^{2+}
pale green	Ba^{2+}
green	Cu^{2+}
blue	Pb^{2+}

Fig. 12.18 flame colours

3. *tests for cations in solids*

i) *action of heat on dry solid*
The residue in the ignition tube from test 1(i) is examined.
A solid which is yellow when hot and white when cold indicates zinc(II) oxide; an orange-yellow solid indicates lead(II) oxide.

ii) *addition of aqueous sodium*(I) *hydroxide to solid*
2 cm³ of 2M aqueous sodium(I) hydroxide is added to about 0·5 g of dry solid and warmed. Ammonia is liberated from an ammonium salt:

$$NH_4^+(aq) + OH^-(aq) \rightarrow NH_3(g) + H_2O(l)$$

iii) *flame tests*
A platinum wire is cleaned by dipping it in concentrated hydrochloric acid, then it is touched into a little of the powdered solid and held at the edge of a Bunsen flame. Colours characteristic of certain cations are shown in Fig. 12.18.

4. *tests for cations in solution*

i) *addition of aqueous sodium*(I) *hydroxide*
Many cations form insoluble hydroxides, the colours of which help to identify the cations. Some of these hydroxides are amphoteric and react with excess hydroxide ions, so that the precipitate disappears.
An aqueous solution of the substance is prepared as in (*2*). To 2 cm³ of solution, aqueous sodium(I) hydroxide is added a few drops at a time until there is no further change. The colour of the precipitate and whether it is soluble (sol.) or insoluble (insol.) in excess alkali, together with the appropriate inference, is shown in Fig. 12.19.

precipitate colour	excess $OH^-(aq)$	
white	insol.	Ca^{2+} Mg^{2+}
white	sol.	Al^{3+} Pb^{2+} Zn^{2+}
blue	insol.	Cu^{2+}
green	insol.	Fe^{2+}
red-brown	insol.	Fe^{3+}
grey-brown	insol.	Ag^+
black	insol.	Hg^+
yellow	insol.	Hg^{2+}

Fig. 12.19 NaOH(aq) on aqueous cations

precipitate colour	excess $NH_3(aq)$	
white	insol.	Mg^{2+} Al^{3+} Pb^{2+}
white	sol.	Zn^{2+}
pale blue	sol. → deep blue solution	Cu^{2+}
green	insol.	Fe^{2+}
red-brown	insol.	Fe^{3+}
brown	sol.	Ag^{+}
black	insol.	Hg^{+}

Fig. 12.20 $NH_3(aq)$ on aqueous cations

ii) *addition of aqueous ammonia*

2M aqueous ammonia contains a much lower concentration of hydroxide ions than does 2M aqueous sodium(I) hydroxide (see section 24.9). Fewer hydroxides dissolve in excess aqueous ammonia than in excess aqueous sodium(I) hydroxide and this helps to distinguish certain cations.

12.30 identification of gases

The identity of a gas should be confirmed by as many tests as possible. The order of testing should use non-destructive tests first, before moving on to tests in which fresh gas must be generated for each test. In most cases the gas should be tested at the mouth of the test-tube, not inside, to avoid spray from the reactants. Fig. 12.21 summarises the results of the tests. The tests are:

i) *colour:* more obvious when viewed down length of tube
ii) *smell:* several gases have a distinctive smell but this test should be avoided if possible; too many gases are poisonous
iii) *action with moist litmus paper*
iv) *effect on a glowing or burning splint:* this is a destructive test
v) *tests for soluble reducing agents:* effect with acidified aqueous potassium(I) manganate(VII) or acidified aqueous potassium(I) dichromate(VI), on filter paper held just above the mouth of the tube

gas	colour	smell	action with moist litmus	effect of glowing or burning splint	special tests
hydrogen				'pops' or burns	
oxygen				splint reignites	brown fumes (NO_2) with nitrogen oxide NO
chlorine	yellow-green	pungent	bleaches		oxidises $I^-(aq)$ to $I_2(aq)$
hydrogen chloride		pungent	blue → red		fumes with $NH_3(g)$
ammonia		distinctive	red → blue		fumes with HCl(g)
carbon dioxide			blue → slight red		white precipitate with limewater
carbon monoxide				gas burns with blue flame	burn and test product for CO_2
sulphur dioxide		pungent	blue → red		soluble reducing gas
hydrogen sulphide		'bad eggs'	blue → slight red	gas burns – sulphur deposits	soluble reducing gas; black ppt. with $Pb^{2+}(aq)$
dinitrogen oxide		sweet		splint reignites	no reaction with NO
nitrogen dioxide	brown	pungent	blue → red		soluble oxidising agent, oxidises $I^-(aq)$ to $I_2(aq)$

Fig. 12.21 identification of gases

vi) *tests for soluble oxidising agents:* effect with dilute acidified aqueous potassium(I) iodide plus starch, on filter paper

vii) *special test for hydrogen sulphide:* effect with aqueous lead(II) nitrate(V) (or lead(II) ethanoate), on filter paper

viii) *special test for ammonia:* effect with hydrogen chloride

ix) *special test for hydrogen chloride:* effect with ammonia

revision summary: Chapter 12

terms to be understood: solute, solvent, solution, suspension, solubility, saturated solution, supersaturated solution, crystallisation, hydration

solvents: polar solvents (e.g. water) good for ionic solutes, non-polar solvents (e.g. tetrachloromethane) for molecular solutes

methods of preparing salts: direct combination, acid + metal, base, carbonate or alkali (soluble salts); precipitation (insoluble salts)

normal and acid salts: if an acid is polybasic, an acid salt will be formed if excess acid is used

questions 12

1. How would you

a) find the solubility of oxygen in water,

b) find whether or not tap-water contains dissolved solids,

c) find whether or not there is a volume change when a salt such as sodium(I) chloride dissolves in water,

d) show that there is a dynamic equilibrium between dissolved and undissolved solute in a saturated solution?

2. a) The temperature of aqueous ammonia in a beaker is raised to 100 °C and kept there for ten minutes. Chemical tests then show only a very tiny trace of ammonia in the liquid. Does this prove that the solubility of ammonia in water at 100 °C is negligibly small? Justify your answer.

b) After preparation, chlorine is often collected over aqueous sodium(I) chloride rather than over water. Explain the reasons for this.

c) Hydrogen sulphide and several other gases are usually collected over hot water. What does this tell us about the solubility of gases in water and about the energy changes which take place when gases dissolve in water?

3. A solution of a salt X in water is cooled slowly from 50 °C to room temperature at 18 °C and remains quite clear. On stirring with a glass rod the temperature rises rapidly to 30 °C and crystals appear. Explain these observations.

4. You are a manufacturer of a solvent for cleaning stains from clothes. What factors do you take into account in order to make your product an economic success?

5. Explain a) what is happening to particles of solute and solvent while a crystal of potassium(I) chloride is dissolving in water,

b) the observation that a large salt crystal left sealed for a long time in a saturated solution of the same salt is found eventually to have changed in shape but not in mass.

6. What is meant by the term 'solubility of a salt'?

Choose any well-known soluble salt, and describe experiments you would perform in order to investigate the effect of temperature on its solubility in water. Give a very simple sketch of the kind of graph you would expect to get if you plotted your results (i.e. solubility against temperature).

From your results do you deduce that it would be easy to crystallise this salt from its solution? Explain your answer. (S)

7. What is meant by the terms *saturated solution* and *supersaturated solution*?

Give full practical details for the determination of the solubility of a salt such as potassium(I) nitrate(V) at room temperature.

Sketch the solubility curves for two salts which could be efficiently separated by crystallisation. Explain your answer. (S)

8. What do you understand by *a saturated solution*?

Describe how you would find the solubility of sodium(I) nitrate(V) in water at 25 °C. How would you check that your solution is saturated?

The solubilities of sodium(I) chloride in 100 g of water at 15 °C and 70 °C are 36 g and 38 g respectively, and the corresponding solubilities of sodium(I) nitrate(V) are 85 g and 135 g.

A mixture of 30 g of sodium(I) chloride and 110 g of sodium(I) nitrate(V) is dissolved in 100 g of water at 70 °C.

The solution is cooled to 15 °C. What will be the mass and composition of the crystals that settle out?

All the remaining liquid is then poured into an evaporating dish (assume no loss) and 50 g of water are then evaporated away at 70 °C. What will be the mass and composition of

(i) the residue that settles out during the evaporation,

(ii) the crystals that will settle if the solution is then cooled to 15 °C? (C)

9. A pupil is told to prepare crystals of zinc(II) chloride from zinc(II) carbonate and a solution of sulphuric(VI) acid which contains 4·9 g per dm^3.

a) Write the equation for the reaction.

b) What volume of the acid will be needed to react with 6·27 g of the zinc(II) carbonate?

c) In a salt preparation of this type it is usual to have a slight excess of one of the reagents. Which one should be in excess and why?

d) Why is lead(II) sulphate(VI) not made in this way?

Outline how you would prepare lead(II) sulphate(VI) starting from lead(II) carbonate.

10. Use the solubility data in section 12.17 to select the most suitable method for preparing the following compounds:

a) zinc(II) sulphate(VI) from zinc(II) oxide,

b) copper(II) nitrate(V) from dilute nitric(V) acid,

c) magnesium(II) chloride from magnesium,

d) barium(II) carbonate from barium(II) nitrate(V),

e) iron(III) chloride from chlorine,

f) lead(II) hydroxide from lead(II) nitrate(V),

g) calcium(II) chloride from calcium(II) carbonate,

h) sodium(I) nitrate(V) from sodium(I) hydroxide.

Write an ionic equation to represent the reaction in each case and mention the technique used.

11. a) When salts are prepared from acids one problem is to make sure that the acid has been neutralised. How is this problem solved when using (i) sodium(I) hydroxide, (ii) zinc powder, to neutralise dilute sulphuric(VI) acid?

Having obtained your solution of the required salt, explain how to obtain good crystals other than by leaving the solution to evaporate at room temperature. On what property of many salts does the method you describe depend? Why is the method not successful with sodium(I) chloride?

Name **three** salts that are best prepared by precipitation.

b) A saturated solution of sodium(I) chloride was prepared at 70 °C. When 60·0 g of this solution were carefully evaporated to dryness, 16·5 g of sodium(I) chloride were left. Calculate the solubility of sodium(I) chloride at 70 °C. (C)

12. A student found that a white crystalline compound, when heated in a test-tube, gave a vapour that condensed to a colourless liquid and left a white powdery residue. He then set up the apparatus shown in Fig. 12.22 to obtain samples of the colourless liquid and the residue.

(a) Suggest and explain **two** improvements you could make to this apparatus.

(b) What do you think is the purpose of the asbestos wool?

(c) It is suggested that the colourless liquid is pure water. How could you confirm this?

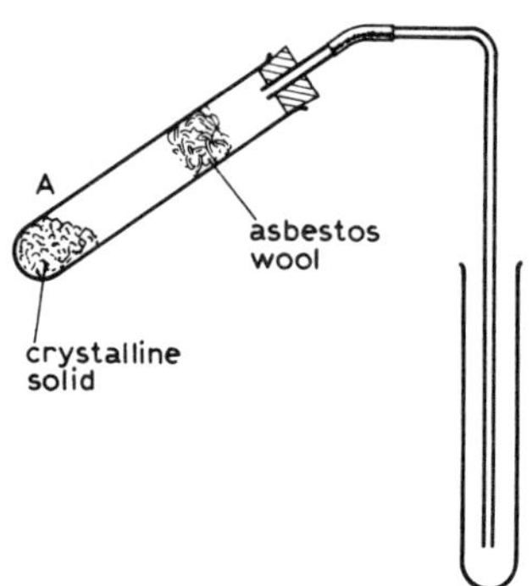

Fig. 12.22 see question 12

(d) At the end of the experiment you are told that the crystalline solid is hydrated magnesium(II) sulphate(VI), $MgSO_4.xH_2O$, which decomposes on heating to give anhydrous magnesium(II) sulphate(VI). Using your apparatus, what weighings would you make to find the percentage by mass of water of crystallisation in the hydrated salt?

(e) What precautions would you take to ensure that all the water of crystallisation had been driven off by the heating?

(f) The compound $MgSO_4.xH_2O$ is found to contain 51·2% by mass of water of crystallisation. Calculate the value of x. (C)

13. Solutions containing 1 mol dm^{-3} of a metal sulphate(VI) and of barium(II) chloride were prepared.

The following mixtures were made in small tubes:

	Vol. of sulphate(VI) solution	Vol. of chloride solution	Vol. of water
(a)	2 cm^3	2 cm^3	8 cm^3
(b)	2 cm^3	4 cm^3	6 cm^3
(c)	2 cm^3	6 cm^3	4 cm^3
(d)	2 cm^3	8 cm^3	2 cm^3
(e)	2 cm^3	10 cm^3	nil

The tubes were centrifuged and the depths of the precipitates formed were measured and recorded as shown in Fig. 12.23.

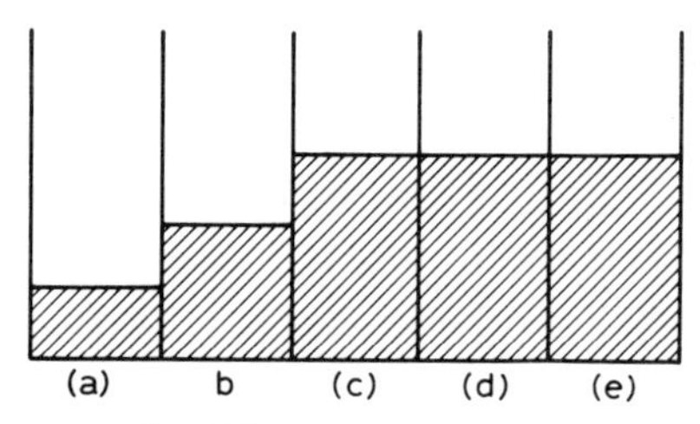

Fig. 12.23 see question 13

(i) Write the ionic equation for the reaction.

(ii) What is the mass of barium(II) chloride in 1 dm^3 of its solution?

(iii) Why did the volume of the precipitate stay constant in (c), (d) and (e)?

(iv) What volume of barium(II) chloride solution reacts with 2 cm^3 of the sulphate(VI) solution?

(v) What volume of barium(II) chloride solution reacts with 1 dm^3 of the sulphate(VI) solution?

(vi) How many moles of barium(II) chloride react with 1 mole of the sulphate(VI)?

(vii) Using the symbol M for the unknown metal, write the formula for the sulphate(VI) and hence the equation for the reaction.

(viii) 1 mole of the metal sulphate(VI) weighs 342 g. What is the relative atomic mass of M?

(ix) How and why was the total volume of the mixture kept constant?

(x) What was the purpose of centrifuging the tubes? What other process might have been used?

(Relative atomic masses: H = 1·0; O = 16·0; S = 32·0; Cl = 35·5; Ba = 137·0.) (JMB)

14. Explain the following observations:

(a) When barium(II) chloride solution is added to a solution of lead(II) nitrate(V), a white precipitate is formed.

(b) When sodium(I) hydroxide solution is added to a solution of zinc(II) sulphate(VI), a white precipitate is formed at first but this dissolves when more alkali is added.

(c) When sodium(I) hydroxide solution is added to a solution of copper(II) sulphate(VI), a bluish precipitate is formed which turns black when the mixture is heated.

(d) When sodium(I) sulphate(VI) crystals are left in an open vessel for some days, a white powder is formed.

(e) When potassium(I) iodide solution is added to a solution of lead(II) nitrate(V), a yellow precipitate is formed which will disappear when water is added and the mixture boiled. (O)

15. How would you distinguish, by test-tube experiments, between the following pairs of substances? Write reaction equations where possible.

a) potassium(I) nitrate(V)	and calcium(II) nitrate(V)
b) ammonium sulphate(VI)	and sodium(I) sulphate(VI)
c) sodium(I) carbonate	and zinc(II) carbonate
d) copper(II) sulphate(VI)	and copper(II) chloride
e) sodium(I) sulphate(VI)	and sodium(I) sulphide
f) iron(II) chloride	and iron(III) chloride
g) aluminium(III) hydroxide	and zinc(II) hydroxide

16. (a) Explain, with full experimental details, how you would prepare in a *pure* condition

(i) zinc(II) oxide, from zinc(II) sulphate(VI) solution,

(ii) copper(II) oxide, from copper.

(b) How would you distinguish between

(iii) zinc(II) oxide and calcium(II) oxide,

(iv) copper(II) oxide and manganese(IV) oxide?

(**One** test is sufficient in (iii) and **one** in (IV) but in **both** parts you must state what happens with **each** oxide).

(c) (v) In preparing lead(II) nitrate(V) crystals starting from concentrated nitric(V) acid, it is essential to dilute the acid. Give **one** reason why this is necessary other than to reduce the vigour of the reaction.

(vi) In preparing sulphates(VI), concentrated sulphuric(VI) acid should also be diluted. Explain what precaution, other than stirring, is essential in diluting sulphuric(VI) acid. (C)

17. Explain the following results, identify **X**, **Y** and **Z** and write equations for all the reactions.

(a) A red solid, **X**, gave a brown residue when warmed with dilute nitric(V) acid. The solid was filtered off and, when warmed with concentrated hydrochloric acid, evolved a gas which turned starch-iodide paper blue.

(b) A metal, **Y**, was warmed with dilute hydrochloric acid and a colourless gas was evolved. When the resulting solution was treated with sodium(I) hydroxide solution, a white precipitate was formed. When excess of the alkali was added, the precipitate dissolved to give a colourless solution.

(c) A black powder, **Z**, gave a blue solution when warmed with dilute sulphuric(VI) acid. When excess of ammonia solution was added to this, the blue colour became much deeper. (JMB)

18. Lead(II) iodide and barium(II) sulphate(VI) are formed as precipitates on mixing solutions of the substances shown on the left in the following equations:

$$Pb(NO_3)_2 + 2KI = PbI_2 + 2KNO_3$$
$$Ba(OH)_2 + H_2SO_4 = BaSO_4 + 2H_2O$$

(A '0·1 M solution' contains one-tenth of a mole of solute per dm^3. Thus 0·1 M HCl contains 3·65 g of HCl – formula weight 36·5 – per dm^3.)

(a) Calculate the formula weights of PbI_2 and $BaSO_4$.

(b) What fraction of a mole of solute is present in 100 cm^3 of a 0·1 M solution?

(c) Calculate the masses of precipitate formed when the following solutions are mixed.

Experiment A.
100 cm^3 of 0·1 M $Pb(NO_3)_2$ + 100 cm^3 of 0·1 M KI

Experiment B.
100 cm^3 of 0·1 M $Pb(NO_3)_2$ + 200 cm^3 of 0·1 M KI

Experiment C.
100 cm^3 of 0·1 M $Ba(OH)_2$ + 100 cm^3 of 0·1 M H_2SO_4

Experiment D.
100 cm^3 of 0·1 M $Ba(OH)_2$ + 200 cm^3 of 0·1 M H_2SO_4.

(d) It is found that the liquids formed in experiments B and D are good conductors of electricity while the liquid formed in experiment C will not conduct. Explain why this is so.

(e) For **each** of the **four** experiments, say whether the liquid formed will be acidic, alkaline or neutral.

(f) If 100 cm^3 of 0·1 M $Ba(OH)_2$ is mixed with 100 cm^3 of 0·1 M HCl, will the liquid formed be acidic, alkaline or neutral? Give reasons for your answer. (C)

19. (a) Starting from copper turnings, describe briefly how you would prepare (i) a solution containing copper(II) nitrate(V), (ii) copper(II) oxide, (iii) crystals of copper(II) sulphate(VI), (iv) copper(II) sulphide.

(b) Give one test for the identification of copper in a dilute solution of copper(II) sulphate(VI). (O)

13.1 effect of electricity on materials

How does electricity affect matter? The answer, which depends partly on what is meant by electricity, has led to some of the most important advances in science. Until Galvani's famous 'frogs'-legs' experiment in 1781 (see section 14.15) and the subsequent development by Volta of the Voltaic Pile in 1800, only static electricity was known. This has no permanent effect upon materials. However, the Voltaic Pile, fore-runner of electric cells, was the key to a source of *moving electricity* (electric current) which, as you will read in this Chapter, can affect materials most profoundly.

13.2 an electric current

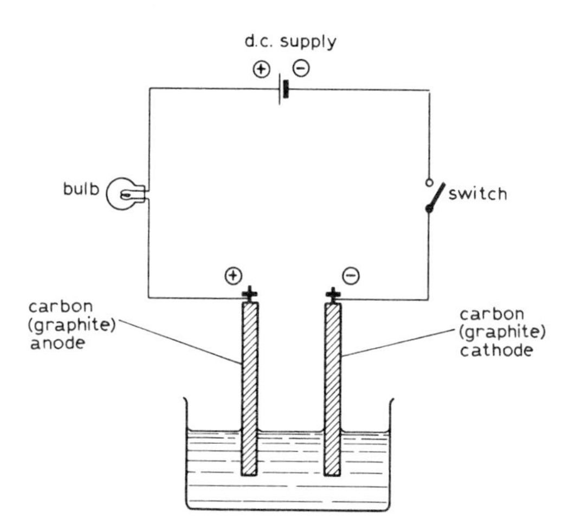

Fig. 13.1 apparatus and circuit for testing conduction by (a) solids, (b) liquids

An electric current is a flow of charge. Before charge flows there must be i) a conductor for it to flow in, ii) a potential difference (voltage) between the ends of the conductor.

In simple experiments which investigate how electric current affects matter, an electric circuit like that shown in Fig. 13.1 may be used. The potential difference is maintained by one or more electric cells. Since currents may only be detected by the effects they produce (particularly heating and magnetic effects), the circuit must include an ammeter or small bulb as current detector. In most electrochemical investigations the circuit is completed by placing the two graphite rods into or across the material under study. Substances other than graphite may be used for these conducting rods, but graphite is unreactive and cheap; irrespective of the material used, the rods are known as **electrodes**.

The electrode connected to the **positive** terminal of the cell or battery is called the **anode.**

The electrode connected to the **negative** terminal of the cell or battery is called the **cathode.**

13.3 conduction of electricity in solids

Using the simple circuit shown in Fig. 13.1 the electrodes are touched on to as many solid materials as possible, including metals, non-metallic elements, salts, sugar, wood, glass and plastics. In which cases does the bulb glow? Does it make any difference if the substances are in powdered form?

It is likely that the bulb glows brightly only when graphite or any metal completes the circuit. This does not prove that the other solids are total non-conductors; it merely shows they are very poor conductors compared with metals and graphite.

13.4 conduction of electricity in liquids

Two types of liquid are investigated:

Pure liquids, often obtained by melting substances. A molten substance is said to be fused.

Aqueous solutions of substances which dissolve well in water.

1. pure liquids

a) *substances which are liquids at room temperature*

About 10–15 cm^3 of the liquids, e.g. mercury, ethanol, water, pure (glacial) ethanoic (acetic) acid and propanone are tested for conductivity in small (50 cm^3) beakers, with the same apparatus as shown in Fig. 13.1. The electrodes must be wiped with tissue paper after each test. The results are summarised in section 13.5.

b) *pure fused substances*

The substances, which should include sugar, sulphur, wax, lead(II) bromide or iodide and potassium(I) iodide, are contained in small crucibles, with the graphite electrodes in an asbestos support. They are warmed gently at first, then more strongly if necessary; they should be fused before the electrodes are inserted. Potassium(I) iodide may take some time to melt, even over a hot flame; the others melt quite quickly over a Bunsen flame. All changes should be noted and the electrodes scraped clean before they are transferred to another substance. The results are summarised in section 13.5.

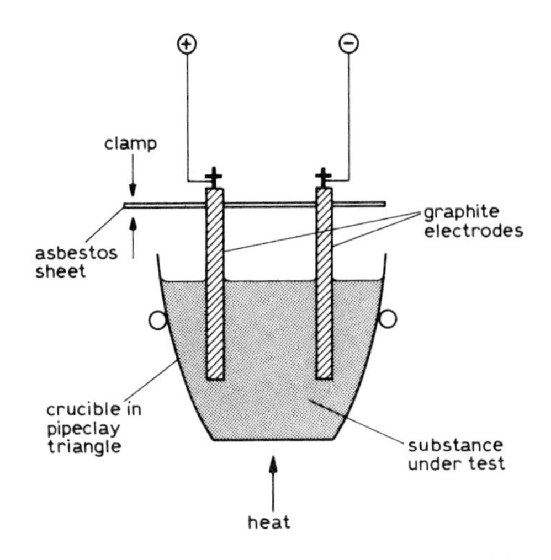

Fig. 13.2 electrolysis of fused solids—this should be conducted in a fume-cupboard or a well-ventilated area, as toxic fumes may be evolved

A more detailed examination of the effect of an electric current on one of the fused salts gives interesting results. Lead(II) bromide melts easily, but the fumes are harmful and the apparatus and circuit should be set up in a fume-cupboard. A large, hard-glass test-tube (or boiling-tube) cut in half, makes an excellent, transparent container for the salt. Longer electrodes (at least 10 cm) may be required, as shown in Fig. 13.3. The white salt lead(II) bromide is placed in the tube to a depth of about 3 cm and is warmed with a small flame until it melts. The graphite electrodes are dipped into the liquid; the glow of the bulb indicates that a current is flowing. The bulb is then removed and the circuit reconnected, since this experiment requires a larger current than the bulb can withstand. While there is a current flowing, the flame is kept just hot enough to maintain the salt in its fused state. After five minutes the electrodes are removed and the apparatus cools. When cold, the solid contents are fairly easy to remove and may be broken up for examination.

During the passage of electricity, a brown gas bubbles out around the anode; this is **bromine.** A small, silvery bead of metal grows below the cathode and **lead** is found in the solid contents after cooling. Clearly **lead(II) bromide decomposes to its elements when conducting electricity in the fused state.** A similar detailed study of lead(II) iodide and potassium(I) iodide affords similar conclusions.

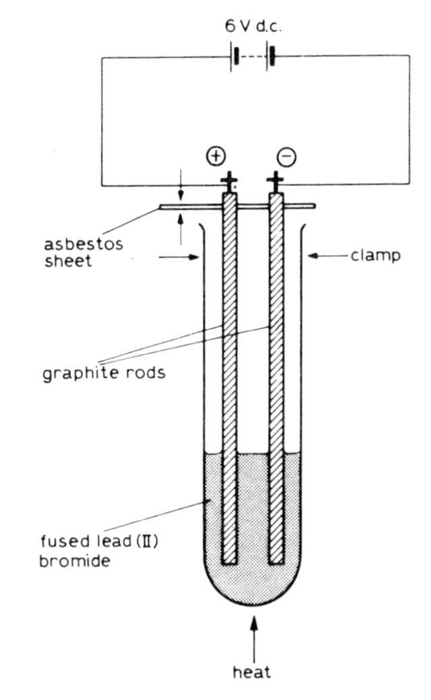

Fig. 13.3 electrolysis of fused lead(II) bromide

2. solutions in water

Conduction in aqueous solutions is tested using the simple circuit of Fig. 13.1 with 20 cm^3 of the liquids in 50 cm^3 beakers. The common dilute acids and alkalis, aqueous solutions of salts, e.g. copper(II) chloride, copper(II) sulphate(VI), potassium(I) iodide and sodium(I) chloride, and aqueous ethanol, propanone and sugar solutions should be examined.

The observations are best summarised in tabular form. Indeed, observations from this and some of the earlier experiments can be condensed into one table as shown in Fig. 13.4. Conclusions derived from these observations are shown in section 13.5 and those referring to the conduction of aqueous solutions are discussed in more detail in section 13.16.

substance	conduction when solid	conduction when fused	other changes when fused	conduction when in solution	other changes when in solution
potassium(I) iodide	×	√	→ elements	√	gas evolved, brown solution

Fig. 13.4 example of tabular summary of results

13.5 some conclusions from the foregoing investigations

a) metals conduct very well when solid or fused;

b) non-metallic elements, e.g. sulphur, do not conduct either when solid or fused;

c) pure compounds which are liquids at room temperature (e.g. ethanol, water, ethanoic (acetic) acid) do not conduct;

d) some compounds (e.g. sugar, plastics, wax) do not conduct either when solid or fused;

e) some compounds (lead(II) bromide, lead(II) iodide, potassium(I) iodide) do not conduct as solids but do conduct when fused–they decompose to their elements when fused and conducting;

f) solutions of some compounds conduct: all acids, alkalis and salt solutions, but not sugar, ethanol or propanone solutions;

g) the changes observed in conducting salt solutions are not always the same as those observed using the same salt in the fused state, e.g. fused potassium(I) iodide decomposes to potassium and iodine, while aqueous potassium(I) iodide gives a gas (hydrogen) at the cathode.

13.6 some terms used in electrochemistry

The conclusions listed in the previous section may well seem confusing at first and we are indebted to two great scientists, Michael Faraday (1791–1867) and Svante Arrhenius (1859–1927), for providing a vocabulary to describe the events and for giving an understanding of the conduction of electricity in liquids and their resulting decomposition.

The important terms are:

electrolysis: the chemical changes in a compound (fused or in solution) when it is conducting electricity

electrolyte: a compound which decomposes when conducting electricity in the fused state or in solution

electrodes: the conductors which connect the electrolyte with the rest of the circuit

anode: the electrode attached to the positive terminal of the cell which supplies the electricity

cathode: the electrode attached to the negative terminal of the cell which supplies the electricity

13.7 ions

An explanation is required for the facts summarised in section 13.5. Why does electrolysis occur only in compounds and not in elements? Why does it occur in some liquids, but not all? Why are the products seen only at the electrodes?

Faraday suggested that electric current in electrolytes is due to a flow of charged particles, which he called ions (*meaning 'wanderers'*). Ions are only free to move in liquids; this explains the absence of conduction in solid compounds. Liquids which do not conduct contain no ions. Since conducting compounds are electrically neutral, they must contain positively and negatively charged ions of such a number and type that the overall total charge in the compound is zero.

13.8 charge and electric current in terms of electrons

It is impossible to explain conduction any further without a knowledge of atomic structure–which Faraday did not possess. Details of atomic structure are given in Chapter 18.

Features relevant to this study are as follows:

a) An atom of any element contains positively and negatively charged parts: the **nucleus (+)** and the **electrons (−).**

b) In a neutral atom the positive charge of the nucleus is exactly balanced by the combined negative charges of the electrons. The charge on a single electron is $1{\cdot}602 \times 10^{-19}$ coulombs (see section 13.13).

c) All charge on anything, from atoms upwards in size, is due to a surplus or deficiency of electrons. Thus:

negative ions are charged particles with one or more electrons in excess of the nuclear charge; **positive ions** are charged particles with one or more electrons fewer than the nuclear charge.

d) An electric current in elements is due to a **flow of electrons.**

By convention, we have become used to considering a movement of electricity from the positive to the negative terminal of a supply. The electrons which are identified with a flow of electricity are negatively charged; hence electrons flow round a circuit in the opposite direction to the conventional current (see Fig. 13.5).

Anode (Gr.: *anodos, way up*) describes the electrode **up which electrons flow**; **cathode** (Gr.: *kathodos, way down*) describes the electrode **down which electrons flow** (see Fig. 13.6).

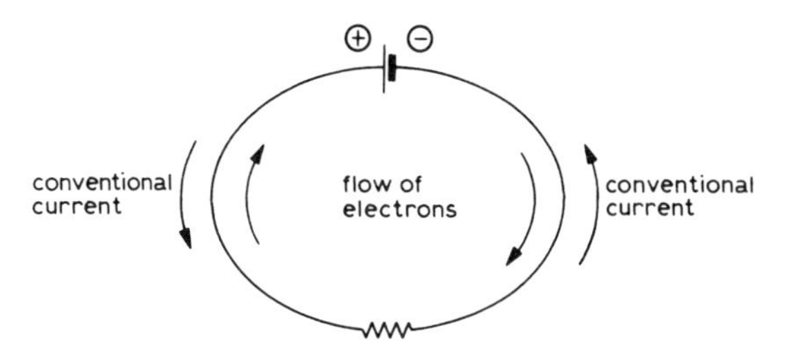

Fig. 13.5 electrons flow in the opposite direction to the conventional representation of current

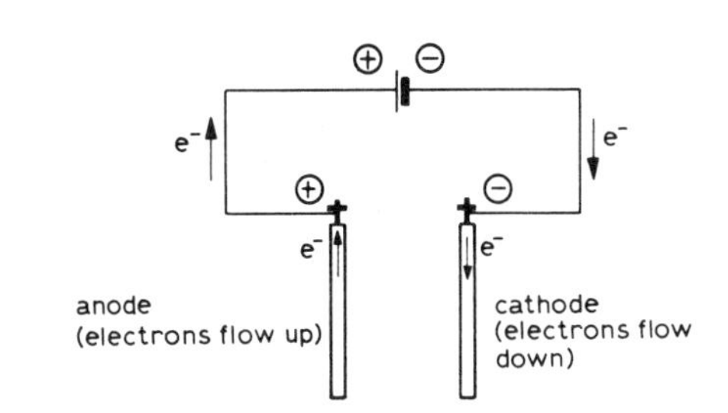

Fig. 13.6 direction of electron flow

13.9 quantitative electrolysis of lead(II) bromide: charge on ions

Careful electrolysis of lead(II) bromide, using a circuit which allows control and measurement of the current, gives the information needed to calculate the charges on single ions of lead and bromide. This is best seen in an example:

When lead(II) bromide undergoes electrolysis with a current of 1·0 A for 30 minutes, 1·931 g of lead is liberated.

$$1{\cdot}0\,\text{A for } 1800\,\text{s} = 1800 \text{ coulombs}$$

$$\therefore \text{ 1 mol (207 g) of lead is liberated by } \frac{207}{1{\cdot}931} \times 1800\,\text{C}$$

$$= 193\,000\,\text{C}$$

$$= \frac{193\,000}{1{\cdot}602 \times 10^{-19}} \text{ electron charges}$$

$$= 12{\cdot}05 \times 10^{23} \text{ electron charges}$$

$6{\cdot}023 \times 10^{23}$ atoms of lead (1 mole) are liberated by the charge of $12{\cdot}05 \times 10^{23}$ electrons

1 atom of lead is liberated by the charge of

$$\frac{12{\cdot}05 \times 10^{23}}{6{\cdot}023 \times 10^{23}} \text{ electrons} = 2 \text{ electrons}$$

The lead ion is Pb^{2+} (see below)

The common unit of charge is the coulomb (C), but when dealing with particles as tiny as ions it is convenient to use a unit of charge of corresponding size. This is the electron charge ($= 1{\cdot}602 \times 10^{-19}$ C), symbolised by *e*.

Since electrons are negatively charged it is often useful in descriptions of chemical processes to show the electron charge as e^- and this convention is followed here.

A negative sign on an ion X^- indicates a surplus of one electron over neutrality;

A positive sign on an ion X^+ indicates a deficiency of one electron from neutrality;

cation with charge of 1+	formula for ion	cations with charge of 2+	formula for ion	cations with charge of 3+	formula for ion
hydrogen	H^+	magnesium(II)	Mg^{2+}	aluminium(III)	Al^{3+}
lithium(I)	Li^+	calcium(II)	Ca^{2+}	iron(III)	Fe^{3+}
sodium(I)	Na^+	barium(II)	Ba^{2+}	chromium(III)	Cr^{3+}
potassium(I)	K^+	copper(II)	Cu^{2+}		
silver(I)	Ag^+	zinc(II)	Zn^{2+}		
copper(I)	Cu^+	iron(II)	Fe^{2+}		
mercury(I)	Hg^+	lead(II)	Pb^{2+}		
ammonium	NH_4^+	mercury(II)	Hg^{2+}		
		tin(II)	Sn^{2+}		
anions with charge of 1−	**formula for ion**	**anions with charge of 2−**	**formula for ion**	**anions with charge of 3−**	**formula for ion**
chloride	Cl^-	oxide	O^{2-}	phosphate(V)	PO_4^{3-}
bromide	Br^-	sulphide	S^{2-}	nitride	N^{3-}
iodide	I^-	sulphate(IV)	SO_3^{2-}		
hydroxide	OH^-	sulphate(VI)	SO_4^{2-}		
nitrate(V)	NO_3^-	carbonate	CO_3^{2-}		
hydrogencarbonate	HCO_3^-				
hydrogensulphate(VI)	HSO_4^-				

Fig. 13.7 charge numbers on ions–positive ions are called cations, negative ions anions (see section 13.11)

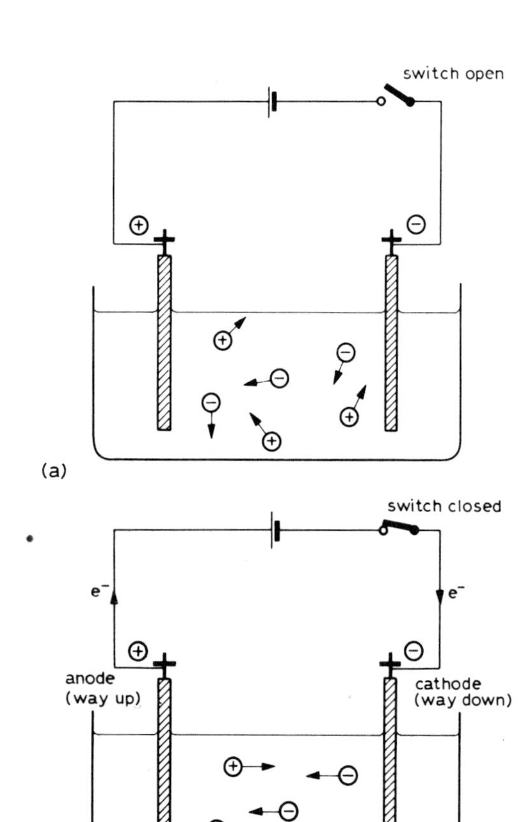

Fig. 13.8 movement of ions (a) before and (b) during electrolysis

A lead ion has two electrons less than a lead atom; the ion is formulated Pb^{2+}. Lead(II) bromide is $PbBr_2$, thus a bromide ion is Br^-.

The symbols and formulae for common ions are shown in Fig. 13.7. Note that **the valency of an element or radical is equal to the charge on the ion** (compare with Figs. 10.10, 10.11).

13.10 environment of ions

a) *in a fused liquid*, ions are shown as

$$M^{n+}(l) \text{ and } X^{n-}(l)$$

e.g. $Pb^{2+}(l)$ and $Br^-(l)$; (fused lead(II) bromide) $Na^+(l)$ and $Cl^-(l)$ (fused sodium(I) chloride)

b) *in water* (*aqueous solution*), ions are surrounded by water molecules–they are said to be hydrated–and are shown as $M^{n+}(aq)$ and $X^{n-}(aq)$,

e.g. $Cu^{2+}(aq)$ and $SO_4^{2-}(aq)$

c) *in solids:* in this book ions are not shown in the formula for a solid. Many substances which exist as ions when fused or in solution are not truly ionic in the solid state and it is misleading to represent them as if they did consist of ions. All solid compounds are represented by formulae of the type MX(s) or M_aX_b(s), even those which X-ray analysis shows to be ionic.

13.11 model for electrolysis of a fused salt

The electrolysis of fused sodium(I) chloride is a good example; for practical details of this electrolysis see section 13.21(1). Electric current in elements (e.g. the connecting wires and electrodes in the circuit) is due to a flow of electrons only; **current in electrolytes is due to a flow of ions.**

Fused electrolytes contain ions, moving about in a random manner before the electric circuit is completed. Immediately the circuit is complete, ions move slowly towards the electrodes, this movement being superimposed upon the general, random motion.

positive ions (cations) move towards the negative electrode (cathode)
negative ions (anions) move towards the positive electrode (anode)

at the cathode: positive ions are discharged by gaining sufficient electrons to become neutral atoms:

e.g. $$Na^+(l) + e^- \rightarrow Na(l)$$

The atoms come together to form metallic sodium (which is fused at the temperature of the fused salt).

at the anode: negative ions are discharged by losing sufficient electrons to become neutral; the liberated species combine to yield molecules of gas:

e.g. $$\left.\begin{array}{l} 2Cl^-(l) \rightarrow 2Cl(l) + 2e^- \\ 2Cl(l) \rightarrow Cl_2(g) \end{array}\right\} 2Cl^-(l) \rightarrow Cl_2(g) + 2e^-$$

Continuous removal of electrons from the cathode and release of electrons at the anode requires a movement of electrons from anode to cathode in the external part of the circuit. The energy for this movement is provided by the electric cell, battery or d.c. source (see section 14.16).

13.12 mole of electrons

In the example in section 13.9 it was calculated that about 193 000 coulombs of electricity are required for discharge of one mole of lead ions, Pb^{2+}. Thus about 96 500 (more accurately 96 487) coulombs are required for the discharge of one mole of singly charged positive ions, e.g. Na^+, Ag^+. Since one mole of singly charged, positive ions is discharged by the addition of one mole of electrons, 96 487 coulombs must be the charge on one mole of electrons.

The quantity of electricity 96 487 C has been referred to as one Faraday of electricity, defined as the quantity of electricity required to liberate one mole (108 g) of silver during electrolysis. This quantity is now referred to as one mole of electrons (units: coulombs) and the symbol F refers to the **Faraday constant** (units: coulombs per mole).

Faraday constant
$F = 96487\,C\,mol^{-1}$
$= 9{\cdot}6487 \times 10^4\,C\,mol^{-1}$

The equation $Na^+(l) + e^- \rightarrow Na(l)$ has one of two meanings, depending on the context:

i) one sodium ion + one electron $\rightarrow$ one sodium atom,
ii) one mole of sodium ions + one mole of electrons $\rightarrow$ one mole of sodium atoms

13.13 discovery of the electron: the 'atom of electricity'

One of the most important conclusions of the early Ionic Theory was that electricity, like matter, is particulate. The name *electron* was given to the tiny 'atom of electricity' which, added to a silver ion, gives a silver atom.

$$\text{electron charge} = \frac{\text{charge required to liberate one mole of silver}}{\text{number of atoms in one mole of silver}}$$
$$= \frac{96487}{6{\cdot}023 \times 10^{23}} \text{ coulombs}$$
$$= 1{\cdot}602 \times 10^{-19} \text{ coulombs}$$

Some years later, cathode rays were identified by J. J. Thomson as streams of electrons, and the electron charge was measured very accurately by R. A. Millikan in his 'oil-drop' experiment (see section 18.1)–the charge on the electron was found to be $1{\cdot}602 \times 10^{-19}$ C.

13.14 electrolysis in solution: strong, weak and non- electrolytes

There are differences in the conducting ability of various solutions. Dilute hydrochloric acid is a good conductor but dilute ethanoic (acetic) acid of the same concentration conducts poorly. Ionic theory tells us that conducting solutions contain ions; dilute ethanoic (acetic) acid must contain fewer ions than dilute hydrochloric acid of the same concentration. Such observations have led to the following classifications:

Strong electrolytes exist completely as ions in water solutions; these are good conductors of electricity. They include 'strong' acids, sulphuric(VI), nitric(V) and hydrochloric; 'strong' alkalis, sodium(I) and potassium(I) hydroxides; and solutions of most soluble salts.

Weak electrolytes exist only partially as ions in water solution; these solutions are generally poor conductors. They include the 'weak' acids, ethanoic (acetic) and tartaric; and the 'weak' alkali, aqueous ammonia. Water itself is a very weak electrolyte due to the partial ionisation:

$$H_2O(l) \rightleftharpoons H^+(aq) + OH^-(aq)$$

Non-electrolytes contain no ions, either when pure or in solution; they do not conduct electricity until the applied potential difference is very high. They include alcohols (ethanol), tetrachloromethane, benzene, sugar, waxes and plastics.

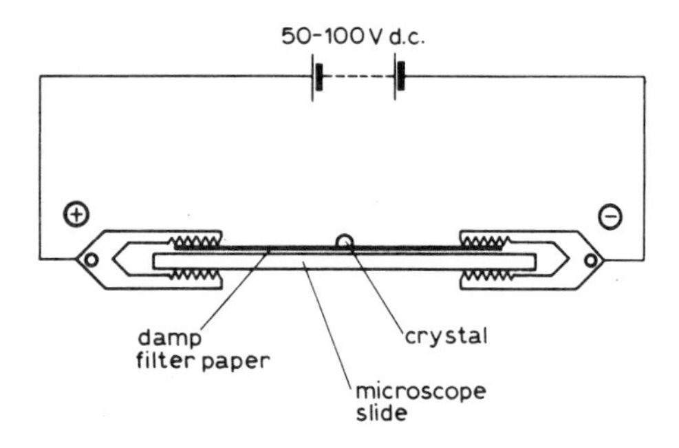

Fig. 13.9 visible migration of ions

13.15 visible migration of ions towards electrodes

An experiment in which the movement of coloured ions may be seen uses the apparatus shown in Fig. 13.9.

A rectangle of filter paper is dampened with water and attached by clip electrodes to a microscope slide. A small crystal of potassium(I) manganate(VII) (permanganate) is placed on the paper and the circuit completed by attaching a source of 50–100 volts d.c.

Quite soon, the purple colour of the manganate(VII) ion $MnO_4^-(aq)$ is seen to move towards the anode (+). No colour is visible on the cathode (−) side; presumably the potassium(I) ion $K^+(aq)$ is colourless. Other coloured salts may be used in this experiment to show movement of the coloured ions.

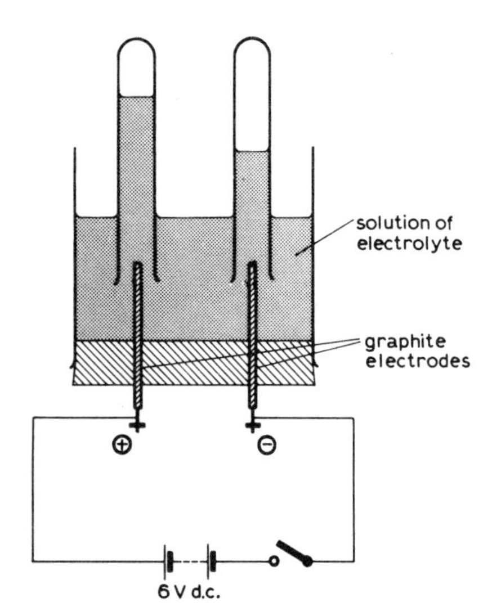

Fig. 13.10 electrolysis in solution

13.16 factors which affect the results of electrolysis in solution

The apparatus in Fig. 13.10 is very effective for collecting information about electrolysis in solution. The main trends become apparent if solutions containing two or more of the following ions are used:

$$H^+(aq),\ Na^+(aq),\ Cu^{2+}(aq),\ SO_4^{2-}(aq),\ Cl^-(aq),\ OH^-(aq)$$

Solutions containing these ions, and their concentrations, are shown in Fig. 13.11. Remember that all solutions in water contain some hydrogen and hydroxide ions from the dissociation

$$H_2O(l) \rightleftharpoons H^+(aq) + OH^-(aq)$$

Liberated gases are collected in small test-tubes and tested with i) damp, blue litmus paper: a change to red indicates an acidic gas; bleaching indicates the presence of chlorine; ii) a glowing splint: a 'pop' indicates hydrogen; re-ignition indicates oxygen.

In most cases the solutions are 2 M but the effect of varying the concentration may be examined by using 0·0001 M sodium(I) chloride and 0·0001 M sulphuric(VI) acid.

The effect of varying the electrodes is studied by using electrodes of graphite, platinum (except with chlorides) and copper.

Electrode reactions in which products appear include:
discharge of hydrogen ions at a cathode

$$2H^+(aq) + 2e^- \longrightarrow H_2(g)$$

discharge of copper(II) ions at a cathode

$$Cu^{2+}(aq) + 2e^- \longrightarrow Cu(s)$$

solution	electrodes	ions present –all (aq)	effect at cathode	ion discharged	effect at anode	ion discharged
0·0001 M sodium(I) chloride	graphite	Na^+, Cl^-, H^+, OH^-	hydrogen	$H^+(aq)$	oxygen and carbon dioxide	$OH^-(aq)$
2M sodium(I) chloride	graphite	Na^+, Cl^-, H^+, OH^-	hydrogen	$H^+(aq)$	chlorine	$Cl^-(aq)$
2M copper(II) chloride	graphite	Cu^{2+}, Cl^-, H^+, OH^-	copper	$Cu^{2+}(aq)$	chlorine	$Cl^-(aq)$
2M copper(II) sulphate(VI)	platinum	Cu^{2+}, SO_4^{2-}, H^+, OH^-	copper	$Cu^{2+}(aq)$	oxygen	$OH^-(aq)$
2M copper(II) sulphate(VI)	copper	Cu^{2+}, SO_4^{2-}, H^+, OH^-	copper	$Cu^{2+}(aq)$	anode dissolves	—
2M sodium(I) hydroxide	platinum	Na^+, H^+, OH^-	hydrogen	$H^+(aq)$	oxygen	$OH^-(aq)$
2M sulphuric(VI) acid	platinum	H^+, SO_4^{2-}, OH^-	hydrogen	$H^+(aq)$	oxygen	$OH^-(aq)$
0·0001 M sulphuric(VI) acid	platinum	H^+, SO_4^{2-}, OH^-	hydrogen	$H^+(aq)$	oxygen	$OH^-(aq)$
2M sodium(I) sulphate(VI)	platinum	Na^+, SO_4^{2-}, H^+, OH^-	hydrogen	$H^+(aq)$	oxygen	$OH^-(aq)$

Fig. 13.11 electrolysis of aqueous solutions

discharge of chloride ions at an anode

$$2Cl^-(aq) \longrightarrow Cl_2(g) + 2e^-$$

discharge of hydroxide ions at an anode

$$4OH^-(aq) \longrightarrow O_2(g) + 2H_2O(l) + 4e^-$$

13.17 interpretation and explanation of the foregoing results

A number of patterns may be found from the investigations described in the preceding section. Their explanation requires a more advanced study than can be given in this book, but in general the processes which occur are those most favoured by energy considerations.

1. substances liberated during electrolysis of solutions

i) at the cathode, only metals or hydrogen are liberated
ii) at the anode, only non-metals are liberated

2. preferential order of discharge of ions

There is a preferential order of discharge for both cations and anions. In many cases the ions derived from water are discharged in preference to those from the electrolyte.

From a solution containing a mixture of cations of equal concentration the order of discharge would be $Cu^{2+}(aq)$ before $H^+(aq)$ before $Na^+(aq)$.

The order for anions is less clear, but again, under conditions of equal concentration, the order of discharge is $OH^-(aq)$ before $Cl^-(aq)$ before $SO_4^{2-}(aq)$.

This topic is dealt with further in a treatment of the electrochemical series [section 14.24(4c)]. *The preferential order of discharge of cations is found to be the reverse of the order of reactivity* as shown in the reactivity series (section 8.14) and the electrochemical series.

3. concentration

An ion in higher concentration may discharge 'out of turn' in the preferential order. A good example is the electrolysis of very dilute (0·0001 M) and more concentrated sodium(I) chloride solutions. Hydroxide ions from the water discharge before the chloride ions in the very dilute solution and oxygen gas is liberated. When the concentration of chloride ions is in very large excess most of the gas liberated is chlorine, though there are still traces of oxygen. The effect of relative concentration on preferential discharge is more commonly observed for anions than for cations. It is, however, also true for cations. Cu^{2+}(aq) is discharged in preference to Zn^{2+}(aq) unless the concentration of Zn^{2+}(aq) is greatly in excess, when copper and zinc are deposited together, as the alloy brass.

4. nature of the electrodes: inert and reactive electrodes

Graphite and platinum are **inert** electrodes in most electrolytes, functioning only as conductors and electron exchangers. Occasionally they do react with the substances liberated, e.g. oxygen attacks graphite (carbon) anodes, forming some carbon dioxide; chlorine attacks platinum anodes. Suitable electrodes must be chosen for a particular solution, e.g. graphite when chlorine is liberated, platinum when oxygen is liberated.

Most metals form **reactive anodes.** No anions are discharged at these anodes. Instead, the material of the anode itself loses electrons, and forms positively charged, hydrated ions which diffuse into the solution. Gradually the anode disappears as electrolysis continues.

A good example is the electrolysis of copper(II) sulphate(VI) with copper electrodes. The solution never loses its blue colour, as copper ions discharged at the cathode are replaced by copper ions from the anode

$$Cu^{2+}(aq) + 2e^- \longrightarrow Cu(s) \quad \text{(cathode)}$$
$$Cu(s) \longrightarrow Cu^{2+}(aq) + 2e^- \quad \text{(anode)}$$

Use is made of this in copper refining [section 13.21(1)].

An extreme example of the influence of the material of an electrode on the preferential discharge of ions is afforded by the electrolysis of aqueous sodium(I) chloride using a mercury cathode. Here sodium is liberated, though hydrogen would be expected from the relative positions in the preferential order of discharge (electrochemical series). This important process is described in section 16.6.

13.18 changes of pH in the vicinity of electrodes during electrolysis

Due to the rapid discharge of hydrogen and/or hydroxide ions during some electrolyses, there are considerable changes in pH round the electrodes. This may be seen, using either the apparatus shown in Fig. 13.10 or a shallow dish on an overhead projector (Fig. 13.12), in the electrolysis of 2 M aqueous sodium(I) sulphate(VI) using platinum electrodes. A few drops of Universal Indicator are added to the solution before electrolysis begins.

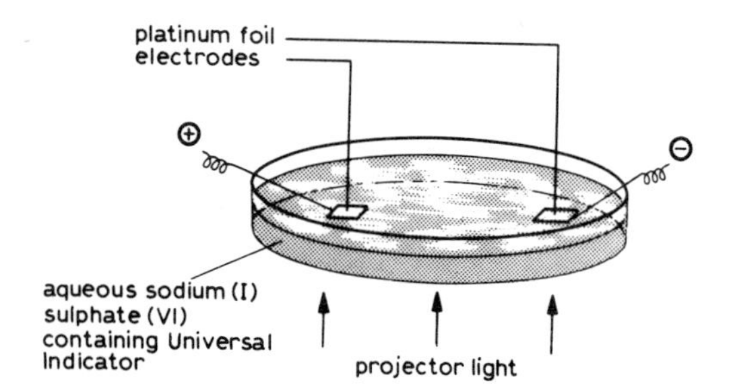

Fig. 13.12 apparatus for overhead projection

at the cathode: hydrogen is liberated and a blue colour indicates an increase of pH (alkaline solution). Discharge of hydrogen ions

$$2H^+(aq) + 2e^- \longrightarrow H_2(g)$$

leaves a local excess of hydroxide ions OH^-(aq), over H^+(aq), which is responsible for the alkaline solution round the cathode.

at the anode: oxygen is liberated and a red colour indicates a decrease of pH (acidic solution). Discharge of hydroxide ions:

$$4OH^-(aq) \longrightarrow O_2(g) + 2H_2O(l) + 4e^-$$

leaves a local excess of hydrogen ions, $H^+(aq)$, over $OH^-(aq)$, and this is responsible for the acidic solution round the anode. When electrolysis is stopped and the solution is stirred, the red and blue colours disappear and the green colour of neutral Universal Indicator is restored. What does this indicate about the local excesses of $H^+(aq)$ and $OH^-(aq)$?

13.19 the masses of elements deposited in electrolysis related to the quantity of electricity passed

Interesting results may be obtained using solutions with the apparatus and circuit shown in Fig. 13.13.

Electrodes of copper and of silver are placed in solutions of 0·05 M aqueous copper(II) sulphate(VI) and 0·05 M aqueous silver(I) nitrate(V) respectively; these systems form **a copper voltameter** and **a silver voltameter.** The copper cathode is conveniently made from copper gauze surrounding the anode; the silver cathode may be of silver-plated copper gauze. The procedure is as follows:

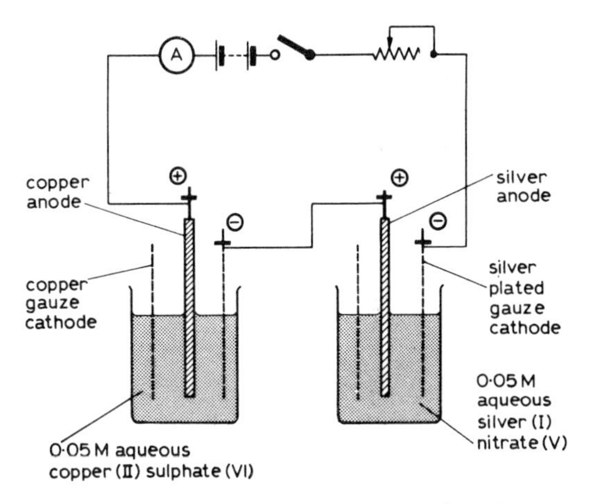

Fig. 13.13 copper and silver voltameters in series

i) The metals deposited should adhere to the cathodes; the cathodes must therefore be carefully cleaned.
ii) The correct operating conditions are arranged by assembling the apparatus completely and adjusting the current to the right order of magnitude. In most experiments, low currents (about 100 mA) give the best results.
iii) The cathodes are removed, washed, dried, weighed and replaced. It is a useful precaution to treat the anodes similarly (see (vi)).
iv) The circuit is completed and a clock started simultaneously.
v) The current is kept steady by operating the variable resistor and after a suitable time (about 30 minutes) the clock and current are stopped.
vi) The cathodes are removed very carefully, in case the metals are not adhering strongly, washed gently with water and propanone, dried in a hot air stream and weighed.

    Silver is notorious for not adhering well; if some does fall into the solution it may be recovered by filtration, washing, etc., its mass being added to the mass of the silver cathode. Alternatively, the *loss* in mass of the anode should equal the *gain* in mass of the cathode. This may be found by the same washing, drying and weighing process used for the cathode.
vii) The cathodes are replaced and further mass readings are taken for different currents and different times.

quantity of electric charge (coulombs)
= current (amperes) × time (seconds)

A carefully conducted experiment gives a graph of the form shown in Fig. 13.14. It is a straight line passing through the origin. Thus the mass of copper or silver liberated is proportional to the quantity of electricity passed.

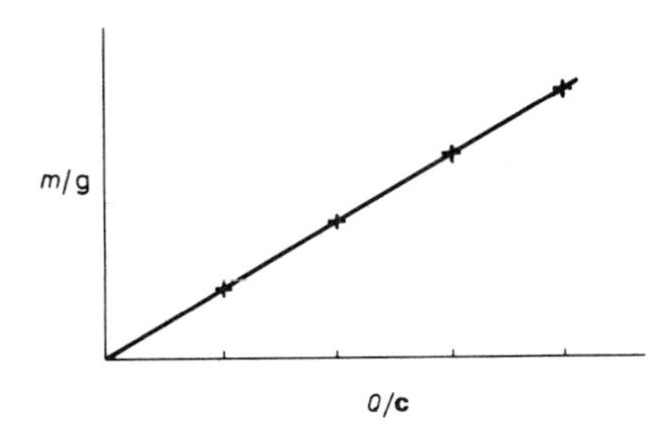

Fig. 13.14 graph of mass of copper deposited at cathode (m) against quantity of electricity passed (Q); since graph is of form $y \propto x$, it follows that $m \propto Q$

13.20 Faraday's Laws of Electrolysis

a) Faraday found the above result to be true for any element liberated in electrolysis and summarised it in his **First Law of Electrolysis:**

The mass of element liberated at an electrode is directly proportional to the quantity of electricity passed.

mass of element liberated ∝ current × time

If m is the mass in grams, I the current in amperes and t the time in seconds,

$$m = I \times t \times \text{a constant}$$

The constant of proportionality is called the **electrochemical equivalent** of the element liberated. It represents the **mass of that element liberated by one coulomb of charge.**

b) **Faraday's Second Law of Electrolysis** was originally derived by comparing the masses of different elements liberated by the same quantity of electricity (number of coulombs). In a modern form it may now be stated:

The quantities of electricity required to liberate one mole of different elements bear a simple relationship to one another.

96487 C are required to liberate one mole of silver
$2 \times$ 96487 C are required to liberate one mole of copper
$3 \times$ 96487 C are required to liberate one mole of aluminium

The Laws of Electrolysis formed the foundation of Faraday's original Ionic Theory and led later to the first measurement of the electron charge, as described in section 13.13.

13.21 uses of electrolysis

Electrolysis is immensely important in a modern economy and accounts for a large proportion of the electrical energy supplied to industry.

1. extraction and purification of metals

Other than by electrolysis, it is extremely difficult to extract the very reactive metals like potassium, sodium and calcium from their compounds, or to obtain magnesium and aluminium in reasonable quantities.

It is interesting to try a simplified Downs Process (see section 16.2) for the extraction of sodium from molten sodium(I) chloride on a laboratory scale. The apparatus required is shown in Fig. 13.15. The melting-point of sodium(I) chloride is brought down to the Bunsen burner temperature range by melting it with an equal mass of calcium(II) chloride in a clay crucible. The anode is a graphite rod, the cathode an iron rod surrounded by a hard-glass tube to trap the sodium produced. Electrolysis is allowed to proceed for 30 minutes using a 6–8 V d.c. supply, then the iron cathode is slowly withdrawn from the fused electrolyte, so that the end is sealed by solidifying salt. Shiny sodium is seen inside the tube.

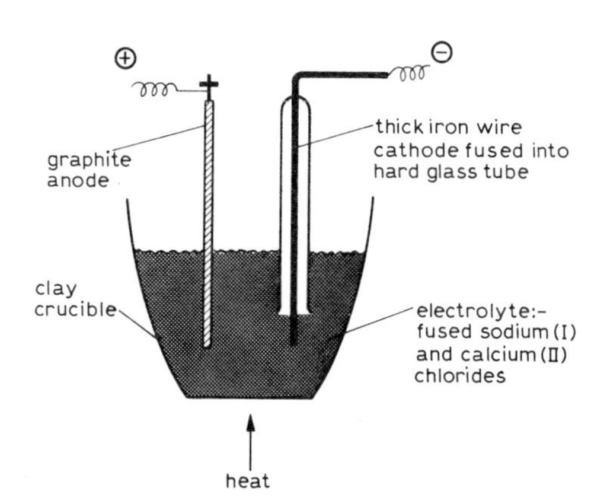

Fig. 13.15 electrolysis of fused sodium(I) chloride

Copper is purified by electrolysis of aqueous copper(II) sulphate(VI) using a cathode of very pure copper and an anode of the impure metal which has been extracted using a chemical reduction process. Effectively copper is 'plated out' of the solution on to the cathode, while the copper in the anode dissolves into the solution:

pure copper cathode: $Cu^{2+}(aq) + 2e^- \rightarrow Cu(s)$
impure copper anode: $Cu(s) \rightarrow Cu^{2+}(aq) + 2e^-$

2. chemical manufacture

Chlorine, hydrogen and many other chemicals as well as metals are prepared by electrolytic processes. Chlorine particularly is in great demand (see Chapter 15) and the Kellner–Solvay process outlined in section 16.6 produces chlorine, hydrogen, sodium(I) hydroxide and, if required, sodium(I) chlorate(I) solution – the main constituent of 'hypochlorite' bleaches.

There are of course many technical difficulties in extraction and refining. Obtaining sufficient electrical energy may be a problem. For this reason the really large users are situated as near to cheap supplies as possible, which partly explains the siting of aluminium extraction industries in Scotland, where cheap hydroelectric power is available, and in Anglesey, near the new nuclear power station. The giant Kariba dam on the Zambesi was built to aid the supply of hydroelectricity to the copper industries of Rhodesia and Zambia.

Maintaining correct operating temperatures, voltages and currents presents difficulties, as does dealing with products other than those particularly desired. Oxygen produced in the extraction of aluminium attacks the graphite anodes and these must be replaced periodically.

3. *electroplating*

A conducting article, made the cathode in a suitable electrolytic cell, may be electroplated with a layer of metal. The conditions of composition of electrolyte, temperature, current and cleanliness must all be carefully controlled, or the metal may not adhere well. Familiar electroplated articles are 'electroplated nickel silver (epns)' cutlery and the thin but strong, rust-resisting, bright chromium finish on car bumpers.

It is easy to plate copper and iron articles with nickel using a solution of ammonium nickel(II) sulphate(VI) as electrolyte, with the object to be plated as the cathode and a nickel anode. Non-metallic objects (even bones or leaves) may be copper-plated if the surface is first made conducting with a layer of graphite or powdered metal. The object is coated first with a thin, dilute solution of adhesive, then shaken in a tin of graphite or powdered metal. When the surface has 'set', the article is immersed as the cathode in aqueous copper(II) sulphate(VI) and electrolysis is conducted using a very low current.

4. *anodising*

The oxide layer always present on aluminium may be made even thicker and stronger if the aluminium is made the anode in electrolysis of dilute sulphuric(VI) acid. By mixing dyes with the electrolyte the layer may be coloured.

5. *electrochemical machining*

Most metal anodes are oxidised during electrolysis [see section 13.17(4)]:

$$M(s) \longrightarrow M^{n+}(aq) + ne^-$$

If a specially shaped cathode is used, the parts of the anode nearest the cathode 'dissolve' more quickly and the anode becomes a reversed copy of the cathode. The cathode may be used many times and complicated shapes can be achieved without the necessity for much additional machining.

revision summary: Chapter 13

conductors: metals, graphite–no decomposition

electrolytes: ionic compounds in liquid state–decomposition

anode: (*way up*) electrons travel up; oxidation occurs

cathode: (*way down*) electrons travel down; reduction occurs

anions: negatively charged ions which migrate to anode

cations: positively charged ions which migrate to cathode

Faraday's Laws: (1) mass of element deposited $\propto$ quantity of electricity passed
(2) quantities of electricity required to liberate one mole of different elements bear simple relation to one another

preferential discharge of ions: reverse of reactivity series, or electrochemical series, for metals

questions 13

1. Sodium conducts electricity well when solid or liquid. Iodine is a non-conductor when solid or liquid. Sodium(I) iodide is a non-conductor when solid but does conduct when fused or dissolved in water.

Explain these observations.

2. Describe the apparatus you would set up to compare the conductances of the following liquids:
(i) distilled water; (ii) benzene; (iii) concentrated sulphuric(VI) acid; (iv) M sulphuric(VI) acid; (v) M ammonium hydroxide; (vi) M ethanoic (acetic) acid.

Classify these liquids as non-conductors, poor conductors, medium conductors, good conductors, giving a brief explanation of the variations in conductance in terms of ions and molecules. (C)

3. What ions may be present in solutions of:
copper(II) chloride, sodium(I) nitrate(V), calcium(II) hydroxide, aluminium(III) sulphate(VI), ammonium carbonate, silver(I) nitrate(V), potassium(I) hydrogencarbonate, zinc(II) hydrogensulphate(IV), barium(II) bromide, tin(IV) iodide, iron(III) sulphate(IV)?

4. Sketch an apparatus for passing an electric current through an aqueous solution of sulphuric(VI) acid, so that any gases liberated may be collected and measured. Label your sketch to show what the electrodes are made of and what their polarity is. Give the names and relative volumes of the gases produced. Name all the ions present in the solution and indicate the direction of their migration. Give electronic equations for the discharge of ions where appropriate, and indicate the direction of flow of electrons in the external circuit.

What mass of chlorine will be liberated by the same quantity of electricity which liberates 23 g sodium during the electrolysis of fused sodium(I) chloride? In what physical state(s) will these elements be liberated in the conditions under which this process is carried out? (O)

5. What do you understand by the term *electrolysis*?

(a) Give an explanation of the fact that hydrochloric acid is a better conductor than ethanoic (acetic) acid of equal concentration, and give the cathode reaction which is common to both solutions.

(b) (i) Give **one** explanation for the fact that metals other than copper are not deposited on the cathode during the electrolytic purification of copper. (ii) Compare the quantity of electricity required to deposit one mole of aluminium with that required to deposit one mole of copper. (O)

6. A solution of copper(II) sulphate(VI), acidified with sulphuric(VI) acid, is electrolysed using copper electrodes.

a) Give the formulae of the ions present in the solution before electrolysis.
b) What changes, if any, are observed at the cathode, at the anode, and in the solution?
c) Explain the reactions which take place at the electrodes.
d) What use is made of this process in industry?

You are asked to investigate the connection between the quantity of electricity passed through a solution of copper(II) sulphate(VI) and the mass of copper deposited.

(i) Sketch the electrical circuit you would use.
(ii) State the measurements you would make.

In such an experiment 1930 coulombs liberated 0·64 g of copper. When the same quantity of electricity was passed through a solution containing silver(I) ions, Ag^+, 2·16 g of silver were liberated. How do you explain these results? (AEB)

7. A steady current is passed through two voltameters connected in series containing copper(II) sulphate(VI) solution and dilute sulphuric(VI) acid respectively. All electrodes are of platinum. What is the total volume of gases liberated in the latter (measured at s.t.p.) when 6·35 g of copper are deposited in the former? (S)

8. When an electric current was passed through dilute sulphuric(VI) acid the volume of hydrogen obtained was 56 cm^3, after correction to s.t.p.

a) Write ionic equations for the reaction at (i) the platinum cathode, (ii) the platinum anode.
b) What volume of oxygen, measured at s.t.p., was liberated in this experiment?
c) What fraction of a mole of water was decomposed by the electric current?
d) How might the oxygen and hydrogen be recombined to form water?

(Relative atomic masses: H = 1·0, O = 16·0; one mole of gas at s.t.p. occupies 22·4 dm^3) (JMB)

9. (a) Give a detailed description of the preparation of a small sample of lead by electrolysis of a fused compound.
(b) Give an outline of the **industrial** preparation of a metal by electrolysis.
(c) A current of 0·5 A, flowing for 6 min 26 s, through two cells in series, was found to deposit 0·216 g of silver on the cathode of the first cell and 0·059 g of nickel on the cathode of the second. The relative atomic masses of silver and nickel are 108 and 59 respectively.
Calculate
(i) the quantity of electricity passed through the two cells,
(ii) the quantity of electricity needed to deposit one mole of silver,
(iii) the quantity of electricity needed to deposit one mole of nickel.

Comment on the results of your calculations in (ii) and (iii). (C)

10. Explain what is meant by (a) an ion, (b) electrolysis.

Describe and explain **any three** of the following electrolytic processes:
(i) the extraction of aluminium from pure aluminium(III) oxide,
(ii) the anodising of aluminium,
(iii) the refining of copper,
(iv) the obtaining of chlorine from sodium(I) chloride. (AEB)

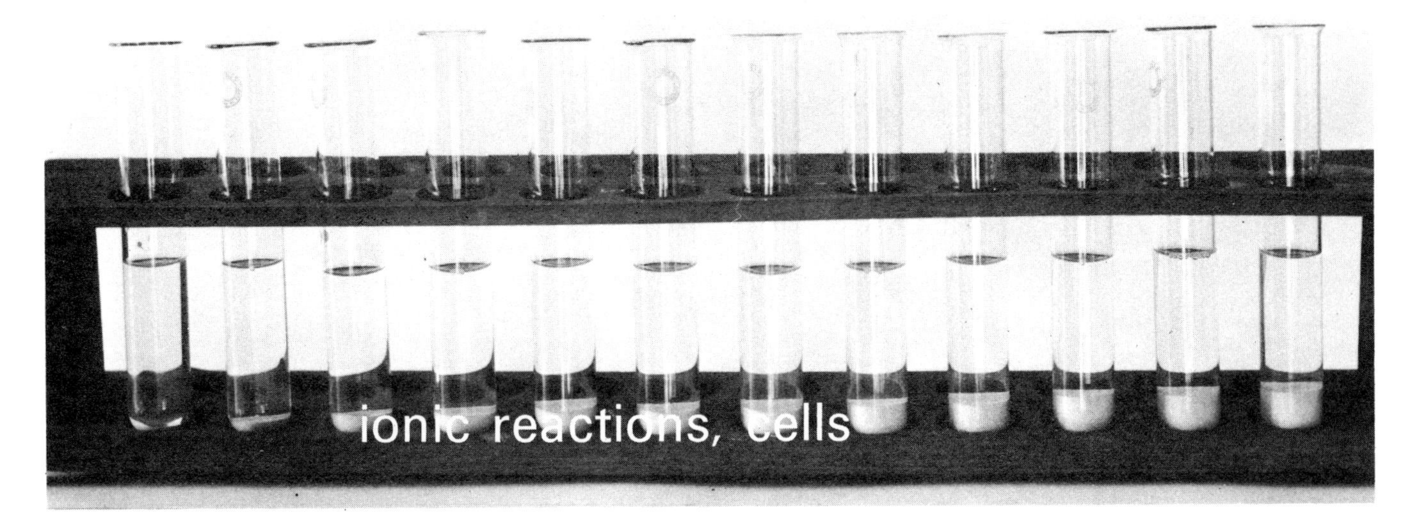

ionic reactions, cells

Many chemical reactions are electrical in nature, in that they involve charged particles. A detailed study of the stability of ions and the nature of the electron and of the proton is deferred to Chapter 18; in this Chapter three important types of chemical reaction encountered in previous work are explained in terms of the charged particles concerned

precipitation reactions, involving positively and negatively charged ions,
oxidation and reduction, which involve transfer of *electrons*, and
acid-base reactions, which involve transfer of *protons*.

Finally, this Chapter deals with the way in which a continuous flow of electrons (an electric current) can be obtained from chemical reactions in **cells.**

14.1 precipitation reactions

Aqueous solutions of salts conduct electricity well because the dissolved salt consists entirely of ions. The ions are surrounded by polar water molecules in indefinite numbers and this is represented by writing the symbol (aq) after the symbol for the ion:

aqueous sodium(I) chloride contains $Na^+(aq)$ and $Cl^-(aq)$,
aqueous silver(I) nitrate(V) contains $Ag^+(aq)$ and $NO_3^-(aq)$.

If these two solutions are mixed, a white precipitate of silver(I) chloride is obtained. The remaining solution contains $Na^+(aq)$ and $NO_3^-(aq)$, these ions – 'spectator ions' – taking no part in the reaction. They are therefore not written in the equation for the reaction, which is

$$Ag^+(aq) + Cl^-(aq) \longrightarrow AgCl(s)$$

The most satisfactory method of writing equations for precipitation reactions is to write down *first* the correct formula for the precipitated compound at the right-hand side (this requires a knowledge of valencies); *then* the symbols for the contributing ions (each with a charge number equal to its valency) at the left-hand side; *finally*, the number of moles of each ion – required is adjusted. The steps in the procedure are exemplified below:

Example 1: precipitation of barium(II) sulphate(VI)

$$BaSO_4(s)$$

$$\mathbf{Ba^{2+}(aq) + SO_4^{2-}(aq) \longrightarrow BaSO_4(s)}$$

Example 2: precipitation of lead(II) iodide

$$PbI_2(s)$$

$$Pb^{2+}(aq) + I^-(aq) \longrightarrow PbI_2(s)$$

$$\mathbf{Pb^{2+}(aq) + 2I^-(aq) \longrightarrow PbI_2(s)}$$

Example 3: precipitation of calcium(II) phosphate(V)

$Ca_3(PO_4)_2(s)$

$$Ca^{2+}(aq) + PO_4^{3-}(aq) \longrightarrow Ca_3(PO_4)_2(s)$$
$$\mathbf{3Ca^{2+}(aq) + 2PO_4^{3-}(aq) \longrightarrow Ca_3(PO_4)_2(s)}$$

It is a useful exercise to practise writing such equations for all the precipitation reactions mentioned in section 12.29.

14.2 oxidation

The following reactions are quoted elsewhere as examples of oxidation:

$$Mg(s) + \tfrac{1}{2}O_2(g) \longrightarrow MgO(s) \quad \text{(section 7.6)}$$
$$Na(s) + \tfrac{1}{2}Cl_2(g) \longrightarrow NaCl(s) \quad \text{(section 15.13)}$$
$$PbO_2(s) + 4HCl(aq) \longrightarrow PbCl_2(s) + 2H_2O(l) + Cl_2(g) \quad \text{(section 15.5)}$$

The reactions appear to be quite different in nature, involving addition of oxygen to magnesium, addition of chlorine to sodium and removal of hydrogen from hydrogen chloride respectively - yet they have all been called 'oxidations'.

Application of the ionic theory soon solves the problem. Let us ask the question: what has been 'oxidised'? *Answer:* magnesium atoms, sodium atoms and chloride ions - for aqueous hydrogen chloride is composed of ions.

$$Mg(s) \longrightarrow ?$$
$$Na(s) \longrightarrow ?$$
$$Cl^-(aq) \longrightarrow ?$$

What sort of particles of magnesium, sodium and chlorine are formed? Magnesium oxide contains Mg^{2+}, sodium chloride contains Na^+ and chlorine gas is composed of neutral molecules. The changes are:

$$Mg(s) \longrightarrow Mg^{2+}(s)$$
$$Na(s) \longrightarrow Na^+(s)$$
$$Cl^-(aq) \longrightarrow \tfrac{1}{2}Cl_2(g)$$

To effect each of these changes, electrons must be removed from the original particles:

$$Mg(s) \longrightarrow Mg^{2+}(s) + 2e^-$$
$$Na(s) \longrightarrow Na^+(s) + e^-$$
$$Cl^-(aq) \longrightarrow \tfrac{1}{2}Cl_2(g) + e^-$$

Oxidation
rem**O**val
Of
electr**O**ns

The three apparently dissimilar reactions all involve loss of electrons.

A species (atom, molecule or ion) is said to be oxidised if electrons are removed from it.

14.3 oxidation number: naming inorganic compounds

When a metal forms a compound, it does so by losing one or more electrons from each atom to form a positively charged ion and it is therefore oxidised in the process. The resulting ion is said to have an **oxidation number** of I if the cation is singly charged, II if doubly charged and III if triply charged. This oxidation number is always written in the name for the compound:

	oxidation number	*example of name*
$Na \longrightarrow Na^+ + e^-$	I	sodium(I) bromide
$Ca \longrightarrow Ca^{2+} + 2e^-$	II	calcium(II) oxide
$Al \longrightarrow Al^{3+} + 3e^-$	III	aluminium(III) chloride

The compound is sometimes said to contain the metal in the **oxidation state** of I, II or III. Some metals, like iron, can exist in more than one oxidation state and the conversion of an iron(II) compound to an iron(III) compound is an oxidation:

$$Fe^{2+} \longrightarrow Fe^{3+} + e^-$$

Such oxidation numbers apply *whether the compound formed is purely ionic or not.*

Non-metals which gain electrons (to form simple negative ions) exist in negative oxidation states in such compounds, this is not written in the name:

	oxidation number	*example of name*
$\frac{1}{2}Cl_2 + e^- \rightarrow Cl^-$	–I	ammonium chloride
$\frac{1}{2}O_2 + 2e^- \rightarrow O^{2-}$	–II	sodium(I) oxide
$\frac{1}{2}N_2 + 3e^- \rightarrow N^{3-}$	–III	magnesium(II) nitride

When a non-metal occurs in an anion containing oxygen, the non-metal is in a *positive* oxidation state, as will be seen by regarding the oxygen in the ion as oxide ion, O^{2-}:

	oxidation number	*example of name*
ClO^-; $(Cl^{+1} + O^{2-})$	I	sodium(I) chlorate(I)
NO_2^-; $(N^{+3} + 2O^{2-})$	III	sodium(I) nitrate(III)
SO_3^{2-}; $(S^{+4} + 3O^{2-})$	IV	sodium(I) sulphate(IV)
NO_3^-; $(N^{+5} + 3O^{2-})$	V	sodium(I) nitrate(V)
SO_4^{2-}; $(S^{+6} + 4O^{2-})$	VI	sodium(I) sulphate(VI)
CO_3^{2-}; $(C^{+4} + 3O^{2-})$	IV	sodium(I) carbonate

Carbon exists in no positive oxidation state other than IV and this number is therefore not mentioned in naming carbonates; in all other cases of oxy-anions the oxidation number is included in the name as shown above.

These ions do not contain positively charged non-metal ions, like S^{6+}, because the oxide ions form covalent bonds with the sulphur particles (see Chapter 18) which greatly modifies the charge on the sulphur particles. The theoretical nature of the oxidation number in these cases is usually shown by writing the symbol for the species as S^{+6} rather than S^{6+}. The designation M^{x+} is reserved for real ions bearing full charges.

14.4 reduction

Reduction is the opposite of oxidation; it is often manifested in the removal of oxygen or addition of hydrogen:

$$CuO(s) + H_2(g) \rightarrow Cu(s) + H_2O(g) \quad (CuO \text{ reduced})$$
$$Cl_2(g) + H_2(g) \rightarrow 2HCl(g) \quad (Cl_2 \text{ reduced})$$

If the product of the second reaction is dissolved in water, producing ions, it will be appreciated that the fundamental changes are:

$$Cu^{2+} + 2e^- \rightarrow Cu$$
$$Cl_2 + 2e^- \rightarrow 2Cl^-$$

A species (atom, molecule or ion) is said to be reduced if electrons are added to it.

In a chemical change, any electrons removed from one species (oxidation) are absorbed by another (reduction). When an oxidation occurs, a corresponding reduction must occur and we therefore speak of **oxidation-reduction,** or **'redox'** processes.

14.5 displacement reactions

1. metal–metal ion reactions

Dip the iron blade of a knife into copper(II) sulphate(VI) solution and it comes out coated with copper metal. A sheet of zinc in a lead(II) salt solution 'grows' handsome crystals of lead; beautiful long crystals of silver appear to grow from a blob of mercury in a silver(I) salt solution. Measurements show that the iron, zinc and mercury have lost mass in each case; evidently these metals are displacing the copper, lead and silver from their salt solutions.

Mg
Zn
Fe
Pb
Cu
Ag

Fig. 14.1 displacement order for metals

Evidence for displacement, or other reactions, may be collected by placing small pieces of various metals in solutions of salts of other metals. Magnesium, zinc, iron, lead, copper and silver are suitable metals and the nitrate(V) is a suitable salt to use in each case. Results are not always very clear but an order of displacement may be distinguished, as shown in Fig. 14.1.

Metals only displace other metals below them in this order. With magnesium and zinc, hydrogen may also appear.

Consider the reaction of zinc with copper(II) sulphate(VI) solution: copper exists in the salt solution as hydrated cations. When it is displaced as a metal, these cations must have gained electrons:

$$Cu^{2+}(aq) + 2e^- \longrightarrow Cu(s)$$

The electrons can only have come from the zinc metal, which then forms hydrated cations; gradually the solid zinc disappears:

$$Zn(s) \longrightarrow Zn^{2+}(aq) + 2e^-$$

The two half-reaction equations may be combined into one ionic equation:

$$Zn(s) \longrightarrow Zn^{2+}(aq) + 2e^-$$
$$Cu^{2+}(aq) + 2e^- \longrightarrow Cu(s)$$
$$\mathbf{Zn(s) + Cu^{2+}(aq) \longrightarrow Zn^{2+}(aq) + Cu(s)}$$

Confirmation of the 'balance' of this equation is obtained by the following experiment:

About 0·5 g of zinc filings is weighed accurately into a weighed test-tube. Half a test-tube full of hot, concentrated aqueous copper(II) sulphate(VI) is added in two successive portions. Thorough mixing is ensured by shaking the tube well for a few minutes after each addition. Copper is displaced and allowed to settle, or is separated using a centrifuge. The solution is removed with a dropping pipette and the copper is washed, first with water then with propanone. When all the propanone has evaporated (helped by warming the tube in boiling water) the dry tube containing copper is weighed.

Calculate i) the masses of zinc and copper involved in the reaction,
ii) the number of moles of each metal,
iii) the ratio of moles of the metals; this confirms the balance of the reaction equation.

2. metal–acid reactions

The reaction between a metal and dilute acid may be classified as a displacement, since hydrogen is displaced from the acid solution. To understand this it is necessary to accept that the particle responsible for acid character in aqueous solutions is $H^+(aq)$. This idea is developed in section 14.10 and more fully in section 18.23.

The reaction between zinc and dilute sulphuric(VI) acid is:

$$Zn(s) \longrightarrow Zn^{2+}(aq) + 2e^-$$
$$2H^+(aq) + 2e^- \longrightarrow H_2(g)$$
$$\mathbf{Zn(s) + 2H^+(aq) \longrightarrow Zn^{2+}(aq) + H_2(g)}$$

Mg
Zn
Fe
Pb
H
Cu
Ag

Fig. 14.2 displacement order including hydrogen

It is not easy to place hydrogen precisely in the displacement series by simple experimental observations, but an order including hydrogen appears to be that shown in Fig. 14.2. Some displacement reactions expected to produce hydrogen, such as lead with dilute acids, are so slow that the gas is not observed.

14.6 oxidation–reduction (redox) reactions

Oxidation of one substance must be accompanied by reduction of another. Redox reactions are reactions in which electron exchange takes place between the reactants. The classification covers a wide range of reactions including displacements (metal–metal ion and metal–acid, section 14.5), electrode processes during electrolysis (sections 13.11, 13.16), the reactions in electric cells (section 14.16), and many others, notably in halogen chemistry (Chapter 15).

Oxidation and reduction processes are conveniently written as separate **half-reaction equations.** These are then added to give a full, ionic reaction equation. As with all reaction equations, it is important to be sure that the reaction actually occurs and what the products are; the 'balance' of reacting moles may be checked by experiment.

The number of electrons gained in reduction must equal the number lost in oxidation.

Examples:

i) reaction between sodium metal and oxygen gas

oxidation half-reaction:

$$4Na(s) \longrightarrow 4Na^+(s) + 4e^-$$

reduction half-reaction:

$$O_2(g) + 4e^- \longrightarrow 2O^{2-}(s)$$

full reaction equation:

$$4Na(s) + O_2(g) \longrightarrow 4Na^+(s) + 2O^{2-}(s)$$

or $$\mathbf{4Na(s) + O_2(g) \longrightarrow 2Na_2O(s)}$$

ii) reaction between iron and hydrochloric acid

oxidation: $Fe(s) \longrightarrow Fe^{2+}(aq) + 2e^-$

reduction: $2H^+(aq) + 2e^- \longrightarrow H_2(g)$

full: $\mathbf{Fe(s) + 2H^+(aq) \longrightarrow Fe^{2+}(aq) + H_2(g)}$

iii) reaction between zinc and aqueous copper(II) sulphate(VI)

oxidation: $Zn(s) \longrightarrow Zn^{2+}(aq) + 2e^-$

reduction: $Cu^{2+}(aq) + 2e^- \longrightarrow Cu(s)$

full: $\mathbf{Zn(s) + Cu^{2+}(aq) \longrightarrow Zn^{2+}(aq) + Cu(s)}$

iv) reaction between chlorine and aqueous iron(II) sulphate(VI)

oxidation: $2Fe^{2+}(aq) \longrightarrow 2Fe^{3+}(aq) + 2e^-$

reduction: $Cl_2(g) + 2e^- \longrightarrow 2Cl^-(aq)$

full: $\mathbf{2Fe^{2+}(aq) + Cl_2(g) \longrightarrow 2Fe^{3+}(aq) + 2Cl^-(aq)}$

v) reaction between bromine and sodium(I) iodide solution

oxidation: $2I^-(aq) \longrightarrow I_2(aq) + 2e^-$

reduction: $Br_2(aq) + 2e^- \longrightarrow 2Br^-(aq)$

full: $\mathbf{Br_2(aq) + 2I^-(aq) \longrightarrow I_2(aq) + 2Br^-(aq)}$

vi) electrode processes are redox reactions in which the oxidation and reduction half-reactions take place at separate electrodes and the electrons are transferred through the external circuit:

oxidation of anions occurs at the anode:

$$Cl^-(aq) \longrightarrow \tfrac{1}{2}Cl_2(g) + e^-$$

reduction of cations occurs at the cathode:

$$H^+(aq) + e^- \longrightarrow \tfrac{1}{2}H_2(g)$$

overall: $\mathbf{H^+(aq) + Cl^-(aq) \longrightarrow \tfrac{1}{2}H_2(g) + \tfrac{1}{2}Cl_2(g)}$

If the anode dissolves it is oxidised:

e.g. $Cu(s) \longrightarrow Cu^{2+}(aq) + 2e^-$

14.7 oxidising agents (oxidants) and reducing agents (reductants)

An oxidising agent or oxidant must remove electrons from another substance; in absorbing these electrons it is itself reduced. Conversely, a reducing agent or reductant must supply electrons to another substance; in doing so it is itself oxidised.

A substance is usually termed an oxidising agent if it is reduced sufficiently easily to enable it to oxidise a variety of other substances; similarly the term reducing agent is reserved for substances which are oxidised very easily.

1. tests for soluble oxidants

An oxidant removes electrons from another substance, therefore a reagent which will reveal the presence of an oxidant must give an **observable change** when electrons are removed from it. Two useful test reagents for soluble oxidants are

a) *acidified aqueous potassium*(I) *iodide*, which changes from colourless to brown when oxidised:

$$\underset{\text{colourless}}{I^-(aq)} \longrightarrow \underset{\text{brown}}{\tfrac{1}{2}I_2(aq)} + e^-$$

If a little starch solution is added to the resulting brown solution, a deep blue colour is produced. This starch-iodine test is very sensitive and may be carried out using 'starch-iodide paper'.

b) *acidified aqueous iron*(II) *sulphate*(VI), which changes from pale green to pale yellow when oxidised:

$$\underset{\text{pale green}}{Fe^{2+}(aq)} \longrightarrow \underset{\text{pale yellow}}{Fe^{3+}(aq)} + e^-$$

If this reagent is used, the colour change can be made more obvious by adding aqueous sodium(I) hydroxide after addition of the oxidant. If oxidation has occurred, red-brown iron(III) hydroxide is precipitated:

$$Fe^{3+}(aq) + 3OH^-(aq) \longrightarrow \underset{\text{red-brown}}{Fe(OH)_3(s)}$$

If no oxidation has occurred, dark green iron(II) hydroxide is formed:

$$Fe^{2+}(aq) + 2OH^-(aq) \longrightarrow \underset{\text{dark green}}{Fe(OH)_2(s)}$$

2 cm^3 of 0·5 M aqueous potassium(I) iodide acidified with 2 cm^3 of M sulphuric(VI) acid and 1 cm^3 of M aqueous iron(II) sulphate(VI) acidified with 1 cm^3 of M sulphuric(VI) acid are suitable quantities of test reagents. 2 cm^3 of a solution containing a soluble oxidant should be added to this quantity of each reagent. Common soluble oxidants which give positive results in these tests are: aqueous chlorine, aqueous bromine, acidified solutions of hydrogen peroxide, sodium(I) chlorate(I), potassium(I) manganate(VII), potassium(I) dichromate(VI) and concentrated nitric(V) acid.

2. half-equations for reduction of oxy-anions

These can be worked out simply using the oxidation number principle. Acidified aqueous potassium(I) manganate(VII) is a soluble oxidant which gives positive results with the reagents described in section (1); when it acts, the purple manganate(VII) ion is reduced to colourless manganese(II) ion. We shall use this example to work out the half-equation.

First write down the symbol for the ion which is reduced and the ion produced on its reduction (*this must be known; it cannot be worked out*):

$$MnO_4^-(aq) \longrightarrow Mn^{2+}(aq)$$

The oxygen removed from the ion *must be regarded as oxide ion*, O^{2-}; two moles of $H^+(aq)$ are required for each mole of O^{2-} removed–this is why the solution is acidified. The next step is to write down

$$MnO_4^-(aq) + 8H^+(aq) \longrightarrow Mn^{2+}(aq) + 4H_2O(l)$$

Finally, the manganese in MnO_4^- changes its oxidation number from VII to II in Mn^{2+}; this requires $5e^-$:

$$MnO_4^-(aq) + 8H^+(aq) + 5e^- \longrightarrow Mn^{2+}(aq) + 4H_2O(l)$$

The full equation for the oxidation of aqueous iodide by acidified manganate(VII) can now be derived:

$$MnO_4^-(aq) + 8H^+(aq) + 5e^- \rightarrow Mn^{2+}(aq) + 4H_2O(l)$$
$$5I^-(aq) \rightarrow 2\tfrac{1}{2}I_2(aq) + 5e^-$$
$$\mathbf{MnO_4^-(aq) + 8H^+(aq) + 5I^-(aq) \rightarrow Mn^{2+}(aq) + 4H_2O(l) + 2\tfrac{1}{2}I_2(aq)}$$

Using this technique, it is possible to work out half-equations for the action of the soluble oxidants listed at the end of section (1):

$$\tfrac{1}{2}Cl_2(aq) + e^- \rightarrow Cl^-(aq)$$
$$\tfrac{1}{2}Br_2(aq) + e^- \rightarrow Br^-(aq)$$
$$H_2O_2(aq) + 2H^+(aq) + 2e^- \rightarrow H_2O(l) + H_2O(l)$$
$$ClO^-(aq) + 2H^+(aq) + 2e^- \rightarrow Cl^-(aq) + H_2O(l)$$
$$Cr_2O_7^{2-}(aq) + 14H^+(aq) + 6e^- \rightarrow 2Cr^{3+}(aq) + 7H_2O(l)$$
$$NO_3^-(aq) + 2H^+(aq) + e^- \rightarrow NO_2(g) + H_2O(l)$$

The overall equation for the action of any of these oxidants with either of the test reagents of section (1) may be obtained by combining the appropriate half-equations, ensuring that the number of electrons gained is equal to the number lost.

3. *tests for soluble reductants*

The oxidants from the list in section (1) which give the clearest observable changes when they act, can be used as test reagents for soluble reductants. Perhaps the best are acidified aqueous manganate(VII), which changes from purple to colourless, and acidified aqueous dichromate(VI), which changes from orange to green. Aqueous solutions of hydrogen sulphide, sulphur dioxide, iodide ion and hydrogen peroxide should be tested with these two reagents; thc appropriate half-equations are given below. *Hydrogen gained or lost should always be treated as H^+, oxygen gained or lost should always be treated as O^{2-}.*

$$H_2S(aq) \rightarrow S(s)$$
$$H_2S(aq) \rightarrow S(s) + 2H^+(aq)$$
$$\mathbf{H_2S(aq) \rightarrow S(s) + 2H^+(aq) + 2e^-}$$

$$SO_2(aq) \rightarrow SO_4^{2-}(aq)$$
$$SO_2(aq) + 2H_2O(l) \rightarrow SO_4^{2-}(aq) + 4H^+(aq)$$
$$\mathbf{SO_2(aq) + 2H_2O(l) \rightarrow SO_4^{2-}(aq) + 4H^+(aq) + 2e^-}$$

$$\mathbf{I^-(aq) \rightarrow \tfrac{1}{2}I_2(aq) + e^-}$$

$$H_2O_2(aq) \rightarrow O_2(g)$$
$$H_2O_2(aq) \rightarrow O_2(g) + 2H^+(aq)$$
$$\mathbf{H_2O_2(aq) \rightarrow O_2(g) + 2H^+(aq) + 2e^-}$$

4. *hydrogen peroxide*

The term 'peroxide' is reserved for substances containing two oxygen atoms linked together in their structure. The structure of hydrogen peroxide is shown in Fig. 14.3. The instability of this structure causes hydrogen peroxide to decompose easily and to behave either as an oxidant or as a reductant according to the reagents with which it reacts, as seen in sections (1), (2) and (3). It is best known as a propellant, due to its decomposition to oxygen, and as a bleach, due to its properties as an oxidant.

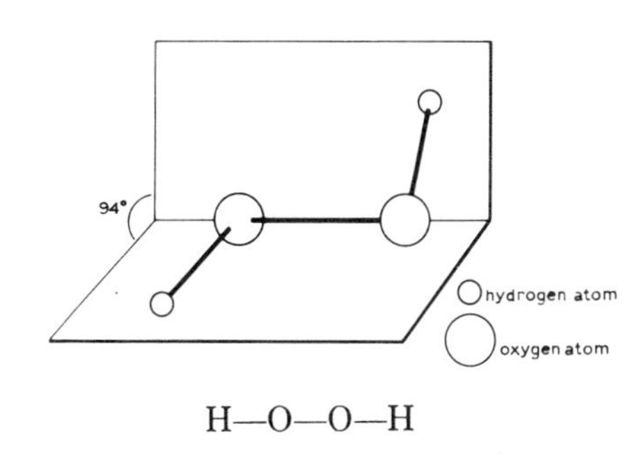

H—O—O—H

Fig. 14.3 molecular structure of hydrogen peroxide

a) *decomposition*

The activation energy for the decomposition process is low and hydrogen peroxide decomposes at room temperature, at measurable speeds, to oxygen and water.

$$H_2O_2(l) \rightarrow H_2O(l) + \tfrac{1}{2}O_2(g)$$

The process is catalysed by light and by a wide variety of materials including powdered metals, manganese(IV) oxide, enzymes in blood and even by dust; the catalysed decomposition of pure hydrogen peroxide may be explosive.

For this reason the pure liquid is used as a rocket fuel. In more dilute solutions, the catalysed decomposition may be used to obtain a supply of oxygen in the laboratory (see sections 7.1–7.3), or as the foaming agent for porous concrete or foam rubber. Hydrogen peroxide should be stored in dark, clean bottles and is usually sold as '10-volume' or '20-volume' solution. This means that one volume of the solution liberates 10 or 20 volumes of oxygen (measured at s.t.p.), on decomposition (section 7.3).

b) *as an oxidant*

In acidic solutions hydrogen peroxide oxidises chloride, bromide, iodide and iron(II) ions. It is reduced to water in each case.

It oxidises i) concentrated hydrochloric acid to chlorine:

$$H_2O_2(aq) + 2H^+(aq) + 2Cl^-(aq) \longrightarrow Cl_2(g) + 2H_2O(l)$$

ii) iodide ions in potassium(I) iodide solution to iodine:

$$H_2O_2(aq) + 2H^+(aq) + 2I^-(aq) \longrightarrow I_2(aq) + 2H_2O(l)$$

iii) iron(II) ions in iron(II) sulphate(VI) solution to iron(III) ions:

$$H_2O_2(aq) + 2H^+(aq) + 2Fe^{2+}(aq) \longrightarrow 2Fe^{3+}(aq) + 2H_2O(l)$$

Hydrogen peroxide has long been known to restorers of old paintings for its ability to restore the brightness to oil paintings which have become dark with time. This too is an oxidation process. Many old paints contain lead compounds; when these absorb hydrogen sulphide from a polluted atmosphere, black lead(II) sulphide forms. Hydrogen peroxide oxidises lead(II) sulphide to white lead(II) sulphate(VI), which allows the colour of the pigments in the paint to show brightly once more:

$$PbS(s) + 4H_2O_2(l) \longrightarrow PbSO_4(s) + 4H_2O(l)$$

The bleaching action of hydrogen peroxide is caused by oxidation of pigments to colourless products. It is used to bleach delicate materials such as hair, silk, wool or feathers which may be damaged by other bleaches, e.g. aqueous sodium(I) chlorate(I) ('hypochlorite bleach').

Its antiseptic properties result from its action in destroying coagulated blood and providing the oxidising environment in which germs do not multiply. It should be used in dilute form or it may damage flesh tissues by too powerful an oxidation.

c) *as a reductant*

In alkaline solution, or in the presence of more powerful oxidising agents, hydrogen peroxide shows reducing properties. It is oxidised to oxygen.

$$H_2O_2(aq) + 2OH^-(aq) \longrightarrow O_2(g) + 2H_2O(l) + 2e^-$$

Hydrogen peroxide reduces: i) iron(III) ions to iron(II) ions in alkaline conditions

$$2Fe^{3+}(aq) + H_2O_2(aq) + 2OH^-(aq) \longrightarrow 2Fe^{2+}(aq) + 2H_2O(l) + O_2(g)$$

(N.B. it oxidises iron(II) ions to iron(III) ions in acid solution)

ii) silver(I) oxide to silver metal

$$Ag_2O(s) + H_2O_2(aq) \longrightarrow 2Ag(s) + H_2O(l) + O_2(g)$$

iii) the very powerful oxidising agent potassium(I) manganate(VII).

The purple manganate(VII) ions change to colourless manganese(II) ions and the change may be used in a titration to find the concentration of hydrogen peroxide in solutions.

$$2MnO_4^-(aq) + 6H^+(aq) + 5H_2O_2(aq) \longrightarrow 2Mn^{2+}(aq) + 8H_2O(l) + 5O_2(g)$$

14.8 insoluble oxidants

A number of oxides, notably lead(IV) oxide and manganese(IV) oxide, behave as oxidants. This behaviour is not due to the oxygen in the oxides (in contrast with hydrogen peroxide) but to the *high oxidation state of the metal* in the oxide. Under conditions in which such substances act as oxidants (usually in the presence of acid) the lower oxidation state of the metal is more stable and the fundamental changes are

$$Pb^{(+4)} + 2e^- \rightarrow Pb^{(+2)}$$
$$Mn^{(+4)} + 2e^- \rightarrow Mn^{(+2)}$$

1. thermal decomposition of lead(IV) oxide

If the dark brown powder is heated strongly, a gas is evolved which rekindles a glowing splint and a yellow residue remains which can be shown on analysis to be lead(II) oxide. (What analytical method would you use?) The equation for the reaction is obviously

$$PbO_2(s) \rightarrow PbO(s) + \tfrac{1}{2}O_2(g)$$

What has been oxidised and what has been reduced? If the oxygen in the oxides is treated as O^{2-}, the answer becomes clear:

$$\text{in } PbO_2(s) \quad \begin{cases} O^{2-} \rightarrow \tfrac{1}{2}O_2(g) + 2e^- \\ Pb^{+4} + O^{2-} + 2e^- \rightarrow Pb^{+2} + O^{2-} \end{cases}$$
$$\mathbf{PbO_2(s) \rightarrow PbO(s) + \tfrac{1}{2}O_2(g)}$$

Oxide ion is oxidised to oxygen gas; lead(IV) is reduced to lead(II).

2. oxidation of hydrochloric acid

If a little lead(IV) oxide is warmed with concentrated hydrochloric acid a white solid remains–lead(II) chloride–and a greenish-yellow gas is evolved which bleaches moist litmus paper–chlorine (see Chapter 15). Redox and precipitation reactions are involved:

(red.) $PbO_2(s) + 4H^+(aq) + 2e^- \rightarrow Pb^{2+}(aq) + 2H_2O(1)$
redox
(ox.) $2Cl^-(aq) \rightarrow Cl_2(g) + 2e^-$
precipitation $Pb^{2+}(aq) + 2Cl^-(aq) \rightarrow PbCl_2(s)$

Manganese(IV) oxide will similarly oxidise concentrated hydrochloric acid to chlorine, but no corresponding precipitation reaction occurs, as manganese(II) chloride is soluble.

14.9 an unusual redox reaction

Warm sodium inflames in an atmosphere of hydrogen gas, leaving the crystalline product sodium(I) hydride:

$$Na(s) + \tfrac{1}{2}H_2(g) \rightarrow NaH(s)$$

In this reaction, what has been oxidised and what has been reduced? You may find the answer a little surprising in view of more elementary definitions of oxidation and reduction than those which we now use. (*Clue:* molten sodium(I) hydride conducts electricity; electrolysis yields sodium metal at the cathode and hydrogen gas at the anode.)

14.10 acid–base reactions

In Chapter 11 the term 'acid' was given to a group of substances having certain properties in common and it was noted that the similarity in properties is most obvious when water is present. The essential structural feature of acids is explained in detail in sections 18.23 and 18.24, after the nature of bonding in chemical compounds has been studied; further evidence for the definition of acids developed in those sections is quoted in section 20.6. In this Chapter it is convenient to use the definition of an acid developed in these later Chapters in order to compare acid–base reactions with the precipitation and redox reactions already studied.

an acid is a substance capable of donating protons

e.g.
$$HCl \rightarrow H^+ + Cl^-$$
$$H_2SO_4 \rightarrow H^+ + HSO_4^-$$
$$HSO_4^- \rightarrow H^+ + SO_4^{2-}$$

The term 'base' is given to a substance capable of counteracting the effect of an acid, which leads to the definition

a base is a substance capable of accepting protons

e.g.
$$O^{2-} + H^+ \rightarrow OH^-$$
$$OH^- + H^+ \rightarrow H_2O$$
$$NH_3 + H^+ \rightarrow NH_4^+$$

The protons donated by an acid are ***never left uncombined*** but are always accepted by another substance, which therefore acts as a base. We therefore speak of **acid–base reactions,** which are characterised by the **transfer of a proton** from one substance (the acid) to another (the base).

e.g. i) reaction between hydrogen chloride gas and ammonia gas:

$$HCl(g) \rightarrow H^+ + Cl^-$$
$$H^+ + NH_3(g) \rightarrow NH_4^+$$
$$\mathbf{\underset{acid}{HCl(g)} + \underset{base}{NH_3(g)} \rightarrow NH_4Cl(s)}$$

ii) neutralisation of hydrogen chloride by a metal hydroxide:

$$HCl \rightarrow H^+ + Cl^-$$
$$H^+ + OH^- \rightarrow H_2O$$
$$\mathbf{\underset{acid}{HCl} + \underset{base}{OH^-} \rightarrow Cl^- + H_2O}$$

iii) dissolving hydrogen chloride in water:

$$HCl \rightarrow H^+ + Cl^-$$
$$H^+ + H_2O \rightarrow H_3O^+$$
$$\mathbf{\underset{acid}{HCl} + \underset{base}{H_2O} \rightarrow Cl^- + H_3O^+}$$

In any reaction involving water, the ions formed are surrounded by an indeterminate number of water molecules and this is represented by writing the symbol (aq) after the symbol for the ion. The last equation written in full would be

$$HCl(g) + H_2O(l) \rightarrow Cl^-(aq) + H_3O^+(aq)$$

The resulting **oxonium ion,** $H_3O^+(aq)$, often simply written as a hydrated proton, $H^+(aq)$, is the particle responsible for the characteristic reactions of acids in water solution.

14.11 familiar and less familiar bases

We should already be aware that metal oxides and solutions of hydroxides (alkalis) can neutralise acids, thus acting as bases:

$$O^{2-} + 2H^+(aq) \rightarrow H_2O(l)$$

e.g. $CuO(s) + 2H^+(aq) \rightarrow H_2O(l) + Cu^{2+}(aq)$

and $OH^-(aq) + H^+(aq) \rightarrow H_2O(l)$

Some acids are very slightly ionised in water and are called weak acids. An example is hydrogen cyanide, HCN:

$$\underset{acid}{HCN(aq)} + \underset{base}{H_2O(l)} \rightleftharpoons CN^-(aq) + H_3O^+(aq)$$

Since this ionisation is very slight, if potassium(I) cyanide is dissolved in water, the cyanide ion will react with water forming hydrogen cyanide molecules:

$$\underset{\text{base}}{CN^-(aq)} + \underset{\text{acid}}{H_2O(l)} \rightleftharpoons HCN(aq) + OH^-(aq)$$

A solution of potassium(I) cyanide in water is alkaline to litmus and smells strongly of hydrogen cyanide. Thus although potassium(I) cyanide is a salt, the cyanide ion is here acting as a base: this idea may well be unfamiliar.

A commoner example is afforded by sodium(I) carbonate. Carbonic acid is a weak, dibasic acid:

$$H_2CO_3(aq) + H_2O(l) \rightleftharpoons HCO_3^-(aq) + H_3O^+(aq)$$
$$HCO_3^-(aq) + H_2O(l) \rightleftharpoons CO_3^{2-}(aq) + H_3O^+(aq)$$

If sodium(I) carbonate is dissolved in water an alkaline solution results; carbonate ion acts as a base towards water:

$$\underset{\text{base}}{CO_3^{2-}(aq)} + \underset{\text{acid}}{H_2O(l)} \longrightarrow HCO_3^-(aq) + OH^-(aq)$$

More surprisingly still, chloride and nitrate(V) ions can act as bases towards concentrated sulphuric(VI) acid:

$$\underset{\text{base}}{Cl^-(s)} + \underset{\text{acid}}{H_2SO_4(l)} \longrightarrow HCl(g) + HSO_4^-(s) \qquad \text{(section 15.5)}$$

$$\underset{\text{base}}{NO_3^-(s)} + \underset{\text{acid}}{H_2SO_4(l)} \xrightarrow{\text{heat}} HNO_3(g) + HSO_4^-(s) \qquad \text{(section 24.12)}$$

14.12 correct classification of reactions of acids

Dilute sulphuric(VI) acid reacting with an alkali or with a carbonate is in each case involved in proton transfer. These are **acid–base** reactions:

$$H^+(aq) + OH^-(aq) \longrightarrow H_2O(l)$$
$$2H^+(aq) + CO_3^{2-}(aq) \longrightarrow H_2O(l) + CO_2(g)$$

However, when the acid reacts with a metal, electrons and not protons are transferred. This is a **redox,** not an acid–base reaction:

$$Zn(s) \longrightarrow Zn^{2+}(aq) + 2e^- \quad \textit{(oxidation)}$$
$$2H^+(aq) + 2e^- \longrightarrow H_2(g) \quad \textit{(reduction)}$$
$$\mathbf{Zn(s) + 2H^+(aq) \longrightarrow Zn^{2+}(aq) + H_2(g)}$$

14.13 amphoteric nature of water

Pure water is very slightly ionised:

$$H_2O(l) \rightleftharpoons H^+(aq) + OH^-(aq)$$

These two ions are responsible for acidic and alkaline character respectively in water solution. Rather than refer to water as a neutral substance, it is more proper to view it as equally acidic and alkaline. It is an example of an amphoteric substance.

a substance which can act either as an acid or as a base is said to be amphoteric

In contact with an acid, water behaves as a base:

$$\underset{\text{acid}}{H_2SO_4(l)} + \underset{\text{base}}{H_2O(l)} \longrightarrow HSO_4^-(aq) + H_3O^+(aq)$$

In contact with a base, water behaves as an acid:

$$\underset{\text{acid}}{H_2O(1)} + \underset{\text{base}}{CO_3^{2-}(aq)} \longrightarrow HCO_3^-(aq) + OH^-(aq)$$

Other examples of amphoteric substances will be encountered elsewhere in the text.

14.14 electron transfer at long range: electrical energy from chemical reactions

In displacement reactions (section 14.5) electrons are exchanged by direct

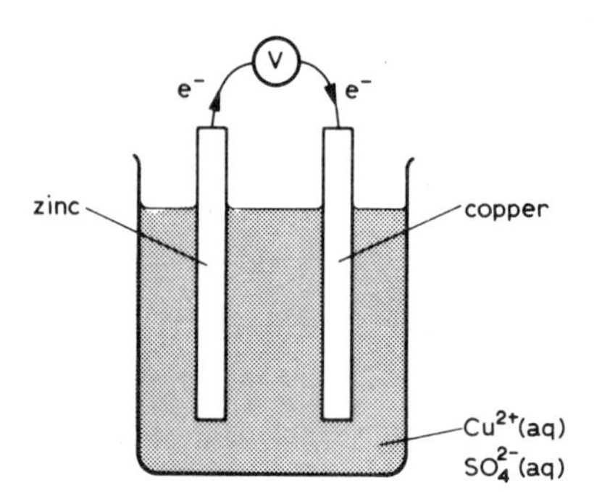

Fig. 14.4 a simple cell

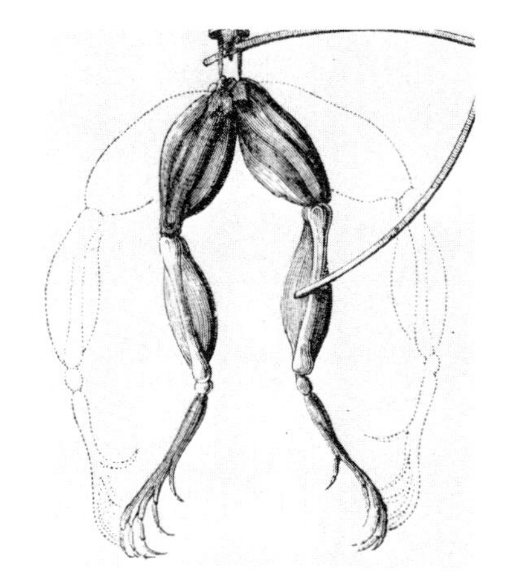
Fig. 14.5 Galvani's experiment with frogs' legs

Fig. 14.6 early voltaic pile in the basement of the Royal Society

contact between cations and metals. Since electrons can travel in metals, it seems reasonable to suppose that electron exchange may take place, without direct contact, through a conductor. This is the basis of **electric cells.**

Sheets of copper and zinc are placed in dilute aqueous copper(II) sulphate(VI) and connected externally through a voltmeter as shown in Fig. 14.4. Deflection of the needle shows that electron flow is taking place in the external circuit. This is a very simple electric cell. Similar simple cells may be made with any two different metals and any electrolyte solution. These conditions create a potential difference between the two different metals. Pairs of metals used in this way are sometimes called **galvanic couples.**

14.15 Galvani and Volta

Luigi Galvani (1737–1798) is credited with the discovery of current electricity. He found, quite accidentally, that the muscles of a frog's leg contracted violently when touched simultaneously by two different metals the other ends of which were in contact. After further experiments, he proposed that the effect was due to electricity originating in the frog's tissues. Until that time only static electricity, produced by rubbing various materials together, was known. Alessandro Volta (1745–1827) at first agreed with Galvani but, on extending the experiments, decided that the electricity somehow came from the metals. Eventually he produced one of the most influential inventions of science–the **Voltaic Pile.** This was the first electric 'battery' or collection of cells.

14.16 how a cell works

The sheets of metal in a simple cell are often called the **poles** or **terminals** of the cell and when the cell is working, electrons flow *from the negative to the positive pole* through the external circuit. From the direction of deflection of the voltmeter needle using the zinc/copper/copper(II) sulphate(VI) cell, it is clear that zinc is the negative pole and copper the positive pole. While electron flow continues, the reaction occurring at the zinc pole (−) releases electrons into the external circuit and hydrated zinc(II) ions into solution; the zinc pole loses mass. This is a process of oxidation:

$$Zn(s) \rightarrow Zn^{2+}(aq) + 2e^- \text{ (to circuit)}$$

Meanwhile, at the copper pole (+), copper(II) ions from the solution take up electrons and are discharged as copper metal; the copper pole gains mass. This is a process of reduction:

$$Cu^{2+}(aq) + 2e^- \text{ (from circuit)} \rightarrow Cu(s)$$

These two processes allow a continuous flow of electrons in the external circuit from the zinc pole (−) to the copper pole (+). (However, remember that conventional electric current is regarded as a flow of positive charge from (+) to (−) pole.)

In this working cell the only permanent changes are the loss of zinc from the negative pole and of copper ions from the solution. Effectively this is a redox (displacement) reaction:

$$Zn(s) + Cu^{2+}(aq) \rightarrow Zn^{2+}(aq) + Cu(s)$$

In fact all electric cells function by *separating a redox reaction into its two half-reactions and allowing the electron transfer to take place through the external circuit.*

14.17 movement of ions inside a cell: cells and electrolysis

Confusion sometimes arises between an electric cell, which generates an electric current from a chemical change, and the process of electrolysis, in which chemical changes in an electrolyte are induced by an electric current supplied from an external source. In particular, the polarity (positive and negative signs) of the terminals or electrodes in the two systems may be confused.

'Anode' always means the same thing: **the place where electrons move up** (*ana* = *up*), i.e. out of the liquid into the external circuit. **'Cathode'** (*cata* = *down*) is always **the place where electrons move down,** from the external circuit into the liquid. Thus the anode is positively charged *in relation to the liquid* and negative ions (anions) in the liquid move towards it; the cathode is negatively charged *in relation to the liquid* and positive ions (cations) move towards it. This is true both for an electrolysis (Fig. 14.7(a)) and a cell producing electricity (Fig. 14.7(b)). The movement of ions in the liquid of a cell constitutes the internal current of a cell circuit.

Since electrons move up the anode, leaving it positively charged with respect to the liquid, the anode will be *negatively* charged *with respect to the external circuit*–this is true both for an electrolysis and a cell (see Fig. 14.7(c)). This fact is largely ignored in electrolysis, as we are mainly concerned with changes in the liquid and therefore think of an anode as a positive pole–its negative end, joined to the wire, is irrelevant. In a cell, the charge with respect to the external circuit is all-important: **the anode is the negative terminal of the cell and the cathode is the positive terminal *in relation to the external circuit.***

The following summary may be helpful. The features of little importance which are easily overlooked are in brackets, those of importance are in bold print.

	ELECTROLYSIS	CELL
anode:	electrons move up	electrons move up
	+ in relation to liquid	(+ in relation to liquid)
	anions move towards it	anions move towards it
	(− in relation to circuit)	**− in relation to circuit**
cathode:	electrons move down	electrons move down
	− in relation to liquid	(− in relation to liquid)
	cations move towards it	cations move towards it
	(+ in relation to circuit)	**+ in relation to circuit**

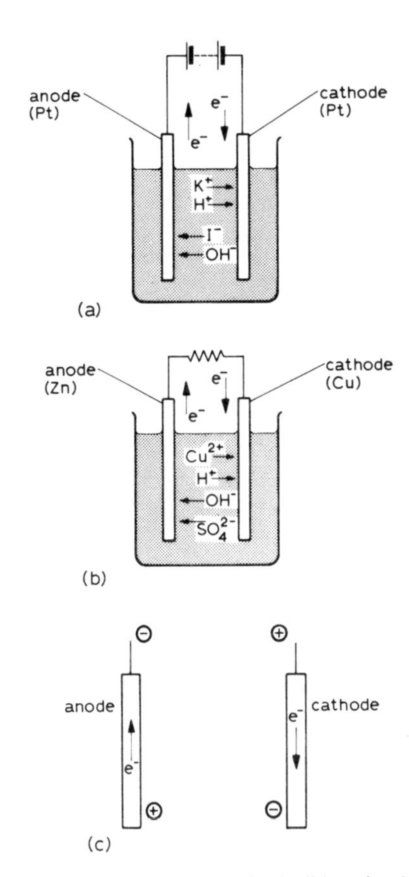

Fig. 14.7 (a) electrolysis of KI(aq), (b) a simple cell, (c) the two polarities of each electrode in both electrolysis and cell

14.18 electromotive force (e.m.f.)

The maximum potential difference (p.d.) between the terminals of a cell is obtained when no current is delivered and is called the **electromotive force (e.m.f.)** of the cell. When a current is delivered, there is a flow of electricity through the liquid in the cell (internal current) and this requires a potential drop within the cell; the greater the current flowing, the greater is this internal potential drop. It follows that the potential difference between the terminals through the external part of the circuit when a current is delivered is lower than the e.m.f. of the cell.

Even a high-resistance voltmeter needs a little current to make it function, but for most purposes in chemistry such an instrument records a p.d. sufficiently close to the cell e.m.f. to be acceptable. More precise measurements require a potentiometer circuit.

14.19 energy changes in cells: electrical energy

Energy changes during direct displacement reactions appear mainly as **heat** of reaction; when the same reaction is used to move electrons in a circuit it is producing **electrical** energy. The moving electrons do work, e.g. in causing electric motors, voltmeters, ammeters, light bulbs, electric heaters, or electromagnets to function. Only a limited amount of the total chemical energy of the reaction can be converted to useful electrical energy, even in the most efficient cells. The rest is 'wasted', mainly as heat. Much of the electrical energy eventually appears as heat in the machines mentioned above.

14.20 polarisation

The simple zinc–copper cell described in section 14.14 does not work efficiently and soon after the circuit is complete, a drop in p.d. is noticed on a voltmeter connected between the terminals. Examination shows layers of copper deposited on *both* copper and zinc poles; the displacement of copper from solution takes place at the zinc surface as well as at the copper surface:

$$Zn(s) + Cu^{2+}(aq) \longrightarrow Zn^{2+}(aq) + Cu(s)$$

Electrons which should have been released into the external circuit, if the cell functioned perfectly, are taken up by copper ions around the zinc pole. The build-up of reaction products on the poles of a cell changes the nature of the poles, causing a reduction in cell efficiency, the process being known as **polarisation.**

14.21 prevention of polarisation

If cells are to be useful and maintain a constant potential difference across the external circuit, polarisation must be avoided or greatly reduced. All cell technology is much exercised by this problem.

1. half-cells

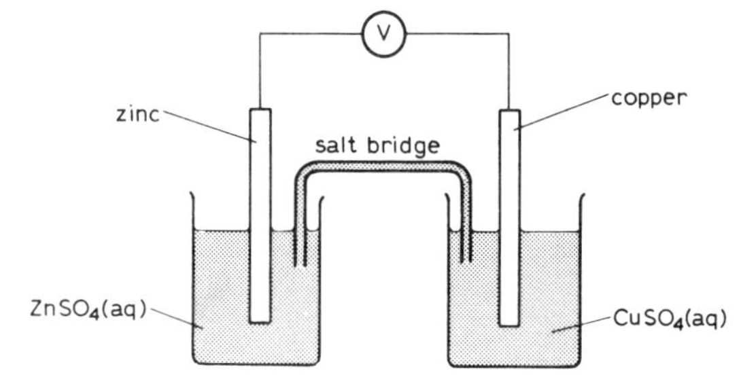

Fig. 14.8 zinc and copper half-cells

Polarisation may often be prevented by dividing a cell into two **half-cells,** with the poles in different solutions of suitable ions. The separate oxidation and reduction half-reactions take place in the half-cells. A zinc–copper cell contains a zinc half-cell of zinc metal in aqueous zinc(II) sulphate(VI) and a copper half-cell of copper metal in aqueous copper(II) sulphate(VI). The half-cells are connected internally by a **salt bridge**–a tube of electrolyte such as aqueous potassium(I) chloride in a jelly to restrict diffusion. Polarisation does not occur, but unfortunately the internal resistance is high and only small currents may be drawn from this type of cell. However, this arrangement is the most suitable for accurate measurement of the possible e.m.f. of a cell system.

2. Daniell cell

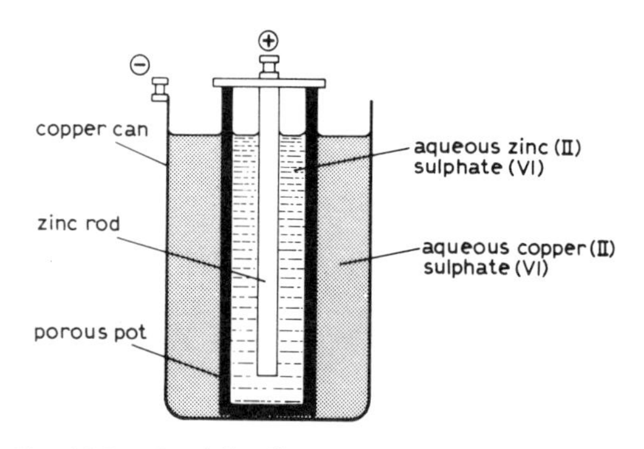

Fig. 14.9 a Daniell cell

A more satisfactory working cell was invented by J. F. Daniell (1790–1845) and the Daniell cell still finds a place in laboratories. Daniell made the copper container the positive terminal (cathode) and filled it with aqueous copper(II) sulphate(VI). Into this he placed a porous pot containing aqueous zinc(II) sulphate(VI) (or dilute sulphuric(VI) acid) and a zinc rod as the negative terminal (anode). Internal resistance is fairly low as the solutions are in contact in the pores of the pot. Copper(II) ions are kept away from the zinc surface unless the cell is left unused for a long time. Daniell cells should therefore be dismantled after use. Other cells may be manufactured using the same principle.

3. removal of hydrogen

When sulphuric(VI) acid is the electrolyte in cells, hydrogen is usually liberated at one or both poles and may cause polarisation.

$$2H^+(aq) + 2e^- \longrightarrow H_2(g)$$

Two fairly effective means of preventing polarisation in this case are:

a) *mechanical removal of hydrogen* by automatically brushing or agitating the poles, so that hydrogen bubbles do not collect on their surfaces;

b) *adding a soluble oxidant* to the electrolyte solution, e.g. potassium(I) dichromate(VI), to oxidise the hydrogen to water as it forms.

14.22 primary and secondary cells (storage cells)

1. primary cells

The Daniell cell will function until the zinc metal or copper ions disappear; fresh chemicals are then needed. Such cells are called **primary cells.** Another primary cell is the familiar **dry cell** used in torches. This is a dry form of the Leclanche cell, first developed in 1868, which still has many advantages–cheap materials, unspillable, capable of being made into almost any shape and being built into compact batteries. The negative pole is zinc, the positive pole is a 'cathodic mix' of manganese(IV) oxide and carbon, with a carbon rod conductor and an electrolyte of ammonium chloride paste. Each cell has an e.m.f. of about 1·5 volt and batteries of any desired voltage may be manufactured by connecting the appropriate number of cells in series.

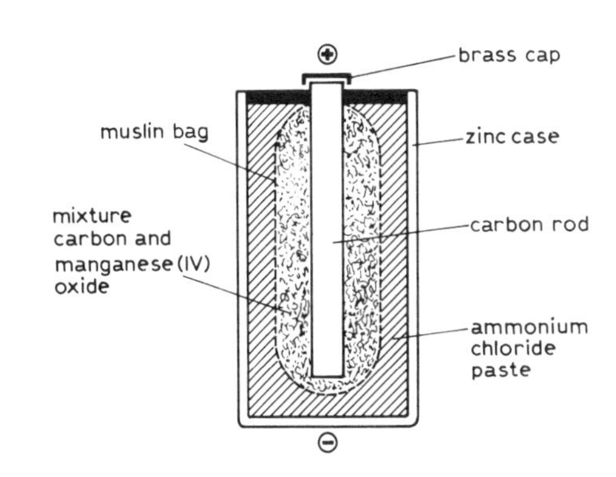

Fig. 14.10 a dry cell

Other important primary cells include the **Reuben–Mallory (R.M.)** cell, using zinc and mercury(II) oxide, which has a very stable, long-lasting e.m.f. of about 1·35 volt. It is most useful when very small cells are required, as in watches, hearing-aids or heart pace-makers.

Fig. 14.11 some uses of dry cells

2. secondary cells

Cells which may be re-charged by simply attaching the run-down cell to another source of potential difference such as a larger cell, power-pack or battery charger, are called **secondary cells.** The larger potential difference drives electrons in the *reverse* of the usual direction, also reversing the reactions which occur at the poles when the cell is working normally. This is a process of electrolysis using the cell poles as electrodes.

The best known secondary cell is the lead–acid cell used to provide the ignition spark for almost all petrol-driven engines and as the source of power in electric battery transport, e.g. milk-floats. A simple lead–acid cell is easily constructed.

If a voltmeter is attached across two sheets of lead dipping into dilute sulphuric(VI) acid, no potential difference is observed. The voltmeter is removed and a 6–9 volt battery or power-pack is attached across the lead sheets and electrolysis is allowed to occur for several minutes. A brown layer of lead(IV) oxide appears on the anode in this electrolysis and hydrogen is liberated at the cathode.

When the charger is removed and the voltmeter replaced, a p.d. of about 2V is registered, showing that the lead–acid system is functioning as a cell,

Fig. 14.12 car battery on charge

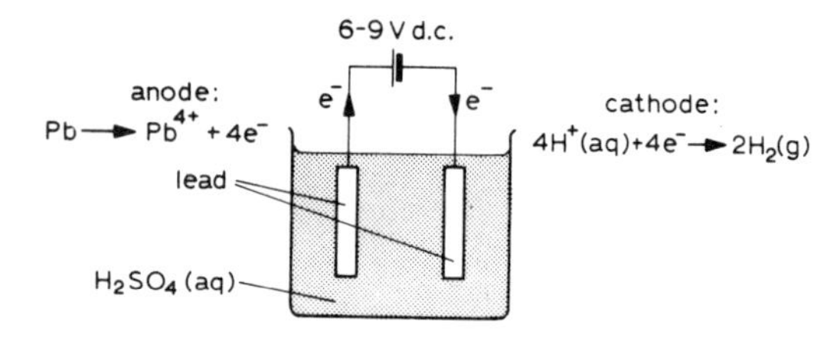

Fig. 14.13 developing a lead–acid cell by electrolysis ('charging'); the subsequent reaction at the anode is

$$Pb^{4+}(s) + 2H_2O(l) \rightarrow PbO_2(s) + 4H^+(aq)$$

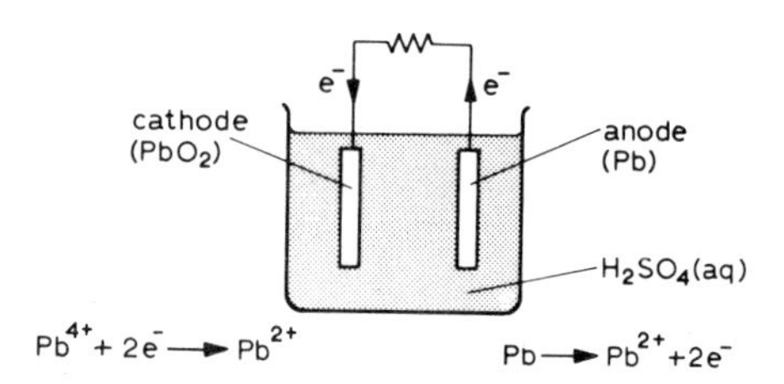

Fig. 14.14 drawing a current from a lead–acid cell ('discharging') – note that the anode and cathode are reversed compared with Fig. 14.13

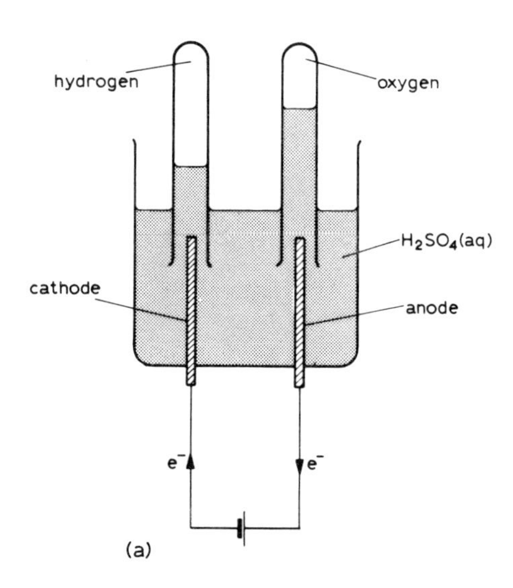

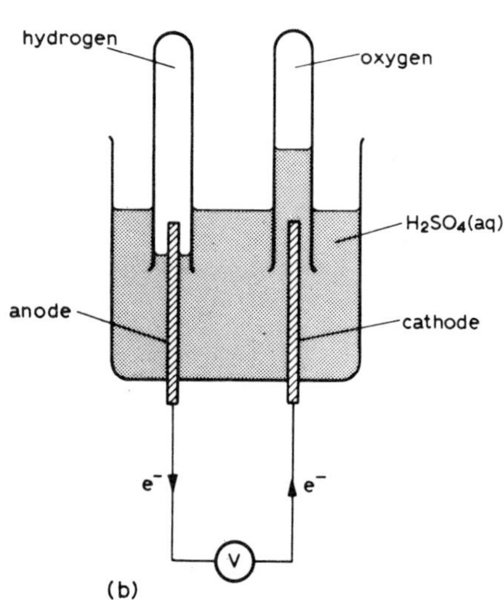

Fig. 14.15 (a) production and (b) use of a hydrogen–oxygen fuel cell

producing a flow of electrons in the reverse direction to that of the electrolytic circuit.

When the cell is working, lead(IV) oxide gradually disappears. It is thought that the cell reactions are:

negative terminal (*anode*):

$$Pb(s) \longrightarrow Pb^{2+}(aq) + 2e^-$$

followed by $$Pb^{2+}(aq) + SO_4^{2-}(aq) \longrightarrow PbSO_4(s)$$

positive terminal (*cathode*):

$$PbO_2(s) + 4H^+(aq) + 2e^- \longrightarrow Pb^{2+}(aq) + 2H_2O$$

followed by $$Pb^{2+}(aq) + SO_4^{2-}(aq) \longrightarrow PbSO_4(s)$$

Despite the production of water in the cell reaction, there is usually a small overall loss of water, due to electrolysis and some evaporation. For this reason, distilled water is added occasionally to car batteries, using a density-measuring hydrometer to decide on the exact dilution.

14.23 fuel cells

Metal–metallic ion reactions form only one type of redox process. Metals are becoming increasingly expensive and research into methods of using cheap materials for cells has been intense in recent years. Oxidisable materials like petroleum and coal are relatively abundant and inexpensive. It is preferable to oxidise these materials in cells, converting their chemical energy directly into electricity, rather than to oxidise them by burning, and using the heat energy evolved to change water to steam for driving electric generators. The use of the latter many-stage process gives an inefficient conversion of chemical energy from fuels to electricity – less than 50%. Direct conversion in an electrochemical **fuel cell** should give high efficiency.

The principle of a hydrogen–oxygen fuel cell may be illustrated in the following way. Electrolysis of dilute sulphuric(VI) acid or aqueous sodium(I) hydroxide in the apparatus shown in Fig. 14.15 produces hydrogen at the cathode and oxygen at the anode. When electrolysis is stopped, a voltmeter attached across the electrodes shows a p.d. of about 1 V. The cell reactions are:

positive terminal (*cathode*): reduction of oxygen

$$\tfrac{1}{2}O_2(g) + H_2O(l) + 2e^- \longrightarrow 2OH^-(aq)$$

negative terminal (*anode*): oxidation of hydrogen

$$H_2(g) \longrightarrow 2H^+(aq) + 2e^-$$

The overall reaction is

$$H_2(g) + \tfrac{1}{2}O_2(g) \longrightarrow H_2O(l); \quad \Delta H = -285{\cdot}8\,kJ\,mol^{-1}$$

Clearly it would be most desirable if a cell could convert continuously the large chemical energy of this reaction efficiently into electrical energy, particularly as:

a) hydrogen and oxygen are abundant,
b) the products of reaction are harmless,
c) there are no moving parts to break,
d) the generating process is noiseless.

Such cells are being developed (pioneered by F. T. Bacon at Cambridge) and improved versions of 'Bacon cells' have found their place in the spacecraft of the Apollo Moon Programme. Hydrogen is still fairly expensive and for really cheap fuel cells, other oxidisable materials must be employed, e.g. petroleum, alcohols and natural gas (methane). Most research goes into finding effective catalysts which work at average atmospheric temperatures, and this is partly why we do not yet see silent, non-polluting cars, powered by the oxidation of petroleum in a fuel cell. It is a hope for the future.

14.24 the electrochemical series

1. reference electrodes and electrode potentials

In a simple zinc–copper cell, the zinc rod acts as the anode and (since it releases electrons to the external part of the circuit) as the negative terminal of the cell:

$$Zn(s) \longrightarrow Zn^{2+}(aq) + 2e^-$$

The potential difference between the copper (positive) and zinc (negative) terminals of the cell, which can be measured using a voltmeter, is the difference in electrical potential between the copper and zinc terminals. Electricity is formally regarded as flowing from a place of high potential (positive) to a place of low potential (negative).

The more reactive a metal is, the more readily it will form cations, releasing electrons to the external circuit, and the more negative will be the value of its electrical potential. Magnesium is more reactive than zinc

$$Mg(s) \longrightarrow Mg^{2+}(aq) + 2e^-$$

so that, if magnesium and copper are connected in a cell, the potential difference registered is greater than that for a zinc–copper cell. The magnesium pole, or electrode, is said to have a higher negative **electrode potential** than zinc.

Very useful information may be obtained by comparing the e.m.f.'s of cells made from different pairs of metals. For convenience, one metal is used as a 'reference' pole (often called a **reference electrode**) throughout the comparison. Copper is a good choice and the e.m.f. measurements give the electrode potentials of the other metals compared with copper. Using the apparatus shown in Fig. 14.16, even very reactive metals like sodium and calcium can be compared.

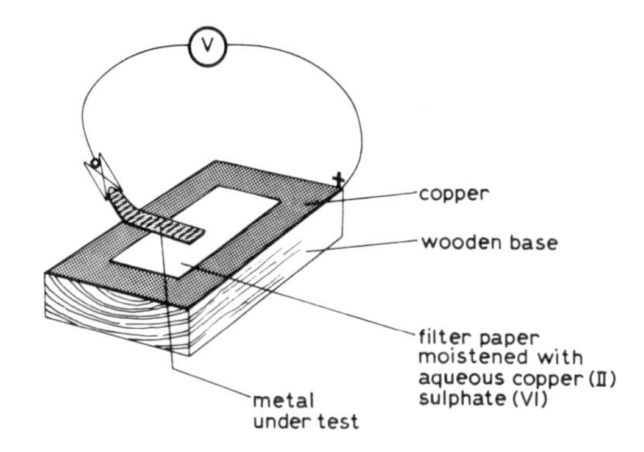

Fig. 14.16 apparatus for measuring the e.m.f. of copper–metal cells

All metals should be cleaned or freshly cut. A clean copper sheet 4 cm × 4 cm is stuck to a wooden base with a screw terminal and used as the reference electrode. A piece of filter paper is moistened with aqueous copper(II) sulphate(VI) and laid on the copper sheet. It must not be too wet. A piece of the metal under study (the other pole of the cell) is connected via a clip and wire to a voltmeter already attached to the copper terminal. The metal is then pressed down firmly on the damp filter paper. The cell and circuit are now complete: the voltmeter reading is noted as the cell e.m.f. It may be necessary, with some metals, to interchange the connections to the voltmeter. The positive pole of the cell (the *less* reactive metal) must be connected to the positive terminal of the voltmeter if the instrument is to function correctly. This therefore indicates whether the metal is the positive or negative pole of the cell with copper.

Sodium must be treated carefully and quickly and not touched with the hands. For aluminium, a little dilute hydrochloric acid in the electrolyte helps to remove the problem of the strongly adhering oxide layer on the surface of the metal. A table of e.m.f. values determined in this way should be drawn up, showing the metals in order of their electrode potentials compared with copper, from the most negative to the most positive.

2. standard electrode potentials and the electrochemical series

For various reasons (which may be found in more advanced texts) a hydrogen electrode is normally used rather than copper as the standard reference electrode. Fig. 14.17 shows a circuit for measuring the e.m.f. of a cell which includes the **standard hydrogen electrode.** The measured e.m.f. gives the value of the standard electrode potential of the other electrode in the cell, as the potential of the standard hydrogen electrode is assigned a value of zero. When the metals and hydrogen are placed in order of their standard electrode potentials, as in Fig. 14.18, they form part of an important series called the

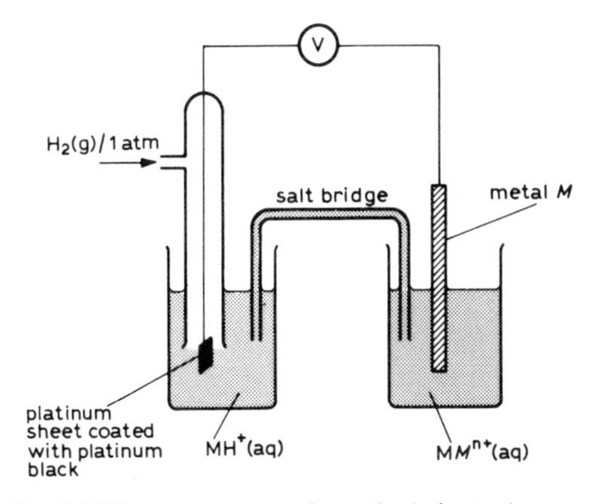

Fig. 14.17 measurement of standard electrode potential of a metal M

element	$E^{\ominus}$/V
K	−2·9
Ca	−2·87
Na	−2·71
Mg	−2·4
Al	−1·66
Zn	−0·76
Fe	−0·44
Sn	−0·14
Pb	−0·13
H	0·00
Cu	+0·34
Hg	+0·80
Ag	+0·81
Pt	+1·2
Au	+1·5

Fig. 14.18 the electrochemical series

Electrochemical Series. The full Electrochemical Series contains many other chemical species (see (5), later in this section).

It is important to remember that for any pair of elements chosen from this series, the element *higher* in the series (that with the more negative electrode potential) forms the *negative* pole of the cell, and the *lower* element is the *positive* pole.

e.g. zinc is the negative pole of a cell with copper;
copper is the negative pole of a cell with silver;
silver is the negative pole of a cell with gold.

3. electropositive metals

When a cell is working, the negative pole releases electrons and forms hydrated cations:

$$M(s) \longrightarrow M^{n+}(aq) + ne^-$$

The electrochemical series allows us to compare how readily the process of forming hydrated cations occurs. Metals high in the series form hydrated cations more readily than those below them. The former metals are said to be highly **electropositive**, since they form positive ions easily during reactions. Conversely, metals low in the electrochemical series (the less electropositive metals) have hydrated cations which are more easily reduced to the metal than those of metals above them in the series.

4. the electrochemical series and other reactivity series

It is hardly surprising that other reactivity series of metals, derived by experiment, are closely related to the electrochemical series. Most reactions of metals in solution involve the loss of electrons (oxidation) from metal atoms, with the formation of hydrated cations. The more electropositive metals are good reducing agents and reducing ability decreases down the series.

a) *metal–metallic ion displacement reactions* (section 14.5)

Metals are displaced from solutions of their salts only by other metals above them in the electrochemical series. Gold is displaced from gold salt solutions by all other metals, since gold cations are the most easily reduced of all.

b) *displacement of hydrogen from water and acids* (sections 8.14, 8.15)

Only the metals which are above hydrogen in the electrochemical series (more electropositive than hydrogen) reduce hydrogen ions to the gas:

$$2H^+(aq) + 2e^- \longrightarrow H_2(g)$$

Metals placed far above hydrogen in the series (the most electropositive metals – potassium, sodium and calcium) displace hydrogen from cold water. Metals nearer hydrogen displace it from steam and acids (magnesium, iron and zinc). Aluminium, normally protected by its tough oxide layer, displaces hydrogen from acids and some alkalis, but not from water. Metals below hydrogen in the electrochemical series (copper, mercury, silver) do not displace hydrogen from water, acids or alkalis.

K, Ca, Na } react with cold water
Mg, Al, Zn, Fe } react with steam
Sn
Pb
H
Cu, Hg, Ag, Pt, Au } do not react with non-oxidising acids

Fig. 14.19 displacement of hydrogen from water and acids

c) *preferential discharge of cations during electrolysis* (section 13.17)

Cations of metals low in the electrochemical series accept electrons and are reduced more easily than cations of metals higher in the series. Thus one criterion for predicting which cation in a mixture is most likely to discharge first during electrolysis is the position in the electrochemical series; the lower in the series the more likely it is to discharge first.

$$M^{n+}(aq) + ne^- \longrightarrow M(s)$$

d) *discovery of metals*

Historically, metals were discovered in approximately the reverse order of the electrochemical series. All metal extractions are based on reduction techniques (section 27.3) and the most easily reduced compounds are those of metals low in the series. Gold, silver and mercury have been known since the earliest times; copper, lead and tin started the Bronze Age. The Iron Age followed, when more powerful reduction techniques, using carbon at high temperatures, were discovered. The very electropositive metals, potassium, sodium and calcium, are highly reactive and their ions are among the most stable known. Only when the most powerful reduction technique of all–electrolytic reduction at a cathode–became available, were these metals liberated from their compounds.

5. redox series

The electrochemical series described in this section is only a part of a much larger series covering all elements, ions and molecules capable of losing or gaining electrons. Electrochemical cells may be produced from combinations of the standard hydrogen electrode with a half-cell containing any oxidation or reduction system. The e.m.f. values of these cells give electrode potentials for all the redox systems and are most useful in many aspects of chemistry–particularly predictions of oxidising or reducing ability. A detailed treatment of the redox series is beyond the scope of this book, but may be found in more advanced texts.

14.25 the rusting of iron and its prevention

Iron, one of the most important manufacturing materials, is a fairly reactive metal and its oxidation to powdery, reddish, hydrated iron(III) oxide (rust) is unfortunately all too familiar. Rusting and its prevention cost the nation many millions of pounds every year.

1. conditions which promote rusting

These conditions are easily demonstrated. Oxygen and water are the chief suspects and sets of test-tubes are arranged to contain two or three clean iron nails each, in a variety of environments:

i) ordinary atmospheric conditions (damp air)
ii) air plus water (Fig. 14.20(a))
iii) dry air (Fig. 14.20(b))
iv) water from which all air has been boiled (Fig. 14.20(c))
v) dry oxygen
vi) oxygen plus a little air-free water.

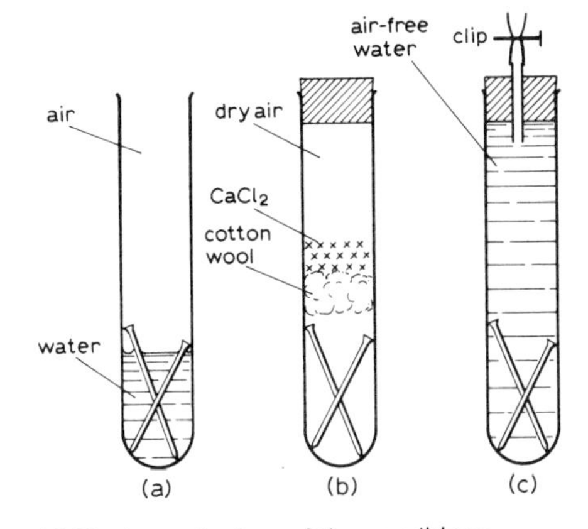

Fig. 14.20 determination of the conditions necessary for rusting

Other gases which are constituents of air may be tried as controls–both dry and damp nitrogen, carbon dioxide and noble gases if available. However, only in (i), (ii) and (vi) does any appreciable rusting take place. **Rusting requires the presence of both oxygen and water**; individually they are not effective. If the water contains some dissolved ionic compound such as sodium(I) chloride, extensive rusting takes place much faster. This is why cars rust more quickly near the sea or after driving over 'salted' roads in winter.

2. electrochemical nature of the rusting process

Rusting involves several stages and two factors about the surface of iron under a water layer help to start the process. The water usually contains some dissolved electrolyte. The two factors are:

a) imperfections in the iron caused by impurities or mechanical strain,
b) differences in the amount of dissolved oxygen in contact with the surface.

Both of these factors cause some areas of the iron to become **anodic** where there is a **lower oxygen concentration** and some areas to become **cathodic** where there is a **higher oxygen concentration.**

In the anodic areas, iron atoms are oxidised by losing electrons to the cathodic areas. Iron(II) ions are formed and migrate into the solution:

$$Fe(s) \rightarrow Fe^{2+}(aq) + 2e^-$$

The iron(II) ions are oxidised by dissolved oxygen to iron(III) ions and these combine with hydroxide ions to give a layer of hydrated iron(III) oxide (rust).

In the cathodic areas, dissolved oxygen takes up the electrons coming in from the anodic areas and is reduced in the presence of water to hydroxide ion:

$$\tfrac{1}{2}O_2(aq) + H_2O(l) + 2e^- \rightarrow 2OH^-(aq)$$

This should explain why rusting requires both water and oxygen. Rust itself does not adhere well to the metal surface and affords little protection of the metal surface to further attack.

3. prevention of rusting

a) *using grease and paint*

Simple protection is given by any coating which prevents both air and water from reaching the iron surface. Oils, greases and paints are quite effective, but need renewing frequently, paint in particular may flake off in unsuspected places. Other commercial 'rust-preventors' combine with the rust itself to form a hard, water-resistant layer. If breaks do occur, rust formation is greatest in the anodic areas *under* the grease or paint layer (i.e. remote from oxygen). The exposed surface is the cathodic area where the oxygen concentration is higher.

b) *galvanising: sacrificial corrosion*

The important stage of rusting to prevent is the oxidation of iron to iron(II) ions in the anodic areas. Perhaps the most effective method involves coating the surface with a layer of zinc; this is known as **galvanising.** If the zinc layer is broken and the two metals are exposed to a conducting solution (such as impure rain-water), the conditions for a short-circuited simple cell are present. As seen from their relative positions in the electrochemical series, zinc is the negative pole of the cell, iron the positive. The likely cell reactions are:

negative pole: $Zn(s) \rightarrow Zn^{2+}(aq) + 2e^-$
positive pole: $2H^+(aq) + 2e^- \rightarrow H_2(g)$

Electrons are transferred from the zinc to the iron, where they are taken up by hydrogen ions. **It is the zinc which is oxidised, not the iron.** This is sacrificial corrosion of the zinc and is widely used for the protection of iron articles. During the process the solution becomes alkaline at the positive pole and the iron surface becomes coated with a protective layer of basic zinc(II) carbonate.

Magnesium is another suitable metal for sacrificial corrosion and bars of magnesium are bolted to the sides of iron ships or to underground iron pipes.

c) *tinning of iron*

'Tin' cans consist of a thin layer of tin upon an iron can. Tin is more resistant to the chemical attack of acids in fruit and vegetable juices than iron. However, tin is below iron in the electrochemical series; if a tin can is scratched, exposing the two metals to the electrolyte solution, iron becomes the *negative* pole of the iron–tin cell and is oxidised (i.e. it rusts) even more quickly than if it had not been tin-plated. Fortunately iron(II) ions are harmless to humans. Zinc compounds are poisonous and zinc-coated cans are never used for containing foodstuffs.

revision summary: Chapter 14

precipitation reactions:	write formula for precipitated compound *first*, then work back to contributing ions
an acid:	a substance which can act as a source of protons
a base:	a substance which can accept protons
acid–base reaction:	transfer of proton(s) from acid to base
oxidation + reduction:	a substance is oxidised if it loses electrons reduced if it gains electrons oxidation and reduction occur together write each half-equation, then combine
cells:	method of converting chemical $\rightarrow$ electrical energy directly work by separating redox reaction into two half-reactions so that electrons transferred through circuit
electrode potential:	e.m.f. of a metal–metallic ion system when connected to a standard hydrogen electrode

questions 14

1. What is present in all solutions of acids in water? Complete the following reaction equations between acids and bases:

a) $HI(aq) + H_2O(l) \rightleftharpoons ? + ?$
b) $NH_4^+(aq) + ? \rightleftharpoons H_2O(l) + ?$
c) $HBr(aq) + CN^-(aq) \rightleftharpoons ? + ?$
d) $CH_3CO_2H(aq) + ? \rightleftharpoons CH_3CO_2^-(aq) + ?$
e) $? + ? \rightleftharpoons CO_3^{2-}(aq) + H_2O(l)$

2. Write ionic reaction equations for:
a) the reaction between any dilute acid and aluminium;
b) neutralisation of ethanoic (acetic) acid by any alkali;
c) the action of sodium(I) carbonate solution with nitric(V) acid;
d) the precipitation of calcium(II) carbonate.

3. 'Metals have basic oxides; non-metals have acidic oxides.' Is this statement true for all cases? Explain your answer. Write ionic reaction equations for the reactions with water of the oxides of elements in Period 3 (sodium to argon) showing how the acid–base properties of oxides change across a period.

4. a) How do chemists distinguish between the terms 'strength' and 'concentration' of an acid? If solutions of hydrochloric and ethanoic (acetic) acids both have a pH of 4, does this mean that they are equally concentrated? Explain your answer.
b) Why are aqueous sodium(I) hydroxide and aqueous ammonia said to be 'strong' and 'weak' bases respectively?
c) Why does aqueous sodium(I) carbonate give a blue colour with Universal Indicator?
d) What is meant by an 'amphoteric' hydroxide?

5. 20 cm³ of a solution containing 0·06 g of a metal hydroxide (XOH) were exactly neutralised by 25 cm³ of 0·1 M hydrochloric acid.
(a) What is the formula weight of the metal hydroxide?
(b) What is the relative atomic mass of X?
(c) Write an ionic equation for the reaction.
(Relative atomic masses: H = 1·0, O = 16·0) (JMB)

6. Write half-reaction equations to show the following conversions. State whether the reaction is an oxidation or a reduction in each case.
a) chlorine molecules to aqueous chloride ions;
b) zinc atoms to aqueous zinc(II) ions;
c) aqueous iron(III) ions to aqueous iron(II) ions;
d) oxide ions to oxygen molecules;
e) bromide ions to bromine molecules;
f) acidified hydrogen peroxide to water;
g) acidified aqueous chlorate(I) ions to aqueous chloride ions.

7. Define the term *reduction*.
The following are classified as 'redox' reactions:
a) the action of carbon monoxide on heated lead(II) oxide;
b) the reaction between ethene (ethylene) and hydrogen;
c) the action of manganese(IV) oxide on warm, concentrated hydrochloric acid;
d) the result of bubbling chlorine into a solution of iron(II) chloride.
Justify the classification of each of these reactions, stating clearly why you consider the substances concerned to be oxidised or reduced. (O)

8. Write ionic equations *only* for each of the following:
(a) the reaction occurring when chlorine is passed into sodium(I) bromide solution,
(b) what occurs at the anode, and the cathode, when a solution of copper(II) sulphate(VI) is electrolysed using copper electrodes.

In (b) contrast the reactions at the anode and the cathode from the point of view of oxidation and reduction. (W)

9. Define the terms *oxidation* and *reduction*.

$$Cl_2 + 2H_2O + SO_2 = 2HCl + H_2SO_4;$$
$$Cl_2 + H_2S = 2HCl + S;$$
$$SO_2 + 2H_2S = 2H_2O + 3S.$$

From the above reaction equations, deduce which of the three gases chlorine, sulphur dioxide, and hydrogen sulphide, is (a) the strongest oxidising agent, (b) the strongest reducing agent. State clearly how you arrive at your answers.

Describe how you would carry out **two** reactions, the one using nitric(V) acid and the other sulphuric(VI) acid, to demonstrate that each of these acids has oxidising properties. In each case, explain why you classify the reaction as oxidation, and indicate how you would recognise the oxidised product of the reaction. (W)

10. Equation *A*: $Ag^+(aq) + Cl^-(aq) \longrightarrow AgCl(s)$
Equation *B*: $2H_2(g) + O_2(g) \longrightarrow 2H_2O(l)$

In the above equations, the physical states of the substances are shown in brackets. Thus, $Ag^+(aq)$ means an aqueous solution containing silver(I) ions, whilst (s), (g) and (l) signify solid, gaseous and liquid states respectively.

(a) Write *ionic* equations on similar lines for the following reactions:

(i) The addition of magnesium ribbon to copper(II) sulphate(VI) solution.
(ii) The addition of zinc to dilute hydrochloric acid.

(b) What simple test-tube reaction would you use to illustrate the reaction in Equation *A*? (W)

11. Explain the following observations. In each section write one equation for a reaction which occurs.

(a) When zinc is added to dilute sulphuric(VI) acid, only a slight reaction occurs. When a few drops of copper(II) sulphate(VI) solution are added, a rapid effervescence begins.

(b) When air-free solutions of sodium(I) hydroxide and iron(II) sulphate(VI) are mixed, a pale green precipitate is formed. When left to stand in contact with air, the precipitate slowly darkens and eventually becomes brown.

(c) When solutions of sodium(I) chloride and silver(I) nitrate(V) are mixed, a white precipitate is formed. On being left to stand in sunlight, the precipitate slowly darkens in colour.

(d) When two pieces of cotton wool, one soaked in concentrated hydrochloric acid and another in concentrated ammonia solution, are placed in the opposite ends of a long glass tube mounted horizontally and closed to prevent draughts, a white ring appears in the tube. The ring is formed about one-third of the way from the end of the tube containing the cotton wool soaked in hydrochloric acid. (JMB)

12. What do you understand by the term *reduction*?

Outline how you could convert (one method in each case):
(a) iron(II) ions (Fe^{2+}) into iron(III) ions (Fe^{3+});
(b) ammonium ions (NH_4^+) into ammonia;
(c) copper(II) ions (Cu^{2+}) into copper;
(d) zinc into zinc(II) ions) (Zn^{2+}).

Hydrogen peroxide solution (H_2O_2) reacts with acidified potassium(I) iodide according to the following equation:

$$H_2O_2 + H_2SO_4 + 2KI \longrightarrow K_2SO_4 + 2H_2O + I_2$$

Calculate the mass of iodine liberated when a solution containing 13·6 g of hydrogen peroxide is added to an excess of potassium(I) iodide solution acidified with dilute sulphuric(VI) acid. (S)

13. (a) Write an equation for the decomposition of hydrogen peroxide in the presence of manganese(IV) oxide, and calculate the volume of oxygen liberated (measured at s.t.p.) from a solution containing one mole of hydrogen peroxide.

(b) When potassium(I) manganate(VII) solution is added to hydrogen peroxide solution in the presence of dilute sulphuric(VI) acid, the reaction can be represented by the equation

$$2MnO_4^- + 5H_2O_2 + 6H^+ = 8H_2O + 5O_2 + 2Mn^{2+}$$

State what will be observed, and calculate the volume of oxygen (measured at s.t.p.) released from a solution containing one mole of hydrogen peroxide.

(c) State what occurs when a solution of iron(II) sulphate(VI) acidified with dilute sulphuric(VI) acid is gently warmed with an excess of hydrogen peroxide solution and suggest an equation for the reaction. What reagent would you apply to the original and final solutions to confirm your statement and what would you expect to observe? (O)

14. Describe what you would see and state whether hydrogen peroxide is acting as an acid, oxidising agent, or reducing agent in each of the following:

(i) the reaction with silver(I) oxide

$$Ag_2O(s) + H_2O_2(l) \longrightarrow 2Ag(s) + H_2O(l) + O_2(g)$$

(ii) the reaction with barium(II) hydroxide solution

$$Ba(OH)_2(aq) + H_2O_2(aq) \longrightarrow BaO_2(s) + 2H_2O(l)$$

(JMB, part question)

15. Two plates, one of zinc and one of copper, held apart by a separator and connected to a small light bulb, are dipped into dilute sulphuric(VI) acid. The bulb lights up but the light soon becomes dim.

(a) What would be observed at the copper plate? Write an ionic equation for the reaction occurring.
(b) What would happen at the zinc plate? Write an ionic equation for the reaction.
(c) Explain why the light fades after a short time.
(d) If the zinc plate were replaced by an iron plate, would the lamp glow more or less brightly?
(e) If the zinc plate were retained but the copper plate were replaced by a silver plate, would the lamp glow more or less brightly? (JMB)

16. Carbon rods are placed in each of two beakers, one containing a solution of iron(II) chloride, the other 'chlorine water'. The solutions are connected by a salt-bridge and the carbon rods connected through a sensitive voltmeter. A reading is noted on the voltmeter.

(a) What reaction is being used in this cell?
(b) What half-reactions are taking place at each pole?

(c) In which direction do electrons flow in the external circuit?
(d) Which is the positive pole and which the negative pole of this cell?
(e) The salt-bridge contains potassium(I) chloride solution in jelly. In which directions do the potassium(I) and chloride ions move when the circuit is complete?

17. The standard electrode potentials of a number of metals in solutions of their doubly charged ions are:

	A	B	C	D	E
$E^{\ominus}$/V:	$-2{\cdot}6$	$+0{\cdot}3$	$-0{\cdot}5$	$+0{\cdot}8$	$-1{\cdot}2$

(a) Which two metals used in a cell gives the largest e.m.f.? Which is the positive and which the negative pole?
(b) What do the positive and negative signs of the electrode potentials mean?
(c) Which metal probably reacts most vigorously with dilute acids?
(d) What is likely to happen when (i) a piece of B is placed in a solution containing ions of D, (ii) a piece of C is placed in a solution containing ions of E, (iii) a piece of A is placed in a solution containing ions of B?

18. The following is a list of the symbols of some of the elements in order of an 'activity series': K, Mg, Al, Zn, Fe, H, Cu, Ag.
(a) Which of these elements will not displace hydrogen from a dilute acid?
(b) Which of these elements has the most stable hydroxide?
(c) A piece of zinc is placed in iron(II) sulphate(VI) solution and a piece of iron is placed in zinc(II) sulphate(VI) solution. In which solution would there be a reaction and why? Give the equation for the reaction.
(d) From these elements name (i) a metal which reacts with cold water, (ii) a different metal which reacts with hot water, but only very slowly with cold, (iii) any other metals which will react when heated in a current of steam.
(e) Name any metals in the list whose heated oxides can be reduced by hydrogen. For one of these metals give an equation for the reaction.
(f) If mixtures of aluminium(III) oxide and iron and of iron(III) oxide and aluminium are heated, in which mixture is there a reaction and why? Give the equation for the reaction. (AEB, part question)

19. What is meant by 'the electrochemical series of metals? How is this series related to the observations:
(a) during electrolysis, copper is deposited at a cathode in copper(II) sulphate(VI) solution but hydrogen is evolved at a cathode in sodium(I) sulphate(VI) solution;
(b) calcium reacts fairly vigorously with cold water, magnesium very slowly with cold water, zinc only with steam and copper with neither water nor steam;
(c) silver and copper were used as metals long before iron; calcium and sodium were discovered much later still.

20. Use the electrochemical series of metals to predict what happens when:
(a) zinc is placed in aqueous lead(II) nitrate(V);
(b) copper is placed in dilute hydrochloric acid;
(c) lead is placed in aqueous calcium(II) nitrate(V);
(d) copper is placed in aqueous silver(I) sulphate(VI);
(e) calcium is placed in aqueous copper(II) chloride.
Write ionic equations where appropriate.

21. (a) What is sacrificial corrosion? What qualities should a metal possess if it is to be sacrificed in preventing corrosion of iron on a large scale?
(b) Why is it important not to connect copper plates using iron bolts? May iron plates be safely connected using brass bolts?

22. (a) Design and describe a quantitative experiment to find out whether a new alloy is oxidised by the atmosphere. What results would you expect to observe if the alloy were
(i) easily oxidised,
(ii) rust resistant?
(b) If the alloy were attacked, what further experiments would you set up to discover which parts of the air cause the reaction?
(c) How is air prevented from attacking iron on
(i) the blades of a lawn mower,
(ii) a dustbin,
(iii) cutlery? (JMB)

23. Oxidation may be defined in terms of electron transfer. Discuss the application of this definition to the following:
(a) dissolving a metal in acid,
(b) burning lead in chlorine,
(c) the change in oxidation state of a metal ion,
(d) the rapid corrosion of a galvanised iron water tank when a copper water pipe is connected to it.
L (NUFFIELD)

24. 50 cm^3 of a solution of iron(III) chloride A containing 0·1 mol of Fe^{3+} per litre were heated until nearly boiling and then a solution of tin(II) chloride B (containing mol of Sn^{2+} per litre) was added slowly: when 2·5 cm^3 of the latter solution had been added the yellow colour of iron(III) solution had just been removed. At this stage it may be assumed that all the iron(III) chloride has been converted to iron(II) and all the tin(II) has been converted to an ion with a higher formal charge.
(a) How many moles of iron(III) were present in the 50 cm^3 of solution A?
(b) How many moles of tin(II) were present in the 2·5 cm^3 of solution B?
(c) How many ions of Fe^{3+} have reacted with one ion of Sn^{2+}?
(d) What is the formula of the tin ion which has been formed?
(e) Write an equation for the reaction.
(f) What alteration in structure takes place when Fe^{3+} changes to Fe^{2+}?
(g) The reaction does not seem to go at room temperature. Why does heating to near boiling cause it to take place?
(h) If some of solution B is electrolysed with a current of 1 A for 160 mins (i.e. 2·67 hrs), how many moles of tin would you expect to be deposited at the cathode?
(the Faraday constant = 96 500 coulombs = 26·7 A-hours)
(i) How do you explain the fact that when solution A is electrolysed it loses its colour but there is no immediate deposition of iron on the cathode?
L (NUFFIELD)

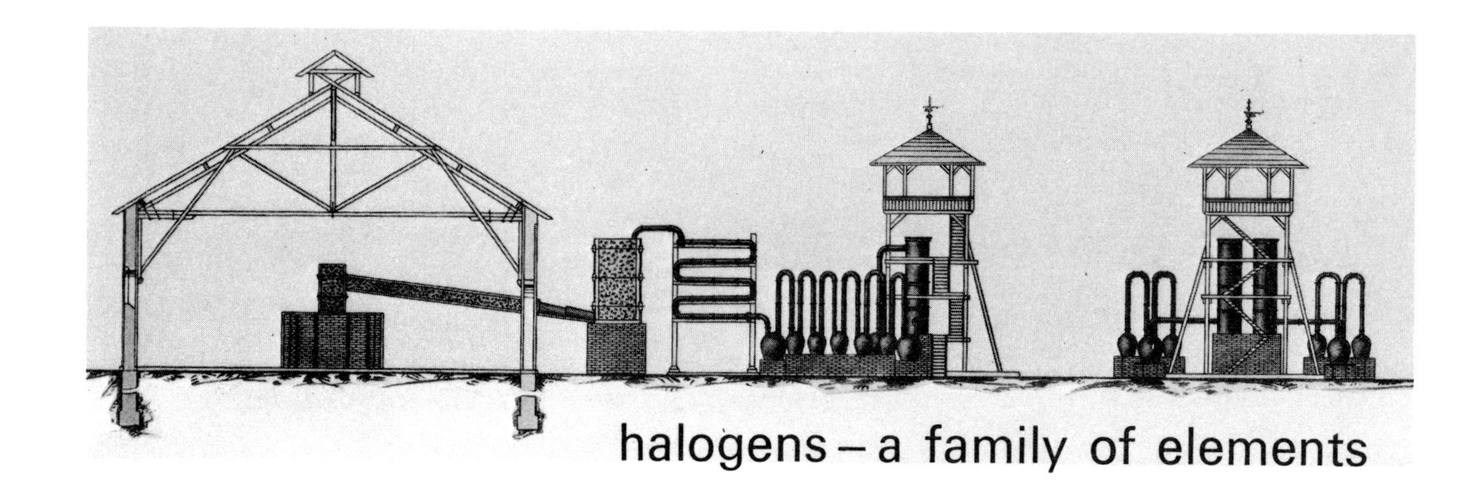

halogens – a family of elements

15.1 gas from salt

If a little common salt is sprinkled into a gas jar and a few drops of concentrated sulphuric(VI) acid are added, steamy fumes are seen at the mouth of the jar – even though the gas contained in the jar is quite clear and not at all steamy. A clue to this strange behaviour is found if the jar is inverted in a trough of water: the water soon rises into the jar.

Apparently the product of the reaction is a colourless gas which dissolves in water. It is so soluble in water that it will absorb water from moist air, forming tiny droplets of solution: this accounts for the appearance of steamy fumes.

By experiment and deduction, it is possible to find out quite a lot about this gas. We will refer to it as **salt gas** until its identity has been established.

The gas has a sharp smell and this gives a further clue, which is confirmed if a piece of moist, blue litmus paper is held in the gas: **it is acidic.** Since acids react with metals, *salt gas* may react with a metal.

15.2 action of *salt gas* on iron

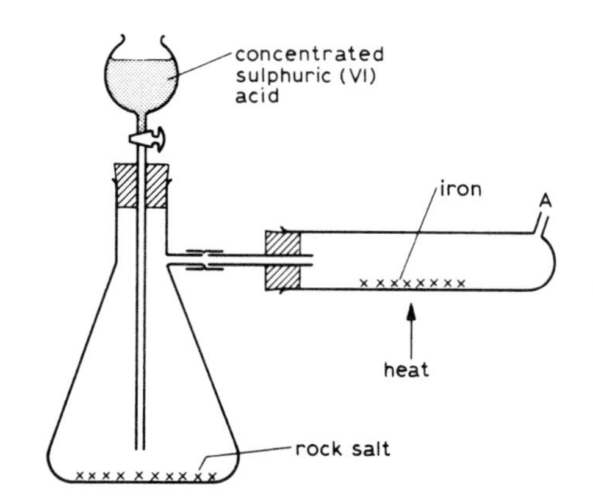

Fig. 15.1 action of 'salt gas' on iron

Concentrated sulphuric(VI) acid is added to rock salt (which produces less frothing than common salt) and the resulting *salt gas* is passed over heated iron filings as shown in Fig. 15.1. Application of a lighted splint at A shows the presence of an inflammable gas. A gas which burns and which is obtained from an acid substance by the action of a metal is surely hydrogen (see Chapter 8).

Thus *salt gas* contains hydrogen.

Some interesting crystals, which can be scraped out and tested, remain in the tube. They dissolve in water and addition of aqueous sodium(I) hydroxide to the resulting solution produces a green gelatinous (jelly-like) precipitate. Our knowledge of hydroxides (section 12.29) helps us to piece together a little more information about *salt gas*:

$$\textit{salt gas} + \text{iron} \longrightarrow \text{an iron(II) compound} + \text{hydrogen}$$

What else does *salt gas* contain? Perhaps if the hydrogen in it were removed, this question could be answered. Consider which substances react readily with hydrogen: such a substance might be used to remove hydrogen from *salt gas*.

15.3 action of *salt gas* on lead(IV) oxide

The apparatus shown in Fig. 15.1 may be used again for this experiment, using lead(IV) oxide in place of the iron filings. A pale green gas with a choking smell (*care–it is poisonous*) is formed and this gas will bleach moist litmus paper. Meanwhile, the brown lead(IV) oxide is coated white.

The name **chlorine** (meaning 'light green') is given to the gas and it can be shown to be an element. What evidence could be gathered to justify this statement?

We now know that

***salt gas* contains hydrogen and chlorine.**

If a jar of hydrogen and a jar of chlorine are placed mouth to mouth and the gases mixed, then a light is applied to the mixture, a violent explosion occurs and a colourless acidic gas is formed which produces steamy fumes in moist air–*salt gas*. Since *salt gas* can be made (synthesised) from the elements hydrogen and chlorine and nothing else, ***salt gas* contains hydrogen and chlorine only,** and it is named **hydrogen chloride.**

15.4 the formula for hydrogen chloride

Tubes A and B (Fig. 15.2) are of equal volume. Tube A is filled with hydrogen and tube B with chlorine, both at atmospheric pressure. Taps 1 and 3 are closed and tap 2 is opened, allowing the gases to mix. The mixture should not be exposed to direct sunlight or an explosion could occur.

After several days, if tap 1 is opened under mercury, no mercury enters the apparatus and no gas escapes: the contents of the apparatus are thus shown to be at atmospheric pressure and therefore if a reaction has occurred it has not caused a change in volume. Tap 1 is closed and reopened under water containing a little potassium(I) iodide. The water fills the apparatus completely, but no brown colour appears in the solution. (Chlorine will give a brown colour with aqueous potassium(I) iodide–see section 15.16(5).) The following deductions can be made from the observations described and it is instructive to trace which observation leads to each conclusion:

i) no hydrogen remains
ii) no chlorine remains
iii) only hydrogen chloride remains
iv) 1 volume of hydrogen reacts with 1 volume of chlorine, producing 2 volumes of hydrogen chloride.

	hydrogen +	chlorine $\rightarrow$	hydrogen chloride
	1 volume +	1 volume $\rightarrow$	2 volumes
(*Avogadro*)	n molecules +	n molecules $\rightarrow$	2n molecules
	1 molecule +	1 molecule $\rightarrow$	2 molecules
	H_2 +	$Cl_2 \rightarrow$	$2HCl$

Thus the formula for hydrogen chloride is HCl.

Additional evidence in favour of this formula is obtainable from measurement of the vapour density of the gas (see section 10.27), which is about 18, and from its mass spectrograph (see Chapter 9). After reading Chapter 9, the student should be able to sketch the sort of chart to be expected if hydrogen chloride were used in a mass spectrometer.

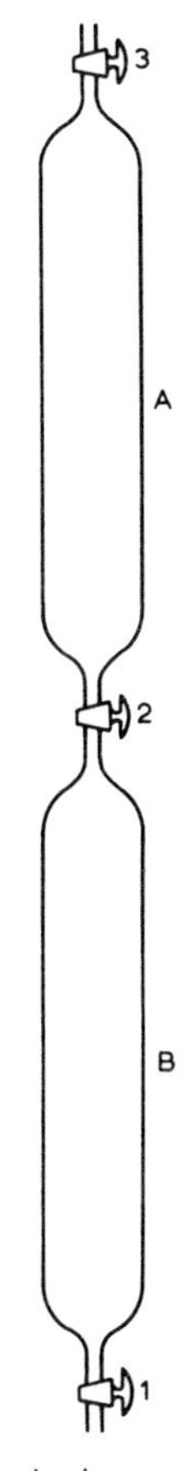

Fig. 15.2 formula for hydrogen chloride

15.5 the production and reactions of *salt gas* explained

It is now possible to write equations for the reactions quoted in this Chapter:

formation of salt gas

$$NaCl + H_2SO_4 \rightarrow NaHSO_4 + HCl$$

(*Why?* see section 12.19)

$$(\text{or } Cl^- + H_2SO_4 \rightarrow HSO_4^- + HCl)$$

$$Fe + 2HCl \rightarrow FeCl_2 + H_2$$

(*Why?* note hydroxide test)

action of salt gas on lead(IV) *oxide*

$$PbO_2 + 2HCl \rightarrow PbO + H_2O + Cl_2$$

but also:

$$PbO + 2HCl \rightarrow PbCl_2 + H_2O$$

overall:

$$PbO_2 + 4HCl \longrightarrow \underset{\text{(white)}}{PbCl_2} + 2H_2O + Cl_2$$

Lead(IV) oxide is an **oxidising agent**: it can add oxygen on to other substances. In this reaction it is oxidising hydrogen chloride **by removing hydrogen from it.** This extension of the idea of oxidation is discussed in Chapter 14.

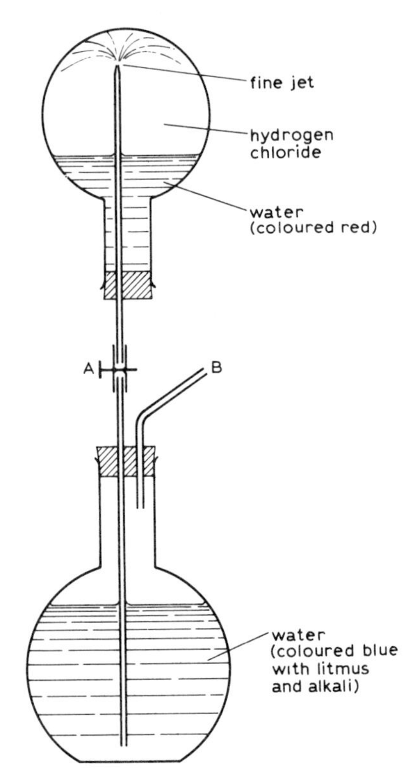

Fig. 15.3 the Fountain Experiment

15.6 solubility of hydrogen chloride: the fountain experiment

The solubility in water of hydrogen chloride can be demonstrated convincingly by filling a large flask with the gas and exposing it to water as illustrated in Fig. 15.3. Further insight into the process is obtained by colouring the water blue with litmus solution and a little aqueous sodium(I) hydroxide.

When the spring clip A is opened, little change is seen. It is usually necessary to blow gently through tube B until the water rises into the upper flask. As soon as the water passes the fine jet of the delivery tube, blowing should be discontinued: the water produces a vigorous fountain as it is drawn into the upper flask.

It is usually possible to see that the liquid which first rises into the upper flask is turned red, whereas that which enters last remains blue. The hydrogen chloride in the upper flask dissolves rapidly in the water which first enters, forming hydrochloric acid; a partial vacuum is created which draws up more water. Since the hydrogen chloride has all been dissolved in the first rush of water, no further reddening of the liquid is seen in the latter part of the experiment.

If the upper flask is completely filled with hydrogen chloride before the experiment, the flask at the end of the experiment should be full of liquid. This is rarely achieved in practice; any gas remaining in the upper flask will be air, the quantity present being a measure of the inefficiency with which the flask was 'filled' with hydrogen chloride. Why is it rarely possible to fill the flask with complete efficiency? What process is used to fill the flask?

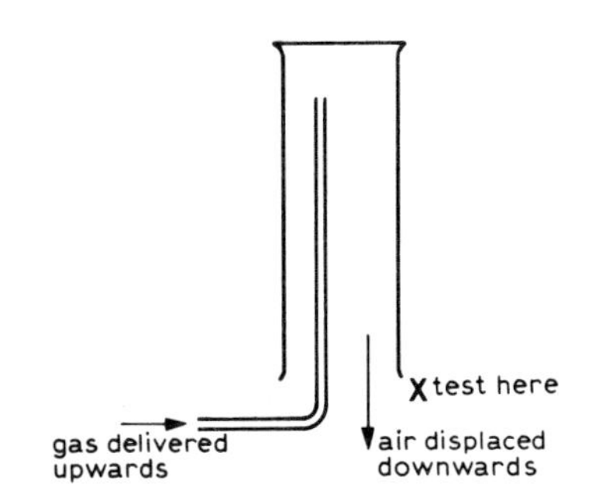

Fig. 15.4 downward displacement of air

15.7 collection of gases by displacement of air

Obviously hydrogen chloride cannot be collected over water, as is hydrogen (section 8.20) or oxygen (section 7.3). Mercury is often suggested as a liquid which will not dissolve the gas and which could be displaced in its collection: this is perfectly reasonable in theory, but suffers from the drawback that a volume of mercury similar to the volume of water used in the collection of hydrogen (section 8.20) would cost about £350; mercury is also very poisonous.

If a gas is soluble in water and its density differs significantly from that of air, it is usually collected by **displacement of air.** A less dense gas is delivered upwards and the more dense air is displaced downwards (Fig. 15.4, for example ammonia–see section 24.5). A more dense gas is delivered downwards and the less dense air displaced upwards (Fig. 15.5). This latter method is used for hydrogen chloride.

Note that it is always necessary when collecting a gas by displacement of air to include a test to show when the jar is 'full' of the gas. Some diffusion will always occur (see Chapter 3) and the jar will always contain some air: herein lies the explanation of the air remaining in the 'fountain' flask.

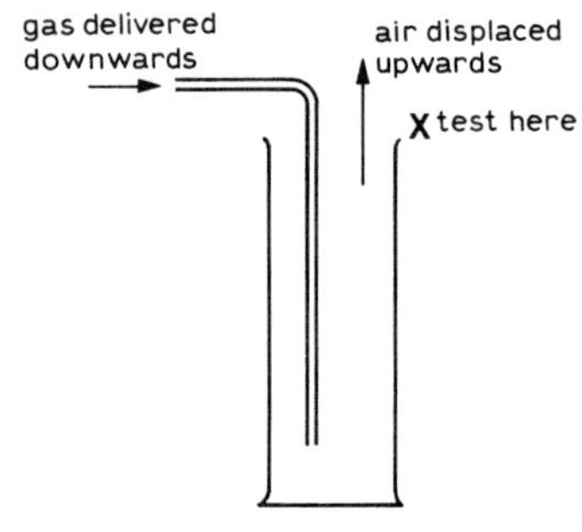

Fig. 15.5 upward displacement of air

A quick calculation can show whether the method of displacement of air can be used for the collection of a gas:

hydrogen:	density $\frac{1}{14} \times$ that of air	Downward displacement

ammonia:	density $\frac{8\cdot5}{14} \times 0\cdot6 \times$ that of air	Downward displacement
carbon dioxide:	density $\frac{22}{14} \approx \frac{3}{2} \times$ that of air	Upward displacement
hydrogen chloride:	density $\frac{18}{14} \approx 1\cdot3 \times$ that of air	Upward displacement (unlikely to be very efficient)
carbon monoxide:	density equal to that of air	Displacement of air cannot be used

gas	M_r	R.V.D.
air	$\approx$28	$\approx$14
H_2	2	1
NH_3	17	8·5
HCl	$\approx$36	$\approx$18
CO_2	44	22
CO	28	14

collection of hydrogen chloride and hydrochloric acid

Fig. 15.6 shows an apparatus which may be used for the collection of hydrogen chloride. The gas is generated from rock salt and concentrated sulphuric(VI) acid: the long rubber tube R permits the raising and lowering of the delivery tube in the jar, a glass or card cover restricts diffusion at the mouth of the jar and L shows the position in which moist blue litmus paper should be held so that it will turn red when the jar is full.

If a solution of hydrochloric acid is required, the delivery tube is disconnected at A and replaced by the arrangement shown in Fig. 15.7. The inverted funnel should dip just below the surface of the water to prevent 'sucking back' of liquid into the reaction flask as the gas dissolves. Visualise what would happen if the water did rise inside the funnel and contrast this with what would happen if the gas was passed into water through a delivery tube without using an inverted funnel. Finally, consider the consequences of water passing back into the reaction flask (refer to section 25.12 if necessary).

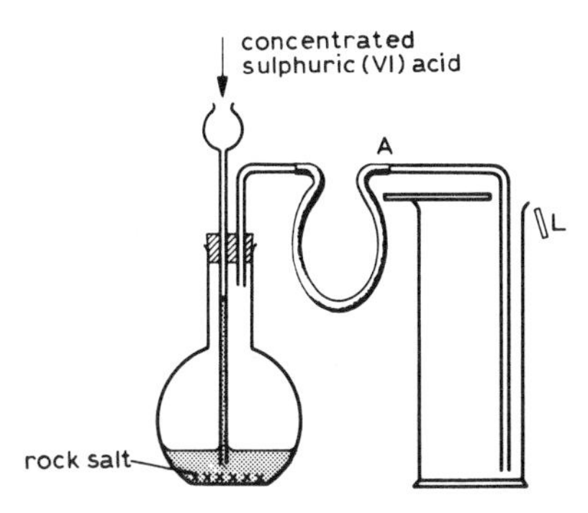

Fig. 15.6 preparation and collection of hydrogen chloride

15.8 physical properties of hydrogen chloride

Hydrogen chloride is a colourless gas with a choking smell; it is very soluble in water (*q.v.*) and fumes is moist air. It is rather denser than air (*q.v.*) and is readily liquefied under pressure.

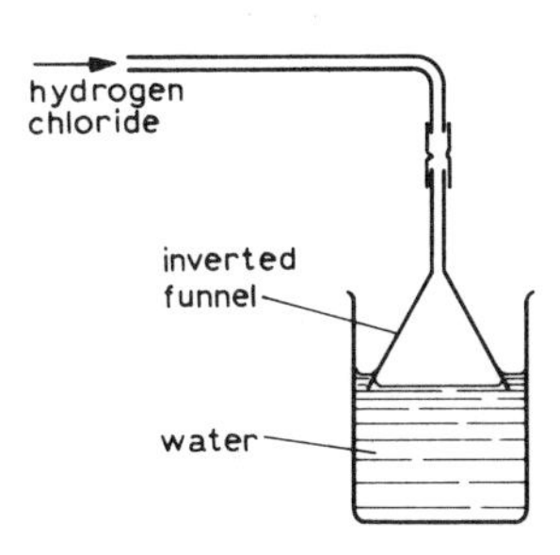

Fig. 15.7 collection of hydrochloric acid

15.9 reactions of hydrogen chloride

a) A lighted splint lowered into a jar of hydrogen chloride is extinguished: the gas will not support the combustion of the splint, neither will it burn. The bond joining the hydrogen and chlorine atoms in the molecule is a strong bond: further evidence for this is available from the energy released when hydrogen chloride is formed from its elements (section 15.13(2)). Hydrogen chloride is difficult to oxidise, though powerful oxidising agents (like lead(IV) oxide) can accomplish the oxidation.

b) Hydrogen chloride will react with certain metals giving chlorides and releasing hydrogen: e.g. magnesium gives magnesium(II) chloride:

$$Mg + 2HCl \rightarrow MgCl_2 + H_2$$

If a reactive metal has more than one valency state, **the chloride of the metal in its lower valency state is formed:** e.g. iron gives iron(II) chloride, not iron(III) chloride:

$$Fe + 2HCl \rightarrow FeCl_2 + H_2$$

Contrast this behaviour with the action of chlorine on metals (section 15.13(a)).

Metals below hydrogen in the reactivity series (section 8.15) or electrochemical series (section 14.24) will not liberate hydrogen from either hydrogen chloride or hydrochloric acid.

c) *Is hydrogen chloride an acid?*

Hydrogen chloride may be regarded as the 'parent acid' of hydrochloric acid; it will, however, dissolve in solvents other than water (e.g. tetrachloromethane), giving a solution with no acidic properties: this problem is discussed in detail in section 18.23.

15.10 reactions of hydrochloric acid

An aqueous solution of hydrogen chloride contains positively charged hydrogen ions (aquated) and negatively charged chloride ions (aquated). The following reactions of hydrochloric acid are interpreted in terms of these ions, following the principles outlined in Chapter 14; in view of varying practice, both designations $H^+(aq)$ and $H_3O^+(aq)$ are given for the aquated hydrogen ion. The authors prefer the designation $H^+(aq)$ and this is given first in each case.

a) blue litmus paper is reddened by the acid.

b) magnesium reacts vigorously at room temperature, liberating hydrogen. The magnesium is oxidised to magnesium(II) ions and the hydrogen ions are reduced to hydrogen:

$$\left.\begin{array}{r} Mg \longrightarrow Mg^{2+}(aq) + 2e^- \\ 2H^+(aq) + 2e^- \longrightarrow H_2 \end{array}\right\} Mg + 2H^+(aq) \longrightarrow Mg^{2+}(aq) + H_2$$

or

$$\left.\begin{array}{r} Mg \longrightarrow Mg^{2+}(aq) + 2e^- \\ 2H_3O^+ + 2e^- \longrightarrow H_2 + 2H_2O \end{array}\right\} \begin{array}{l} Mg + 2H_3O^+ \\ \quad \longrightarrow Mg^{2+}(aq) + 2H_2O + H_2 \end{array}$$

c) Sodium(I) carbonate effervesces in the acid at room temperature, giving carbon dioxide:

$$CO_3^{2-} + 2H^+(aq) \longrightarrow H_2O + CO_2$$

or

$$CO_3^{2-} + 2H_3O^+ \longrightarrow 3H_2O + CO_2$$

d) Copper(II) oxide dissolves to give a blue solution containing aquated copper(II) ions:

$$O^{2-} + 2H^+(aq) \longrightarrow H_2O, \quad \text{or more fully}$$
$$(Cu^{2+} + O^{2-}) + 2H^+(aq) \longrightarrow Cu^{2+}(aq) + H_2O$$

or

$$O^{2-} + 2H_3O^+ \longrightarrow 3H_2O, \quad \text{or more fully}$$
$$(Cu^{2+} + O^{2-}) + 2H_3O^+ \longrightarrow Cu^{2+}(aq) + 3H_2O$$

e) Manganese(IV) oxide, like lead(IV) oxide, will oxidise hydrochloric acid to chlorine; an acid–base and an oxidation–reduction reaction are involved:

$$(Mn^{4+} + 2O^{2-}) + 4H^+(aq) \longrightarrow Mn^{4+} + 2H_2O \qquad \textit{(acid–base)}$$
$$Mn^{4+} + 2e^- \longrightarrow Mn^{2+} \qquad \textit{(reduction)}$$
$$2Cl^-(aq) \longrightarrow Cl_2 + 2e^- \qquad \textit{(oxidation)}$$

overall:

$$MnO_2(s) + 4H^+(aq) + 2Cl^-(aq) \longrightarrow Mn^{2+}(aq) + 2H_2O(l) + Cl_2(g)$$

Hydrochloric acid is a typical strong acid in aqueous solution: it will redden litmus and react with carbonates and other bases, and with reactive metals liberating hydrogen. It is not an oxidising agent, as are nitric(V) and sulphuric(VI) acids under appropriate conditions. It can itself be oxidised by powerful oxidising agents, giving chlorine.

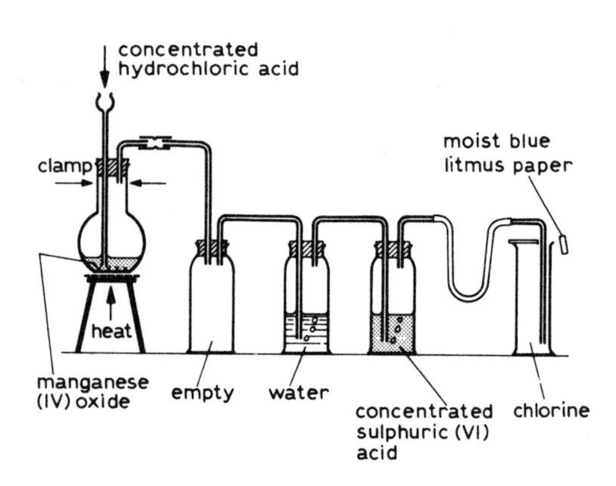

Fig. 15.8 preparation and collection of chlorine

15.11 preparation of chlorine in quantity

Granular manganese(IV) oxide is preferable to powder in this experiment, as the powder can prevent the glass from becoming 'wetted' by the liquid and the glass may then crack on heating.

Chlorine is generated when concentrated hydrochloric acid is warmed with manganese(IV) oxide as shown in Fig. 15.8; the gas is washed free from hydrogen chloride by passing it through water (which also dissolves some chlorine) and it is then dried by passing it through concentrated sulphuric(VI) acid. As the calculation opposite shows, it is possible to collect the chlorine by **upward displacement of air** (see section 15.7). The moist blue litmus paper is bleached when the jar is full; since chlorine is poisonous the gas should not be allowed to escape into the laboratory in quantity. The poisonous nature of the gas, combined with its high density, caused it to be used as a 'poison gas' in the First World War.

The first washbottle in the line is a safeguard against water from the second washbottle 'sucking back' into the reaction flask. The equation for the reaction is given in section 15.10(e).

Other oxidising agents may be used in this preparation; potassium(I) manganate(VII), for example, is powerful enough to oxidise hydrochloric acid to chlorine at, or just above, room temperature:

$$MnO_4^-(aq) + 8H^+(aq) + 5e^- \rightarrow Mn^{2+}(aq) + 4H_2O(l)$$

$$5Cl^-(aq) \rightarrow 2\tfrac{1}{2}Cl_2(g) + 5e^-$$

overall:

$$MnO_4^-(aq) + 8H^+(aq) + 5Cl^-(aq) \rightarrow Mn^{2+}(aq) + 4H_2O(l) + 2\tfrac{1}{2}Cl_2(g)$$

relative vapour densities:

chlorine	= 35·5
air (approx.)	= 14
density chlorine : air	= 35·5 : 14
	= 2·5 : 1

15.12 physical properties of chlorine

Chlorine is a yellow-green (the name means 'light green') gas, with a choking smell. It is poisonous. The gas is moderately soluble in water and is easily liquefied under pressure; it is often transported as a liquid in steel containers; it is much denser than air (*q.v.*).

15.13 reactions of chlorine

1. with metals

The action of chlorine on warm sodium is described in section 4.2. It is strongly exothermic and the product is crystalline sodium(I) chloride; sodium is oxidised to sodium(I) ions and chlorine is reduced to chloride ions:

$$\left.\begin{array}{r} Na \rightarrow Na^+ + e^- \\ \tfrac{1}{2}Cl_2 + e^- \rightarrow Cl^- \end{array}\right\} Na + \tfrac{1}{2}Cl_2 \rightarrow (Na^+ + Cl^-)$$

Magnesium ribbon will continue to burn in chlorine, forming magnesium(II) chloride:

$$\left.\begin{array}{r} Mg \rightarrow Mg^{2+} + 2e^- \\ Cl_2 + 2e^- \rightarrow 2Cl^- \end{array}\right\} Mg + Cl_2 \rightarrow (Mg^{2+} + 2Cl^-)$$

If a jar of chlorine is inverted over a watch-glass containing Dutch metal (a copper–zinc alloy), the metal inflames and a pale green residue of copper(II) chloride and zinc(II) chloride remains:

$$\left.\begin{array}{r} Cu \rightarrow Cu^{2+} + 2e^- \\ Zn \rightarrow Zn^{2+} + 2e^- \\ 2Cl_2 + 4e^- \rightarrow 4Cl^- \end{array}\right\} Zn + Cu + 2Cl_2 \rightarrow ZnCl_2 + CuCl_2$$

Heated iron wool inflames when lowered into a jar of chlorine. The dense brown 'smoke' settles to a solid which, if dissolved in water (or dilute hydrochloric acid) and tested with aqueous sodium(I) hydroxide, gives a brown precipitate of iron(III) hydroxide. Thus chlorine oxidises iron to iron(III) ions:

$$\left.\begin{array}{r} Fe \rightarrow Fe^{3+} + 3e^- \\ 1\tfrac{1}{2}Cl_2 + 3e^- \rightarrow 3Cl^- \end{array}\right\} Fe + 1\tfrac{1}{2}Cl_2 \rightarrow (Fe^{3+} + 3Cl^-)$$

also $$Fe^{3+}(aq) + 3OH^-(aq) \rightarrow Fe(OH)_3(s)$$

It is a general rule that if a metal has more than one valency, it will react with chlorine to form the **chloride of the metal in its highest valency state.** Contrast this with the action of hydrogen chloride on metals (section 15.9(b)). Even the unreactive metal mercury combines directly with chlorine at room temperature. This can be shown by dropping a small bead of mercury into a jar of chlorine and swirling the jar vigorously so that the mercury forms a 'mirror' on the inside of the jar. In ten or fifteen minutes the 'mirror' is eaten away and a white residue of mercury(II) chloride is seen. This can be dissolved and identified using aqueous sodium(I) hydroxide.

$$\left.\begin{array}{l} Hg \longrightarrow Hg^{2+} + 2e^- \\ Cl_2 + 2e^- \longrightarrow 2Cl^- \end{array}\right\} Hg + Cl_2 \longrightarrow HgCl_2$$

also $$Hg^{2+}(aq) + 2OH^-(aq) \longrightarrow HgO(s) + H_2O(l)$$

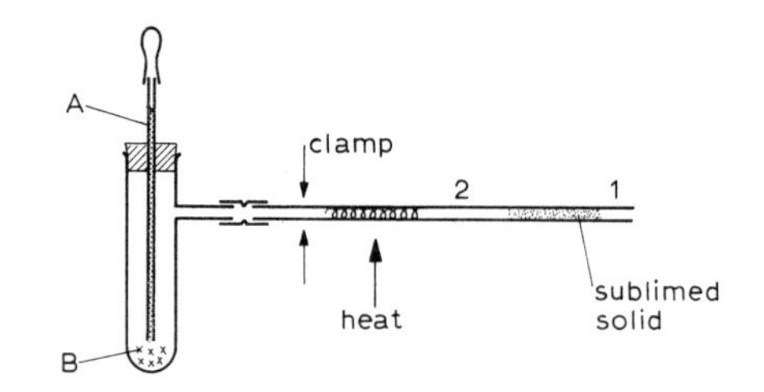

Fig. 15.9 preparation of anhydrous chlorides

A	c. H_2SO_4	c. HCl(aq)
B	NaCl	$KMnO_4$
gas	HCl	Cl_2
product	$FeCl_2$	$FeCl_3$

Fig. 15.10

preparation of anhydrous chlorides

Chlorides prepared from solution by the action of a metal, a base or a carbonate on hydrochloric acid (Chapter 12) almost always contain water of crystallisation. In some cases (e.g. barium(II) chloride) the anhydrous chloride can be obtained simply by heating the hydrated chloride:

$$BaCl_2.2H_2O \xrightarrow{heat} BaCl_2 + 2H_2O$$

In most cases, however, such treatment results in the decomposition of the chloride by the water (**hydrolysis**):

e.g. $$FeCl_3 + 3H_2O \xrightarrow{heat} Fe(OH)_3 + 3HCl$$

The preparation of anhydrous chlorides may be illustrated using iron (steel wool acts well) in the apparatus illustrated in Fig. 15.9. The iron is heated in the side tube and a stream of gas is generated by dropping liquid A on to solid B. The crystalline anhydrous chloride can be driven along the tube by sublimation on heating and a good specimen obtained, out of contact with moist air, by removing the stopper from the reaction tube and sealing the side-tube at 1 and 2.

The product of the reaction depends upon the gas employed, which in turn depends upon the identity of A and B, as shown in Fig. 15.10.

Chlorine reacts directly with all metals; indeed, it reacts directly with all elements except carbon, nitrogen and oxygen–and compounds of even these three elements with chlorine can be made by indirect methods.

Chlorine is often formed in electrolysis and obviously metal electrodes cannot be used where chlorine is being produced. Fortunately, one of the three elements named above conducts electricity; this is used for the anode in the electrolysis of chlorides (see Chapters 13, 16 and 27).

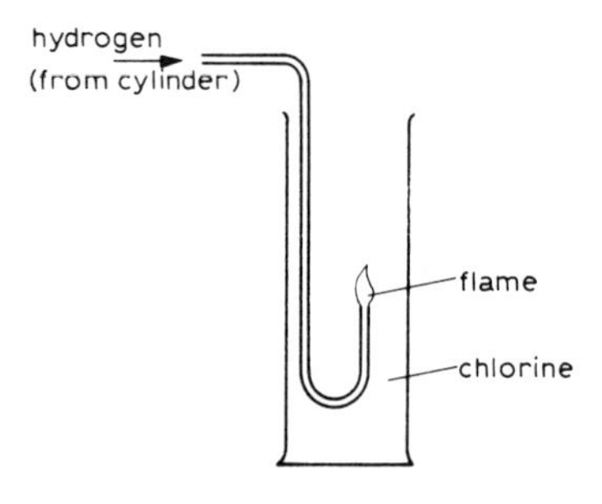

Fig. 15.11 hydrogen burning in chlorine

2. *with hydrogen*

A jet of hydrogen burning in air continues to burn when lowered into a jar of chlorine. The colour of the chlorine disappears and white fumes of hydrogen chloride in moist air appear at the mouth of the jar:

$$H_2 + Cl_2 \longrightarrow 2HCl$$

Ignition of equal volumes of hydrogen and chlorine produces a violent explosion; such a mixture should not be exposed to direct sunlight, which could provide the activation energy E_A (Fig. 15.12) for the exothermic reaction and an explosion could result.

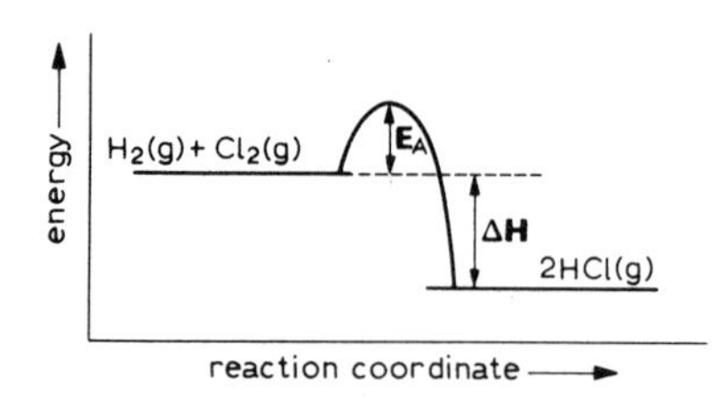

Fig. 15.12 energy changes in formation of hydrogen chloride

3. *with compounds containing hydrogen*

Chlorine will oxidise compounds containing hydrogen by removing the hydrogen from them. Several examples are given:

Turpentine ($C_{10}H_{16}$). Glass wool soaked in warm turpentine (*not* turpentine 'substitute') inflames when lowered into a jar of chlorine, with the formation of fumes of hydrogen chloride and particles of soot:

$$C_{10}H_{16} + 8Cl_2 \longrightarrow 10C + 16HCl$$

A wax candle (C_xH_y), burning in air, burns in chlorine with the formation of much soot:

$$C_xH_y + \frac{y}{2}Cl_2 \longrightarrow xC + yHCl$$

Acetylene (C_2H_2) reacts with dangerous vigour, as described in section 19.12:

$$C_2H_2 + Cl_2 \longrightarrow 2C + 2HCl$$

Hydrogen sulphide (H_2S) reacts with chlorine if jars of the two gases are brought mouth to mouth and shaken. The presence of a little water in the jars accelerates this reaction. A yellow deposit of sulphur is seen:

$$H_2S + Cl_2 \longrightarrow S + 2HCl$$

Ammonia (NH_3) reacts in a most interesting way with chlorine, as is described in section 24.8(3). The ammonia is first oxidised to nitrogen and the chlorine is reduced to hydrogen chloride; as ammonia is alkaline it then reacts with the acidic hydrogen chloride, giving dense white fumes of ammonium chloride:

$$2NH_3 + 3Cl_2 \longrightarrow N_2 + 6HCl$$

then

$$NH_3 + HCl \longrightarrow \underset{\text{(white fumes)}}{(NH_4^+ + Cl^-)}$$

4. with other non-metals

Chlorine will react with non-metals other than hydrogen, as can be demonstrated in the case of phosphorus: a piece of white phosphorus inflames spontaneously if lowered into a jar of chlorine, forming a mixture of its two chlorides:

$$P_4 + 6Cl_2 \longrightarrow 4PCl_3 \quad \text{phosphorus trichloride}$$
$$P_4 + 10Cl_2 \longrightarrow 4PCl_5 \quad \text{phosphorus pentachloride}$$

5. with water and alkalis: bleaching action

Chlorine dissolves in water to give a green solution containing molecular chlorine, Cl_2. In addition, some reaction with water takes place, giving an acidic solution:

$$Cl_2 + H_2O \rightleftharpoons H^+(aq) + Cl^-(aq) + HClO(aq) \quad \text{chloric(I) acid}$$

If chlorine is dissolved in cold, dilute aqueous alkali, a similar reaction takes place and a solution containing chlorate(I) ions is formed:

$$Cl_2 + 2OH^-(aq) \rightleftharpoons H_2O + Cl^-(aq) + ClO^-(aq) \quad \text{chlorate(I) ion}$$

Chlorine gas will **bleach** coloured substances, but only in the presence of water. This may be shown by painting a design with water on a piece of dry coloured fabric and lowering it into a jar of chlorine which has been allowed to stand for some time with a little concentrated sulphuric(VI) acid in it. The design appears as the damp fabric, but not the dry fabric, is bleached.

A solution of chlorine in water has good bleaching properties, especially if acid is added to it. This may be interpreted as resulting from the powerfully oxidising chloronium ion:

$$Cl_2 \rightleftharpoons Cl^- + Cl^+ \quad \text{(chloronium ion)}$$

in water: $Cl^+ + H_2O \longrightarrow H^+(aq) + HClO(aq)$

in alkali: $Cl^+ + 2OH^-(aq) \longrightarrow H_2O + ClO^-(aq)$

When acid is added to aqueous chlorate(I) ions, the last reaction is reversed and the chloronium ion is available for oxidising purposes (bleaching):

$$ClO^-(aq) + 2H^+(aq) \longrightarrow H_2O + Cl^+$$
$$Cl^+ + 2e^- \longrightarrow Cl^-(aq)$$

At the same time, the addition of acid to aqueous chlorate(I) ions with chloride ions present forms chlorine once more:

$$ClO^-(aq) + 2H^+(aq) \longrightarrow H_2O + Cl^+$$
$$Cl^+ + Cl^-(aq) \longrightarrow Cl_2(aq)$$

overall:

$$ClO^-(aq) + Cl^-(aq) + 2H^+(aq) \longrightarrow Cl_2(aq) + H_2O$$

Solutions of chlorate(I) and the solid chlorate(I) 'Bleaching Powder', which contains some calcium(II) chlorate(I), are used domestically and industrially as bleaches.

15.14 manufacture and uses of chlorine

Chlorine is used extensively for bleaching certain fabrics and wood-pulp, for sterilising water supplies by killing bacteria and for the production of hydrochloric acid and chlorides; the petrochemicals industry utilises much chlorine in forming organic compounds containing chlorine which are used as solvents, synthetic fibres and plastics (see Chapter 19).

Chlorine is produced on the large scale during the electrolysis of chlorides, in which it is a very valuable by-product. Some examples of electrolytic processes generating chlorine are given in Chapters 16 and 27.

15.15 bromine and iodine

Bromine, Br_2, is a deep red liquid which readily gives a brown vapour on warming. Iodine, I_2, is a blue-black solid which sublimes to a purple vapour on heating. Bromine and iodine are closely related to chlorine, as the following experiments reveal. The three elements are members of the chemical family called **halogens** and their compounds are called **halides.**

synthesis of hydrogen chloride, hydrogen bromide and hydrogen iodide

Jars of chlorine gas, bromine vapour and iodine vapour are separately mixed with an equal volume of hydrogen gas and the mixtures are ignited. Combination is explosive in the case of chlorine, less so with bromine and rather feeble with iodine: in each case steamy acid fumes are produced:

$$Cl_2 + H_2 \longrightarrow 2HCl \quad \text{hydrogen chloride}$$
$$Br_2 + H_2 \longrightarrow 2HBr \quad \text{hydrogen bromide}$$
$$I_2 + H_2 \longrightarrow 2HI \quad \text{hydrogen iodide}$$

These hydrogen halides are all soluble in water, giving strongly acidic solutions.

The difficulty of oxidising hydrogen chloride to chlorine is explained by the strength of the hydrogen–chlorine bond in the molecule (section 15.9, see also Fig. 18.21) and this is in turn related to the energy released when hydrogen chloride is formed. The observations in the formation of hydrogen bromide and hydrogen iodide allow us to predict successively weaker bonds in these molecules; this idea is developed in the succeeding sections of this Chapter.

15.16 chlorides, bromides and iodides

Bromine and iodine form salts (bromides and iodides) similar in many ways to chlorides. Experiments on these salts reveal interesting similarities and gradations in properties between the three elements in the family of halogens.

1. action of aqueous silver(I) nitrate(V) on aqueous halides

A little sodium(I) chloride is dissolved in water. A few drops of dilute nitric(V) acid are added to the solution, followed by aqueous silver(I) nitrate(V). A *white* precipitate of silver(I) chloride, soluble in aqueous ammonia, is produced:

$$Ag^+(aq) + Cl^-(aq) \longrightarrow AgCl(s)$$
$$AgCl(s) + 2NH_3(aq) \longrightarrow \underset{\text{diamminesilver(I) ion}}{[Ag(NH_3)_2]^+(aq)} + Cl^-(aq)$$

This is one of the tests for a soluble chloride.

The above experiment repeated with sodium(I) bromide in place of sodium(I) chloride gives a *cream* precipitate of silver(I) bromide, sparingly soluble in aqueous ammonia:

$$Ag^+(aq) + Br^-(aq) \longrightarrow AgBr(s)$$
$$AgBr(s) + 2NH_3(aq) \rightleftharpoons [Ag(NH_3)_2]^+(aq) + Br^-(aq)$$

Sodium(I) iodide gives a *yellow* precipitate of silver(I) iodide, insoluble in aqueous ammonia:

$$Ag^+(aq) + I^-(aq) \longrightarrow AgI(s)$$

2. action of manganese(IV) oxide and concentrated sulphuric(VI) acid on solid halides

A little of each solid sodium(I) halide is separately mixed with an equal bulk of manganese(IV) oxide and a little concentrated sulphuric(VI) acid is added. The test-tube may be warmed gently if necessary to produce an observable reaction. Green, brown and purple vapours are obtained from the chloride, bromide and iodide respectively. Careful observation in this experiment will indicate the relative ease of oxidation of each halide to the halogen:

i) $$MnO_2 + 4H^+ + 2e^- \longrightarrow Mn^{2+} + 2H_2O$$

ii) $$\begin{cases} 2Cl^- \longrightarrow Cl_2 + 2e^- \\ 2Br^- \longrightarrow Br_2 + 2e^- \\ 2I^- \longrightarrow I_2 + 2e^- \end{cases}$$

Combination of (i) with (ii) gives the overall equation for the reaction.

3. action of concentrated sulphuric acid(VI) alone on solid halides

Hot concentrated sulphuric(VI) acid is itself an oxidising agent, though it is not so powerful as manganese(IV) oxide. It has an observably different effect upon the three halides.

A little of each solid sodium(I) halide is placed in an ignition tube or test-tube. To each tube a few drops of concentrated sulphuric(VI) acid are added; the tube may be warmed if necessary to give an observable reaction. Sodium(I) chloride gives hydrogen chloride, but no chlorine:

$$Cl^- + H_2SO_4 \longrightarrow HSO_4^- + HCl$$

..phuric(VI) acid will *not* oxidise chloride ion

Sodium(I) bromide gives a mixture of hydrogen bromide and bromine, together with sulphur dioxide:

$$Br^- + H_2SO_4 \longrightarrow HSO_4^- + HBr$$

also:

$$SO_4^{2-} + 4H^+ + 2e^- \longrightarrow SO_2 + 2H_2O \quad (\textit{reduction})$$
$$2Br^- \longrightarrow Br_2 + 2e^- \quad (\textit{oxidation})$$

overall: $$SO_4^{2-} + 4H^+ + 2Br^- \longrightarrow SO_2 + Br_2 + 2H_2O$$

sulphuric(VI) acid oxidises bromide ion *partially*; bromide ion reduces sulphuric(VI) acid to sulphur dioxide

Sodium(I) iodide gives iodine vapour, but little if any hydrogen iodide; a smell of 'bad eggs' is also apparent as hydrogen sulphide is formed:

$$SO_4^{2-} + 10H^+ + 8e^- \longrightarrow H_2S + 4H_2O$$
$$8I^- \longrightarrow 4I_2 + 8e^-$$

overall: $$SO_4^{2-} + 10H^+ + 8I^- \longrightarrow H_2S + 4I_2 + 4H_2O$$

sulphuric(VI) acid oxidises iodide ion *completely*; iodide ion reduces sulphuric(VI) acid to hydrogen sulphide

Iodide ion is thus the most easily oxidisable halide and is therefore the strongest reducing agent of the three halides.

Hydrogen bromide and iodide can **not** be prepared by the action of concentrated sulphuric(VI) acid on bromides or iodides, in a way analogous to the method for the preparation of hydrogen chloride. Hydrogen bromide and iodide are prepared by hydrolysis of the appropriate phosphorus halide.

4. recognition of halogens in solution

The three halogens give solutions in water with pale colours which are not readily distinguishable. Each halogen dissolves more readily in organic solvents (e.g. tetrachloromethane or trichloromethane) than in water; the

Colours of halogens in tetrachloromethane:

chlorine	green
bromine	orange
iodine	purple

colour of the halogen in the organic solvent is more intense than in water and is a colour characteristic of the halogen in its molecular state.

This can be shown by adding a little tetrachloromethane (*care: the vapour is toxic*) to each of three tubes, containing respectively aqueous chlorine, dilute aqueous bromine and a dilute solution of iodine in aqueous potassium(I) iodide. Gentle shaking of each tube shows the characteristic colour of each halogen in the lower (organic) layer.

This test is used for the elements chlorine, bromine or iodine in the following experiments.

5. displacement reactions

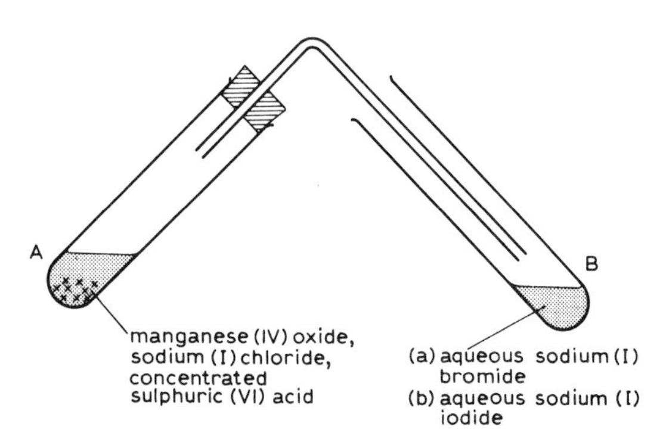

Fig. 15.13 displacement of one halogen by another

Chlorine is generated in tube A shown in Fig. 15.13; a bung and short delivery tube is attached to tube A so that the gas is delivered into tube B but the delivery tube does not enter the liquid. Tube B contains aqueous sodium(I) bromide. On shaking tube B a colour change is seen. This may be interpreted more readily if a little tetrachloromethane is added to the solution in tube B: an orange colour is observed in the organic layer.

If the experiment is repeated using aqueous sodium(I) iodide in place of aqueous sodium(I) bromide in tube B, a brown solution results which gives a purple colour in the organic layer.

chlorine displaces bromine from a bromide:

$$\left.\begin{array}{r}\tfrac{1}{2}Cl_2 + e^- \longrightarrow Cl^- \\ Br^- \longrightarrow \tfrac{1}{2}Br_2 + e^-\end{array}\right\} \tfrac{1}{2}Cl_2 + Br^-(aq) \longrightarrow \tfrac{1}{2}Br_2(aq) + Cl^-(aq)$$

chlorine displaces iodine from an iodide:

$$\left.\begin{array}{r}\tfrac{1}{2}Cl_2 + e^- \longrightarrow Cl^- \\ I^- \longrightarrow \tfrac{1}{2}I_2 + e^-\end{array}\right\} \tfrac{1}{2}Cl_2 + I^-(aq) \longrightarrow \tfrac{1}{2}I_2(aq) + Cl^-(aq)$$

A similar experiment in which bromine is generated in tube A and passed successively into aqueous sodium(I) iodide and aqueous sodium(I) chloride reveals that *bromine displaces iodine from an iodide:*

$$\left.\begin{array}{r}\tfrac{1}{2}Br + e^- \longrightarrow Br^- \\ I^- \longrightarrow \tfrac{1}{2}I_2 + e^-\end{array}\right\} \tfrac{1}{2}Br_2 + I^-(aq) \longrightarrow \tfrac{1}{2}I_2(aq) + Br^-(aq)$$

bromine will not displace chlorine from a chloride

the halogen family

chlorine
bromine
iodine

i) the higher displaces the lower from the halide
ii) strongest oxidant at the top
iii) strongest reducing halide at the bottom
iv) most stable hydrogen halide at the top

The three elements are found, in this order, in the same vertical group of the Periodic Table (Chapter 17). A fourth halogen, **fluorine**, is found in this group, above chlorine. Many of its properties may be predicted successfully in the light of the foregoing experiments. Reasons for the observed gradation in behaviour may be found by examining the atomic structure of the halogen atoms (Chapter 18) and by comparing the relative sizes of the halogen atoms with each other and of the halide ions with each other (section 27.6).

revision summary: Chapter 15

hydrogen chloride:	*preparation:*	conc. H_2SO_4 on a chloride
		$Cl^-(s) + H_2SO_4(l) \longrightarrow HSO_4^-(s) + HCl(g)$
	collection:	upward displacement of air
	test:	reddens litmus, fumes with ammonia
	reactions:	neither burns nor supports combustion
		metals give hydrogen + lower valency chloride
		'parent acid' of hydrochloric acid
		oxidised to chlorine

hydrochloric acid: see 'Acids', revision summary, Chapter 11

chlorine:

preparation: MnO_2 on hot conc. HCl(aq)
$MnO_2(s) + 4H^+(aq) + 2Cl^-(aq) \rightarrow Mn^{2+}(aq) + 2H_2O(l) + Cl_2(g)$

collection: upward displacement of air

test: smell (poisonous!), bleaches moist litmus paper

reactions: metals give higher valency chloride
powerful oxidant, bleaches (damp) by oxidation

chlorides:

test: white precipitate with acidic $AgNO_3$(aq), soluble in NH_3(aq)

bromine, iodine:

similar to chlorine but successively less reactive
hydrogen halides successively less stable
halides successively easier to oxidise
displacement order: Cl_2 displ. Br_2 displ. I_2

questions 15

1. Hydrogen chloride can be prepared from sodium(I) chloride and sulphuric(VI) acid.
(a) State the conditions necessary to obtain a steady supply of the gas in the laboratory.
(b) Give the equation for the reaction which occurs under these conditions.
(c) How is the gas collected?
(d) Give a diagram of the apparatus you would use.
(e) What precaution would you take when dissolving the gas in water?
(f) What reaction occurs when hydrogen chloride molecules dissolve in water?

Describe reactions (**one** in each case) in which hydrochloric acid
(i) is oxidised;
(ii) reacts to form a gaseous compound;
(iii) reacts to give a precipitate of an insoluble salt. (AEB)

2. Chlorine is often prepared in the laboratory by the reaction between manganese(IV) oxide and hydrochloric acid.
(a) State the conditions necessary to obtain a reasonable quantity of chlorine.
(b) What is oxidised in this reaction?
(c) Name the other products of the reaction, and give the equation.
(d) What is the main impurity in the gas, other than moisture? State how you would remove it.
(e) Give a diagram of the apparatus you would use for preparing, purifying and collecting the chlorine.

The rest of this question refers to an experiment performed to find the formula of hydrogen chloride. 40 cm^3 of hydrogen and 30 cm^3 of chlorine were mixed in subdued light. When the mixture was passed over hot platinum it lost its colour. On cooling to the original temperature the total volume was found to be unchanged. The remaining gas was pumped into water, when all but 10 cm^3 dissolved.
(i) Why was the hydrogen mixed with chlorine in *subdued* light?
(ii) What was the purpose of the platinum?
(iii) Show clearly how the results of this experiment enable you to calculate the formula of hydrogen chloride. (You may assume that hydrogen and chlorine are diatomic gases.)
(iv) State the chemical law you use in (iii). (AEB)

3. How would you prepare and collect a few jars of damp, but otherwise pure, chlorine (not by electrolysis)? What particular impurity has to be removed?

Give **four** properties, physical or chemical, in which chlorine differs from hydrogen chloride. State whether the properties you choose are physical or chemical.

Give **two** large-scale uses for chlorine and **one** large-scale use for a solution of hydrogen chloride. (C)

4. (a) Draw a labelled diagram of the apparatus you would use to prepare and collect reasonably pure and dry chlorine from hydrochloric acid (**not** by electrolysis). State the conditions and write an equation for the reaction used.
(b) Describe the reactions of chlorine with
(i) a **named** metallic element,
(ii) a **named** non-metallic element,
(iii) a **named** alkali,
(iv) a **named** organic compound.
The equation for **each** reaction should be given. (C)

5. (a) Describe the preparation of moist but otherwise pure chlorine. (**No** diagram is required.)

(b) What would you observe, and what is formed, when
(i) a lighted taper, or candle, is introduced into chlorine,
(ii) a small piece of white phosphorus is put into chlorine,
(iii) chlorine is passed through a solution of sulphur dioxide?
(c) How is cloth bleached industrially, and how is it treated after bleaching? (C)

6. Describe a method of manufacturing chlorine by an electrolytic process. Name the electrolyte and give the name, sign and material of each electrode. Name also the products of the electrolysis and write ionic equations for the reactions occurring at the electrodes.

When dry hydrogen burns at a jet in a gas jar of dry chlorine the following changes are observed: the colour of the chlorine disappears and a gas is formed which fumes in moist air and turns moist blue litmus red. Explain these observations.

Give the name of the structural formula for the product formed when one g-molecule (mole) of hydrogen chloride reacts with one g-molecule (mole) of acetylene (ethyne). Give **one** use of this product. (JMB)

7. Describe how you would prepare reasonably pure samples of
(a) lead(II) chloride given lead, concentrated nitric(V) acid and sodium(I) chloride,
(b) anhydrous iron(III) chloride given iron, concentrated sulphuric(VI) acid, sodium(I) chloride and manganese(IV) oxide. (O&C)

8. (a) Chlorine, bromine and iodine are members of the halogen family of elements. Illustrate similarities between these elements by reference to
(i) the electronic structures of their atoms;
(ii) their compounds with hydrogen;
(iii) their compounds with lead.
(b) Give **one** example of a reaction in which chlorine is converted to chloride ions. State, with the reason, whether you need to use an oxidising or a reducing agent to bring about this change.
(c) Compare the reactions which occur when concentrated sulphuric(VI) acid is added in turn to solid specimens of sodium(I) chloride and sodium(I) bromide. (W)

9. The halogens have the following atomic numbers: fluorine 9, chlorine 17, bromine 35, iodine 53.
(a) Give the electronic configuration for
(i) an atom of fluorine,
(ii) an atom of chlorine.
(b) How many valency electrons has an iodine atom?
(c) Write down the names of the four halogens and after each state whether it is a solid, a liquid or a gas under ordinary laboratory conditions.
(d) How, and under what conditions, does chlorine react with (i) hydrogen, (ii) iron, (iii) potassium(I) bromide?
(e) State whether iodine is more or less reactive than chlorine. Give an example to illustrate your answer. (JMB)

10. What is meant by a **group** in the Periodic Table? The halogens form a group and include the element astatine (At).
(a) Suggest in what state astatine exists at room temperature.
(b) What is the most likely colour of astatine?
(c) Is astatine more likely to be soluble in water or in tetrachloromethane?
(d) What is the charge on an astatide ion? Write its formula.
(e) How does chlorine react with a solution of sodium(I) astatide? Write an ionic equation, if possible.
(f) What action, if any, does astatine have with aqueous sodium(I) bromide?
(g) How does astatine react with iron?
(h) How would you obtain a pure dry sampe of an insoluble astatide? Give brief experimental details.
(i) What would be the most likely effect of concentrated sulphuric(VI) acid on dry sodium(I) astatide?
(j) What brief statement can you make about the electron configuration of astatine?

11. Give an account of the experimental evidence which might be collected in a school laboratory which indicates that the halogens (chlorine, bromine and iodine) belong to the same group in the Periodic Table of the elements and that each of these halogens has a particular position in the group.
L (NUFFIELD)

12. Describe how chlorine can be made from sodium(I) chloride (a) by electrolysis, and (b) without using electrolysis.

In (a) a diagram is required, with the electrodes carefully labelled. In (b), no diagram is required, but you should give the names of other chemicals used and also explain the reactions which take place, giving equations.

13. This is a question about the element fluorine, symbol F. It is a member of the halogen family, and is more reactive than chlorine. You will not be familiar with the chemistry of fluorine, but you should be able to predict some of its important properties from what you know about the properties of the other halogens.
(a) What physical properties would you expect fluorine to have?
(b) How would you expect fluorine to occur in nature?
(c) Suggest a method which might be used to obtain fluorine as the free element.
(d) What would you expect to see if fluorine was reacted with aqueous solutions of potassium(I) chloride, potassium(I) bromide and potassium(I) iodide? Write an equation for any one of these reactions.
(e) How would you expect fluorine to react with water? Predict the approximate pH of the resulting liquid.
(f) How would you expect fluorine to react with hydrogen? Name the product of this reaction and write an equation for its formation. What properties would you expect this product to have?
(g) Fluorine reacts with sodium to form sodium(I) fluoride. What properties would you expect this compound to have?

L (NUFFIELD)

In previous Chapters, the chemistry of two 'families' of elements has been considered in some detail; that of the **noble gases** (Chapter 6) and the **halogens** (Chapter 15). 'Families', or groups, of elements are sets of elements which have many properties in common. The reason for the similarity in properties of the members of a 'family' or group is explained in Chapters 17 and 18: the atoms of such elements all have the same number of electrons in their highest energy levels.

One very broad classification divides elements into metals and non-metals; metals have certain properties in common, but they are by no means similar in reactivity. The different ways in which sodium, aluminium and copper behave with water exemplify conspicuous differences between these elements, although they are all metals. This Chapter is concerned with a group of elements known as the **alkali metals:** lithium, sodium, potassium, rubidium and caesium. They are all very reactive metals and the atom of each element contains one electron in the highest energy level. The chemistry of one element typical of the group (**sodium**) is first considered in detail and the similarities and differences between the chemistry of sodium and that of the other alkali metals is then outlined.

Sodium compounds have a wide variety of uses, many of which are very familiar. In the home, sodium(I) chloride, $NaCl$ (common salt), is well known, sodium(I) carbonate-10-water, $Na_2CO_3.10H_2O$ (washing soda), is used for softening water, sodium(I) hydroxide, $NaOH$ (caustic soda), is a powerful and corrosive heavy-duty cleaning material, sodium(I) hydrogencarbonate, $NaHCO_3$ (bicarbonate of soda), is used in baking, sodium(I) silicate(IV), Na_2SiO_3 (water-glass), is a preservative for eggs and sodium(I) sulphide, Na_2S, is used as a setting lotion for hair.

16.1 sodium metal

Metallic sodium is usually encountered in the laboratory in the form of small white lumps of solid, stored under oil. In this form, it does not have a very metallic appearance, but if a piece is removed from the oil, dried on filter paper and cut with a pen-knife (it is very soft) a surface is exposed which has a characteristic metallic lustre. This shiny surface very soon tarnishes as a result of reaction with substances in the atmosphere, giving the metal once more its familiar dull white appearance.

1. reaction with water

Sodium reacts with cold water liberating hydrogen and leaving an aqueous solution of sodium(I) hydroxide, as is described in section 8.12:

$$Na(s) + H_2O(l) \longrightarrow Na^+(aq) + OH^-(aq) + \tfrac{1}{2}H_2(g)$$

The reaction generates sufficient energy to melt the sodium, which floats on the surface of the water as a small, shiny, rapidly moving globule. Sodium atoms reduce hydrogen ions from the water to hydrogen atoms, which then form molecules; the sodium atoms are themselves oxidised to sodium(I) ions:

$$H_2O(l) \rightleftharpoons H^+(aq) + OH^-(aq)$$
$$Na(s) \longrightarrow Na^+(aq) + e^- \qquad (\textit{oxidation})$$
$$H^+(aq) + e^- \longrightarrow \tfrac{1}{2}H_2(g) \qquad (\textit{reduction})$$

2. reaction with oxygen

If a cleaned piece of sodium is warmed gently in air in a deflagrating spoon and the hot metal is then immersed in a gas jar of oxygen, it burns with a bright yellow flame leaving a yellow residue which consists mainly of sodium(I) peroxide:

$$2Na + O_2 \longrightarrow Na_2O_2$$

At first sight, it appears that this compound shows sodium to be exerting a valency of two. This is not so; the product is an ionic solid which contains peroxide ions ($2Na^+ + O_2^{2-}$).

3. reaction with air

A cleaned piece of sodium heated in air on a deflagrating spoon burns readily with a yellow flame, but not so intensely as it does in oxygen. The residue is a mixture of sodium(I) peroxide, Na_2O_2, and sodium(I) oxide, Na_2O.

As previously described, sodium reacts with substances in air at room temperature. The components of air principally involved in the reaction are *water vapour* and *carbon dioxide*; the chain of changes which takes place if a piece of sodium is left exposed to the atmosphere for a few days is described in section 8.17.

4. reaction with chlorine

The reaction between sodium and chlorine is described in section 4.2. The product is sodium(I) chloride:

$$Na(s) + \tfrac{1}{2}Cl_2(g) \longrightarrow NaCl(s)$$

Sodium atoms reduce chlorine molecules to chloride ions and are themselves oxidised to sodium(I) ions:

$$Na \longrightarrow Na^+ + e^- \qquad (\textit{oxidation})$$
$$\tfrac{1}{2}Cl_2 + e^- \longrightarrow Cl^- \qquad (\textit{reduction})$$

The synthesis of sodium(I) chloride from sodium and chlorine is difficult to demonstrate by lowering a piece of warm sodium on a deflagrating spoon into a jar of chlorine, as dense black fumes appear as a result of reaction between the chlorine and the iron spoon.

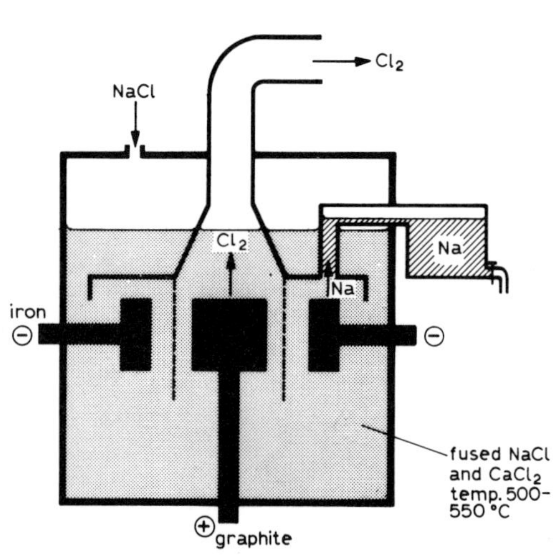

Fig. 16.1 diagram of Downs cell for manufacture of Na from molten NaCl (+ $CaCl_2$)

16.2 industrial production of sodium

The preceding sections show sodium to be a very reactive metal which readily forms compounds. The considerable energy released when sodium forms compounds must be supplied to the sodium compounds in order to produce the metal. Chemical reduction processes, e.g. using carbon or hydrogen, are inadequate to supply this energy and the only feasible method for producing sodium from its compounds is by the process of electrolysis. Aqueous solutions of sodium compounds cannot be used, since they contain aqueous hydrogen ions as well as aqueous sodium(I) ions; the hydrogen ions are discharged at the cathode in preference to the sodium(I) ions (see section 13.17).

Sodium is manufactured by the **electrolysis of molten sodium(I) chloride** in the Downs cell as illustrated in Fig. 16.1. Calcium(II) chloride is added to the sodium(I) chloride to reduce its melting point and the working temperature of the cell is lowered from over 800 °C to about 550 °C (see section 4.6). The electrical resistance of the cell maintains the temperature once the unit is in operation. The products of the electrolysis are sodium and chlorine:

anode: $Cl^- \rightarrow \frac{1}{2}Cl_2 + e^-$ (*oxidation*)
cathode: $Na^+ + e^- \rightarrow Na$ (*reduction*)

The sodium ion is discharged at the cathode in preference to the calcium ion as calcium is above sodium in the electrochemical series (but not in the reactivity series for metals). It is necessary to keep separate the sodium and chlorine produced in the electrolysis and this is effected using a nickel hood over the anode, as shown in Fig. 16.1. The anode is made of graphite, the only non-metallic electrical conductor; chlorine would corrode a metallic electrode. The molten sodium produced at the cathode is less dense than the molten electrolyte; it floats to the surface and is run off through a tap as illustrated.

An acceptable description of an industrial electrolytic process ought to include the following features: material of the containing vessel, nature of the electrolyte, cathode material, anode material, polarity (charge sign) of the electrodes, products of the reaction, temperature of the process and siting, or relative positions, of the important parts of the plant. A useful mnemonic for remembering what to include in such a description is shown opposite. Other aids to memory are given in section 27.1.

Very	**V**essel
Earnest	**E**lectrolyte
Chemists	**C**athode
And	**A**node
Physicists	**P**olarity
Produce	**P**roducts
Top	**T**emperature
Scientists	**S**iting

16.3 uses of sodium

Metallic sodium is not widely used in industry, in contrast with its compounds. It is used in the manufacture of sodium(I) peroxide; as a powerful reducing agent in the manufacture of certain metals, e.g. titanium; and in the liquid state as a coolant in some nuclear power stations. It is quite widely used as a powerful reducing agent in laboratory work.

16.4 sodium(I) oxides

a) *sodium*(I) *oxide*

Sodium heated in a limited supply of air produces sodium(I) oxide:

$$4Na(s) + O_2(g) \rightarrow 2Na_2O(s)$$

It is a white solid which reacts violently with cold water producing an aqueous solution of sodium(I) hydroxide:

$$O^{2-}(s) + H_2O(l) \rightarrow 2OH^-(aq)$$

It reacts explosively with acids forming salts and on heating in oxygen it is converted into sodium(I) peroxide:

$$Na_2O(s) + \frac{1}{2}O_2(g) \rightarrow Na_2O_2(s)$$

b) *sodium*(I) *peroxide*, Na_2O_2

This is produced as a yellow solid when sodium is heated in a plentiful supply of oxygen (*q.v.*).

If sodium(I) peroxide is added to water at room temperature, a gas is evolved which rekindles a glowing splint and an alkaline solution remains:

$$O_2^{2-}(s) + H_2O(l) \rightarrow 2OH^-(aq) + \frac{1}{2}O_2(g)$$

If this simple test-tube experiment is repeated using a test-tube cooled in a mixture of ice and water, no bubbles are produced and the clear solution which is formed contains hydrogen peroxide:

$$O_2^{2-}(s) + 2H_2O(l) \rightarrow 2OH^-(aq) + H_2O_2(aq)$$

The interesting properties of this solution are described in section 14.7(4).

Sodium(I) peroxide gradually changes colour from yellow to white on exposure to air as it reacts with carbon dioxide:

$$O_2^{2-}(s) + CO_2(g) \rightarrow CO_3^{2-}(s) + \tfrac{1}{2}O_2(g)$$

Because of this property, sodium(I) peroxide is used to refresh the air in confined spaces, e.g. in submarines and underground tunnels.

Sodium(I) peroxide is an extremely powerful oxidising agent. If it is carefully mixed with dry sugar on a metal dish supported on a tripod and a drop of water is subsequently added, the mixture suddenly inflames. The oxidation of the sugar by sodium(I) peroxide, catalysed by water, is sufficiently exothermic to ignite the mixture.

The peroxides of lithium, potassium, rubidium and caesium are formed in much the same way, and they are all strong oxidising agents.

16.5 sodium(I) hydroxide (caustic soda)

Aqueous sodium(I) hydroxide is formed when sodium reacts with cold water (*q.v.*). If the resulting solution is evaporated carefully to dryness, pure sodium(I) hydroxide remains as a deliquescent, white, waxy solid. It should not be touched with the fingers; being caustic, it attacks skin. It is now manufactured largely by the electrolysis of aqueous sodium(I) chloride as described in section 16.6.

1. aqueous sodium hydroxide

If two or three pellets of solid sodium(I) hydroxide are added to a boiling-tube half-full of water and the tube is gently agitated, the solid dissolves readily and heat is evolved. The resulting solution turns red litmus blue; sodium(I) hydroxide is a 'strong' alkali, being completely ionised in aqueous solution. The same features are observed when the reaction is carried out using the hydroxides of lithium, potassium, rubidium and caesium.

2. action of heat on sodium(I) hydroxide

Sodium(I) hydroxide is not decomposed on heating, unlike hydroxides of almost all other metals. If the solid is heated, it melts at 591 K, but continued heating of the liquid produces no further change. On cooling, the original solid is re-formed. The only other hydroxides not decomposed are those of potassium, rubidium and caesium.

3. action of carbon dioxide on sodium(I) hydroxide

Both solid and aqueous sodium(I) hydroxide readily absorb carbon dioxide forming either sodium(I) carbonate:

$$CO_2 + 2OH^- \rightarrow CO_3^{2-} + H_2O$$

or, if excess carbon dioxide is used, sodium(I) hydrogencarbonate:

$$CO_2 + OH^- \rightarrow HCO_3^-$$

Sodium(I) hydroxide is, however, rarely used to remove carbon dioxide from gas mixtures because the solubility of sodium(I) carbonate in aqueous sodium(I) hydroxide is not high; potassium(I) hydroxide is used instead for this purpose as it does not suffer from the same disadvantage.

Other references to the reactions of sodium(I) hydroxide may be found as follows: with acids, Chapters 11 and 12; as a reagent in analysis, section 12.29; with metals, sections 26.2, 26.29; with chlorine, section 15.13(5).

16.6 industrial production of sodium(I) hydroxide

Electrolysis of fairly concentrated aqueous sodium(I) chloride in the Kellner–Solvay cell produces aqueous sodium(I) hydroxide. The problem solved by this ingenious process is the isolation of sodium(I) hydroxide from an aqueous solution containing sodium(I) chloride.

The aqueous sodium(I) chloride is electrolysed using a graphite anode (as chlorine is produced) and a cathode consisting of a thin, flowing layer of mercury. The anode product is chlorine:

$$Cl^-(aq) \rightarrow \frac{1}{2}Cl_2(g) + e^- \quad (oxidation)$$

The high overpotential of hydrogen at a mercury surface leads to the preferential discharge of aqueous sodium(I) ions rather than hydrogen ions at the mercury cathode:

$$Na^+(aq) + e^- \rightarrow Na(l) \quad (reduction)$$

The discharged sodium dissolves in the mercury and is transferred to a second tank; here it reacts with water, producing hydrogen and aqueous sodium(I) hydroxide. The mercury, freed from sodium in this way, is returned to the first tank for re-use and the aqueous sodium(I) hydroxide, which is free from chloride ions, is evaporated to dryness to give the solid product.

Sodium(I) hydroxide is used in the manufacture of many important organic compounds and in the manufacture of soap, paper, artificial silk and dyes. It is important in the purification of aluminium ore (see section 26.1) and as a cleansing agent, as it attacks greases and oils readily.

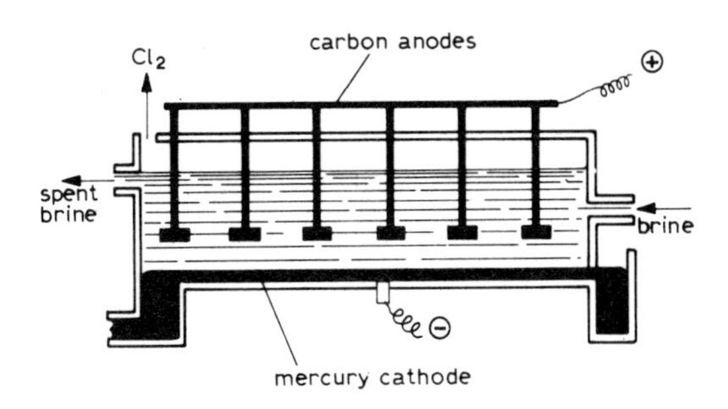

Fig. 16.2 diagram of Kellner–Solvay cell for manufacture of NaOH from NaCl(aq) using moving mercury cathode

16.7 sodium(I) carbonate, Na_2CO_3

1. laboratory preparation

Aqueous sodium(I) carbonate is produced by bubbling carbon dioxide into aqueous sodium(I) hydroxide until the solution has increased in mass by an amount sufficient to indicate completion of the change:

$$CO_2(g) + 2OH^-(aq) \rightarrow CO_3^{2-}(aq) + H_2O(l)$$

Crystallisation of the resulting solution produces transparent, colourless crystals of sodium(I) carbonate-10-water. On exposure to air, this solid *effloresces* (see section 8.17) to the monohydrate, whereas all of the water of crystallisation is lost on heating and anhydrous sodium(I) carbonate is produced. Unlike the carbonates of most metals, but like those of potassium, rubidium and caesium, anhydrous sodium(I) carbonate does not decompose on heating.

2. aqueous sodium(I) carbonate

Aqueous sodium(I) carbonate turns red litmus blue; the solution is alkaline as it contains an excess of aqueous hydroxide ions over aqueous hydrogen ions. The hydroxide ions are produced by reversible *hydrolysis* of carbonate ions:

$$\underset{\text{(base)}}{CO_3^{2-}(aq)} + \underset{\text{(acid)}}{H_2O(l)} \rightleftharpoons OH^-(aq) + HCO_3^-(aq)$$

A standard solution of alkali for use in volumetric analysis can be prepared by dissolving a carefully measured mass of anhydrous sodium(I) carbonate in water and diluting the resulting solution to a known volume; this is because anhydrous sodium(I) carbonate can be obtained in a fairly high state of purity. Sodium(I) hydroxide cannot be used in this way for producing standard solutions of alkali: why not?

Like sodium(I) carbonate, the carbonates of potassium, rubidium and caesium are basic.

3. action of acids on sodium(I) carbonate

Like other carbonates, sodium(I) carbonate reacts with acids producing an effervescence of carbon dioxide:

$$CO_3^{2-}(s) + H^+(aq) \rightleftharpoons HCO_3^-(aq)$$
$$HCO_3^-(aq) + H^+(aq) \rightleftharpoons H_2O(l) + CO_2(g)$$

ion in solution	precipitate colour
K^+	no precipitate
Ca^{2+}	white
Mg^{2+}	white
Al^{3+}	white
Zn^{2+}	white
Cu^{2+}	green
Fe^{2+}	green
Fe^{3+}	red-brown
Pb^{2+}	white

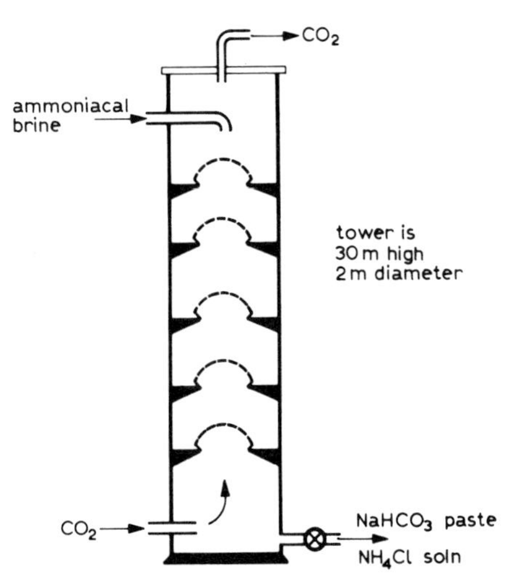

Fig. 16.3 diagram of carbonating tower in Solvay process for Na_2CO_3, showing bubble caps and opposite flow of $CO_2(g)$ and ammoniacal brine

4. aqueous sodium(I) carbonate in analysis

Aqueous sodium(I) carbonate produces characteristically coloured precipitates when added to solutions containing aqueous metal ions, but the precipitate never redissolves in excess of the reagent (contrast this with the action of aqueous sodium(I) hydroxide, section 12.29). The colours of precipitates obtained on adding aqueous sodium(I) carbonate to solutions containing different metal ions are listed here. The precipitate is usually a 'basic carbonate', $MCO_3.M(OH)_2$, since aqueous sodium(I) carbonate contains hydroxide as well as carbonate ions; in some cases, e.g. aluminium, the hydroxide is precipitated.

16.8 industrial production of sodium(I) carbonate

The Solvay, or ammonia-soda, process for the manufacture of sodium(I) carbonate illustrates well the economic principles underlying an efficient industrial operation. The raw materials (sodium(I) chloride, ammonia and limestone) are readily available and the by-products are re-used in the process with such ingenuity that the only waste product is calcium(II) chloride.

Brine (concentrated aqueous sodium(I) chloride) is saturated with ammonia, obtained from the Haber process (see section 24.11). Carbon dioxide is bubbled through the 'ammoniacal brine' in a 'carbonating tower' as shown in Fig. 16.3; the solution emerging from the base of the tower contains many ionic species, among them $Na^+(aq)$, $HCO_3^-(aq)$ and $NH_4^+(aq)$:

$$CO_2(g) + H_2O(l) \longrightarrow HCO_3^-(aq) + H^+(aq)$$
$$NH_3(aq) + H^+(aq) \longrightarrow NH_4^+(aq)$$

Although sodium(I) hydrogencarbonate is soluble in water, it is not very soluble in brine; it is precipitated as a solid:

$$Na^+(aq) + HCO_3^-(aq) \longrightarrow NaHCO_3(s)$$

The precipitate is filtered off and heated, when a residue of anhydrous sodium(I) carbonate ('soda ash') is obtained:

$$2NaHCO_3(s) \longrightarrow Na_2CO_3(s) + H_2O(g) + CO_2(g)$$

Sodium(I) carbonate-10-water may be obtained from this by a process of solution and crystallisation.

Carbon dioxide for the carbonating tower is obtained from the thermal decomposition of sodium(I) hydrogencarbonate (above) and by heating calcium(II) carbonate (limestone):

$$CaCO_3(s) \longrightarrow CaO(s) + CO_2(g)$$

The calcium(II) oxide formed in this reaction, being alkaline, is used to liberate ammonia from the filtrate emerging from the process of separation of 'insoluble' sodium(I) hydrogencarbonate:

$$2NH_4^+(aq) + (Ca^{2+} + O^{2-})(s) \longrightarrow 2NH_3(g) + Ca^{2+}(aq) + H_2O(l)$$

The whole process is summarised diagrammatically in Fig. 16.4 opposite.

Sodium(I) carbonate is used in glass-making (section 21.13), water-softening (section 22.9) and in the manufacture of other sodium compounds.

16.9 sodium(I) hydrogencarbonate, $NaHCO_3$

1. preparation

An aqueous solution is obtained by the action of excess carbon dioxide on aqueous sodium(I) hydroxide:

$$CO_2(g) + OH^-(aq) \longrightarrow HCO_3^-(aq)$$

or on aqueous sodium(I) carbonate:

$$CO_3^{2-}(aq) + H_2O(l) \longrightarrow HCO_3^-(aq) + OH^-(aq) \quad (i)$$
$$CO_2(g) + OH^-(aq) \longrightarrow HCO_3^-(aq) \quad (ii)$$

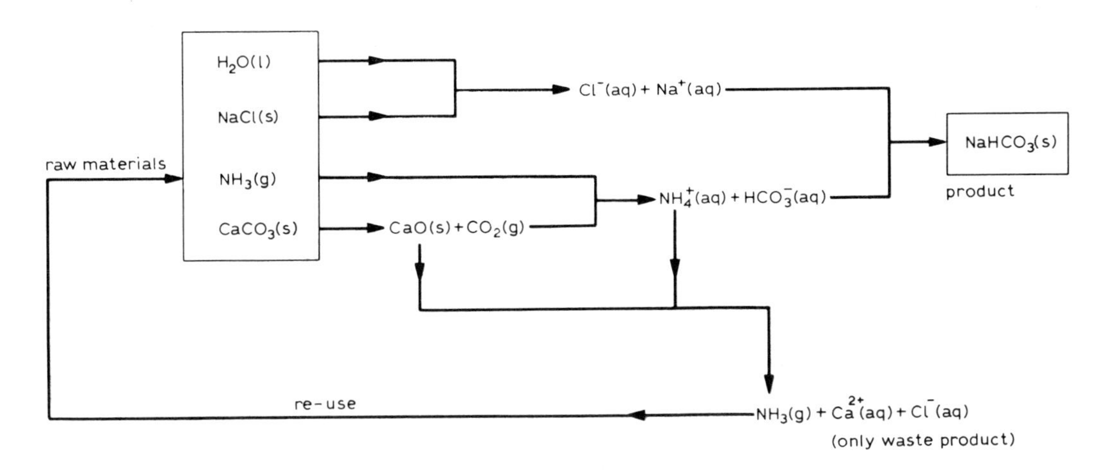

Fig. 16.4 flow sheet for the Solvay process

combining (i) and (ii):

$$CO_3^{2-}(aq) + H_2O(l) + CO_2(g) \longrightarrow 2HCO_3^-(aq)$$

2. thermal decomposition

If solid sodium(I) hydrogencarbonate is heated in an ignition tube, a gas is evolved which turns 'lime water' milky:

$$2NaHCO_3(s) \longrightarrow Na_2CO_3(s) + H_2O(l) + CO_2(g)$$

Because of this, solid sodium(I) hydrogencarbonate cannot be isolated from the aqueous solution prepared as described in (1); a similar decomposition occurs when the aqueous solution is boiled:

$$2HCO_3^-(aq) \longrightarrow CO_3^{2-}(aq) + H_2O(l) + CO_2(g)$$

3. use of 'baking soda

Sodium(I) hydrogencarbonate causes cakes to 'rise' because of the production of carbon dioxide, either by heating in the oven (*q.v.*) or by interaction with acids in the mixture:

$$HCO_3^-(s) + H^+(aq) \longrightarrow H_2O(l) + CO_2(g)$$

16.10 the family of alkali metals

Li
Na
K
Rb
Cs
Fr

Some similarities between the alkali metals which form Group I of the Periodic Table have been mentioned in the preceding sections. They form unipositive ions and the elements therefore show a valency of one; they tarnish in air and react rapidly with cold water; their hydroxides are strongly basic; they have a greater range of soluble salts than do any other metals.

However, there is a gradation in properties from lithium to caesium, as the following examples show.

1. formation of chlorides

All alkali metals react directly with chlorine, forming crystalline chlorides, having an ionic lattice similar to that shown in Fig. 18.22. The heat evolved on the formation of one mole of each chloride under standard conditions ($\Delta H_f^\ominus$) increases down the Group, as is consistent with the increase in metallic character down the Group (see section 17.6(3)).

2. action of the alkali metals on water

Potassium inflames when dropped on to cold water, sodium melts but usually does not inflame, while lithium floats, liberating hydrogen smoothly, but does not melt. The order of reactivity towards water is **Li < Na < K**, exemplifying once more the increasing metallic character down the Group.

$$Li(s) + H_2O(l) \rightarrow Li^+(aq) + OH^-(aq) + \tfrac{1}{2}H_2(g)$$
$$Na(s) + H_2O(l) \rightarrow Na^+(aq) + OH^-(aq) + \tfrac{1}{2}H_2(g)$$
$$K(s) + H_2O(l) \rightarrow K^+(aq) + OH^-(aq) + \tfrac{1}{2}H_2(g)$$

3. ionisation of alkali metals

The change common to reactions (1) and (2) is

$$M \rightarrow M^+ + e^-$$

Since lithium has a small atom, a relatively large quantity of energy is required to ionise it; potassium, having a large atom, requires less energy to ionise it (see section 18.10). Hence potassium atoms form positive ions more readily than do lithium atoms–potassium is the more reactive metal.

Further study will reveal that other factors (e.g. lattice energy evolved when crystals are formed, energy evolved when ions are hydrated) tend to complicate the above relatively simple argument. In particular, these factors render the chemistry of lithium somewhat atypical of the Group of alkali metals; some of its salts have comparatively low solubilities and its carbonate and hydroxide decompose on strong heating. In spite of this, the Group of alkali metals shows many more similarities than differences in the behaviour of its members.

revision summary: Chapter 16

alkali metals: Li Na K Rb Cs

preparation: electrolysis of fused chloride

reactions: burn in air, chlorine
form alkaline oxides and hydroxides
form crystalline chlorides
liberate hydrogen from cold water
reactivity increases Li → Cs

compounds: generally very stable, e.g. hydroxides, carbonates thermally stable
generally all soluble in water

questions 16

1. Francium is the last member of the family of alkali metals. From your knowledge of the other members of this family predict the properties of francium and its compounds. You should consider particularly (a) the solubility of francium compounds, (b) the reaction of francium with water, (c) the reaction of francium with chlorine, (d) the effect of francium(I) hydroxide solution on indicators and (e) the most likely method of extracting metallic francium from its chloride.

Do you think that the reaction of francium with water will be more or less vigorous than that of sodium with water?

2. Outline briefly one method by which sodium(I) hydroxide is manufactured on a large scale. Give two large-scale uses of this compound.

Describe what you would see when sodium(I) hydroxide solution is added to (a) colourless phenolphthalein solution, (b) aqueous copper(II) sulphate(VI), (c) aqueous iron(III) chloride, (d) aqueous aluminium(III) sulphate(VI). Explain the reactions occurring and give ionic equations for the reactions occurring in (b) and (c).

3. Which of the following contains most sodium?
a) one kilogram of sodium(I) carbonate,
b) one kilogram of sodium(I) hydrogencarbonate,
c) one kilogram of sodium(I) chloride.

4. Describe, in outline, how the following may be obtained starting from sodium(I) chloride: (a) metallic sodium, (b) silver(I) chloride, (c) hydrochloric acid, (d) aqueous sodium(I) hydroxide. Give equations for all reactions.

5. What volume of 2M aqueous sodium(I) hydroxide is required to exactly neutralise 50 cm^3 of a solution containing 49 g of sulphuric(VI) acid per dm^3?

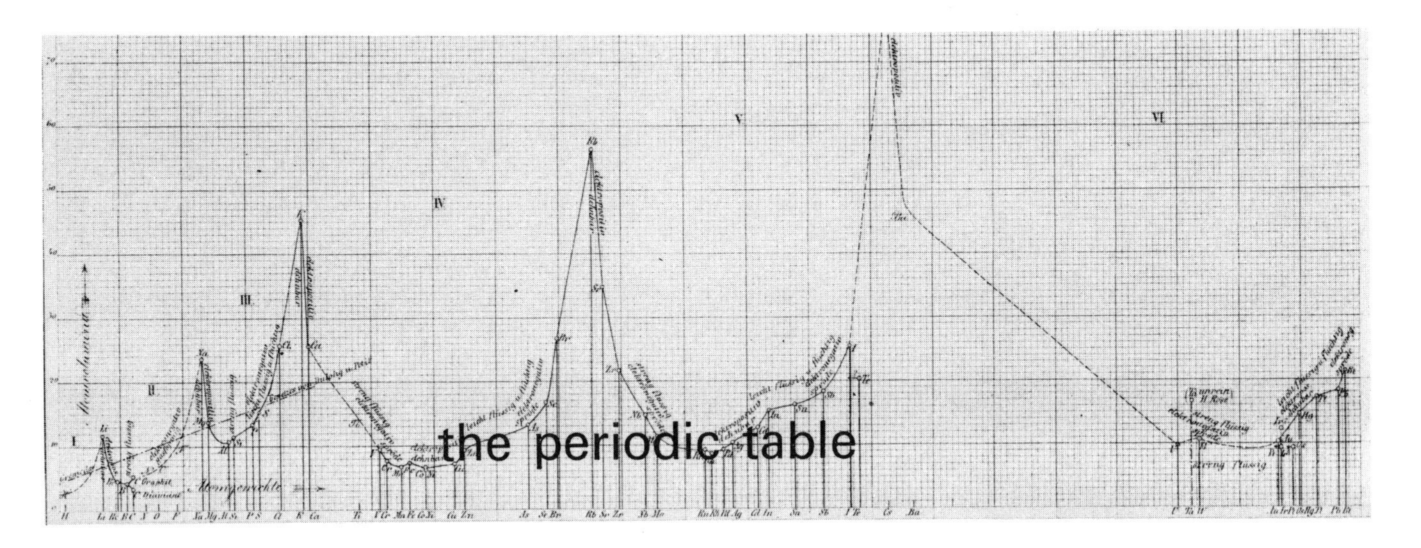

the periodic table

17.1 patterns among elements

Natural philosophers have always looked for patterns in the behaviour of matter. An early classification of elements was that into metals and non-metals (see section 7.10). An element was classified as a metal if it possessed certain physical properties in common with other such elements; examples are a shiny surface when freshly cut, malleability, ductility, high density, the ability to conduct heat and electricity well. Metals were found to have certain chemical properties in common, such as a solid hydride, a crystalline chloride and a basic oxide. Some non-metals share a number of the properties listed above and a clear dividing line is hard to find. Indeed, few classifications in nature are very clear-cut; the patterns may be difficult to discern and careful collection and summary of evidence is essential before judgments are made.

17.2 patterns in properties and relative atomic mass: Döbereiner's 'triads'

Certain groups of elements show strong similarities in their chemical properties; examples of such 'families' of elements encountered in earlier chapters are chlorine, bromine and iodine (Chapter 15) and lithium, sodium and potassium (Chapter 16). By about 1820 Dalton had popularised the atomic theory of matter and relative atomic masses (then known as 'atomic weights') were being estimated. J. W. Döbereiner (1780–1849) discerned an interesting pattern in the relative atomic masses of the groups of elements mentioned above. He arranged the elements into **triads**, or groups of three.

The chemical properties of the 'middle' element in each triad were seen to be *intermediate* between those of the outer two members of the triad. An example of such a pattern is depicted in Fig. 17.1 for the triad calcium, strontium, barium. Döbereiner also claimed that the relative atomic mass of the middle element in the triad was the *average* of the relative atomic masses of the other two elements. Examples of this are shown in Fig. 17.2. This relationship in relative atomic mass was not apparent for every group of three elements of similar properties and in any case the values for the relative atomic masses (or 'atomic weights') which were in use at the time were acknowledged to be somewhat unreliable; as a consequence, Döbereiner's ideas received only limited attention.

	Ca	**Sr**	**Ba**
metal on cold water	smooth	rapid	vigorous
type of hydroxide	sparingly soluble, weakly alkaline	soluble, fairly alkaline	more soluble, strongly alkaline
temperature of decomposition of carbonate(K)	1100	1400	1700
water-solubility of sulphate(VI)	moderate	fairly insoluble	insoluble

Fig. 17.1 properties of the Ca–Sr–Ba triad

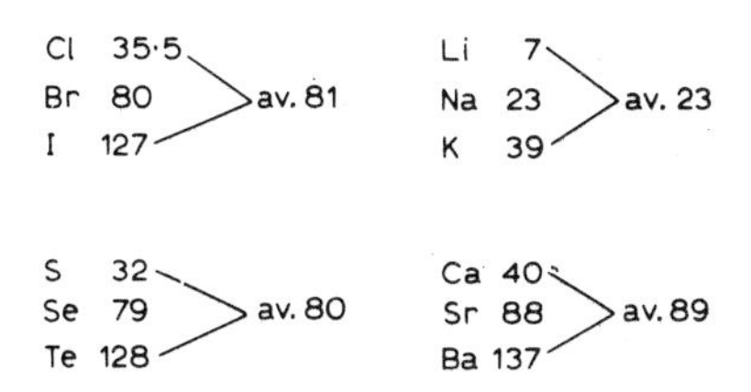

Fig. 17.2 the average A_r of the outer elements is close to the A_r of the middle elements

17.3 periodicity: Newland's law of octaves

By about 1860, more elements were known (about sixty) and the values for relative atomic masses ('atomic weights') were less subject to dispute. J. A. R.

Newlands arranged the elements known at the time in order of increasing relative atomic mass, and in doing so claimed to observe a regular, repeating pattern, or **periodicity.** The word is used in the sense that a regularly repeating and predictable event is said to be periodic; as a weekly, monthly or quarterly journal is termed a 'periodical'. Newlands used a musical analogy: he said **'the eighth element, starting from a given one, is a kind of repetition of the first, like the eighth note in an octave of music'.** His first three 'octaves' are shown in the key of G major in Fig. 17.3.

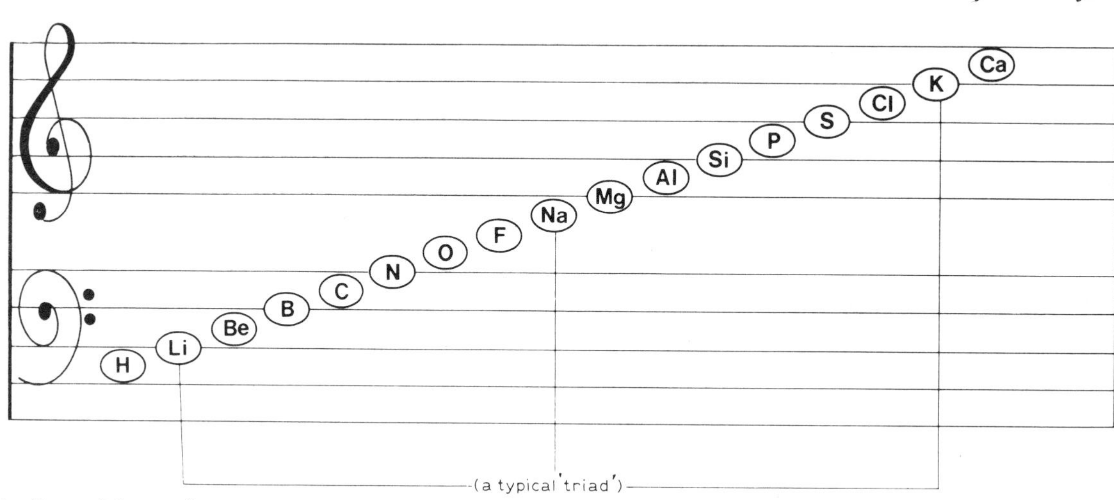

Fig. 17.3 the 'Law of Octaves'

This arrangement undoubtedly shows a regular pattern in chemical behaviour, but to the scientific minds of the time it suffered from certain disadvantages: it only worked satisfactorily for the first sixteen elements, and the idea that atomic properties were periodic suggested a repeating structure within the atom, which contradicted the firmly held belief in the indivisible, unchanging atom. Further, the 'octaves' left no room for elements yet to be discovered. When Newlands read his paper on the topic, it was greeted with laughter and scorn; indeed, Newlands was asked whether he had considered arranging the elements in alphabetical order of the initial letters of their names. In the face of such a reception, Newlands did not pursue his ideas further. However, other scientists were working along similar lines; in particular, Lothar Meyer in Germany and Dmitri Mendeleeff in Russia, independently pronounced a **'periodic law'** describing the properties of elements. Mendeleeff had attended the important meeting of scientists in 1858 at which Cannizzaro outlined his new method for the determination of 'atomic weights' (see section 10.27); much impressed by what he heard, Mendeleeff returned to St Petersburg to work on his ideas for thirteen years before publishing his results.

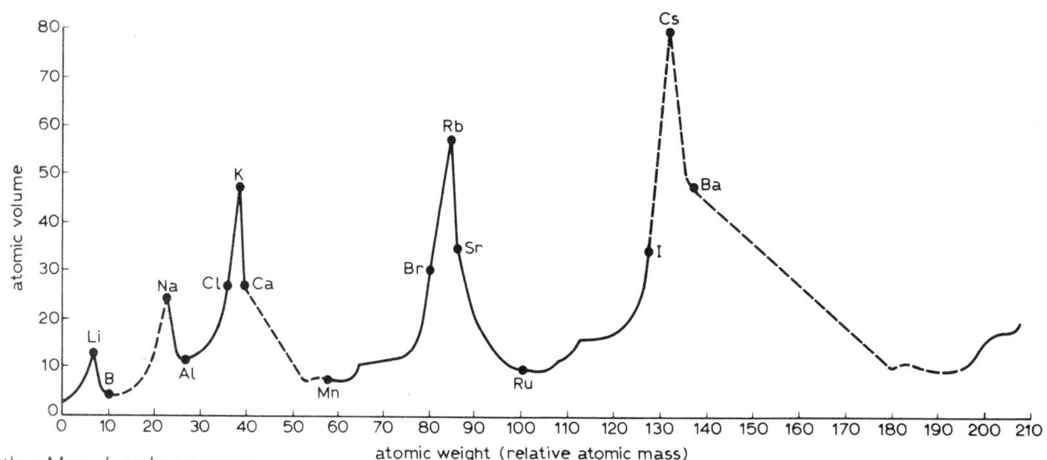

Fig. 17.4 Lothar Meyer's volume curves

17.4 Lothar Meyer (1830–1895) : atomic volumes

In this section, non-standard chemical terms are used in the interests of reasonable historical accuracy.

Meyer observed a periodicity in properties among the elements when he plotted a graph of 'atomic volume' against 'atomic weight', where atomic volume = atomic weight/density. Fig. 17.4 shows the nature of Meyer's results, the peaks and troughs being the important characteristic feature. A particular family of elements, the alkali metals, appear along the peaks and the other families of elements occupy corresponding positions on the curve. Such results afforded clear evidence of periodicity in atomic properties.

17.5 Mendeleeff and the Periodic Table

Mendeleeff assembled the most comprehensive and conclusive proof of chemical periodicity and summarised it in his **Periodic Law:**

The chemical elements, as characterised by their physical and chemical properties, fall into a repeating pattern, if they are arranged in order of increasing 'atomic weight'.

In his **Periodic Table**, Mendeleeff arranged the elements in ascending order of their relative atomic masses (atomic weights) and ordered his arrangement into *vertical groups*, containing elements of similar or related properties (e.g. halogens, alkali metals), and *horizontal periods* or rows, in which the elements of each row show a gradual change in properties. Such a change is exemplified by the elements of the third period:

Na Mg Al Si P S Cl

Mendeleeff was quite prepared to leave gaps in his Periodic Table, if suitable elements did not exist to fit into the pattern; he suggested that elements would eventually be discovered which would fill the gaps. Interpolating from the observed patterns of physical and chemical properties of members of the vertical groups, he predicted many of the properties of such unknown elements and their compounds. The best-known case is that of germanium, to which Mendeleeff gave the name *eka-silicon* before it was discovered. Fig. 17.5 shows how accurate Mendeleeff's predictions proved to be.

The value of the Periodic Table, in rationalising the mass of facts known about the chemistry of the elements and their compounds, and in stimulating research and discovery, led to its immediate and widespread acceptance. Its influence on the growth of understanding of chemistry since about 1870 (an appropriate time for unifying procedures) is difficult to overestimate. The place of an element in the Periodic Table assumed enormous importance as the central feature in the interpretation of its chemistry and each element was given a number describing its place in the order. These numbers were integers, unlike the relative atomic masses on which the order of the Table was based, which were not whole numbers. Hence the first element, hydrogen, was given the number 1, helium 2, lithium 3, etc. The name **atomic number** was given to this numerical order of an element in the Table.

The classification had its imperfections: according to a strict order of increasing relative atomic mass, potassium ($A_r = 39{\cdot}1$) should be element 18 and argon ($A_r = 39{\cdot}9$) element 19. This would group potassium with the noble gases and argon with the alkali metals. Consequently, these two elements (and some other anomalous pairs) were placed out of strict A_r order; argon was recognised as the true element 18 and potassium as 19. Prophetically, atomic number was seen to be of profounder chemical significance than relative atomic mass; the wisdom of this decision was fully revealed when knowledge of the structure of the atom gave a new and fundamentally important meaning to the term atomic number.

	(1) *Es*	(2) Ge
colour	light grey	dark grey
nature	metal	metal
A_r	72	72·3
$\rho/\mathrm{g\,cm^{-3}}$	5·5	5·36
oxide	EsO_2	GeO_2
oxide: $\rho/\mathrm{g\,cm^{-3}}$	4·7	4·7
chloride	$EsCl_4$	$GeCl_4$
chloride: b.p./K	373	356
chloride: $\rho/\mathrm{g\,cm^{-3}}$	1·9	1·88

Fig. 17.5 properties (1) predicted for *eka-silicon*, (2) confirmed on discovery of germanium

Group O

He	(2)
Ne	(2,8)
Ar	(2,8,8)
Kr	(2,8,18,8)
Xe	(2,8,18,18,8)
Rn	(2,8,18,32,18,8)

Group I

Li	(2,1)
Na	(2,8,1)
K	(2,8,8,1)
Rb	(2,8,18,8,1)
Cs	(2,8,18,18,8,1)
Fr	(2,8,18,32,18,8,1)

Group VII

F	(2,7)
Cl	(2,8,7)
Br	(2,8,18,7)
I	(2,8,18,18,7)
At	(2,8,18,32,18,7)

Fig. 17.6 electron configurations for elements in Group O (noble gases), Group I (alkali metals) and Group VII (halogens)

17.6 the modern Periodic Table

A full understanding of the succeeding sections in this Chapter depends upon a knowledge of atomic structure as outlined in Chapter 18. It is therefore advisable to delay further consideration of the modern Periodic Table until the material in Chapter 18 has been read and understood.

1. atomic number and electron configurations

Following the work of H. J. G. Moseley (section 18.4) the **atomic number** of an element is seen to represent the number of units of positive charge on the nucleus of the atom; this is in turn equal to the **number of protons** in the nucleus and hence, for a neutral atom, to the **number of electrons** outside the nucleus. The discovery of the electron configuration of elements (section 18.10) gave a fundamentally satisfying explanation for the periodicity in properties of the elements in the Table. Two important features of the Periodic Table in terms of electron configurations emerge clearly:

(a) Atoms of adjacent elements in the Table differ by one electron in the highest energy level.

(b) Atoms of elements in the same vertical Group all contain the same number of electrons in the highest energy level.

The number of electrons in the highest energy level of an atom critically influences the chemical and physical properties of the element and its compounds. The electron configurations of elements in Group O, Group I and Group VII are shown in Fig. 17.6. Detailed consideration of the correlation between properties and electron configurations in atoms and molecules is given in sections 18.12 to 18.25.

2. new elements

Many new elements have been discovered since Mendeleeff published his Periodic Classifications between 1869 and 1872 The noble gases were incorporated easily as a complete Group and indeed form a keystone in the understanding of the relationship between electron configuration and properties, as is shown in the next two sections (3) and (4) of this Chapter.

Periodic Tables published before 1940 end with element 92, uranium; since then nuclear physicists and chemists have prepared a series of artificial elements, including element 101 which is named mendelevium (Md) in commemoration of the founder of the Table. The production of such elements is extremely expensive and most only exist as short-lived radioactive isotopes. There has been some controversy between East and West over the naming of a number of these elements, but element 105 is generally agreed as hahnium, in honour of Otto Hahn, who discovered nuclear fission in 1939. In theory there is no foreseeable limit to the number of elements which may be prepared artificially.

The modern Periodic Table may be displayed in a number of ways, but most show vertical Groups of elements having similar chemical properties and horizontal rows or Periods of elements having different chemical properties (but see 'transition elements').

3. group relationships in the table

With the exception of the 'transition elements', which should be considered separately (see the next section), elements in the vertical Groups usually show similarity in chemical properties and always show a gradation in physical properties. The similarities are most marked in Groups I and II (which are metals) and in Groups VI and VII (which are non-metals). In explaining Group similarities, as well as the differences apparent across horizontal rows, the most important principle to remember is that stated in section 18.13; that

all elements react in such a way as to achieve completely filled electron energy levels.

Elements in Group I achieve filled energy levels by losing one electron per atom to form positively charged ions; they are all metals with a valency of I. Elements in Group II lose two electrons to form ions bearing two units of positive charge: they are all metals with a valency of II. The common valency number of elements in the same Group is a marked feature of the Table.

Elements in Group VII achieve filled electron energy levels most easily by gaining one electron to form a negatively charged ion: they are non-metals with a valency number of I. Elements in Group VI similarly form negatively charged ions, but such ions bear two negative charges and the common valency number of elements in Group VI is II.

Since a 'valency' electron is more easily removed from a high energy level (as in caesium) than from a low energy level (as in lithium) and since metallic properties are associated with the relative ease of removal of such an electron from the atom of the element, **metallic character increases down a vertical Group of the Table;** caesium and francium are the most metallic elements, whilst barium (Group II) is chemically more similar to sodium than it is to calcium, in terms of metallic nature. Conversely, the extra electron in a negatively charged ion will be more tightly bound to the nucleus if it enters a low energy level (as in fluorine) than it will if it enters a higher energy level (as in iodine). Consequently, fluorine is the most markedly non-metallic halogen with the strongest oxidising properties in the Group, whereas iodine has only weak oxidising properties (see Chapter 15).

Elements in the middle Groups of the Table (III, IV and V) show a similar gradation in properties from non-metallic at the top of the Group to metallic at the bottom of the Group; it is hardly surprising that these Groups include both metals and non-metals.

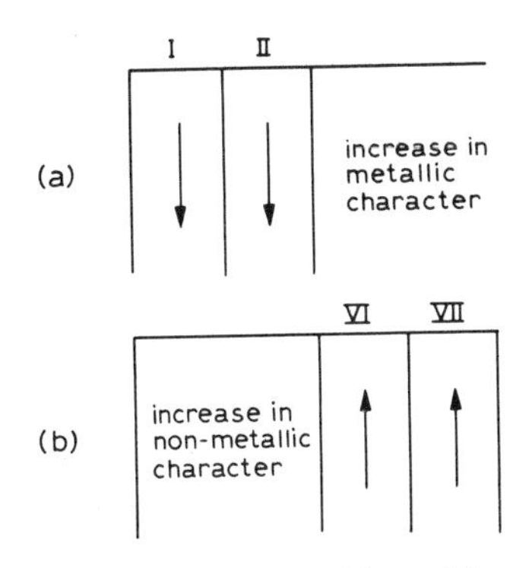

Fig. 17.7 change of (a) metallic or (b) non-metallic character down a Group

4. relationships in horizontal rows or Periods

The Periods vary in length, each Period beginning with an alkali metal and ending with a noble gas.

Period 1 contains only two elements, hydrogen and helium.
Periods 2 and 3 each contain eight elements (Li to Ne and Na to Ar).
Periods 4 and 5 each contain eighteen elements (K to Kr and Rb to Xe).
Period 6 contains thirty-two elements (Cs to Rn) and
Period 7 is not yet complete.

Fig. 17.8 change in character across a horizontal row (m = metal, n-m = non-metal)

The gradation in properties across a horizontal row or period are clearly exemplified in Periods 2 and 3; the elements at the **left-hand side** lose electrons to form positively charged ions and are **metals**, whereas those at the **right-hand side** of the row gain electrons to form negatively charged ions and are **non-metals**. This is shown diagrammatically in Fig. 17.8. When this gradation in chemical character is seen in conjunction with that which occurs on proceeding down a vertical Group of the Table, the existence of **diagonally adjacent elements having similar chemical properties** may be understood (Fig. 17.9). Beryllium resembles aluminium more closely than it does magnesium, while boron and silicon have much in common. This diagonal relationship does not extend beyond adjacent members–a geographical analogy which may be helpful is that many Cumbrians have a pattern of life similar to that which is followed on the Yorkshire Moors, but life in Cumberland is very different from that in East Anglia or Kent. The most marked differences in the Table are represented in Fig. 17.10–Northumbrians are very different from Cornishmen.

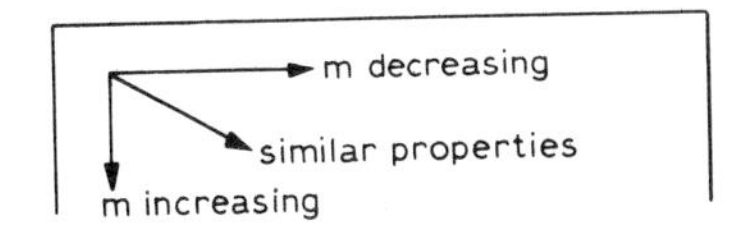

Fig. 17.9 similarity of diagonal neighbours (m = metallic character)

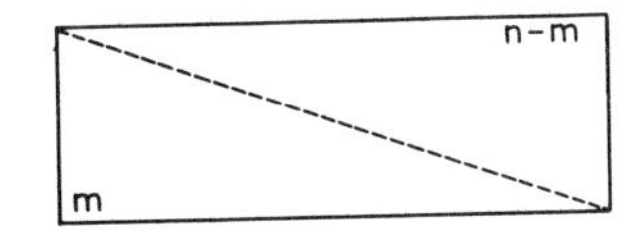

Fig. 17.10

The elements in the first horizontal row of the Table (Li–Ne) are sometimes referred to as **'typical elements'**, i.e. elements having a chemistry

property	Na	Mg	Al	Si	P	S	Cl	Ar
relative atomic mass (A_r)	23	24·3	27	28	31	32	35·5	40
atomic number	11	12	13	14	15	16	17	18
electron configuration	(2,8,1)	(2,8,2)	(2,8,3)	(2,8,4)	(2,8,5)	(2,8,6)	(2,8,7)	(2,8,8)
main valencies	1	2	3	4	3,5	2,4,6	1	0
type and structure	metal crystalline	metal crystalline	metal crystalline	non-metal crystalline giant structure	non-metal molecular solid	non-metal molecular solid	non-metal gas	non-metal gas
ion (if any)	Na^+	Mg^{2+}	Al^{3+}	—	—	S^{2-}	Cl^-	—
chloride	$NaCl$	$MgCl_2$	$AlCl_3$	$SiCl_4$	PCl_3, PCl_5	S_2Cl_2	—	—
structure	ionic crystal	ionic crystal	molecular (covalent) crystal	molecular (covalent) liquid	molecular (covalent) liquid, solid	molecular (covalent) liquid		
solution in water acidic/alkaline	neutral	slightly acidic	acidic	acidic	acidic	acidic		
oxide	Na_2O	MgO	Al_2O_3	SiO_2	P_2O_5	SO_2, SO_3	—	—
structure	ionic crystal	ionic crystal	ionic crystal	covalent giant lattice	molecular (covalent) solid	molecular (covalent) gas, solid		
acid-base properties	basic (soluble)	basic (slightly) soluble	amphoteric (insoluble)	acidic (insoluble)	acidic (soluble)	acidic (soluble)		
hydroxide	$NaOH$	$Mg(OH)_2$	$Al(OH)_3$	?$Si(OH)_4$	H_3PO_4	H_2SO_4	—	—
solubility in water	high	low	very low	moderate	high	high		
acidic/alkaline solution	very alkaline	mildly alkaline	amphoteric	weakly acidic	acidic	very acidic		
hydride	NaH	MgH_2	?(AlH_3)	SiH_4	PH_3	H_2S	HCl	—
physical state	solid	solid	polymerised solid (?)	gas	gas	gas	gas	
bonding	ionic	ionic	covalent (?)	covalent	covalent	covalent	covalent	—

Fig. 17.11 gradation in chemical properties across the second horizontal row of the Periodic Table

typical of the elements in their vertical Groups. This is rather a misnomer, as there are sharper differences between the chemistry of lithium and sodium than between sodium and potassium; similarly, beryllium resembles magnesium rather less than magnesium resembles calcium. The explanation for this slightly anomalous behaviour is beyond the scope of this book, but it is for this reason that the detailed trends in properties of elements in a horizontal row are represented in Fig. 17.11 by the elements sodium to argon rather than lithium to neon.

In Period 4 there is a break in the trends outlined above after element 20 (calcium), which is only resumed at element 31 (gallium). The ten elements

21–30 (Sc to Zn) all show distinct horizontal similarities, particularly in metallic properties, and are called **transition elements.** Transition elements also occur in the rest of the Periods and it is because of the existence of these elements that it is difficult to display a Periodic Table which shows vertical Groups without showing gaps in the first three Periods. This difficulty, however, is only one of paper representation and has no significance in chemistry. The usual form of the modern Periodic Table is shown at the end of this book.

5. periodicity in physical properties

It is interesting to note how the physical properties of elements are related to their atomic numbers. The periodicity of the molar volumes of elements (V_m) may be displayed in a very clear way by cutting and painting pieces of wooden blocks in proportions representing the molar volumes, and sticking the bits to cards representing the elements; such a display is depicted in Fig. 17.12.

Fig. 17.12 'molar volumes' of elements

Other properties may be studied by plotting graphs of atomic number against density, melting-point, boiling-point, specific latent heats of fusion and vaporisation, electrical resistivity and thermal conductivity. It is instructive to note the elements which occupy similar positions on the curves which result.

6. use of the Periodic Table in the study of chemistry

The Periodic Table is exceptionally important in the development of the study of chemistry as it introduces a degree of order into the enormous fund of factual information derived from observation and experiment. It encourages the appreciation of relationships between elements and their compounds; such appreciation stimulates further investigations. The Periodic Table printed at the end of this book should be referred to constantly as an accompaniment to the study of the detailed chemistry of the elements described in subsequent Chapters.

questions 17

1.

$_{1}$G							
			$_{6}$J	$_{7}$L	$_{8}$M	$_{9}$Q	$_{10}$R
$_{11}$T	$_{12}$X				$_{16}$Y	$_{17}$Z	

The diagram above represents a portion of the Periodic Table of the elements. The elements are represented by letters which are not the usual symbols of the elements, but the atomic numbers are correct.

a) Which is the element of lowest density?
b) Which is the least reactive halogen?
c) Which is the most reactive metal?
d) Which element is most likely to have a valency of three?
e) Which element is likely to form an ion carrying two positive charges?
f) Which is the least reactive element? (JMB)

2. Sodium and aluminium have atomic numbers of 11 and 13 respectively. They are separated by one element in the Periodic Table and have valencies of one and three respectively. Chlorine and potassium are also separated by one element in the Periodic Table (they have atomic numbers of 17 and 19 respectively) and yet they both have a valency of one. Explain the difference.

(b) The halogens, fluorine, chlorine, bromine and iodine, show a gradation in properties. Illustrate this by reference to their ease of combination with hydrogen and the ease of replacement of one halogen by another. (JMB)

3. Period 3 contains the elements:
sodium, magnesium, aluminium, silicon, phosphorus, sulphur, chlorine and argon.

Describe how

(a) the elements change in structure, atomic volume, electron configuration and type of ion formed, across the period;
(b) the oxides change in structure and acid–base behaviour in water;
(c) the chlorides change in structure;
(d) the hydrides change in structure and acid–base behaviour in water.

4. What evidence can you find from your studies of the Periodic Table that metallic properties and behaviour appear to increase down a group and decrease across a period?

5. A new, metallic-looking solid, discovered on the Moon, is given the name Lunium. What experiments and measurements may be carried out to find whether or not Lunium is an element and if so, in what position it should be placed in the Periodic Table?

6. Give a brief account of the work of Döbereiner, Newlands, Meyer and Mendeleeff in the understanding of the periodic properties of elements. Include answers to the following questions.

a) Why did Döbereiner observe only a very limited number of Triads?
b) How did Mendeleeff manage to convince a majority of his fellow scientists when Newlands had failed with a fairly similar idea?

7. In his Periodic Law, Mendeleeff suggested that the properties of the elements are a periodic function of their atomic masses. It has also been said that as a result of H. J. G. Moseley's work, the modern Periodic Table 'substituted for Mendeleeff's somewhat romantic classification a completely scientific accuracy'.

What does the author mean by this statement and do you think he has sound reasons for making it?

8. (a) Justify the inclusion of chlorine, bromine and iodine in the same chemical family of elements by considering,

(i) the electronic structures of their atoms,
(ii) their compounds with hydrogen,
(iii) their compounds with silver.

(b) Describe and explain what you would expect to observe in each of the following experiments:

(i) Aqueous lead(II) nitrate(V) is added to separate solutions of sodium(I) chloride and sodium(I) iodide.
(ii) Concentrated sulphuric(VI) acid is added to crystalline sodium(I) bromide. (W)

9. The element barium (Ba) has a relative atomic mass of 137 and the element selenium (Se) has a relative atomic mass of 79.

Barium oxide contains 89·5% by mass of barium. The oxide is a white solid with a very high melting-point. It reacts with water with the evolution of heat and the product of this reaction gives an alkaline solution in water.

Selenium oxide contains 71·2% of selenium. This oxide is a white solid that sublimes at 300 °C and reacts with water to give an acidic solution.

(i) Calculate the empirical formulae of barium oxide and selenium oxide.
(ii) What are the valencies of barium and of selenium in these oxides?
(iii) Name a familiar oxide that closely resembles barium oxide in physical and chemical properties.
(iv) Write an equation for the probable reaction between barium oxide and water and the formulae for the ions present in the resulting solution.
(v) How, and under what conditions, would you expect the element barium to react with water?
(vi) Write a possible formula for the acid formed when selenium oxide reacts with water.
(vii) Describe in terms of electrons the difference between the chemical bonds you would expect to be present in barium and selenium oxides. (C)

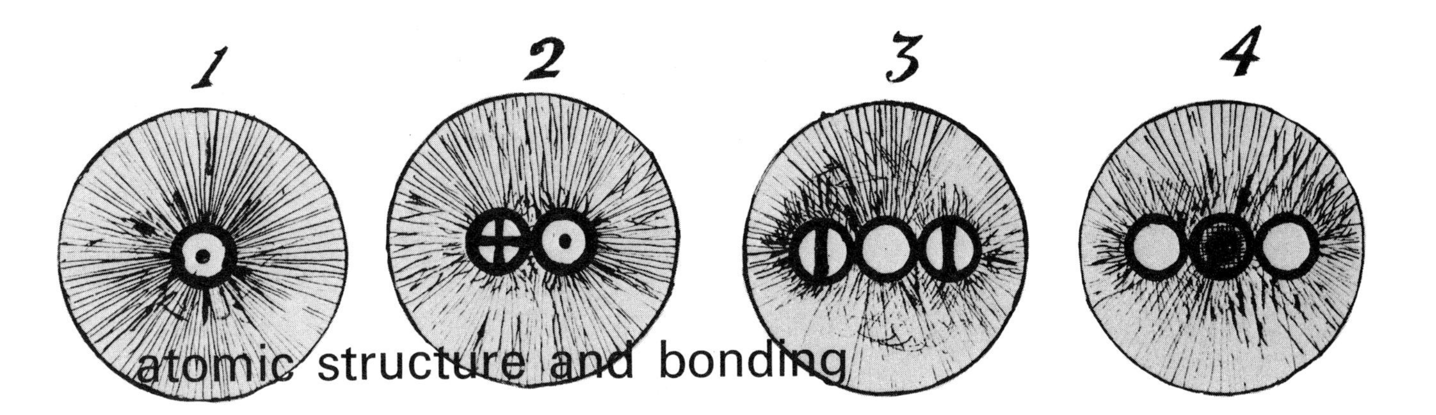

atomic structure and bonding

The Ancient Greeks (particularly Democritus, 460 B.C.) are credited with the earliest known ideas about atoms. However, an atomic theory of matter became widely accepted only in the nineteenth century, when it was propagated by John Dalton and others. The following two fundamental propositions were common to all early atomic theories:

a) atoms are indivisible and indestructible minute particles,
b) all atoms of the same element are identical (though the meaning of the term *element* tended to vary).

Such theories were forced to give way to other models of the atom as the science and technology of the nineteenth and twentieth centuries produced new insights: the model of the atom now proposed by modern physics is very complicated. The value of a model, or theory, lies in what it explains or predicts and the most modern atomic models are not necessary for explanation and prediction in most aspects of our study of chemistry. This Chapter describes some of the particles which are thought to make up the structure of an atom and some of the evidence which was adduced in favour of their existence and nature. The atom is described only in terms of such particles as are necessary to enlighten our chemical observations. The description cannot give details of the chance discoveries, laborious collection of data, design of experiments, testing, rejection and improvement of theories, or of the sheer imagination which resulted in the very valuable model of the atom which we shall use. The story is fascinating and its full flavour can only be captured by reference to more specialised works.

18.1 discovery of sub-atomic particles: the electron

The first idea of an 'atom of electricity' emerged from investigations of electrolysis, such that:

one silver ion + one 'atom of electricity' ⟶ one silver atom
(electron)

By this method an electron charge was measured at $1{\cdot}602 \times 10^{-19}$ coulombs.

1. the production of cathode rays

Gases at normal pressures are usually insulators, but at very low pressures they will permit the passage of electricity through them. Physicists studying the effect caused when electricity was passed through gases at low pressures found that the glass of the containing vessel opposite the **cathode** could be made to glow if the applied potential difference was sufficiently high. A solid object, placed between the cathode and the glow, was found to cast a shadow.

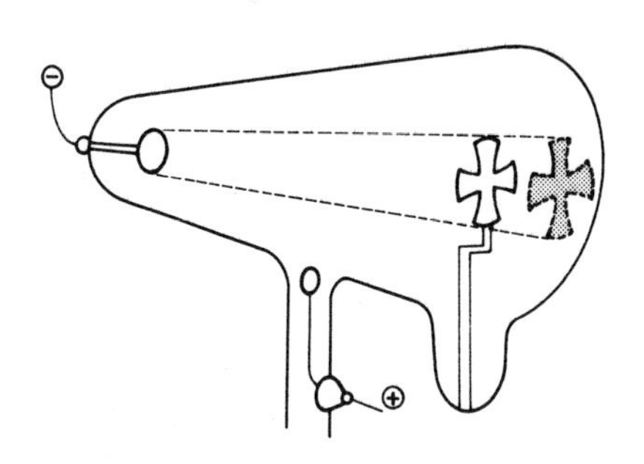

Fig. 18.1 vacuum tube with Maltese cross which casts shadow

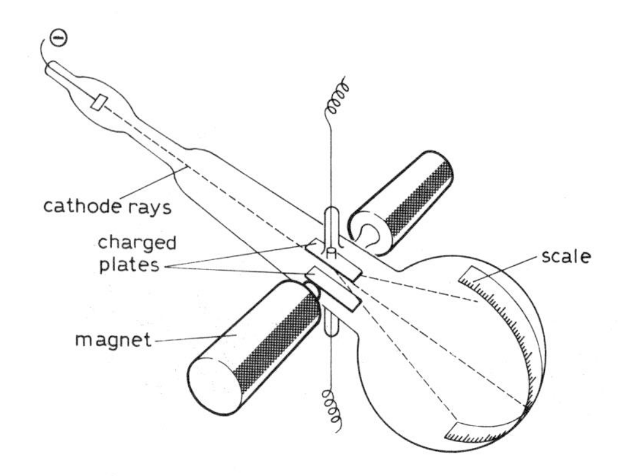

Fig. 18.2 J. J. Thomson's *e/m* tube

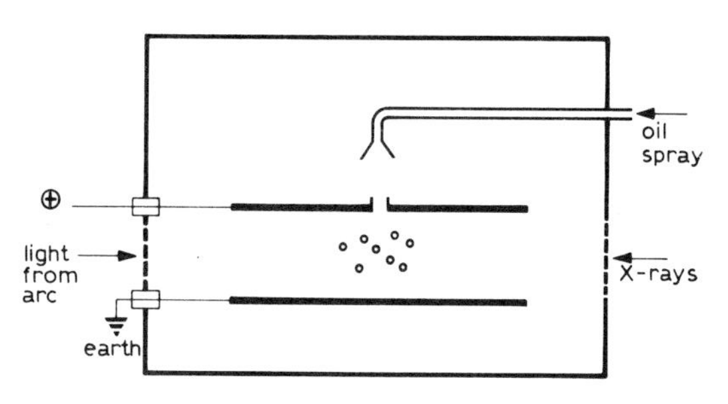

Fig. 18.3 Millikan's oil-drop experiment

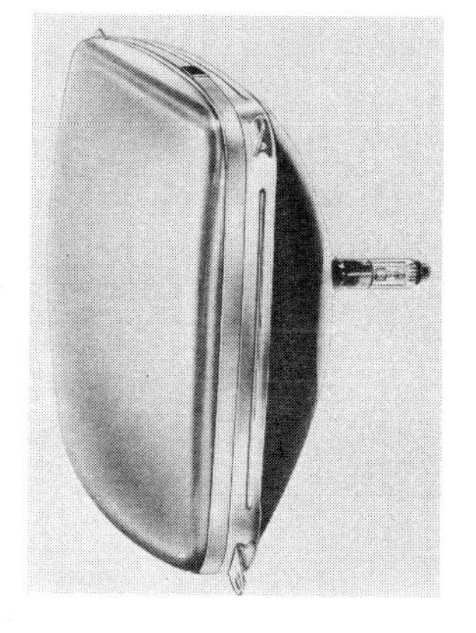

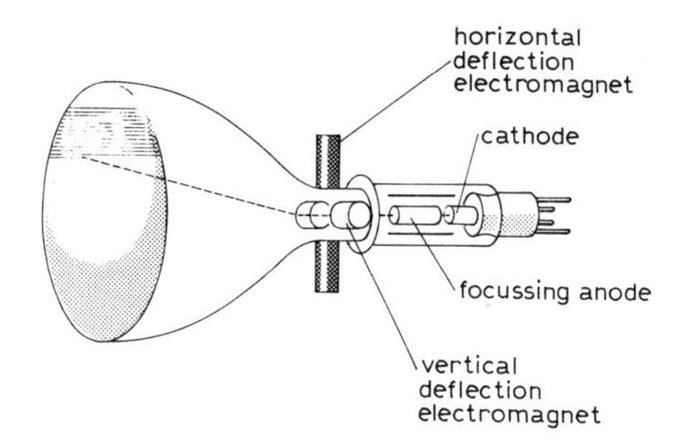

Fig. 18.4 photograph and diagram of a television tube

The glow was therefore believed to be caused by rays coming from the cathode; these rays were called **cathode rays.** At one time it was thought that cathode rays were similar to light rays, but their deflection in a magnetic field subsequently showed them to be streams of electrically charged particles. The direction of the deflection showed that the particles were **negatively** charged.

2. nature of cathode rays: J. J. Thomson's charge/mass (e/m) measurements

In 1897, Professor J. J. Thomson, working in the Cavendish Laboratory, Cambridge, measured the deflection of a narrow beam of cathode rays in both magnetic and electric fields. The results afforded a method of calculation of the charge/mass ratio and of the velocity of the particles: the charge/mass ratio was found to be exactly the same, whatever gas or type of electrodes were used in the experiment. Further experiments on their charge caused Thomson to decide that the particles had a mass very much less than that of the lightest atom known–approximately 1/2000 of the mass of a hydrogen atom. The particles in cathode rays were then called **electrons**–the name already suggested for the 'atom of electricity' to which reference has already been made.

In 1909, R. A. Millikan measured the electron charge accurately in his famous 'oil-drop' experiment; details of this experiment are given in most text-books of physics. The charge was found to be $1{\cdot}602 \times 10^{-19}$ C, identical with that measured earlier in electrolysis experiments. The mass of an electron was found to be $9{\cdot}109 \times 10^{-31}$ kg, which is 1/1837 of the mass of a hydrogen atom.

Beams of electrons in cathode-ray tubes are widely used in laboratories today and the familiar television tube is a further development of the same principle.

Fig. 18.5 Madame Curie

18.2 radioactivity

While Thomson was investigating cathode rays, a French physicist, Henri Becquerel, discovered a strange new phenomenon which rapidly became known as **radioactivity.** His assistant, Marie Curie, together with her husband Pierre, found that radioactivity is the continuous and spontaneous emission of radiations from certain 'radioactive' elements such as radium, uranium and thorium. The rate of emission was found to be unlike that in any known chemical change, since it was completely unaffected by temperature, pressure or the presence of catalysts. The radiations were found to affect photographic plates, ionise gases, produce fluorescence in other substances and cause damage to cells in plants and animals. During radioactive changes energy is also released in large amounts relative to the small quantities of material involved in the change. Mass for mass, the energy liberated is about one million times as great as that liberated in most chemical reactions–such energy may be harnessed for destructive purposes, if released rapidly in the explosion of a nuclear bomb, or more usefully when its release is controlled in a nuclear reactor.

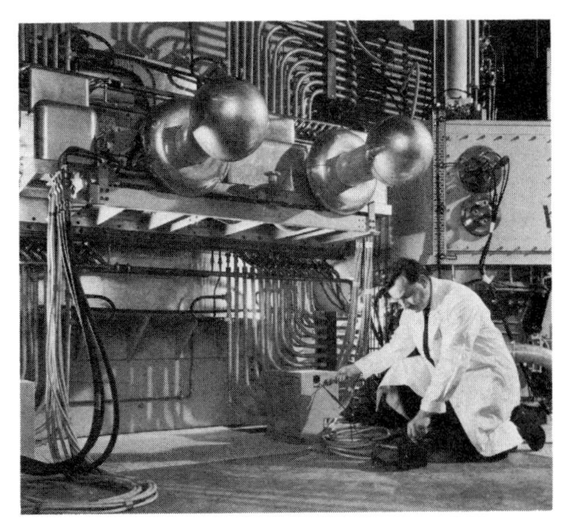

Fig. 18.6 a nuclear pile

1. types of radiation emitted from radioactive materials

The fairly simple experiment represented in Fig 18.7 gives evidence of three types of radiation. The three types behave differently when passed through an electric field and they clearly have different masses and charges. The three types of radiation are called α-, β- and γ-radiation.

i) *α-radiation* consists of particles which have a positive charge exactly twice that of the negative charge of an electron and the same mass as an atom of helium–about four units on the relative atomic mass scale.

α-particles are slower-moving and less penetrating than the other two radiations emitted.

ii) *β-radiation* is identified as consisting of very rapidly moving streams of electrons–particles possessing one unit of negative charge ($1{\cdot}602 \times 10^{-19}$ C) and negligible mass.

iii) *γ-radiation* is not affected by electric or magnetic fields. All the evidence points to this being a highly energetic, very short wavelength electromagnetic radiation, similar to, but more penetrating than, X-rays.

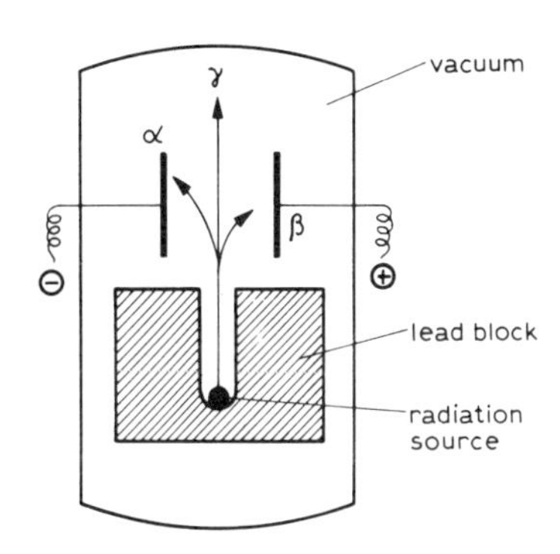

Fig. 18.7 identifying nature of radiation by passing between charged plates

2. radioactive disintegration

Ernest Rutherford and Frederick Soddy suggested in 1902 a theory to explain radioactivity which completely contradicted all previous ideas about atoms. They proposed that when a radioactive substance emits an α- or a β-particle it changes into a different element; the radioactive atom disintegrates into a different atom with the ejection of a sub-atomic particle.

Such a transmutation of elements had been the dream of the alchemists centuries earlier, but had later been rejected by all 'right-thinking' scientists. In fact, radium, uranium, thorium and certain other metals *are* changing continuously, as part of the process of Nature, into other elements.

18.3 the nuclear atom

The flood of new discoveries about sub-atomic particles in the early years of this century demanded new theories describing atoms. Thomson's discoveries of negatively charged electrons in all atoms necessarily implied the existence of a compensating positively charged part; for some time, the most favoured atomic model was Thomson's own 'plum-pudding', in which electrons (the plums) were considered to be embedded in a 'pudding' of positive charge. Experiments carried out by Hans Geiger and Ernest Marsden, two members of Rutherford's research team at Manchester University, on the scattering of α-particles by thin metal foils, completely changed the model in 1909.

Fig. 18.8 Lord Rutherford (1871–1937)

A stream of α-particles was directed at a thin sheet of platinum or gold. Since α-particles are rapidly moving, massive (in atomic terms) particles possessing a great deal of energy, the observation that the vast majority of particles passed through the foil with only slight deflections in their path caused no surprise. The paths taken by the particles were detected by the small flashes of light (called 'scintillations') which they caused on impact with a screen. Such slight deflections were easily explained as being caused by the small scattering effect resulting from random encounters with atoms in the foil, as predicted by the 'plum-pudding' model of the atom (see Fig. 18.9). They were not inconsistent with the notion that atoms were fairly solid entities.

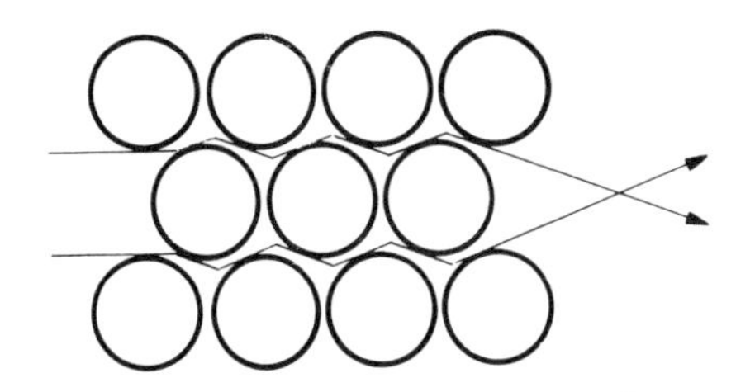

Fig. 18.9 Thomson model of atoms predicts small-angle scattering of α-particles

However, Geiger and Marsden noticed some large deflections and very occasionally scintillations were found to occur on a screen placed on the *same side* of the metal foil as the course of α-particles. Some particles apparently failed to penetrate the foil at all. This was most unexpected. Rutherford said 'it was almost as incredible as if you had fired a 15-inch shell at a piece of tissue paper and it came back and hit you'.

Rutherford explained these surprising observations by proposing his **nuclear model** of the atom. This supposed that atoms consist largely of empty space and that the mass of an atom is concentrated into a very small, positively charged central core called the **nucleus.** The nucleus is thought to occupy only about one ten-thousandth of the diameter of the atom – rather like a marble placed at the centre of a football pitch. A positively charged particle, such as an α-particle, would be strongly repelled on close approach to such a nucleus and would be deflected from its original path through large angles.

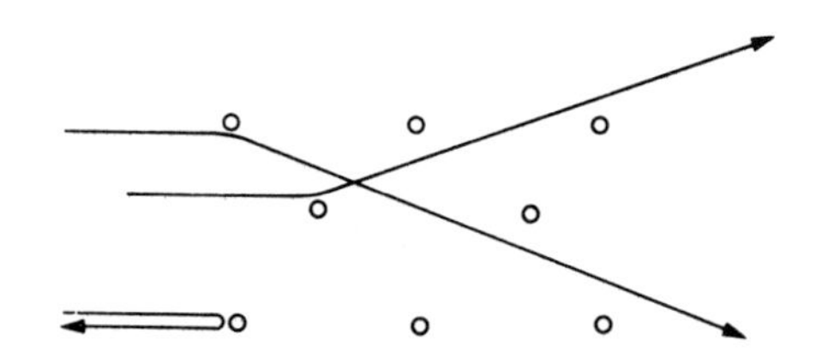

Fig. 18.10 Rutherford nuclear model explains wide-angle scattering of α-particles

18.4 relation between nuclear charge and atomic number

Rutherford proposed that radioactivity is due to a partial disintegration of the nucleus of an atom and that in the case of emission of an α- or a β-particle, the nucleus of another element is formed. A radioactive element which loses one α-particle from the nucleus of its atom changes into an element *two positions earlier* in the Periodic Table (i.e. having an atomic number two units less – see section 17.5). Since loss of an α-particle means loss of two positive charges from the nucleus, it seems likely that successive positions in the Periodic Table are associated with successive unit increases in the nuclear charge.

This idea was proved to be correct by a brilliant series of experiments conducted by H. J. G. Moseley when working as a member of Rutherford's team in 1913. He established a relationship between the frequency of the X-rays emitted by an element and its position in the Periodic Table (atomic number). Moseley said of his results 'We have here a proof that there is in the atom a fundamental quantity which increases by regular steps as we pass from one element to the next. This quantity can only be the charge on the central positive nucleus.' The main conclusion to be drawn from Moseley's work is that the **atomic number** of an element is a fundamental property of each atom of that element, which represents the **magnitude of the positive charge on the nucleus** of the atom.

The Periodic Table (Chapter 17) which Mendeleeff had devised by arranging elements in increasing order of their 'atomic weights' (now called relative atomic masses), could now be explained and modified slightly by arranging the elements in it in ascending order of their atomic numbers. Thus Moseley was able to verify argon and potassium as elements 18 and 19 respectively, although their relative atomic masses are 39·9 and 39·1 respectively. He also revised the positions of nickel and cobalt in the original table.

18.5 the proton

Research on the positive rays which were found to be emitted from discharge tubes and further experiments in α-particle bombardment led Rutherford to propose the existence of a particle in the nucleus of all atoms, which is responsible for the positive nuclear charge. He called it the **proton.** The proton carries a positive charge of $1{\cdot}602 \times 10^{-19}$ C, equal in magnitude but opposite in nature to the charge on an electron, and has a mass of $1{\cdot}672 \times 10^{-27}$ kg. It is convenient in the subsequent discussion to refer to the mass of a proton as one unit and the charge on a proton as one unit of charge, equal and opposite to the unit of charge on the electron. Rutherford further proposed that the **atomic number** of an element represents the **number of protons in the nucleus** of an atom of that element. The fundamental chemical differences between atoms of different elements can therefore be traced to one single difference in the composition of atoms: the number of protons in the nucleus of each atom. Each electrically neutral atom will, of course, possess the same number of electrons outside the nucleus as there are protons within the nucleus.

Atomic number is now taken to describe **three** features of an element:

a) the position of the element in the Periodic Table,
b) the number of protons in the nucleus of the atom,
c) the number of electrons outside the nucleus in a neutral atom of the element.

18.6 the neutron

The scale of relative atomic mass described in section 9.4 referred the masses of all atoms to the mass of a hydrogen atom. A hydrogen atom can be shown to consist of one proton and one electron; since the mass of an electron is negligible, it follows that such a scale of relative atomic mass is identical with a scale which refers the masses of atoms to the mass of a proton taken as one unit.

The mass of an atom is essentially that of its nucleus and the atomic number of an element is seen to be, very approximately, half of its relative atomic mass. The mass of an atom, which is concentrated in its nucleus, cannot be accounted for by the existence of protons alone. Rutherford therefore proposed the existence of a particle in the nucleus having a mass equal to that of a proton but with zero electric charge; he thought of this particle as a proton and an electron in very close association. The existence of such a particle was very hard to detect and not until twelve years after Rutherford's suggestion did Sir James Chadwick decide that he had sufficient evidence for the existence of a nuclear particle with a mass similar to that of the proton but with no electrical charge. The particle was named the **neutron.**

18.7 isotopes

1. nature of isotopes

According to the Rutherford model of the atom, the nucleus consists of protons and neutrons, each having a mass of one unit on the proton scale. The relative atomic masses of elements should therefore be whole numbers. The fact that chlorine has a relative atomic mass of approximately 35·5 seemed at first to be inconsistent with this model of the atom.

This question had been considered before the discovery of the proton and the neutron. In 1913 Frederick Soddy, from his investigations into the breakdown of radioactive elements, reinforced even earlier ideas that atoms of the same element are not all identical, in sharp contrast to the postulates stated at the very beginning of this Chapter. Soddy proposed that atoms of the same element, having the same chemical properties, could have different atomic masses; for such atoms he coined the term **isotopes.** The word means 'equal place'–occupying the same place in the Periodic Table and having the same atomic number.

2. isotopes and atomic mass

Support for Soddy's theory came with the development of accurate mass spectrometers (see Chapter 9). For example, the mass spectrograph of naturally occurring lead (Fig. 18.11) shows four peaks indicating four lead isotopes of different relative abundance:

isotopic mass:	204	206	207	208
% relative abundance:	2	24	22	52

The **relative atomic mass** of an element is the statistically weighted average of the isotopic masses; natural lead has a relative atomic mass of 207·2.

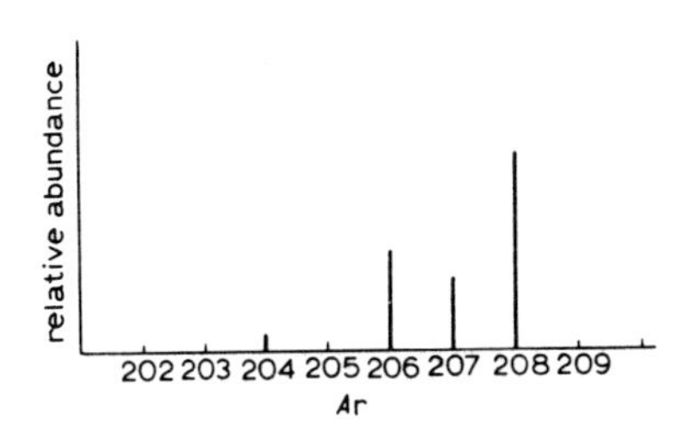

Fig. 18.11 mass spectrograph diagrams showing four isotopes of lead

Most elements have isotopes: chlorine, which has a relative atomic mass of approximately 35·5, has atoms of mass 35 (75·5%) and 37 (24·5%).

3. *international standard of atomic mass*

Reference to the relative masses of atoms and sub-atomic particles in preceding sections of this book has necessarily been made in approximate terms. Relative atomic masses have been referred to the mass of a hydrogen atom taken as unity or to the mass of a proton taken as unity; the masses of these two species are very nearly identical if hydrogen having an isotopic mass of 1 unit is considered. However, isotopes of hydrogen having isotopic masses of 2 and 3 units also exist and this means that the relative atomic mass of naturally occurring hydrogen atoms, referred to the mass of a proton taken as unity, is approximately 1·008. Further, the masses of a proton and a neutron are not quite identical:

$$m_{\mathrm{p}} = 1{\cdot}672 \times 10^{-27}\,\mathrm{kg}$$
$$m_{\mathrm{n}} = 1{\cdot}675 \times 10^{-27}\,\mathrm{kg}$$

For all but the most exact statements it is convenient to refer to the masses of protons and neutrons in terms of the 'unified atomic mass unit, u', where

$$1\,\mathrm{u} \approx 1{\cdot}66041 \times 10^{-27}\,\mathrm{kg}$$

It is sufficiently precise for our purposes to say, for example, that in the case of the chlorine isotope of mass 35, the nucleus contains 17 protons of total mass 17 u and 18 neutrons of total mass 18 u, the total isotopic mass being 35 u.

However, the relative atomic masses (old name: 'atomic weights') must be referred to a scale other than 'H = 1'.

Relative atomic mass is now defined as the ratio of the average mass per atom of the natural nuclidic composition of an element to 1/12 of the mass of an atom of nuclide ^{12}C. This means that the standard to which the masses of all atoms are compared is the mass of an atom of carbon containing six protons and six neutrons in its nucleus taken to be twelve units *exactly*. In these terms, the relative atomic mass of chlorine, $A_r(\mathrm{Cl}) = 35{\cdot}453$.

4. *isotopes, protons and neutrons*

The discovery of protons and neutrons gave an excellent explanation for the existence of atoms with the same atomic number but different atomic masses. In isotopes of one element the number of protons must be the same but the number of neutrons may be different.

A species of atoms which are identical as regards atomic number (number of protons in the nucleus) and mass number (sum of numbers of protons and neutrons, or nucleons, in the nucleus) is called a **nuclide.** Nuclides are symbolised conventionally as follows:

$$^{\text{mass number}}_{\text{atomic number}}\mathrm{X} \quad \text{or} \quad ^{A}_{Z}\mathrm{X}$$

If the atomic number Z and the mass number A of a nuclide are both known, it is easy to calculate the number of protons and neutrons in its nucleus; the number of protons is equal to the atomic number and the number of neutrons is equal to the difference between the mass number and the atomic number.

number of protons $= Z$
number of neutrons $= A - Z$

Examples:

hydrogen has three isotopes:	$^{1}_{1}\mathrm{H}$	$^{2}_{1}\mathrm{H}$	$^{3}_{1}\mathrm{H}$
	protium	deuterium	tritium
protons:	1	1	1
neutrons:	0	1	2

uranium, with the heaviest naturally occurring atom, consists mainly of three isotopes:

	$^{234}_{92}U$	$^{235}_{92}U$	$^{238}_{92}U$
protons:	92	92	92
neutrons:	142	143	146

Fig. 18.12 a particle accelerator

5. radioactive isotopes

Some naturally occurring isotopes, such as $^{3}_{1}H$, $^{40}_{19}K$, $^{14}_{6}C$, are radioactive. All of the isotopes of uranium are radioactive. Other radioactive isotopes may be obtained artificially; by neutron bombardment in a nuclear reactor, or by bombardment of certain 'targets' with accelerated, electrically charged particles such as protons, α-particles or even quite massive ions. Almost all elements now have a number of these artificially made isotopes; they are always radioactive. Several elements, particularly those of higher atomic number than uranium (92), only exist as tiny quantities of their artificial isotopes. The search for new elements and new isotopes of known elements still goes on, limited only by the enormous expense of the technology.

6. half-life of radioactive isotopes

The stability of a radioactive isotope is measured by its **half-life.** This is the length of time taken for the radioactivity to be reduced to half of its original value, i.e. the time taken for half of the number of radioactive atoms to disintegrate.

For a particular radioactive isotope, half-life is a constant; no matter how much radioactive material is present initially, the time taken for half of it to disintegrate is always the same. Half-lives vary from millions of years to fractions of a second.

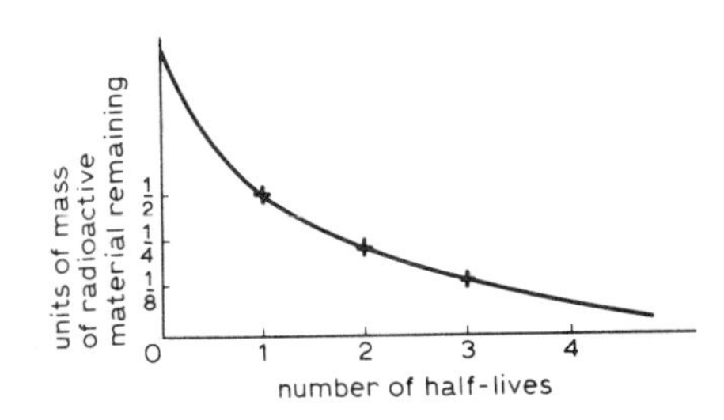

Fig. 18.13 radioactive decay in terms of 'half-life

Uranium 238 ($^{238}_{92}U$) is an α-emitter with a half-life of $4{\cdot}51 \times 10^9$ years. For this reason there is still a good deal of uranium on Earth and it is much sought after for processing for the manufacture of nuclear fission bombs and reactors.

From an initial mass of x g of uranium 238, after one half-life ($4{\cdot}51 \times 10^9$ y), $x/2$ g remain; after two half-lives ($9{\cdot}02 \times 10^9$ y), $x/4$ g remain and after three half-lives ($13{\cdot}53 \times 10^9$ y), $x/8$ g remain, etc.

Carbon 14 ($^{14}_{6}C$), a β-emitter, has a half-life of 5570 y which enables it to be used for dating carbon-containing materials in archaeological and geological remains.

Strontium 90 ($^{90}_{38}Sr$), one of the dangerous components in radioactive fall-out from nuclear bomb explosions, has a half-life of 27 y.

The only known isotope of element 103, lawrencium, ($^{257}_{103}Lr$) has a half-life of 8 seconds and polonium 212 ($^{212}_{84}Po$) has a half-life of 3×10^{-7} seconds.

18.8 changes during radioactive decay

a) *α-emission* An α-particle is a helium nucleus, $^{4}_{2}He^{2+}$; loss of this particle decreases the mass of the nuclide by four units and the atomic number by two units:

e.g. $$^{238}_{92}U \xrightarrow{-\alpha} {}^{234}_{90}Th$$

Uranium 238 changes to thorium 234.

b) *β-emission* A β-particle is an electron emitted from the nucleus when a neutron changes to a proton. The mass of the nuclide remains the same on such emission but the atomic number increases by one unit.

$$^{14}_{6}C \xrightarrow{-\beta} {}^{14}_{7}N$$

Carbon 14 changes to nitrogen 14.

Fig. 18.14 Niels Bohr (1885–1962)

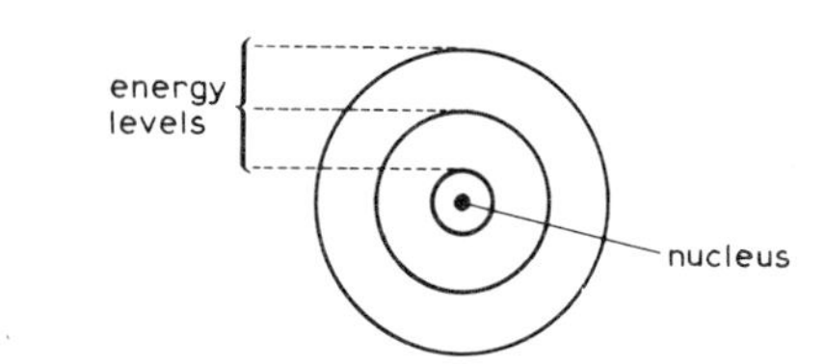

Fig. 18.15 simplified representation of energy levels for electrons

18.9 arrangement of electrons outside the nucleus

The problem of how electrons are arranged in atoms has exercised scientists ever since electrons were discovered. One of the best of the earlier theories was proposed by Niels Bohr (1885–1962) and the combination of his ideas about electrons and Rutherford's proposals concerning the nature of the nucleus is called the 'Rutherford–Bohr' model of the atom.

In 1901, Max Planck (1858–1947) had proposed that energy, like matter, is particulate. Energy can be transferred only in multiples of certain units which he called **quanta.** Bohr applied this idea to electrons, suggesting that as electrons have energy, they must possess it in certain definite amounts. This means that **electrons can only exist in a stable way in particular energy levels in the atom,** i.e. at particular distances from the nucleus.

Fig. 18.15 gives a very convenient, though highly over-simplified, pictorial representation of the energy levels in an atom. If we imagine a negatively charged electron adhering closely to the positively charged nucleus, energy needs to be supplied to the electron to separate it from the nucleus and promote it to the innermost energy level. This energy is best regarded as **potential energy**, similar to that which must be supplied to an object to raise it above the earth in opposition to the force of gravitational attraction represented by its weight. The electron would obviously need even more energy to be supplied to it to promote it to successive energy levels further away from the nucleus. The energy levels therefore represent the ability of the electron to escape from the attractive force of the nucleus; those in the outermost energy levels will escape most easily. We shall see later in this Chapter that **chemical reactions involve the exchange of electrons** between atoms; obviously the electrons in the **outermost** energy levels will be the ones involved in chemical reactions.

Bohr also proposed that electrons can move from a lower to a higher energy level when the atom absorbs quanta of energy, and that when they move from higher to lower energy levels, the atom emits energy. The electron cannot remain between the energy levels. This is rather like the situation when we climb a ladder – we cannot remain in a stable way between the rungs. Bohr's proposals helped to explain many observations, particularly in **spectroscopy** – the study of emission and absorption of radiation. His work earned Bohr the Nobel Prize for Physics in 1922.

18.10 number of electrons in energy levels: evidence from ionisation energies

It is possible to measure how much energy is required to remove electrons one at a time from an atom. This is called **ionisation energy,** since ions are formed when electrons are removed; the energy required to remove one electron is the **first** ionisation energy, that required to remove the second electron from the resulting ion is the **second** ionisation energy, and so on.

Ionisation energies give very valuable information which can be used to gain an insight into the chemistry of different elements and into the pattern of arrangement of electrons within an atom. We will consider first, the *first* ionisation energies of *different* atoms, then *all* of the ionisation energies of *one* atom.

1. first ionisation energies of different atoms

This graph shown in Fig. 18.16 on the page opposite has several significant features closely linked with periodicity:

i) *The peaks* are occupied by members of one group of the Periodic Table: *the noble gases.*
ii) *The troughs* are occupied by members of one group of the Periodic Table: *the alkali metals.*

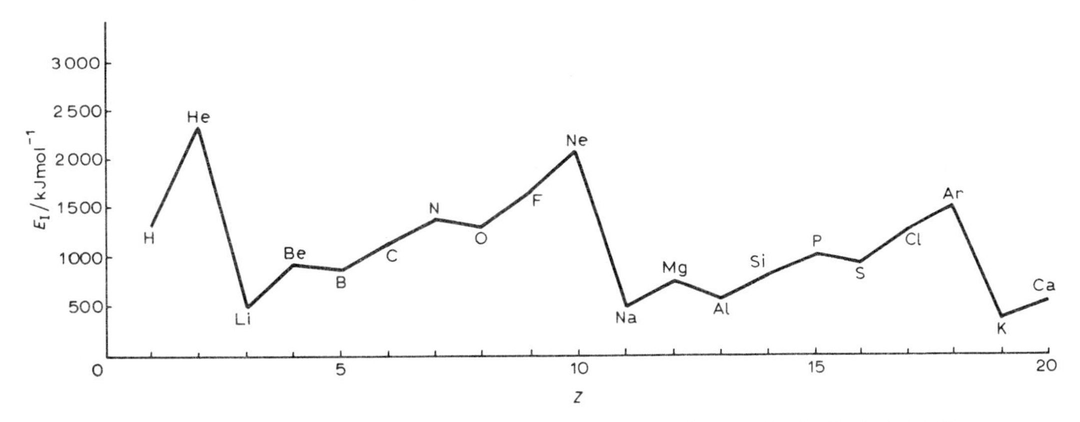

Fig. 18.16 first ionisation energy plotted against atomic number for the first twenty elements in the Periodic Table

iii) *Sharp drops* in ionisation energy occur between a noble gas and the adjacent alkali metal, e.g. He to Li and Ne to Na.
iv) Ionisation energy *increases gradually* from an alkali metal to the next noble gas, e.g. Li to Ne, i.e. across a horizontal row of the Periodic Table.
v) Members of the other 'family' of elements which has been considered in detail (Chapter 15) occupy *corresponding positions* on the graph (F, Cl).

The atom of hydrogen contains one electron; that of helium, two; that of lithium, three. Why is more energy required to remove the first electron from a helium atom (two protons) than from a hydrogen atom (one proton) when much *less* energy is required to remove the first electron from a lithium atom (three protons)?

A reasonable theory is that the hydrogen atom has one electron in the first energy level, attracted by one proton; that of helium has two electrons in the same energy level, attracted by two protons–more energy is required to remove one of these electrons because of the presence of the extra proton in the helium nucleus. If the atom of lithium has two electrons in the first energy level and *one electron in the second energy level,* the greater potential energy possessed by this third electron renders it more easily removed from the nucleus–hence the first ionisation energy of lithium is even lower than that of hydrogen.

If the extra electrons contained in the atoms from lithium to neon are placed in this second energy level, the gradual increase in first ionisation energy is explained by the corresponding increase in the number of protons in the nucleus of each atom. The atom of neon contains eight electrons in the second energy level. If the electrons in the atom of sodium are arranged so that there are two electrons in the first energy level, eight electrons in the second energy level and *one in the third energy level,* the low value of the first ionisation energy of the sodium atom can be explained in terms similar to the argument used above for the case of lithium. The argument may then be repeated for the elements up to argon, which has eight electrons in the third energy level. Potassium, having *one electron in the fourth energy level,* is similar to sodium.

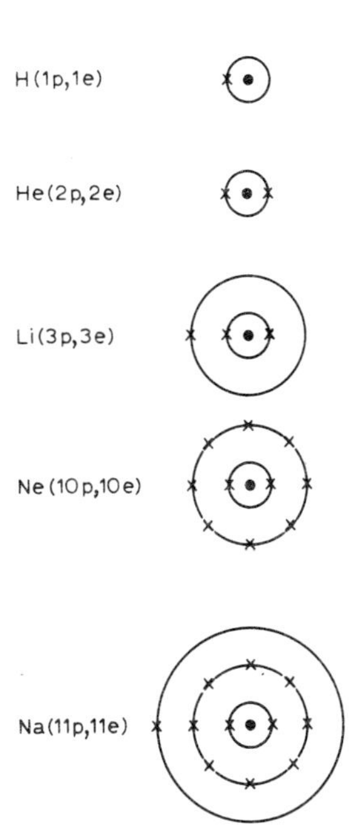

Fig. 18.17 number of electrons in energy levels

Explanations for small irregularities in the graph, e.g. Be–B and N–O, are beyond the scope of this book; however, this information may be found in more advanced texts.

2. *values for all ionisation energies of the potassium atom*

The atom of potassium contains nineteen electrons and Fig. 18.18 shows the energies required to remove all nineteen of these electrons, one at a time. Reading the graph from left to right, the points show the energy required to remove one electron from the neutral K atom, then that required to remove one electron from the K^+ ion, then that required to remove one electron from

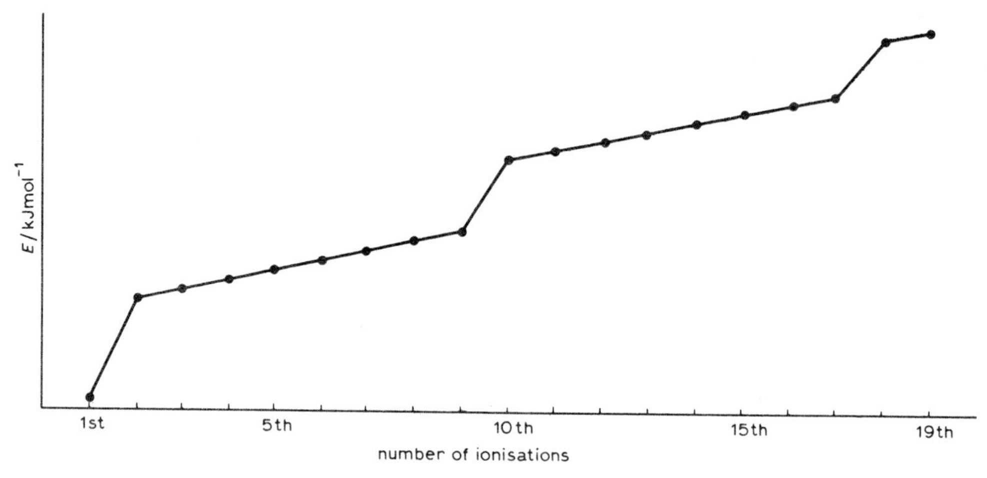

Fig. 18.18 all ionisation energies of the potassium atom plotted against the number of the ionisation

the K^{2+} ion, and so on. The energy required for successive ionisations does not increase regularly; sharp increases occur after the *first, ninth and seventeenth* electrons have been removed. This argues the existence of groups of **one, eight, eight and two** electrons in the potassium atom. If the graph is now read from right to left, these groups of electrons can readily be interpreted as being the two electrons in the first energy level (requiring most energy for their removal), eight electrons in the second energy level, eight electrons in the third energy level and one electron in the fourth energy level (requiring least energy for its removal).

18.11 electron configurations

The evidence obtained by interpretation of Fig. 18.18 supports that obtained on interpretation of Fig. 18.17 and enables us to understand the number of electrons in each energy level of an atom. This is represented in the **electron configuration** of the atom; the electron configuration of potassium is written as (2,8,8,1). Electron configurations for the first twenty elements in the Periodic Table are shown in Fig. 18.19.

H (1)							He (2)
Li (2,1)	Be (2,2)	B (2,3)	C (2,4)	N (2,5)	O (2,6)	F (2,7)	Ne (2,8)
Na (2,8,1)	Mg (2,8,2)	Al (2,8,3)	Si (2,8,4)	P (2,8,5)	S (2,8,6)	Cl (2,8,7)	Ar (2,8,8)
K (2,8,8,1)	Ca (2,8,8,2)						

Fig. 18.19 electron configurations for the first twenty elements in the Periodic Table

A clear periodicity in electron configurations is evident in the above table:

the noble gases have completely filled energy levels,
the alkali metals have one electron in their outermost energy levels,
the 'alkaline-earth' metals have two electrons in their outermost energy levels,
the halogens are one electron short of a completely filled outermost energy level.

If the material in this Chapter has been thoroughly understood, it should be possible to offer reasons for the arrangement of elements in the Periodic Table as described in Chapter 17. The remaining sections in this Chapter are devoted to an explanation of the physical and chemical properties of some important substances, in terms of the principles of atomic structure which have been developed in the preceding sections.

18.12 physical properties and bonding

substance	m.p./K	b.p./K	L_V/J mol^{-1}	molecular (m) or non-molecular (n-m)	bonding
hydrogen	14	20	450	m	covalent
nitrogen	63	77	2800	m	covalent
oxygen	54	90	3400	m	covalent
fluorine	54	85	3200	m	covalent
methane	91	112	8400	m	covalent
hydrogen chloride	158	188	16200	m	intermediate
ammonia	196	240	23500	m	intermediate
sulphur dioxide	200	263	25000	m	intermediate
tetrachloromethane	250	350	30600	m	covalent
water	273	373	41200	m	intermediate
magnesium(II) chloride	987	1680	137300	n-m	ionic
potassium(I) iodide	950	1600	145700	n-m	ionic
potassium(I) chloride	1040	1680	163000	n-m	ionic
sodium(I) chloride	1074	1740	171500	n-m	ionic
graphite	4000	4500	718200	n-m	covalent

Fig. 18.20 physical properties of molecular and non-molecular substances

Examine carefully the first three columns of numbers in Fig. 18.20; the third column gives values for the latent heat of vaporisation of the substances (see sections 2.6 and 2.7). Some of the substances named in the table have strong forces binding the atoms within their molecules, but relatively weak forces between the molecules; they are classed as **molecular substances.** Contrast the energy required to separate the molecules of such substances from each other (L_V) with the energy required to separate the atoms within the molecules (bond energies, Fig. 18.21).

For other substances in Fig. 18.20, the attractive forces between the particles which must be overcome in vaporising the substance (L_V) are of the same order as the strong forces between atoms in molecular substances (bond energies). Such substances do not behave as though they are composed of molecules; they possess binding forces between the particles which are

bond	bond energy/J mol^{-1}
H—H	436800
O=O	495600
F—F	160000
Cl—Cl	243600
C=C	613200
C—H	415800
C—O	361200
C—Cl	340200
H—Cl	432600
H—O	466200

Fig. 18.21 bond energies (energy required to separate atoms in molecules)

strong and continuous throughout the structure. They are called **non-molecular substances**; the classification of substances into molecular and non-molecular was introduced in section 10.6. We must now try to explain how the strong, continuous forces in some non-molecular substances arise.

18.13 ionic compounds

We will make use of the following ideas which are developed in other sections:

a) compounds which conduct electricity in the molten state are composed of ions (section 13.7)

b) the magnitude of the charges on the ions can be determined (section 13.9)

c) when an element is formed from an ion at an electrode, the charge on the ion is neutralised by a transfer of electrons (section 13.11)

e.g. $Na^+ + e^- \longrightarrow Na$ (*cathode*)

$Cl^- \longrightarrow \frac{1}{2}Cl_2 + e^-$ (*anode*)

d) the electron configurations of atoms can be determined (section 18.10)

e.g. Na (2,8,1)

Cl (2,8,7)

When sodium reacts with chlorine forming sodium chloride, the reaction can be represented as a simple electron transfer, the reverse of (*c*) above:

$$Na\ (2,8,1) \longrightarrow Na^+\ (2,8) + e^-$$
$$Cl\ (2,8,7) + e^- \longrightarrow Cl^-\ (2,8,8)$$

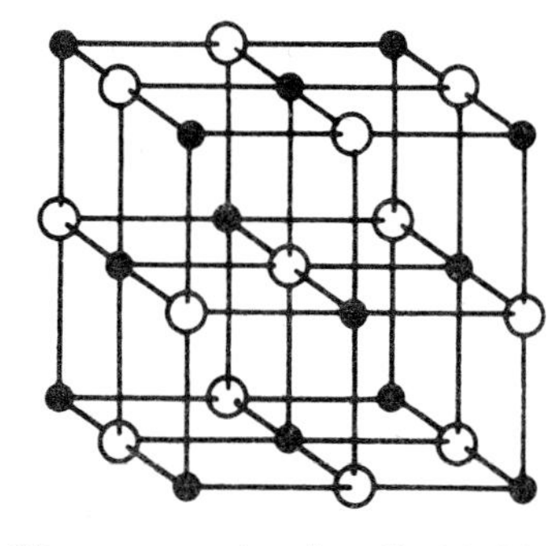

Fig. 18.22 structure of sodium(I) chloride (solid, high m.p.)

The result of this electron transfer is the formation of charged particles or **ions**; these do not merely attract each other in pairs, for the force of attraction between particles bearing unlike charges is exerted in all directions in space. Sodium ions will surround themselves by chloride ions and vice versa, a three-dimensional **lattice** being built up as shown in Fig. 18.22. When the charged particles come together to form such a lattice, much energy is given out; this is called the **lattice energy** of the crystal, and it must of course be supplied if the solid is vaporised.

There is no such thing as a 'molecule' of sodium chloride. The description of sodium chloride as a non-molecular compound, composed of a lattice of oppositely charged ions, is consistent with its high melting-point, boiling-point and latent heat of vaporisation (see Fig. 18.20).

The lack of chemical reactivity of the noble gases and the electron configurations of these gases (sections 6.5 and 18.11 respectively) give evidence for the particular stability of filled electron energy levels. Note that the sodium(I) ion and the chloride ion both have filled electron energy levels. The electron transfer in the formation of sodium(I) chloride can be represented pictorially as follows:

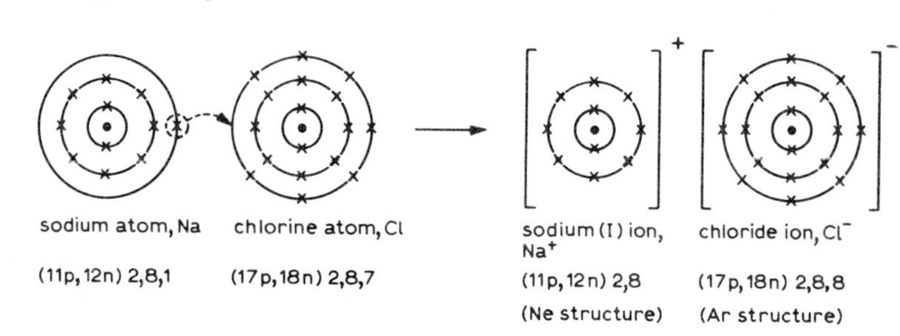

Fig. 18.23

The formation of sodium(I) chloride exemplifies a principle which is of vital importance in explaining chemical reactions:

all elements react in such a way as to achieve completely filled electron energy levels

The noble gases have atoms with filled electron energy levels–they have virtually no chemical reactions. Elements with one or two electrons in the outermost energy level achieve the electron configuration of a noble gas most easily by *losing* electrons, forming *positively* charged ions. Elements with six or seven electrons in the outermost energy level achieve the electron configuration of a noble gas most easily by *gaining* two electrons, or one.

18.14 further examples of ionic compounds

1. potassium fluoride Formula: KF

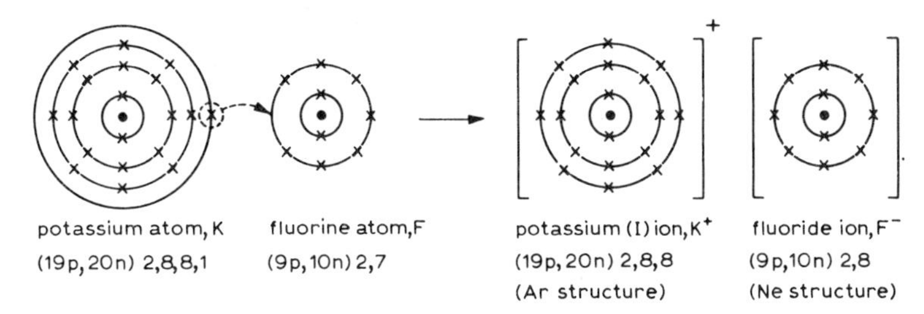

Fig. 18.24

2. magnesium(II) chloride Formula: $MgCl_2$

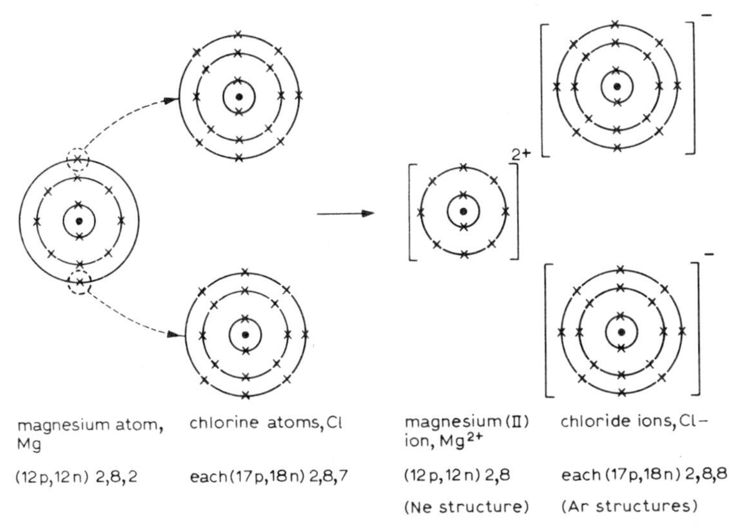

Fig. 18.25

3. calcium(II) oxide Formula: CaO

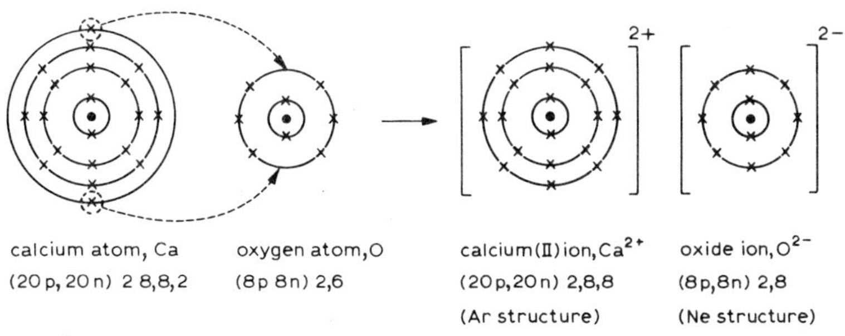

Fig. 18.26

It is now possible to explain the reason for the **valency numbers** of elements which form ionic compounds: the **valency number is equal to the charge number on the ion** in each case. The 'rules' used in constructing formulae from valency numbers (section 10.15) result from the condition that the crystal lattice of an ionic compound must be **electrically neutral.**

18.15 physical properties and bonding in molecular substances

Further reference to Figs. 18.20 and 1821 indicates the existence of some substances having strong forces holding together the atoms in the molecule, but weak forces holding the molecules to each other: such substances are called **molecular substances.**
e.g. hydrogen:

energy necessary to separate *molecules*, $L_V = 450\,J\,mol^{-1}$
energy necessary to separate *atoms* in molecules (bond energy), 436800 $J\,mol^{-1}$

In order to explain the existence of strong binding forces between the atoms in the molecules of molecular substances, we make use of two principles

1. the electron-pairing principle

Although electrons are all negatively charged and hence would be expected to repel each other, they can and do come together to form pairs. The explanation for the stability of such pairs is beyond the scope of this book. Each electron pair will, however, repel other electron pairs according to the normal law describing like charges.

2. the stability of filled electron energy levels

The highest energy level of an atom is filled when it contains 8 electrons (4 pairs), except for the case of the first level, which is filled when it contains 2 electrons (1 pair), as in the structure of helium.

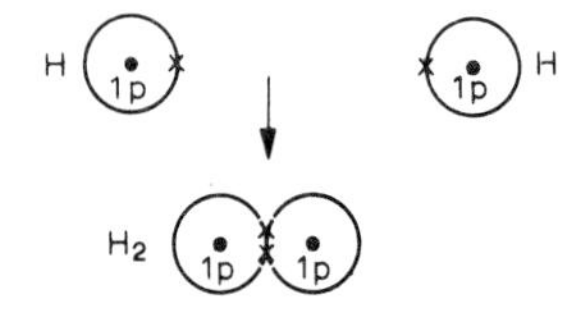

Fig. 18.27 the hydrogen molecule

When two hydrogen atoms join to form a molecule, the two electrons come together to form a pair; this pair is shared between the two nuclei. Each hydrogen atom has thus attained the particularly stable electron structure of helium, the first energy level being 'filled' with two electrons. The electron pair is strongly and equally attracted to each nucleus and this attraction effectively links the nuclei together: it is called a **covalent bond.**

a covalent bond consists of a pair of electrons shared between two nuclei

There is very little electrical attraction between adjacent molecules (in contrast with the ionic lattice of the preceding section), hence the forces between molecules are weak; since the hydrogen molecule also has a very small mass ($M_r = 2$), hydrogen is a gas at normal temperatures.

The 'structural formula' for the hydrogen molecule is often written thus: H—H. The line represents the pair of electrons forming the covalent bond.

18.16 further examples of bonding in molecular compounds

1. methane (the simplest hydride of carbon)

(*look at the properties in Figs. 18.20 and 18.21*)

The carbon atom has attained the electron structure of neon and each hydrogen atom has attained the electron structure of helium. Note that there is no hydride of carbon with the formula CH, or CH_2, or CH_3, or CH_5; such hypothetical 'compounds' would not accord with principle 2 of the preceding section. The valency number IV for carbon and the formula CH_4 are explained.

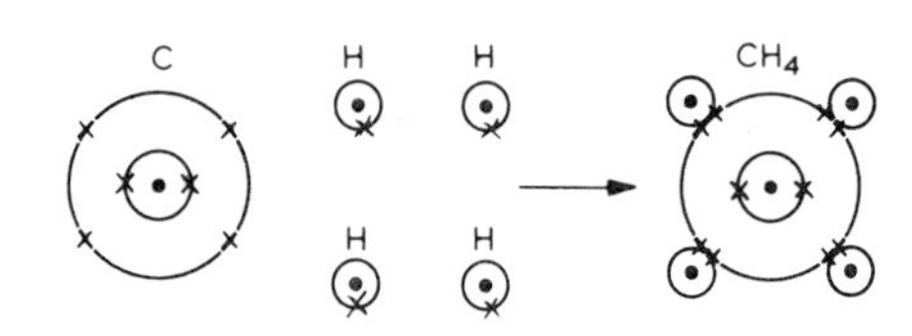

Fig. 18.28 the methane molecule

arrangement in space of the four covalent bonds in a methane molecule

A covalent bond lies in a **straight line joining the nuclei** of the bonded atoms. Thus the atoms in a molecule are arranged in definite positions in space, determined by the mutual repulsion between the electron pairs forming the bonds. The four electron pairs binding the atoms in the methane molecule become separated from each other as far as possible in **three** dimensions, resulting in a **tetrahedral** disposition as shown in Fig. 18.29.

Methane ($M_r = 16$) is a gas at normal temperatures.

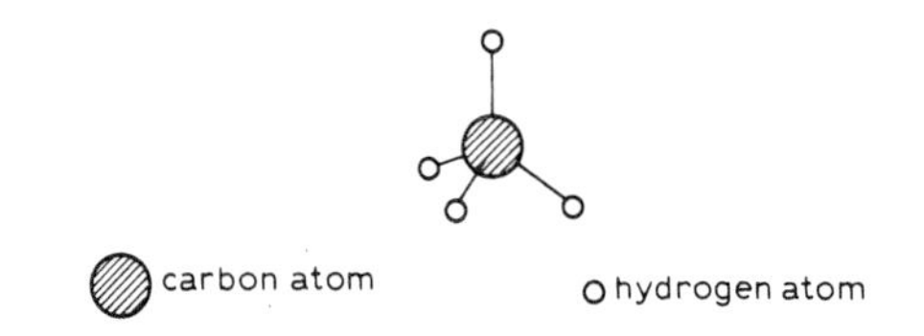

Fig. 18.29

2. tetrachloromethane

(look at the properties in Figs. 18.20 and 18.21)

The carbon atom has attained the electron structure of neon. Each chlorine atom has attained the electron structure of argon. Tetrachloromethane ($M_r = 154$) is a volatile liquid at normal temperatures. The shape of its molecule is **tetrahedral,** like the methane molecule.

Apparently the thermal energy available at room temperature is sufficient to separate molecules of small mass and keep them in the gaseous state. For molecules of greater mass, higher temperatures are needed to provide the energy to separate the molecules.

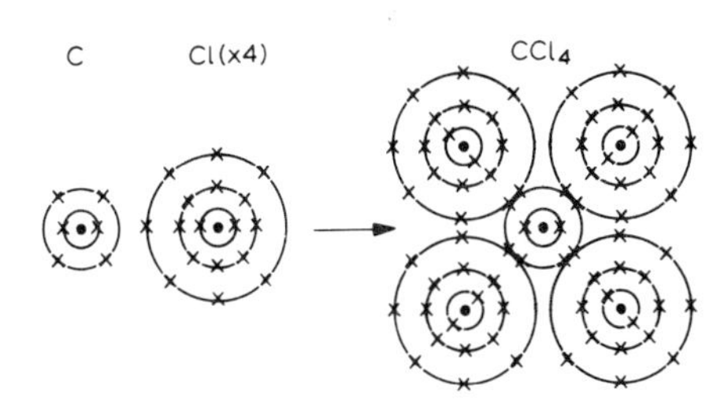

Fig. 18.30 the tetrachloromethane molecule

18.17 physical properties and bonding in some other non-molecular substances (giant-molecule structures)

The physical properties of graphite shown in Fig. 18.20 are those of a non-molecular substance, but it is not ionic. Carbon has four electrons in its highest energy level and it can neither gain nor lose four electrons to achieve a filled energy level; the bonding in all carbon compounds is covalent. The continuity of strong bonding forces arises as shown in Fig. 18.31.

The shared electron pairs (covalent bonds) between the carbon atoms are *repeated right through the structure*, giving a giant-molecule structure in which the forces between the atoms are as strong as those holding together the atoms in molecular compounds.

Silicon(IV) oxide (formula SiO_2) is another example of a compound with a giant-molecule structure; in this case the repeating covalent bonds hold together silicon and oxygen atoms.

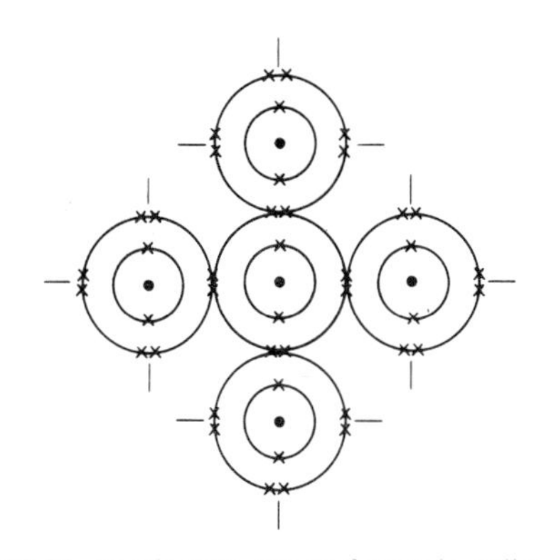

Fig. 18.31 continuous strong forces in solid carbon

18.18 the unequally shared electron pair: polarity of covalent bonds

The preceding descriptions of the strong forces between atoms and weak forces between molecules in molecular compounds, and of the continuous, uniformly strong forces operating in non-molecular compounds are adequate to explain the properties of some of the compounds listed in Fig. 18.20. A number of the other compounds in the table have **intermediate** properties; the properties of hydrogen chloride, for example, indicate much stronger binding forces between molecules than are present in the substances above it in the list. The intermolecular forces in hydrogen chloride are nevertheless much weaker than the force binding hydrogen and chlorine atoms (Fig. 18.21) or the forces in truly non-molecular compounds. This rather complex statement may be clarified by extracting some of the relevant figures:

substance	(a) L_v/J mol^{-1} (represents force between molecules)	(b) bond energy/J mol^{-1} (represents force between atoms)
hydrogen	450	436800
fluorine	3200	160000
graphite	718200	613200
hydrogen chloride	16200	432600

Fig. 18.32 properties representative of forces between (a) molecules, (b) atoms

Hydrogen and graphite are extreme examples of molecular and non-molecular substances. Fluorine is obviously molecular; hydrogen chloride has a relative molecular mass similar to that of fluorine. Why are its intermolecular forces more than *five times as strong* as those in fluorine?

Such intermediate properties are indicative of bonding which is **intermediate between covalent and ionic.** We will now examine how this arises.

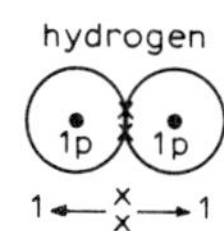

Fig. 18.33 electron pair equally shared

In the hydrogen molecule the electron pair constituting the covalent bond is attracted *equally* by each hydrogen nucleus, since each nucleus contains one proton. There is no great concentration of positive or negative charge in the molecule–it is electrically symmetrical. The absence of electrical attraction between adjacent hydrogen molecules explains the weak intermolecular forces indicated by its physical properties.

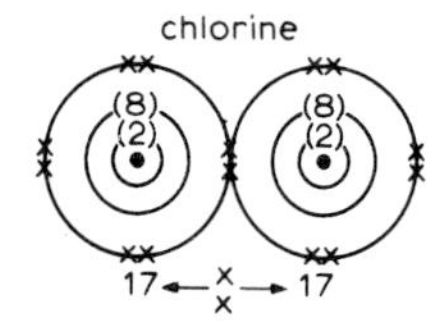

Fig. 18.34 electron pair equally shared

The chlorine molecule is similarly electrically symmetrical, with the bonding electrons attracted *equally* by the two nuclei, which each contain 17 protons. The same argument would hold for fluorine; chlorine and fluorine are truly molecular substances.

In the hydrogen chloride molecule, the bonding electrons are attracted by one proton in the hydrogen nucleus and, in the opposite direction, by 17 protons in the chlorine nucleus. The electron pair is thus *unequally shared* and this leads to a positive pole (at the electron-deficient hydrogen end of the molecule) and a negative pole (at the electron-rich chlorine end). The hydrogen chloride molecule is said to be **dipolar** (see Fig. 18.35).

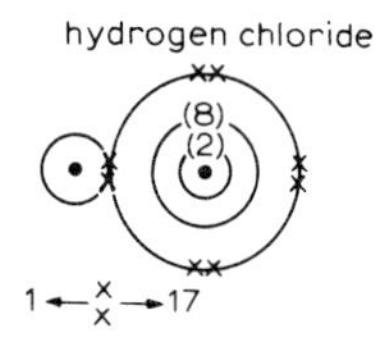

Fig. 18.35 electron pair unequally shared

Since some sharing does take place, the charges on the poles are only a fraction of the full charges occurring on ions. The fractional nature of these charges is represented in this text by writing the positive and negative signs in circles. The hydrogen chloride molecule is electrically unsymmetrical; this leads to electrical attraction *between adjacent molecules*, resulting in stronger intermolecular forces than can arise between electrically symmetrical, or non-polar, molecules like hydrogen, chlorine and fluorine.

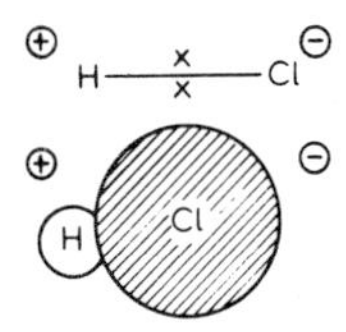

Fig. 18.36 'dipolar' hydrogen chloride molecule

In interpreting the properties of molecular compounds it is important to consider the shape and the polarity of the molecules. We shall now consider two very important compounds, ammonia and water, which like hydrogen chloride are intermediate in their physical properties and bonding (see Fig. 18.20).

18.19 the shape and polarity of the ammonia molecule

The nitrogen atom achieves a filled outermost energy level (neon structure) by the formation of three covalent bonds. Thus the formula for ammonia, the hydride of nitrogen, is NH_3. For a further insight into the properties of ammonia, we must ask:

i) what is the shape of the molecule?
ii) what is the polarity of the N—H bonds?

a) *molecular shape*

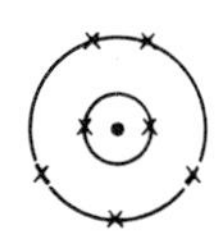
Fig. 18.37 nitrogen atom (7p, 7n) 2,5

The nitrogen atom is attached to three hydrogen atoms, but the hydrogen atoms are not arranged symmetrically: this is because the molecular shape arises from the **mutual repulsion between the electron pairs.** There are **four** electron pairs in the filled outermost energy level of the nitrogen atom in ammonia; these will be arranged **tetrahedrally** around the nitrogen nucleus (compare methane, Fig. 18.29, and tetrachloromethane, section 18.16 (2)). Three of these electron pairs form bonds with hydrogen nuclei; the fourth pair is not involved in bonding and is called a **non-bonding,** or **'lone', pair** of electrons. The tetrahedral shape of the molecule is not quite perfect, as we will see in the next section.

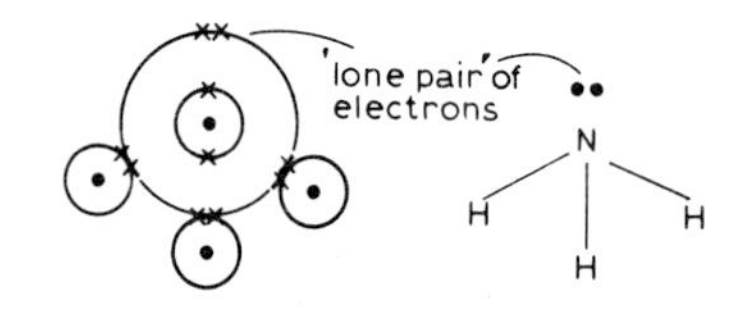

Fig. 18.38 ammonia molecule

b) *polarity of the N—H bonds*

The electron pair constituting the covalent bond is attracted by seven protons in the nitrogen nucleus and one proton in the hydrogen nucleus; the nitrogen atom develops a slight excess and the hydrogen atom a slight deficiency of electrons. The ammonia molecule is **dipolar,** with the negative pole concentrated at the non-bonding electron pair and the positive pole spread over the three hydrogen atoms.

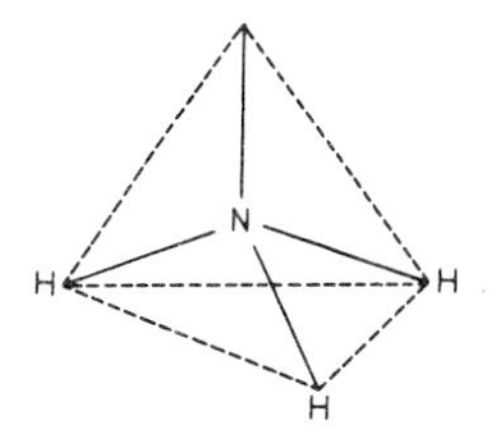

Fig. 18.39 ammonia molecule–distorted tetrahedron

(N) 7p ←— $\overset{\times}{\times}$ —→ 1p (H)

Fig. 18.40 electron pair unequally shared in nitrogen–hydrogen bond

The attraction between the slightly positively charged hydrogen atoms and the negatively charged electrons in adjacent bonds distorts the tetrahedral arrangement slightly. The angle between the N—H bonds is 107°, compared with the C—H bond angle of 109° 28′ in methane.

Understanding of the dipolar nature of the ammonia molecule not only explains the physical properties of the compound (Fig. 18.20); it is also vital in explaining its chemistry. This is considered briefly in section 18.25 and in more detail in Chapter 24.

18.20 the shape and polarity of the water molecule

Reference to the Figs. 18.41 and 18.42 showing the electron structures of the oxygen atom and the water molecule, and application of arguments exactly similar to those used for ammonia in the preceding section, lead to the conclusions:

i) the oxygen atom achieves a filled outermost energy level by the formation of two covalent bonds,
ii) the water molecule has two bonding and two non-bonding electron pairs in the highest energy level of the oxygen atom,
iii) the O—H bonds are polar, the oxygen atom being slightly electron-rich and the hydrogen atoms slightly electron-deficient (Figs. 18.43, 18.44),
iv) the water molecule is dipolar (Fig. 18.44),
v) the shape of the molecule is that of a distorted tetrahedron (H—O\H bond angle 104·5°).

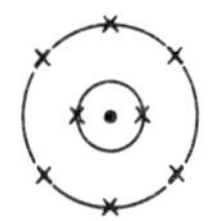

Fig. 18.41 oxygen atom (8p, 8n) 2,6

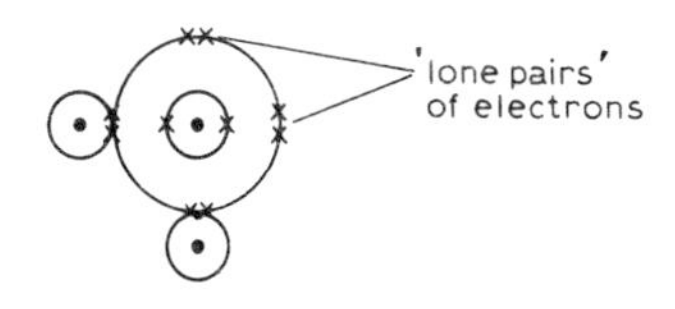

Fig. 18.42 water molecule

(O) 8p ←— $\overset{\times}{\times}$ —→ 1p (H)

Fig. 18.43 electron pair unequally shared in oxygen–hydrogen bond

It will be instructive for you to consider whether the polarity of the O—H bonds in water is greater or smaller than that of the N—H bonds in ammonia, and how the answer to this question throws light upon the differences in physical properties between the two compounds (Fig. 18.20).

The chemical properties of water are of vast significance, since most of the reactions studied in our chemistry course take place in aqueous solution. For example, acids constitute an important class of substances which depends upon the presence of water for its activity, as has been discussed in Chapter 11. We will consider the reason for this in the last part of this Chapter. Before doing this, however, it is as well to summarise some of the very important ideas which we have considered so far.

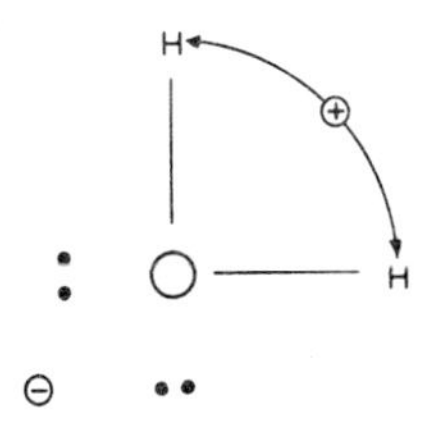

Fig. 18.44 'dipolar' water molecule

shapes of some simple molecules		
methane	tetrahedral	non-polar
tetrachloromethane	tetrahedral	non-polar
hydrogen chloride	linear	polar
ammonia	pyramidal	polar
water	bent	polar

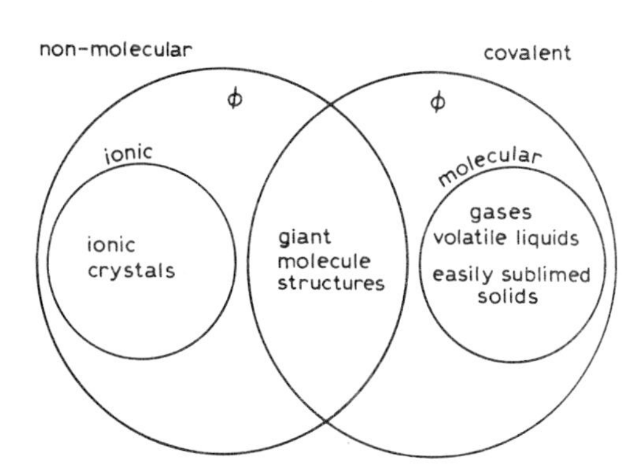

Fig. 18.45 Venn diagram showing molecular/non molecular and ionic/covalent bonding

18.21 summary: properties of ionic and covalent substances

	bonding	properties
molecular	covalent	1. low melting-points and boiling-points, i.e. gases, volatile liquids, easily vaporised solids variation left to right with increasing → *a*) relative molecular mass *b*) polarity of bond 2. soluble in organic solvents rather than water 3. do not conduct electricity in either solid or liquid state
non-molecular	ionic	1. hard, crystalline solids with high melting-points and boiling-points 2. soluble in water rather than organic solvents 3. conduct electricity when molten
	covalent	1. hard, crystalline solids with very high melting-points and boiling-points 2. insoluble both in water and organic solvents 3. do not conduct electricity in solid or liquid state (except graphite)

18.22 the coordinate bond

A covalent bond is a pair of electrons shared between two nuclei; so far we have considered cases in which one electron is supplied by each atom. It is also possible for *both* electrons in the pair to be supplied by *one* atom: in this case the bond formed is called a **coordinate bond,** though once formed it is indistinguishable from a covalent bond.

e.g. in the reaction between ammonia gas and hydrogen chloride gas:

$$\begin{array}{l} H \\ H{-}N: \\ H \end{array} + H\div Cl \rightarrow \left[\begin{array}{l} H \\ H{-}N{\rightarrow}H \\ H \end{array}\right]^+ + Cl^-$$

the nitrogen atom in ammonia supplies both electrons to form the N→H bond in NH_4^+; all four N—H bonds in NH_4^+ are, however, identical.

The resulting ammonium chloride contains covalent, coordinate and ionic bonds, though the four N—H bonds in the NH_4^+ ion are completely identical apart from their method of formation. The coordinate bond is usually represented by an arrow showing the direction in which the electron pair is donated: the **donor atom** must have at least one 'lone pair' of electrons and the **acceptor atom** must be at least one electron pair short of a completely filled energy level.

18.23 comparison of the properties of hydrogen chloride in methylbenzene and in water

We saw in Chapter 11 that water is necessary for acids to show their common properties. This idea is confirmed by carrying out tests on solutions of hydrogen chloride gas in (*a*) methylbenzene, (*b*) water.

Small portions of each solution are tested as follows:

1) a piece of carefully dried Universal Indicator paper (or litmus paper), held in tweezers, is dipped in the solution;
2) a small piece of marble is dropped into a portion of the solution;
3) a piece of magnesium is dropped into a portion of the solution;
4) a pair of electrodes connected to a 6-volt d.c. supply is dipped into the solution contained in a small beaker.

The results of these four tests are summarised in Fig. 18.46.

test	hydrogen chloride in methylbenzene	hydrogen chloride in water
indicator paper	no colour change	turns red
marble chip	no reaction	effervescence
magnesium	no reaction	effervescence
effect of electricity	no reaction	bubbles seen at electrodes

Fig. 18.46

5) If a thermometer is dipped in methylbenzene and is then lowered into a jar of dry hydrogen chloride, no change in temperature is observed. However, if a thermometer is dipped in water and is then lowered into a jar of dry hydrogen chloride, a marked rise in temperature is evident at once.

Certain important conclusions may be drawn from these five tests. They are:

a) The common properties of acids are evident in a solution of hydrogen chloride in water and absent in a solution of hydrogen chloride in methylbenzene (tests (1), (2) and (3)).

b) A solution of hydrogen chloride in water contains ions; a solution of hydrogen chloride in methylbenzene does not (test (4), see also section 13.14).

c) The ions in the aqueous solution of hydrogen chloride are not simply H^+ and Cl^-. We can be confident of this because the change

$$\underset{\text{(molecules)}}{HCl} \rightarrow \underset{\text{(ions)}}{H^+ + Cl^-}$$

would require energy to be supplied to separate the oppositely charged particles. Test (5) shows that when water and hydrogen chloride are mixed, energy is *released*. Positively charged hydrogen ions are, of course, bare protons; a tiny proton carrying a full positive charge would be so intensely charged that it would attract an electron pair from any other chemical species available. 1 gram of protons react with water molecules giving out 1 200 000 J: this is equivalent to saying that a bare proton has no chance of existing in water.

d) In view of (*c*) above, there are very strong grounds for believing that when hydrogen chloride and water react, bonds are formed as well as bonds being broken.

The water molecule possesses lone pairs of electrons and can act as a donor molecule in forming a coordinate bond with a hydrogen ion:

$$HCl \rightarrow H^+ + Cl^- \qquad \text{(energy } \textit{required}\text{)}$$

$$\underset{H}{\overset{H}{\diagdown}}\ddot{O}\!: + H^+ \rightarrow \left[\underset{H}{\overset{H}{\diagdown}}\ddot{O} \rightarrow H\right]^+ \qquad \text{(energy } \textit{released}\text{)}$$

oxonium ion

overall reaction:

$$\underset{H}{\overset{H}{\diagdown}}\ddot{O}\!: + H{-}Cl \rightarrow \left[\underset{H}{\overset{H}{\diagdown}}\ddot{O} \rightarrow H\right]^+ + Cl^- \qquad \text{(i)}$$

The H—O bond is stronger than the H—Cl bond (see Fig. 18.21). Is this consistent with the results of the experiment?

You will probably not have any knowledge of the composition of methylbenzene. What deductions can you make, as a result of these experiments, about the nature of methylbenzene molecules? Methylbenzene is often called a non-polar solvent; others are hexane, benzene and tetrachloromethane.

18.24 acids

An acid is usually now defined thus:

an acid is a substance which can act as a source of protons

The protons liberated by an acid will be absorbed by a second substance with the formation of a coordinate bond. One example of this is the reaction between ammonia and hydrogen chloride (section 18.22); another is the ionisation of sulphuric(VI) acid in water solution:

$$H_2O + H_2SO_4 \rightarrow H_3O^+ + HSO_4^- \quad \text{(ii)}$$
$$H_2O + HSO_4^- \rightarrow \underset{\text{(oxonium ion)}}{H_3O^+} + SO_4^{2-} \quad \text{(iii)}$$

(Interpret the results of the experiment described in section 25.12 in the light of this representation of the reaction between sulphuric(VI) acid and water.)

The particle responsible for acidic properties in water solution is not H^+, but H_3O^+; the concentration of such particles gives a good measure of the strengths of acids. In dilute aqueous solution the following acids are **completely ionised** and are called **'strong acids'**:

a) hydrogen chloride $HCl + H_2O \rightarrow H_3O^+ + Cl^-$
b) nitric(V) acid $HNO_3 + H_2O \rightarrow H_3O^+ + NO_3^-$ (iv)
c) sulphuric(VI) acid $H_2SO_4 + 2H_2O \rightarrow 2H_3O^+ + SO_4^{2-}$

In the case of hydrochloric and nitric(V) acids, one mole of the acid produces one mole of oxonium ions in dilute aqueous solutions. One mole of sulphuric(VI) acid produces, in dilute aqueous solution, two moles of oxonium ions.

Some acids are only **slightly ionised,** even in dilute aqueous solution, and are called **'weak acids'**. An example is ethanoic (acetic) acid:

$$CH_3COOH + H_2O \rightleftharpoons H_3O^+ + CH_3COO^-$$

One mole of acetic acid in excess water produces an equilibrium mixture of molecules and ions, containing only a very small fraction of a mole of oxonium ions. Such equilibria are discussed in Chapter 23.

18.25 bases

a base is a substance which can act as an acceptor of protons

The most familiar base in aqueous chemistry is the hydroxide ion; when this accepts a proton it forms a water molecule:

$$[H\!:\!\ddot{\underset{..}{O}}\!:]^- + H{-}Cl \rightarrow H\!:\!\ddot{O}\!:\!H + Cl^-$$

Sodium(I) and potassium(I) hydroxides are **completely ionised** in dilute aqueous solution and are called **'strong bases'**:

$$\left.\begin{array}{l} NaOH \rightarrow Na^+ + OH^- \\ KOH \rightarrow K^+ + OH^- \end{array}\right\} \text{one mole of each hydroxide produces one mole of hydroxide ions}$$

Ammonia acts as a base towards water by accepting protons from water molecules:

$$H_3N\!: + H{-}O{-}H \rightarrow [H_3N{\rightarrow}H]^+ + OH^- \quad \text{(v)}$$

This ionisation is only partial; one mole of ammonia in excess water produces an equilibrium mixture containing only a very small fraction of a mole of hydroxide ions:

$$NH_3 + H_2O \rightleftharpoons NH_4^+ + OH^-$$

Ammonia is thus an example of a 'weak base'.

18.26 nature of water

Note that in reactions (i), (ii), (iii) and (iv) in sections 18.23 and 18.24, water molecules are accepting protons, whereas in (v) above, water molecules are acting as a source of protons. Water can react as a base or as an acid; it is said to be **amphoteric.**

revision summary: Chapter 18

sub-atomic particles:

	proton	neutron	electron
charge:	1+	0	1−
mass:	1 unit	1 unit	negligible

atomic number, Z: position in Periodic Table
number of protons in nucleus
number of electrons (for a neutral atom)

mass number, A: sum of number of protons and neutrons

isotopes: atoms of different mass occupying same position in Periodic Table,
i.e. having same number of protons but different numbers of neutrons

symbols for nuclides: ${}^{A}_{Z}X$

radioactive emission: α-particles: ${}^{4}_{2}He^{2+}$
β-particles: electrons
γ-rays: high-energy electromagnetic radiation

radioactive decay: α-emission – shift of 2 places to the left in the Periodic Table
β-emission – shift of 1 place to the right in the Periodic Table

ionisation energy: energy required to remove an electron from a particle

stability of particles: all atoms react to achieve filled outer electron energy levels

molecular compound: strong forces between atoms, weak forces between molecules

non-molecular compounds: strong, uniform forces throughout structure

ionic bond: electrostatic attraction between oppositely charged particles (ions) – non-directional

covalent bond: pair of electrons shared between two nuclei (directional)

intermediate bonding: results from polarity of covalent bond

molecular shape:	determined by repulsion of electron pairs (both bonding and non-bonding)
coordinate bond:	covalent bond formed by donation of an electron pair from one atom to another
acid:	a substance which can act as a source of protons
base:	a substance which can accept protons
amphoteric substance:	one which can act as both an acid and a base

questions 18

1. Compare the Dalton, Thomson and Rutherford–Bohr models of the atom, giving details of experimental evidence which caused the older models to be changed.

2. The table shows the mass numbers and atomic numbers of atoms labelled T to Z.

	mass number	atomic number
T	2	1
V	3	1
W	4	2
X	6	3
Y	9	4
Z	11	5

a) How many protons are there in an atom of Y?
b) How many electrons are there in an atom of W?
c) How many neutrons are there in an atom of Z?
d) Which atoms are isotopes of the same element?
e) Which atom would readily form an ion with a single positive charge?
f) Which is an atom of a noble gas?

3. Neon is thought to consist of single atoms, yet the mass spectrometer trace of naturally occurring neon contains more than one peak. Explain this observation.

4. Chlorine has an atomic number of seventeen and exists mainly in isotopic forms A and B of relative atomic masses 35 and 37 respectively. State the number of
a) electrons in each atom of A,
b) protons in each atom of B,
c) neutrons in each atom of B,
d) electrons in each $^{35}_{17}Cl^-$ ion.
Calculate the A : B ratio in ordinary chlorine gas (relative molecular mass = 71). (JMB)

5. In radioactivity, what is
(a) an α-particle,
(b) a β-particle?

$$^{228}_{88}Ra \xrightarrow{(i)} {}^{228}_{89}Ac \xrightarrow{(ii)} {}^{228}_{90}Th \xrightarrow{(iii)} {}^{224}_{88}Ra$$

The first three stages in the radioactive decay of the radium atom, $^{228}_{88}Ra$, are shown above.

State what type of emission occurs in (i), (ii), (iii).

What name is given to such a pair of atoms as the first and last in the above series? (W)

6. (a) $^{212}_{83}Bi \xrightarrow[\text{emission}]{\beta\text{ (Beta)}} Po \xrightarrow[\text{emission}]{\alpha\text{ (Alpha)}} Pb$

The above scheme shows how the radioactive isotope of bismuth, $^{212}_{83}Bi$, changes successively to polonium, Po, and finally to lead, Pb.
Complete the following statements:
The atomic number of polonium is
The atomic number of lead is
The mass number of the polonium atom is
The mass number of the lead atom is

(b) 'The half-life' of a radioactive element is ten years. In what time will 1 g of the element be reduced in mass to 0·25 g? (W)

7. (a) How do measurements of ionisation energies help to show that electrons in atoms are distributed among energy levels and that the energy levels are filled by particular numbers of electrons?

(b) What are the electron configurations of:
(i) an oxygen atom (ii) an argon atom (iii) a potassium(I) ion (iv) a calcium(II) ion (v) a chloride ion (vi) a fluoride ion?

8. State **one** law of chemical combination, and show how it was explained by the atomic theory. How has Dalton's atomic theory had to be modified in more recent times?

Explain the terms nucleus, proton, electron, ion.

Sodium has a relative atomic mass of 23 and its atomic number is 11. Draw simple diagrams to show the particles which make up (a) a sodium atom, (b) a sodium ion, and give a brief explanation of the difference between them. (S)

9. Fluorine (9); Neon (10); Sodium (11); Magnesium (12).

F Ne Na Mg

The following refer to the above four elements, the atomic numbers of which are shown in brackets:
(a) What **two** facts about the structure of the atom of any **one** of these elements can be deduced from its atomic number?
(b) State the numbers of electrons in successive electron-shells of the magnesium atom.
(c) If the symbol for the sodium ion is written Na^+, write similar symbols for the ions of fluorine and magnesium.

(d) Write the chemical formula for
   (i) Magnesium(II) fluoride, (ii) Sodium(I) fluoride.
(e) Name and explain briefly, with the aid of a diagram, the type of chemical bond linking atoms of fluorine in the molecule F_2.
(f) Caesium and sodium are both alkali metals. In what respect do the structures of atoms of these two elements resemble each other?
(g) $^{24}_{12}Mg$, $^{25}_{12}Mg$, $^{26}_{12}Mg$ are the symbols for the three isotopes of magnesium present in the metal as commonly obtained.
   (i) What do these symbols tell you about the nuclei of the atoms of these isotopes?
   (ii) The relative atomic mass of magnesium is 24·32. Which is the most abundant of the above isotopes?
(W)

10. The following symbols refer to atoms of fluorine, nitrogen and aluminium.

$$^{19}_{9}F, \quad ^{14}_{7}N, \quad ^{15}_{7}N, \quad ^{27}_{13}Al.$$

(a) For the aluminium atom **only**, write
   (i) the atomic number
   (ii) the mass number
(b) What information about the nucleus of the aluminium atom do these numbers provide?
(c) Using the symbols for the nitrogen atoms, explain briefly the meaning of 'isotopes'.
(d) Answer the following questions on the elements aluminium and fluorine.
   (i) State the numbers of electrons in successive electron-shells of their atoms.
   (ii) Explain why fluorine forms an ion F^-.
   (iii) Write the symbol for the aluminium ion.
   (iv) Write the formula for the ionic solid formed when these two elements combine.
   (v) Would you expect this compound to have a high or a low melting-point? State briefly the reason for your answer.
(e) Show diagrammatically how the nitrogen and hydrogen atoms in ammonia are linked together, and name the type of chemical bond involved.
(f) Name the element into which the radioactive isotope of carbon, $^{14}_{6}C$, is transformed by β(beta)-ray emission, and write its symbol showing the mass number and the atomic number.
(W)

11. Write the words appropriate to replace the question marks in the following statements:
(a) Substance *A* is a crystalline compound. When heated, it melts and the resultant liquid will conduct electricity.
A probably contains ? bonds. *A* might be ?.
(b) Substance *B* is a gas which is almost insoluble in water. The gas does not react with cold water. It burns readily when ignited, but does not support combustion.
B probably contains ? bonds. *B* might be ?.
(c) Substance *C* is a gas which dissolves readily in water forming a solution which is a ready conductor of electricity.
C probably contains ? bonds. *C* might be ?. (JMB)

12. Use the information in the table to answer the questions which follow. Answer by writing the letter indicating the correct substance.

substance	melting point °C	boiling point °C	electrical conductivity of aqueous solution
A	−119	78	none
B	801	1413	good
C	−112	−84	good
D	186	decomposes below b.p.	none
E	−4	101	good

Indicate the substance which
(a) is likely to be an ionic crystalline solid at room temperature,
(b) is likely to be a covalently bonded solid at room temperature,
(c) is an aqueous solution of a salt,
(d) consists of widely spaced, rapidly moving molecules at room temperature,
(e) could be ethanol (ethyl alcohol),
(f) could be hydrogen chloride. (JMB)

13. (a) What do the numbers in the symbol $^{23}_{11}Na$ indicate about the nucleus of this sodium atom?
(b) Explain why sodium forms an ion Na^+.
(c) In its reactions, sodium forms the Na^+ ion by 'electron transfer'. Explain this statement using **one** of the following reactions as an example:
   (i) the addition of sodium to water,
   (ii) heating sodium in chlorine.
(d) The atomic numbers of nitrogen and phosphorus are 7 and 15 respectively. Explain why a chemist would expect to find some chemical similarity between the elements nitrogen and phosphorus however different these elements may seem to be.
(e) Show diagrammatically, by means of a suitable example, the structure of a covalent bond. (W)

14. (a) Explain what is meant by (i) an electrovalent bond, (ii) a covalent bond. Give the electronic structure of **one** compound containing one or more electrovalent bonds and of **one** compound containing one or more covalent bonds.
(b) Give **two** characteristic properties of covalent compounds.
(c) 'The properties of sodium and potassium are similar to each other and so are the properties of chlorine, bromine and iodine.' For each of the two groups give **two** chemical properties which justify this statement, and explain how the chemical similarity in each group is accounted for by the electronic structure of the elements concerned. (AEB)

15. A solution of hydrogen chloride in methylbenzene (toluene), an organic solvent, does not conduct electricity and has no apparent reaction with a solid metal oxide or a solid metal carbonate. A solution of hydrogen chloride in water conducts electricity and reacts with both a metal oxide and a metal carbonate.

What does this information indicate about the nature of the bonding in hydrogen chloride when it is in solution in (a) methylbenzene (toluene), (b) water?

Write equations, either molecular or ionic, for the reactions of an aqueous solution of hydrogen chloride with (a) magnesium(II) oxide, (b) zinc(II) carbonate. (JMB)

16. (a) (i) What type of bond exists between hydrogen and chlorine in hydrogen chloride gas?

(ii) When this gas is dissolved in water, what particles are to be found in the solution of the acid so produced?

(iii) Give the structure of the particle which is responsible for the acidity, showing the bonding clearly.

(b) (i) Sketch **one** face only of a crystal of sodium(I) chloride. Each edge of the face should show **two** sodium(I) ions (each labelled +) and two chloride ions (each labelled −) and should also show, roughly, the relative sizes of the ions.

(ii) Try to imagine what a crystal of sodium(I) chloride would look like in three dimensions and, with the help of your diagram in (i) suggest how many chloride ions would be surrounding each sodium(I) ion which is not on the face or edge of the crystal.

(iii) Describe what happens when a small crystal of sodium(I) chloride is placed in water.

(c) (i) Write formulae for the ions present in a solution of copper(II) sulphate(VI) showing the charge which each different ion carries.

(ii) Name the precipitate produced when this solution is treated with sodium(I) hydroxide solution. Write the simplest ionic equation for the precipitation, showing only the ions which are involved. (JMB)

17. (a) Describe experiments by which you could demonstrate **three** differences between a solution of hydrogen chloride in water and a solution of hydrogen chloride in methylbenzene (toluene). What explanation can you give for these differences?

(b) Explain why hydrogen and chlorine form diatomic molecules, H_2 and Cl_2, whereas the molecules of helium and argon are monatomic.

(c) Choose **one** substance from **each** of the groups below and explain the essential differences between the crystals of the three substances chosen in terms of the particles present and the forces holding these particles together. (Details of the geometrical arrangements in the crystals are **not** required.)

(i) Sodium(I) chloride, calcium(II) oxide, potassium(I) bromide.

(ii) Naphthalene, solid carbon dioxide, sugar.

(iii) Diamond, graphite, copper. (C)

18. (a) Using diagrams to show electron distributions, describe the bonding in molecules of methane, ammonia, water and ethene. With these examples, explain what is meant by (i) a bonding pair of electrons, (ii) a lone pair of electrons, (iii) a double bond.

(b) Each of the molecules in (a) has its own individual shape. What factors determine these shapes?

(c) Why is it correct to refer to a molecule of methane or ammonia but not to a molecule of sodium(I) chloride?

19. Criticise the following statements:

(a) 'Covalent compounds are all gases, liquids or low-melting solids.'

(b) 'The heat of vaporisation of methane is 8·2 kJ mol^{-1}; the heat of vaporisation of sodium(I) chloride is 171 kJ mol^{-1}. These figures prove that covalent bonds are much weaker than ionic bonds.'

20. (a) (i) How does an ammonia molecule combine with a proton to form the ammonium ion?

(ii) Show the electron distribution inside an ammonium ion.

(iii) Why does the ammonium ion have a positive charge?

(iv) What is the likely shape of the ammonium ion?

(b) The hydrated copper(II) ion in water solution is usually written as $Cu^{2+}(aq)$ or $[Cu(H_2O)_6]^{2+}$. How are the molecules of water of hydration bonded to the copper(II) ion?

What happens when a concentrated aqueous solution of ammonia is added to copper(II) sulphate(VI) solution and what new ion is formed?

21. Use your knowledge of atomic structure to explain the following:

(a) When the atoms of the alkali metals form ions, they do so by the loss of one electron, but when the atoms of the halogens form ions they gain one electron.

(b) The ionic radii of the alkali metals increase down the group from 0·06 nm for the lithium ion to 0·169 nm for the caesium ion.

(c) The successive ionisation energies of an element such as sodium, can be used to predict the arrangement of electrons in the atoms of the element.

L (NUFFIELD)

22. Differentiate clearly between a covalent bond and an electrovalent bond (ionic). State the type of bond in each of the following:

(i) hydrogen (ii) potassium chloride (iii) methane

Make a comparison between covalent compounds and electrovalent compounds under the following headings:

(a) melting point and boiling point

(b) solubility

(c) electrical conductivity

Describe with the help of a simple diagram the shape of a crystal of sodium chloride.

19.1 what is petroleum?

Petroleum, or crude oil, is one of our most valuable natural resources. In the state in which it is found in nature it is of little or no value, but it is possible to obtain from it a wide variety of substances which are vital to our present-day lives.

Petroleum is a thick, dark-coloured, viscous liquid which is found in underground deposits. There are a number of theories about its origin, one of the more likely being that marine life deposited on the sea-bed, subjected to increasing pressures by subsequent deposits, changes chemically over a very long period of time to produce the many closely-related substances of which petroleum is composed. The composition of petroleum can be discovered by laboratory experiments and calculations; an outline of the procedure is given in this Chapter.

19.2 separation of crude oil

About 2 cm^3 of crude oil is added to a loose absorbent plug in a test-tube, using a dropping pipette. The tube is fitted with a side-arm and a thermometer, the bulb of which should be adjacent to the side-arm, as shown in Fig. 19.1. The tube which is to collect the first fraction stands in a beaker of cold water.

The plug is warmed very gently and the fraction distilling below 70 °C is collected. Three further fractions are collected, distilling in the temperature ranges shown in Fig. 19.2. The tubes in which the second and subsequent fractions are collected need not be cooled in water.

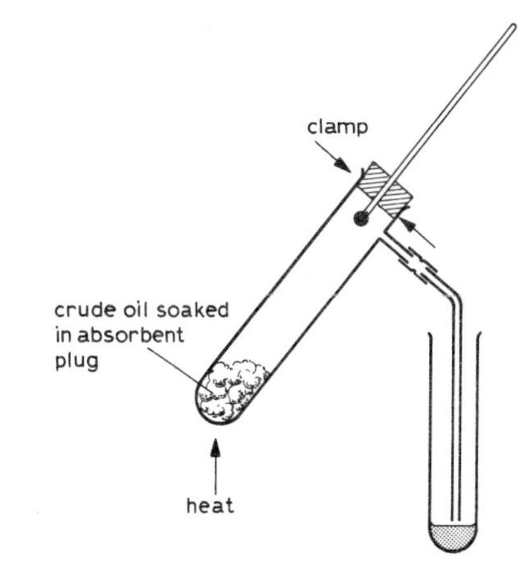

Fig. 19.1 laboratory distillation of crude oil

fraction	colour	viscosity	inflammability
up to 70 °C	colourless	very mobile	burns very rapidly
70 °C–120 °C	colourless	mobile	burns rapidly
120 °C–170 °C	pale yellow	mobile but oily	may just burn alone
170 °C–220 °C	yellow	slightly sticky	burns only with wick
residue in tube	brown	thick	burns with difficulty

Fig. 19.2 properties of fractions from distillation of crude oil

The colour and viscosity of each fraction is examined and each fraction is tested for inflammability by pouring a little into a watch-glass and applying a lighted splint (*not* a wax taper). The residue in the tube, if preserved, serves well for the experiment outlined in section 19.8.

The temperature shown on the thermometer rises continuously during the experiment: this shows that the fractions collected are not pure compounds (see section 4.7); they are mixtures of compounds.

If each fraction is subjected to a more sensitive form of fractional distillation, in the manner described in section 5.2(6), a large number of pure compounds can be obtained. Analysis shows that the vast majority of these compounds contain carbon and hydrogen only: they are called **hydrocarbons.** (Do not confuse *hydrocarbons*, which contain carbon and hydrogen only, with *carbohydrates*, which contain carbon, hydrogen and oxygen.)

Combustion analysis of the individual hydrocarbons gives information from which their molecular formulae can be deduced. A typical example of the results of such an analysis, and the route to the formula, is as follows:

10 cm³ of a gaseous hydrocarbon was mixed with 100 cm³ of oxygen. After explosion and cooling to the original temperature and pressure, the volume of the residual gases was 75 cm³; on shaking with aqueous potassium(I) hydroxide this volume reduced to 35 cm³. The remaining 35 cm³ of gas was absorbed completely by alkaline pyrogallol. Find the formula for the hydrocarbon.

The above results give the *relative volumes* of gases involved in the reaction. Application of Avogadro's law (see section 10.27) to these figures enables us to calculate the *relative number of moles* of the gases involved in the reaction. This enables us to solve the unknown quantities in the equation for the reaction and hence calculate the formula.

	hydrocarbon		oxygen		carbon dioxide
a) volumes:	10 cm³		100 − 35 = 65 cm³		75 − 35 = 40 cm³
b) relative volumes:	1	:	6·5	:	4
c) relative number of moles (*Avogadro*):	1	:	6·5	:	4

Let the formula for the hydrocarbon be C_xH_y.

The equation will be:

$$C_xH_y + \left(x + \frac{y}{4}\right)O_2 \longrightarrow xCO_2 + \frac{y}{2}H_2O$$

relative number of moles: $1 \qquad x + \frac{y}{4} \qquad x$

From (*c*) above, $x = 4$

and $x + \frac{y}{4} = 6{\cdot}5$

since $x = 4$, $4 + \frac{y}{4} = 6{\cdot}5$

$y = 10$

the formula for the hydrocarbon is C_4H_{10}

This hydrocarbon (butane) would occur in the first fraction shown in Fig. 19.2 (below 70 °C). Formulae which may be similarly calculated for hydrocarbons occurring in the other fractions are:

fraction	*typical hydrocarbon formula*
70 °C–120 °C	C_7H_{16}
120 °C–170 °C	C_9H_{20}
170 °C–220 °C	$C_{11}H_{24}$

These hydrocarbons all belong to the group called **alkanes.** Their names all end in '-ane' and the **general formula** covering all members of the group is C_nH_{2n+2}.

Petroleum consists mainly of a mixture of a large number of different alkanes. The pure alkanes may be obtained by repeated fractional distillation, but in practice a single distillation, giving fractions containing mixtures of alkanes, is sufficient to meet industrial and domestic requirements.

19.3 primary distillation of petroleum in the refinery

In the refinery, the crude petroleum is largely vaporised at the base of a steel fractionating column which may be 60 m high and 10 m in diameter. The vapours cool as they rise up the tower and condense as liquids in the trays. Vapours of more volatile constituents bubble through the condensed liquids (the feed pipes are protected by 'bubble caps') and themselves condense higher up the tower. The most volatile components of all escape from the top of the tower as vapour and are artificially condensed in the condenser: the liquid so formed is called 'petrol ether'. By returning some of this liquid to the topmost tray, the temperature at the top of the column is fixed within the boiling range of petrol ether. Fig. 19.3 shows this process diagrammatically and includes the approximate boiling-point ranges of the principal fractions. A typical throughput would be 10 dm^3 of crude oil per second.

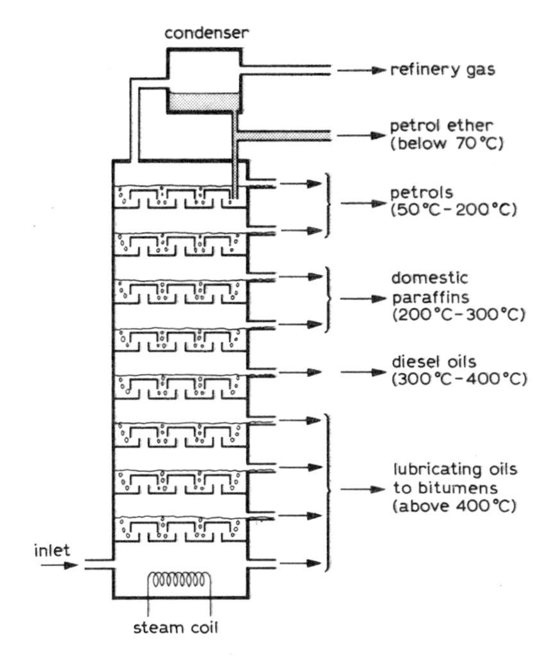

Fig. 19.3 fractionation of petroleum

19.4 molecular structures of the alkanes

All hydrocarbons, including alkanes, are **molecular compounds** and have the properties of such compounds (see section 18.21). The structure of the molecule of methane (the simplest alkane) has already been explained (section 18.16). In the molecule of any alkane, each carbon atom is joined by a covalent bond to each of four other atoms, which may be hydrogen atoms or other carbon atoms, as shown in Fig. 19.4. Each carbon atom has the four atoms to which it is linked arranged **tetrahedrally** around it; all bond angles are thus of the order of 109°. It is strongly recommended that this arrangement is studied thoroughly using three-dimensional models of methane, ethane, propane and butane before proceeding further.

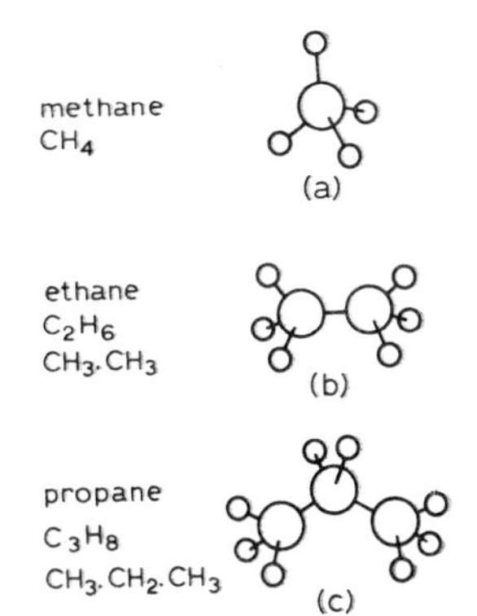

Fig. 19.4 shapes of simple alkane molecules

19.5 the butanes: isomerism

Familiarity with three-dimensional models of butane, C_4H_{10}, will show that there are two, and only two, possible different arrangements of the four carbon and ten hydrogen atoms. These two arrangements are shown diagrammatically in Fig. 19.5.

All arrangements of '*normal*-butane' are structurally equivalent, as the bonds are all capable of 'free rotation'. The arrangement of atoms in '*iso*-butane' (its systematic name is methylpropane) is quite different from that in '*normal*-butane'; the shape of the molecule is different and the physical and chemical properties of *normal*- and *iso*-butane differ from each other.

butane (normal butane)
C_4H_{10} or $CH_3.CH_2.CH_2.CH_3$
(a)

methylpropane (isobutane)
C_4H_{10} or $CH_3.CH(CH_3).CH_3$
(b)

Fig. 19.5 isomers of butane

'*Normal*-butane' and '*iso*-butane' are isomers. Isomers are substances with the same molecular formula which have different structural formulae and different physical and chemical properties. The existence of such compounds is called isomerism.

Structural formulae for the alkanes are often written as follows; remember, however, that all bond angles are about 107°, not 90° and 180° as these two-dimensional drawings imply:

```
                                                                              H
                                                                              |
                                                                            H—C—H
                                                                              |
   H           H  H          H  H  H          H  H  H  H               H      |     H
   |           |  |          |  |  |          |  |  |  |               |      |     |
 H—C—H       H—C—C—H       H—C—C—C—H        H—C—C—C—C—H              H—C——————C—————C—H
   |           |  |          |  |  |          |  |  |  |               |      |     |
   H           H  H          H  H  H          H  H  H  H               H      H     H

 methane      ethane        propane                         butanes
```

alkane	b.p./K
methane	111
ethane	184
propane	231
n-butane	273
(*iso*-butane)	(261)

Fig. 19.6 boiling points of alkanes

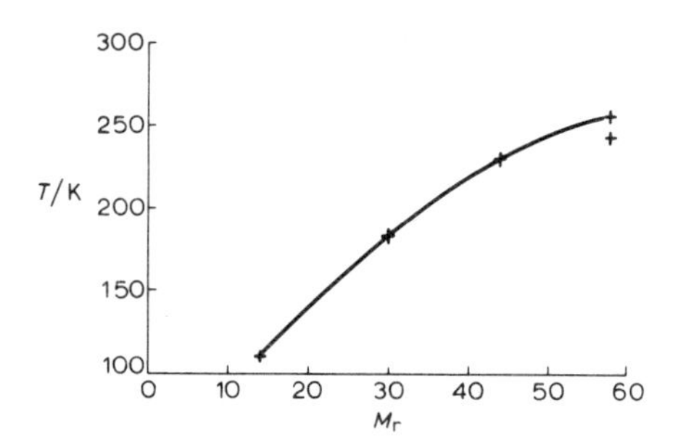

Fig. 19.7 boiling points (T) and relative molecular masses (M_r), methane to butanes

19.6 boiling points of alkanes

The physical properties of alkanes are well exemplified by their boiling points. The boiling points increase as the relative molecular mass increases. This is to be expected, for the more massive the molecule, the more energy will be required to separate it from its neighbours. This is reflected in the energy having to be supplied at a higher level, i.e. at a higher temperature.

Fig. 19.7 shows the relationship between the boiling points of alkanes and their relative molecular masses: it is not a straight line. It should be possible to offer an explanation for the slope of the curve, and for the fact that the boiling point of *iso*-butane is lower than that of *normal*-butane.

19.7 chemical properties and uses of alkanes

1. combustion

All hydrocarbons, including alkanes, burn in air or oxygen forming carbon dioxide and steam:

any hydrocarbon:

$$C_xH_y + \left(x + \frac{y}{4}\right)O_2 \rightarrow xCO_2 + \frac{y}{2}H_2O$$

any alkane:

$$C_nH_{(2n+2)} + \left(\frac{3n+1}{2}\right)O_2 \rightarrow nCO_2 + (n+1)H_2O$$

methane:

$$CH_4 + 2O_2 \rightarrow CO_2 + 2H_2O$$

Methane burns with a blue, non-luminous and non-smoky flame; the combustion is strongly exothermic–alkanes have an extensive use as fuels.

a) *domestic fuels*

'Town gas' (originally 'coal gas') has recently been replaced by 'natural gas', or 'North Sea gas', as a gaseous domestic fuel.

The compositions of 'town gas' and 'natural gas', and the heat evolved by the combustion of one mole of each of the compounds contained in the fuels, are shown in Fig. 19.8. It is possible from these figures to calculate the heat evolved in burning equal volumes of the two fuels:

	(% by volume)
'town gas'	
hydrogen	50
methane	40
carbon monoxide	10
'natural gas'	
methane	100

(a)

	ΔH/kJ mol^{-1}
hydrogen	286
methane	882
carbon monoxide	286

(b)

Fig. 19.8 (a) composition of fuels, (b) heat evolved on combustion of one mole of each component

Consider $2{\cdot}24 \times 10^{-2}\,m^3$ of each fuel at s.t.p.
$2{\cdot}24 \times 10^{-2}\,m^3$ of 'natural gas' contains

$$2{\cdot}24 \times 10^{-2}\,m^3 \text{ of methane}$$
$$= 1 \text{ mole of methane}$$

This evolves 882 kJ

$2{\cdot}24 \times 10^{-2}\,m^3$ of 'town gas' contains $\frac{1}{2} \times 2{\cdot}24 \times 10^{-2}\,m^3$ hydrogen, $\frac{4}{10} \times 2{\cdot}24 \times 10^{-2}\,m^3$ methane, $\frac{1}{10} \times 2{\cdot}24 \times 10^{-2}\,m^3$ carbon monoxide

$= \frac{1}{2}$ mole hydrogen, $\frac{4}{10}$ mole methane, $\frac{1}{10}$ mole carbon monoxide

This evolves $(\frac{1}{2} \times 286) + (\frac{4}{10} \times 882) + (\frac{1}{10} \times 286)$ kJ

$= 143 + 353 + 29$ kJ

$=$ **525 kJ**

Thus 'natural gas' furnishes 60% more heat energy than an equal volume of 'town gas'. The hotter flame is more efficient, though it necessitates changing the burners on domestic appliances. The waste gases contain far less corrosive sulphur dioxide than those from 'town gas', which contains some sulphur-bearing compounds.

'Calor gas' and other similar fuels are mixtures of propane and butane; fuel for cigarette-lighters is largely butane. Propane and butane are gases at atmospheric pressure but they are easily liquefied at higher pressures. They are therefore transported in pressurised containers as liquids and vaporised when the valve is opened: a large volume of gaseous fuel can be conveniently transported as a relatively small bulk of liquid.

'Paraffin' for domestic heaters consists of a mixture of alkanes in the range C_9H_{20} to $C_{16}H_{34}$ with a preponderance of the smaller molecules. These alkanes are liquids at normal temperatures and pressures and are vaporised in the flame.

b) *motor fuels*

These are mixtures of hydrocarbons in the C_5H_{12}–$C_{12}H_{26}$ range, a typical molecule being the well-known **octane**, C_8H_{18}. They are liquids at normal temperatures and pressures (*ease of transportation in petrol tank*) which when mixed with air (*carburettor*) form a mixture which explodes on sparking (*spark-plugs*) in the cylinder. The increase in volume on explosion pushes the piston which drives the motor.

$$C_8H_{18} + 12\tfrac{1}{2}O_2 \longrightarrow 8CO_2 + 9H_2O$$

The increase in volume accompanying this reaction is enormously amplified by the expansion of the waste gases due to the temperature rise: the combustion of one mole of octane liberates 5460 kJ.

c) *energy from foodstuffs*

The fundamental changes in all of the above energy-releasing reactions are the breaking of C–C, C–H and O–O bonds and the formation of C–O and H–O bonds. The reactions are exothermic because the energy released in the formation of the C–O and H–O bonds is much greater than that required to break the C–C, C–H and O–O bonds.

Carbohydrates, like hydrocarbons, contain many C–C and C–H bonds, though unlike hydrocarbons they contain some C–O and H–O bonds as well; when carbohydrates are oxidised in the body, energy is released. This process is discussed more fully later in this Chapter (section 19.13 *et seq*).

2. *reaction with chlorine and other halogens: substitution reactions*

If a mixture of chlorine with an alkane is ignited, a dangerously explosive reaction occurs:

$$C_nH_{2n+2} + (n + 1)Cl_2 \longrightarrow nC + (2n + 2)HCl$$

If the gases are mixed in diffused daylight, without sparking or exposure to direct sunlight, a quiet reaction occurs in which the hydrogen atoms of the alkane are successively **substituted** by halogen atoms. Methane forms the following products:

$$\begin{array}{c}H\\|\\H\text{—}C\text{—}H\\|\\H\end{array} + Cl\text{—}Cl \longrightarrow \begin{array}{c}H\\|\\H\text{—}C\text{—}Cl\\|\\H\end{array} + H\text{—}Cl$$

or: $CH_4 + Cl_2 \longrightarrow HCl + CH_3Cl$ chloromethane

then: $CH_3Cl + Cl_2 \longrightarrow HCl + CH_2Cl_2$ dichloromethane

$CH_2Cl_2 + Cl_2 \longrightarrow HCl + CHCl_3$ trichloromethane (chloroform)

$CHCl_3 + Cl_2 \longrightarrow HCl + CCl_4$ tetrachloromethane (carbon tetrachloride)

The first two products are useful intermediates in the production of other compounds, while trichloromethane is used as an anaesthetic and as a solvent and tetrachloromethane is used as a solvent and has had extensive use in fire extinguishers. This last use depends on the properties of the liquid tetrachloromethane: it boils at 77°C giving a non-inflammable vapour with a density approximately five times that of air. This vapour, produced when the liquid is sprayed on to a fire, 'blankets' the burning material and prevents further access of air. Its disadvantage is that the vapour is toxic and it has now been largely superseded by 'foam' extinguishers.

19.8 the 'cracking' of petroleum

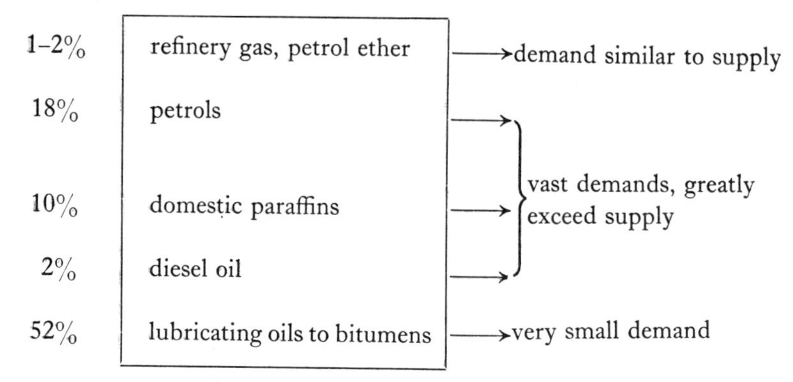

1–2%	refinery gas, petrol ether	→ demand similar to supply
18%	petrols	→ vast demands, greatly exceed supply
10%	domestic paraffins	→ vast demands, greatly exceed supply
2%	diesel oil	→ vast demands, greatly exceed supply
52%	lubricating oils to bitumens	→ very small demand

Fig. 19.9 components of primary distillation: supply and demand

The economic problem posed by the above summary diagram is: how to convert the 52% of 'base stock' into the more valuable, volatile petrols? The problem for the chemist is to convert large molecules into smaller ones. It is answered by the process of **cracking** the molecules by heating the liquid under pressure, or by passing the liquids or vapours over a heated catalyst ('cat-cracking'). The following experiment indicates a laboratory version of the cracking process and shows the nature of some of the products formed.

The pieces of broken pot are heated strongly, then the flame is spread to the 'base stock' (the residue in the tube from the experiment described in section 19.2 serves well for this). A liquid collects in the water-cooled receiver and a gas may be collected over water. Three tubes of gas should be collected. Application of a lighted splint to a tube of the gas shows that the gas burns with a luminous and rather smoky flame. If a few drops of aqueous bromine are added to the second tube of gas and a few drops of dilute acidified aqueous manganate(VII) to the third and the tubes shaken, both reagent solutions are decolourised. This should be contrasted with the behaviour of alkanes: methane burns with a different flame (see section 19.7(1)) and fails to decolourise either of the two test solutions mentioned.

Fig. 19.10 a laboratory 'cracking' process

The liquid product in the experiment also burns with a luminous, smoky flame and decolourises aqueous bromine and acidified aqueous manganate(VII).

Before interpreting these results, you should make an intelligent guess as to the likely number of carbon atoms in the molecules contained in (*a*) the gaseous product and (*b*) the liquid product.

interpretation of the results

Both the gaseous and liquid products contain hydrocarbons which (*i*) are composed of *smaller molecules* than those of the 'base stock', (*ii*) are *not* alkanes. Pure compounds can be isolated from the gas and the liquid and the formulae for these compounds can be determined by a process of analysis and calculation exactly similar to that of section 19.2. Typical formulae are:

(from gas) C_2H_4, C_3H_6
(from liquid) C_8H_{16}
general formula C_nH_{2n}

These hydrocarbons contain insufficient hydrogen atoms to satisfy fully the valency requirements of the carbon atoms; they are said to be **unsaturated hydrocarbons.** Hydrocarbons of this type are called **alkenes**, e.g. C_2H_4 ethene (ethylene), C_3H_6 propene (propylene), C_8H_{16} octene. (N.B. the spelling is very important: the name ends '-ene', not '-ine'.)

the double bond

A single covalent bond consists of one pair of electrons shared between two nuclei: a double bond consists of *two* pairs of electrons shared between two

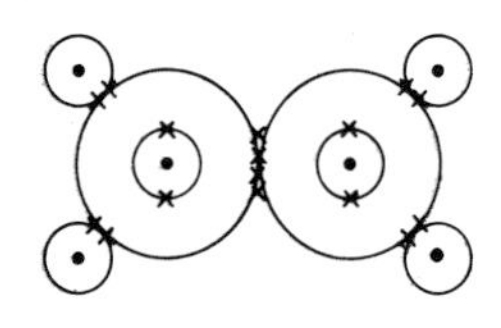

Fig. 19.11 an electron diagram for ethene, C_2H_4

nuclei. Examination of Fig. 19.11 shows how the formation of a double bond enables the carbon and hydrogen atoms in the molecule C_2H_4 to attain electron structures of noble gases. Fig. 19.12 shows how the structural formulae of ethene and propene may be written.

Suppose that the 'base stock' before cracking contained as a typical molecule $C_{18}H_{38}$, and suppose that this molecule 'cracked' at the points shown in the structure below. Rearrangement of hydrogen atoms between the fragments, and formation of two double bonds, gives rise to octane and octene (which would occur in the liquid product) and ethene (which would occur in the gas).

Octene would not itself be an acceptable constituent of petrol. (*Why not?*)

(a) $H_2C{=}CH_2$

(b) $H_3C{-}CH{=}CH_2$

Fig. 19.12 structural formulae for (a) ethene, (b) propene

$C_{18}H_{38}$ → fragments C_8H_{17}, C_2H_4, C_8H_{17} → octane, C_8H_{18}; ethene, C_2H_4; octene, C_8H_{16}

It would need to be hydrogenated (see section 19.9) to octane; this is a relatively simple matter.

The cracking process produces a high yield of molecules in the petrol range, from larger, less useful, molecules; as side-products small, double-bonded molecules are formed as unsaturated gases. These gases are not wasted: they are so reactive that a whole industry (the 'petrochemicals industry') is based on their utilisation. Indeed, it may fairly be claimed that the gaseous alkenes formed as side-products have become even more valuable than the petrols which the process was designed to produce.

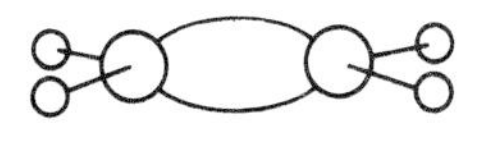

Fig. 19.13 all atoms in ethene lie in one plane

19.9 alkenes: the reactions of ethene (ethylene)

molecular shape

When two electron pairs approach sufficiently closely to form a double covalent bond, the electron pairs can no longer be distributed tetrahedrally about the carbon atoms; inspection of a molecular model will show how a **trigonal** (triangular) distribution of atoms around each carbon atom results. The two carbon atoms and four hydrogen atoms in ethene all lie in one plane.

1. hydrogenation: addition reactions

If one of the two bonds in the double bond is broken, the tetrahedral distribution of the electron pairs can be restored; this occurs when ethene reacts with hydrogen to give ethane. The reaction occurs at room temperature in the presence of a finely divided platinum catalyst; it is called **catalytic hydrogenation.** The reaction for ethene is exothermic to the extent of 139 kJ mol^{-1}.

$$C_2H_4 + H_2 \xrightarrow{Pt} C_2H_6$$

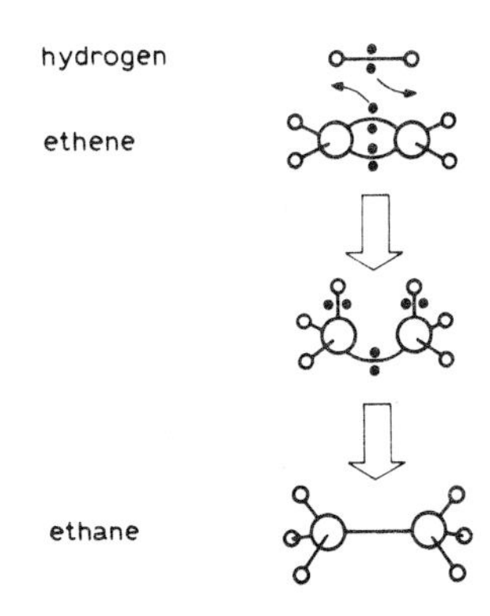

Fig. 19.14 a schematic representation of the hydrogenation of ethene

Hydrogenation is an important process in obtaining petrols from the products of 'cracking' (section 19.8) and in the conversion of liquid vegetable oils (which contain double-bonded molecules) into edible fats such as margarine.

Note that the reaction is quite different from the substitution reactions of alkanes; the hydrogen molecule simply **adds on** to the ethene molecule without the formation of any side-product. Such a reaction is called an **addition reaction:** it is a characteristic reaction of unsaturated compounds and can be represented in general terms by:

$$\begin{matrix} H & & H \\ & C{=}C & \\ H & & H \end{matrix} + X{-}Y \rightarrow \begin{matrix} & X & Y & \\ & | & | & \\ H{-} & C{-} & C & {-}H \\ & | & | & \\ & H & H & \end{matrix}$$

or

$$CH_2{=}CH_2 + XY \rightarrow CH_2X.CH_2Y$$

Since a wide range of substances can be made from ethene by this type of reaction, the unsaturated gases from the 'cracking' plant, once a mere by-product, have become a rich source of many chemical substances.

2. reaction with halogens

Ethene gas is absorbed when bubbled into liquid bromine at room temperature: a colourless volatile liquid, dibromoethane, is formed:

$$CH_2{=}CH_2 + Br_2 \rightarrow CH_2Br.CH_2Br$$

This reaction also occurs when ethene is shaken with aqueous bromine, but the water takes part in the reaction and a mixture of products results.

With chlorine, dichloroethane is formed:

$$CH_2{=}CH_2 + Cl_2 \rightarrow CH_2Cl.CH_2Cl$$

3. reaction with hydrogen chloride

Ethene is absorbed by concentrated hydrochloric acid at room temperature, forming chloroethane:

$$CH_2{=}CH_2 + HCl \rightarrow CH_3.CH_2Cl$$

This is a better method for the production of chloroethane than the substitution reaction between ethane and chlorine described in section 19.7(2). Why?

4. reaction with sulphuric(VI) acid and water

Ethene is absorbed by concentrated sulphuric(VI) acid at room temperature. Dilution of the resulting liquid results in the formation of ethanol (ethyl alcohol). The overall reaction is referred to as the **hydration** of ethene (not to be confused with *hydrogenation*):

$$CH_2{=}CH_2 + H_2O \rightarrow \underset{\text{ethanol}}{CH_3.CH_2OH}$$

This hydration of ethene from the petroleum cracking plant is the main industrial source of ethanol.

5. polymerisation

Ethene (ethylene) can be made to add on to itself producing the long chain molecule poly(ethene) ('polythene' or 'alkathene'):

$$CH_2{=}CH_2 \quad CH_2{=}CH_2 \quad CH_2{=}CH_2 \quad CH_2{=}CH_2$$

$$\downarrow$$

$$|{-}CH_2{-}CH_2{-}| \quad |{-}CH_2{-}CH_2{-}| \quad |{-}CH_2{-}CH_2{-}| \quad |{-}CH_2{-}CH_2{-}|$$

$$\downarrow$$

$$\text{etc.}{-}CH_2{-}CH_2{-}{-}CH_2{-}CH_2{-}{-}CH_2{-}CH_2{-}{-}CH_2{-}CH_2{-}\text{etc.}$$

A similar process of polymerisation, using the closely related compound chloroethene (vinyl chloride), gives poly(chloroethene) or polyvinyl chloride, PVC:

$$CH_2{=}CHCl \quad CH_2{=}CHCl \quad CH_2{=}CHCl \quad CH_2{=}CHCl$$
$$\downarrow$$
$$\text{etc.}{-}CH_2{-}CHCl{-}CH_2{-}CHCl{-}CH_2{-}CHCl{-}CH_2{-}CHCl{-}\text{etc.}$$

6. combustion of ethene

Ethene burns in air with a luminous and smoky flame. The luminosity and smoke are both caused by small particles of carbon resulting from incomplete combustion; this is a characteristic of unsaturated compounds. Complete combustion gives carbon dioxide and steam:

$$C_2H_4 + 3O_2 \rightarrow 2CO_2 + 2H_2O$$

19.10 alkynes: the reactions of ethyne (acetylene)

The hydrocarbons described so far in this Chapter have been either alkanes (C_nH_{2n+2}) or alkenes (C_nH_{2n}). A third family of hydrocarbons exists in which the molecular formulae fit neither of these two patterns. The formula for the simplest member of the series (ethyne, or acetylene) may be deduced by analysis and calculation as outlined below.

10 cm³ of ethyne is exploded with 50 cm³ of oxygen. (*N.B. This is a very dangerous process indeed.*) After cooling to room temperature and pressure, the gases occupy 45 cm³; this volume reduces to 25 cm³ on shaking with aqueous potassium(I) hydroxide. The remaining gas is completely absorbed by alkaline pyrogallol.

	hydrocarbon	oxygen	carbon dioxide
a) volumes:	10 cm³	50 − 25 = 25 cm³	45 − 25 = 20 cm³
b) relative volumes:	1	: 2·5	: 2
c) relative number of moles (*Avogadro*)	1	: 2·5	: 2

Let the formula for ethyne be C_xH_y.
The equation will be:

$$C_xH_y + \left(x + \frac{y}{4}\right)O_2 \rightarrow xCO_2 + \frac{y}{2}H_2O$$

relative number of moles: $1 \qquad x + \frac{y}{4} \qquad x$

From (*c*) above, $x = 2$

and $x + \frac{y}{4} = 2{\cdot}5$

Since $x = 2$, $2 + \frac{y}{4} = 2{\cdot}5$

$y = 2$

the formula for ethyne is C_2H_2

Formulae for other members of this family of hydrocarbons are C_3H_4, C_4H_6, C_5H_8; **the general formula** for the series is C_nH_{2n-2}. The hydrocarbons are called **alkynes** and their molecules are distinguished by the possession of two carbon atoms joined by a **triple bond** (three pairs of electrons shared between two nuclei).

Inspection of a molecular model of ethyne illustrates how the four atoms in the molecule lie in the same straight line. The molecule is said to be **linear** (compare: tetrahedral in alkanes and trigonal in alkenes).

If the reactions and structures of the alkanes and alkenes have been learned and understood, you should be able to make an intelligent guess at the simple

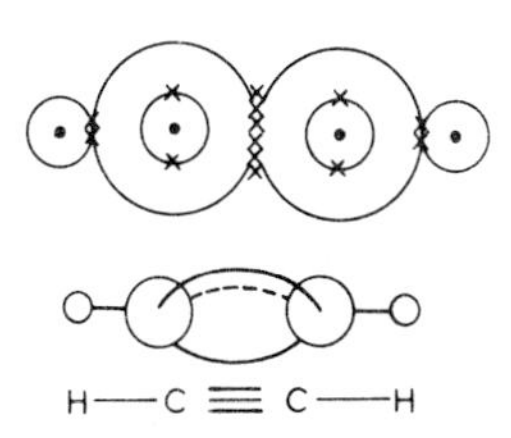

Fig. 19.15 representations of ethyne (acetylene)

$CH_3{-}C{\equiv}C{-}H$
(a)

$CH_3{-}C{\equiv}C{-}CH_3$
(b)

Fig. 19.16 semistructural formulae for (a) propyne, (b) but-2-yne reactions of ethyne in the light of the structures shown in Fig. 19.15. For example, will it burn? If so, with what sort of flame? Will its reactions be similar to those of ethane, or to those of ethene? Which reagents might react with ethyne giving a colour change? What type of reaction would be expected as characteristic of the alkynes?

19.11 small-scale preparation of ethyne and tests

A few drops of water are added to a piece of calcium(II) dicarbide (which should not be handled) in a test-tube. The tube is placed in a rack and a lighted splint is applied *at arm's length* to the mouth of the tube. The smoky nature of the flame should be interpreted in terms of the molecular structure of ethyne.

$$C_2H_2 + 2\tfrac{1}{2}O_2 \longrightarrow 2CO_2 + H_2O$$

The above equation represents the complete combustion of ethyne, but it does not account for the smoky nature of the flame. Incomplete combustion, giving particles of carbon, causes the smoke, but we cannot write a meaningful equation for this. Why not? In answering the question it may be useful to consider why such an equation as

$$C_2H_2 + \tfrac{1}{2}O_2 \longrightarrow 2C + H_2O$$

should not be written in an attempt to describe the combustion of ethyne. Certain other equations written elsewhere in this book are liable to similar criticism.

The complete combustion of ethyne (acetylene), using the correct quantity of oxygen, gives a very hot flame with no soot: this reaction is used in oxy-acetylene blowpipes.

Tubes of the gas may be collected by placing two or three pieces of calcium(II) dicarbide in a test-tube, adding a few drops of water, attaching a bung and delivery tube and collecting the gas in test-tubes over water.

$$CaC_2 + 2H_2O \longrightarrow Ca(OH)_2 + C_2H_2$$

No attempt should be made to ignite the gas collected in this way as it will be mixed with air and is liable to cause a dangerous explosion.

If a tube of gas is shaken with a little dilute aqueous bromine, the colour disappears. Likewise, dilute acidic aqueous manganate(VII) is decolourised. Compare this with the behaviour of ethene; note that alkanes will not decolourise these two reagents.

19.12 addition reactions of ethyne

1. with chlorine

Ethyne and chlorine ignite spontaneously if the gases mix–this reaction can cause dangerous explosions and should not be attempted in practice. The products are mainly carbon and hydrogen chloride. However, if ethyne and chlorine are passed alternately into a solvent, so that they never mix in the gas phase, an *addition* reaction occurs:

$$CH{\equiv}CH + Cl_2 \longrightarrow \underset{\text{dichloroethene}}{CHCl{=}CHCl}$$

then further: $$CHCl{=}CHCl + Cl_2 \longrightarrow \underset{\text{tetrachloroethane}}{CHCl_2{-}CHCl_2}$$

Bromine reacts similarly with ethyne.

2. with hydrogen

Ethyne adds on hydrogen in the presence of a platinum catalyst, forming first ethene, then ethane:

$$CH{\equiv}CH + H_2 \xrightarrow{Pt} \underset{\text{ethene}}{CH_2{=}CH_2}$$

$$CH_2{=}CH_2 + H_2 \xrightarrow{Pt} \underset{\text{ethane}}{CH_3.CH_3}$$

3. with hydrogen chloride

The first addition product is chloroethane (or vinyl chloride), from which poly(chloroethene) or polyvinyl chloride, PVC, is made (see section 19.9(5)):

$$CH{\equiv}CH + HCl \rightarrow CH_2{=}CHCl$$

Further reaction produces dichloroethane:

$$CH_2{=}CHCl + HCl \rightarrow \underset{\text{dichloroethane}}{CH_3.CHCl_2}$$

A wide variety of organic compounds, including useful polymers, can be obtained from ethyne. It is interesting to note that this furnishes a route by which **organic** compounds can be synthesised from simple **inorganic** substances, as calcium(II) dicarbide can be made by heating coke and lime in an electric furnace at 2500°C:

$$CaO + 3C \rightarrow CaC_2 + CO$$

19.13 energy changes in plants and animals

Photosynthesis in plants involves absorption of *carbon* dioxide and evolution of oxygen; **respiration** in plants and animals involves absorption of oxygen and evolution of *carbon* dioxide. Both plants and animals must contain *carbon* compounds, these being made in plants (using the carbon in carbon dioxide) and destroyed in plants and animals (releasing the carbon as carbon dioxide) during the processes of respiration. These carbon compounds are called **carbohydrates**: they are different from the hydrocarbons studied earlier in this Chapter in that their molecules contain oxygen as well as carbon and hydrogen.

However, both hydrocarbons and carbohydrates contain carbon–carbon and carbon–hydrogen bonds; when these bonds are broken and replaced by carbon–oxygen and hydrogen–oxygen bonds, **energy is released.** Hydrocarbons are sources of energy as fuels; carbohydrates are sources of energy as food. Conversions (a) and (b) shown in Fig. 19.18 both require oxygen and they both release energy.

When we eat certain types of food we take into the body energy-rich carbohydrates which have been made by plants. This energy becomes available for use when the carbohydrates are converted into carbon dioxide and water in the body; it may be used for performing mechanical work or for generating heat. In fact, both processes take place simultaneously: when we take exercise, we become hot.

Carbohydrates are made in plants by the process of photosynthesis: the raw materials are carbon dioxide (*from the atmosphere*), water (*from the soil*) and energy (*from the sun*). A catalyst (*chlorophyll*) is necessary. During this process, only some of the oxygen in the carbon dioxide and water is built into the carbohydrates, the rest being released to the atmosphere as oxygen gas. The carbohydrate contains more chemical energy (energy has been absorbed from the sun) than the carbon dioxide and water from which it is made. It is this extra energy which is released for mechanical work and generation of heat when a carbohydrate food is broken down by an animal and reconverted to carbon dioxide and water. The energy changes in the processes of photosynthesis and carbohydrate breakdown are summarised diagrammatically in Fig. 19.19.

19.14 the composition of starch

Starch is the most important carbohydrate synthesised in plants and eaten by animals. The composition of starch is very complex, but the following

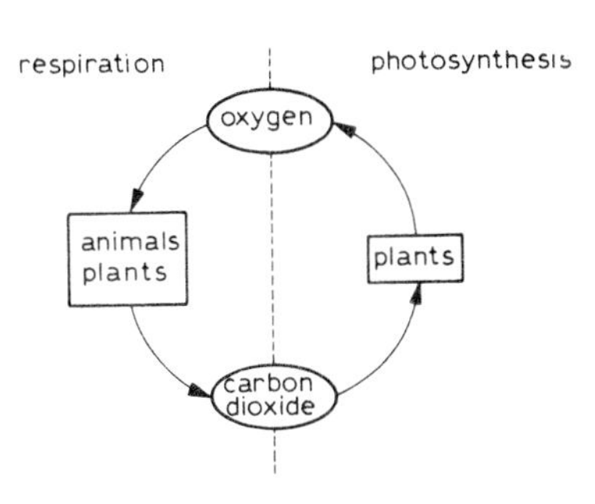

Fig. 19.17 carbon dioxide–oxygen interconversions in photosynthesis and respiration

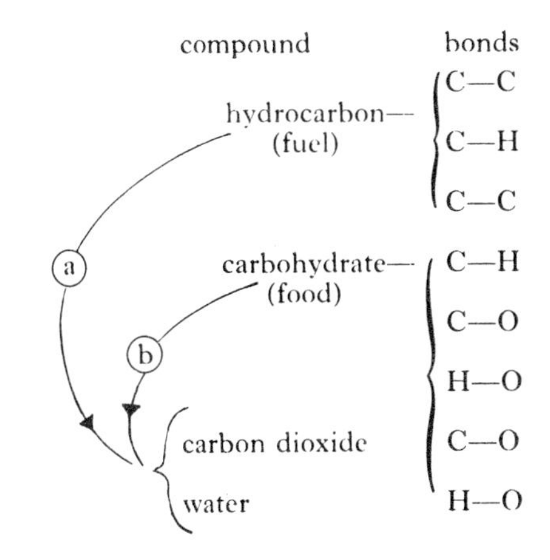

Fig. 19.18 the fundamental changes which release energy from (a) fuels, (b) foods

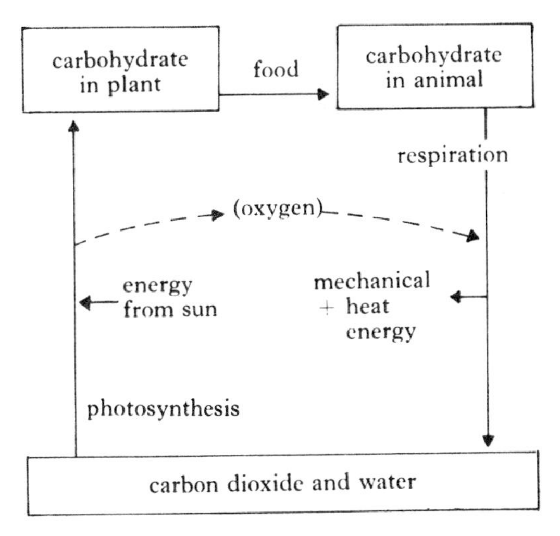

Fig. 19.19 photosynthesis and respiration cycle—the mechanical and heat energy produced originate from solar energy

experiments indicate some important points about its structure and how it is broken down in the body.

Starch forms a deep blue compound with iodine. Aqueous iodine is used in the experiment to test for the presence of starch; when no blue colour is produced, all of the starch has been broken down. The breakdown of the starch is caused by the action of water (hydrolysis), but a catalyst is necessary. The hydrolysis is carried out in duplicate: in one case an **acid catalyst** is used and in the other the **enzymes** in saliva form the catalyst.

1. *hydrolysis of starch*

$10\,cm^3$ of starch solution is placed in each of two boiling-tubes; 10 drops of 2M hydrochloric acid is added to one sample and $1\,cm^3$ of saliva to the other. Both tubes are placed in a beaker of water at 35–40 °C for ten minutes. A drop of each solution is withdrawn and tested with aqueous iodine: the two tubes should be left in the warm water until no blue colour is observed in this test. It may be noted that when the iodine test gives no blue colour, the liquids remaining in the boiling-tubes are *less viscous* than at the outset. The breakdown of starch accomplished so far involves large molecules being converted to smaller ones.

The liquids in the boiling tubes are transferred to evaporating basins and each is evaporated to about $1\,cm^3$.

Fig. 19.20 marking a paper for chromatography

2. *identification of the products by paper chromatography*

A clue to the nature of the products can be obtained by chewing slowly on a piece of dry bread while the experiment is proceeding. By the time the iodine test gives no colour change, the bread will taste quite sweet. The products of the breakdown are in fact **sugars;** just which sugars, may be determined by comparing the hydrolysed liquids with solutions known to contain the sugars **maltose** and **glucose**. The comparison is effected by the technique of **paper chromatography.**

(Note: the solvent and locating agent used in this process will be specified in various laboratory manuals.)

A large sheet of filter paper is marked at one end in pencil as shown in Fig. 19.20. Using a clean capillary tube, a spot of aqueous glucose is placed on mark G. When this has dried, another spot is added in exactly the same place. After drying, a third spot is added. Using clean capillary tubes in each case and following exactly the same procedure, spots are added to the other marked points as follows:

M aqueous maltose
A liquid from acid-catalysed hydrolysis
S liquid from saliva-catalysed hydrolysis

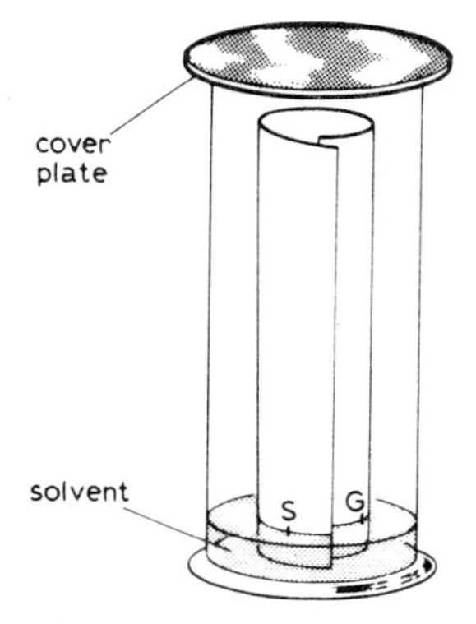

Fig. 19.21 running a chromatogram

When all of the spots are completely dry, the paper is fastened in a cylinder and allowed to stand overnight in a jar containing a 1 cm depth of solvent, the mouth of the jar being closed by a cover-glass, as shown in Fig. 19.21.

The chromatogram is removed from the jar, the solvent front is marked with a pencil line and the paper is air-dried. The paper is sprayed with, or rapidly dipped into, the locating agent and is placed in an oven at 100 °C for two or three minutes. The locating agent reacts with the sugar producing a compound which, on heating in the oven, appears as a coloured spot, or patch. The colour and position of the patch is characteristic of the sugar.

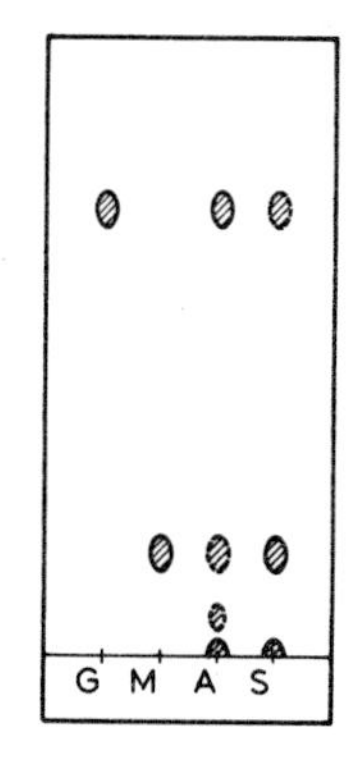

Fig. 19.22 a developed chromatogram

3. *interpretation of the observations*

An illustration of a developed chromatogram is shown in Fig. 19.22. It appears that starch is hydrolysed in the presence of acid largely to glucose, but in the presence of the enzymes in saliva, maltose is the main product. Also, the change in viscosity during the hydrolysis indicates that large molecules in starch are hydrolysed to smaller molecules (glucose and maltose).

Glucose has the formula $C_6H_{12}O_6$ and its molecule consists of a six-membered ring, which may be referred to as a single 'sugar unit'. Glucose is said to be a **monosaccharide.** *Maltose* has the formula $C_{12}H_{22}O_{11}$ and its molecule consists of two six-membered rings, or 'sugar units': it is said to be a **disaccharide.** Maltose is readily hydrolysed to glucose in the presence of acid:

$$C_{12}H_{22}O_{11} + H_2O \longrightarrow 2C_6H_{12}O_6$$

The enzymes in saliva will not catalyse this conversion. It is reasonable to infer that starch contains large molecules which can be broken down on hydrolysis to disaccharide and monosaccharide units. Starch is a mixture of **polysaccharides.** (This is shown on the next page.)

When a piece of bread is chewed slowly, the sweet taste is due to the formation of maltose. The completion of this process and the further hydrolysis of maltose to glucose takes place in the small intestine under the action of other enzymes. This glucose is the main source of energy in the body: it can be broken down into various products in various ways.

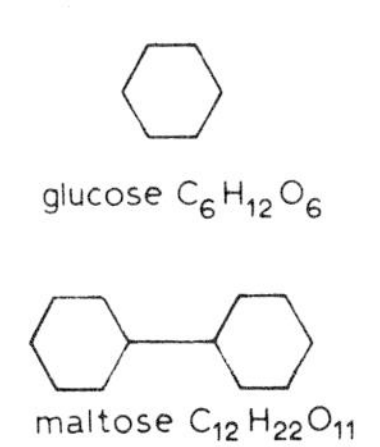

Fig. 19.23 simplified representations for glucose and maltose

19.15 fermentation of glucose

An aqueous solution of glucose and a teaspoonful of yeast (which contains several different enzymes) are enclosed in a flask as shown in Fig. 19.24 for a period of several days. The limewater becomes turbid, indicating the evolution of carbon dioxide. If the contents of the flask are then filtered into a second flask and fractionally distilled, the first product is an inflammable liquid distilling at 78°C. This liquid is **ethanol** (ethyl alcohol).

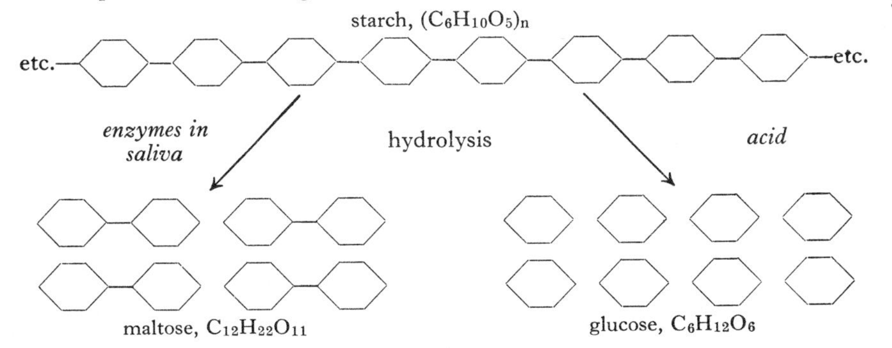

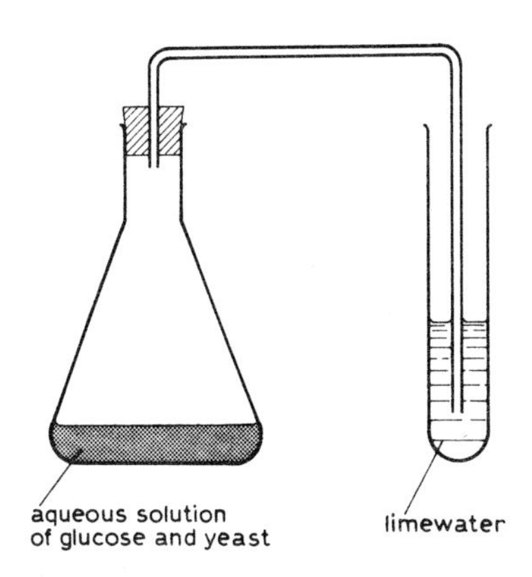

Fig. 19.24 fermentation of glucose

$$\underset{\text{glucose}}{C_6H_{12}O_6} \longrightarrow \underset{\text{ethanol}}{2C_2H_6O} + 2CO_2$$

The production of ethanol from glucose in this way is called **alcoholic fermentation**; it is the basis of the beers, wines and spirits industry. Beers and wines are made by a process of fermentation only, whereas spirits require the additional process of distillation. N.B. Not only is it illegal to carry out the distillation process domestically, it is highly dangerous, as the alcohol product may be sufficiently concentrated to be poisonous, as well as being highly inflammable.

Some of the alcoholic beverages produced from glucose from different sources are shown in Fig. 19.25.

drink	glucose from
beer	barley, malt
wines	grapes
port / sherry	grapes, 'fortified' with added alcohol
whisky	barley, malt
gin	barley, malt with juniper berries
brandy	grapes
rum	molasses

Fig. 19.25 sources of glucose for alcoholic beverages

19.16 breakdown of glucose in the body

In the presence of plenty of oxygen, glucose is oxidised to carbon dioxide and water:

$$C_6H_{12}O_6 + 6O_2 \longrightarrow 6CO_2 + 6H_2O$$

This reaction liberates 2830kJ of energy per mole of glucose.

If very violent or prolonged exercise is taken, energy may be required at a greater rate than that at which oxygen can be supplied to the muscle. In this event the glucose is broken down in the absence of oxygen to lactic acid:

$$\underset{\text{glucose}}{C_6H_{12}O_6} \longrightarrow \underset{\text{lactic acid}}{2C_3H_6O_3}$$

As the lactic acid concentration builds up, it attacks the nerve, causing pain, and eventually prevents the muscle from contracting. The muscle is 'fatigued' and 'stiff' and will not function properly until sufficient oxygen has been supplied to oxidise the lactic acid to carbon dioxide and water:

$$C_3H_6O_3 + 3O_2 \longrightarrow 3CO_2 + 3H_2O$$

The purpose of 'training' is largely to improve the rate at which oxygen can be supplied to the muscle.

The synthesis and breakdown of polysaccharides described in the preceding sections is summarised in Fig. 19.26.

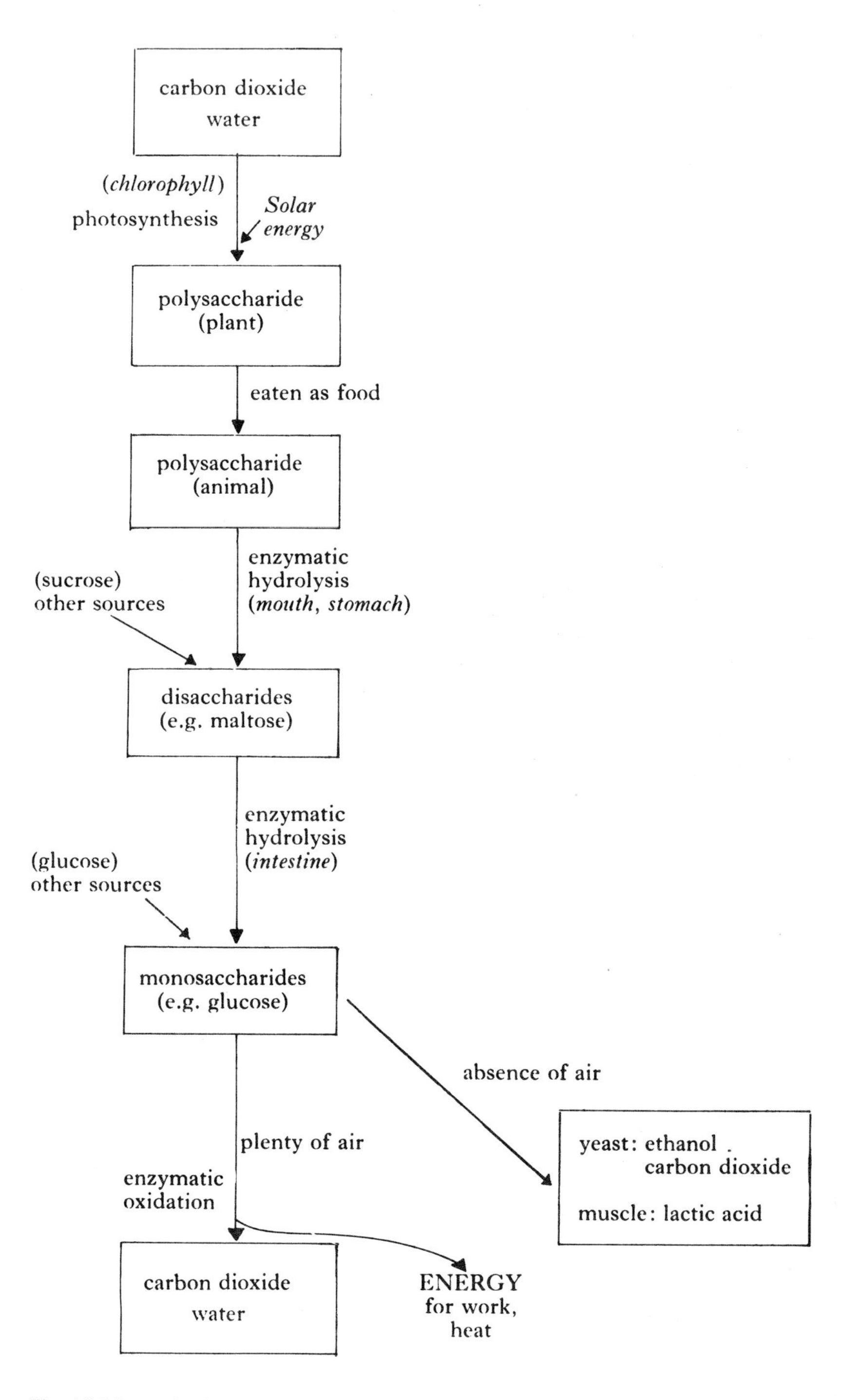

Fig. 19.26 synthesis and utilisation of polysaccharides

revision summary: Chapter 19

petroleum: naturally occurring, complex mixture of alkanes

hydrocarbons: compounds containing carbon and hydrogen only

combustion of hydrocarbons: $C_xH_y + \left(\frac{x+y}{4}\right)O_2 \rightarrow xCO_2 + \frac{y}{2}H_2O$

alkanes: general formula C_nH_{2n+2}, contain single bonds only

isomers: compounds with same molecular formula but different structural formulae

alkenes: general formula C_nH_{2n}, contain C=C

alkynes: general formula C_nH_{2n-2}, contain C≡C

carbohydrates: general formula $C_x(H_2O)_y$, contain oxygen as well as carbon and hydrogen

photosynthesis: energy-absorbing build-up of carbohydrates by certain plants, from CO_2 and H_2O

respiration: energy-releasing oxidation of carbohydrates to CO_2 and H_2O, in plants and animals

questions 19

1. Give the name and formula of (i) an unsaturated hydrocarbon and (ii) a saturated hydrocarbon. What do you understand by the meaning of these terms? How do these hydrocarbons react with chlorine? State the type of reaction involved.

Write the equation for the complete combustion of the hydrocarbon C_xH_y. If $x = \frac{1}{2}y$, to what series does the hydrocarbon belong? What volume of oxygen would be required completely to combust 20 cm³ of the hydrocarbon C_5H_{12}? (Both volumes measured at s.t.p.) (S)

2. 10 cm³ of a gaseous hydrocarbon were mixed with 30 cm³ of oxygen and the mixture was exploded. After the mixture had cooled to room temperature, 20 cm³ of gas remained. After shaking this gas with sodium(I) hydroxide solution, its volume was reduced to 10 cm³. The remaining gas rekindled a glowing splint.

(a) Name the gas remaining at the end.
(b) What were the reacting volumes of the hydrocarbon and oxygen?
(c) Name the gas absorbed by the alkali.
(d) State the volume of the gas named in (c).
(e) Name the other product of the reaction.
(f) Work out the formula for the hydrocarbon. (JMB)

3. Carbon forms a regular series of compounds with hydrogen which can be expressed by the general formula C_nH_{2n+2}. Give the names and formulae of the first four members of this series.

When the structures of the molecules are considered, it is found that two arrangements can be made for the compounds containing four carbon atoms; give these structures, and state the term used to describe this phenomenon.

Compare the reactions which can occur when one molecule of bromine reacts with a molecule of ethane or with a molecule of ethene (ethylene). What general terms are used to describe (a) the reactions, (b) the substances which behave in these ways? (O)

4. Write the equation for the reaction between ethane and oxygen. Find the smallest volume of oxygen needed to burn 1 litre of ethane completely. What volume of carbon dioxide is formed? (All volumes measured at s.t.p.) (S)

5. Write an equation to show the reaction which takes place when propane is burned in an excess of oxygen.

Assuming that all gas volumes are measured at room temperature and pressure, calculate

(a) the minimum volume of oxygen needed for the complete combustion of 20 cm³ of propane,
(b) the volume of the gaseous product of the complete combustion of 20 cm³ of propane. (JMB)

6. A mixture of 400 cm³ of a gas, A, and 400 cm³ of hydrogen reacted completely in the presence of a catalyst to give 400 cm³ of a new gas, B. When 400 cm³ of this new gas, B, were mixed with 400 cm³ of hydrogen in the presence of a different catalyst, $400\,cm^3$ of ethane were produced. (All volumes measured under the same conditions of temperature and pressure.)

A liquid, *D*, was produced when *B* and bromine reacted together. Addition of another catalyst to *B* gave a solid, *E*, which did not react with bromine and had a relative molecular mass of about 20000.

Explain these results, identify *A*, *B*, *D*, *E* and give equations for the reactions.

How might polyvinyl chloride be obtained from *A*?

(JMB)

7. (a) A mixture of $50\,cm^3$ methane with $100\,cm^3$ oxygen maintained at 110 °C was ignited by a spark, after which the resulting mixture occupied $150\,cm^3$ so long as the temperature and pressure remained unaltered; when cooled to room temperature (but with unchanged pressure) the gas volume fell to $50\,cm^3$ and it was also found that this final residue could be absorbed completely in sodium(I) hydroxide.

What substances were produced when the original mixture was ignited?

What were the respective volumes of these substances?

What principle (or rule) have you assumed?

What chemical law has been illustrated?

Account for the changes in volume which occurred (i) on cooling, (ii) when sodium(I) hydroxide was added.

(b) Equations representing reactions of equimolar proportions of chlorine with ethane and ethene (ethylene) are respectively

$$C_2H_6 + Cl_2 \rightarrow C_2H_5Cl + HCl$$

and

$$C_2H_4 + Cl_2 \rightarrow C_2H_4Cl_2$$

Name the substances formed and point out the essential difference between these reactions. (O)

8. Write structural formulae for the two gaseous hydrocarbons of molecular formulae C_3H_8 and C_3H_6.

If you were supplied with a gas cylinder of **one** of the above gases describe how you would proceed,

(a) to show experimentally that the gas contains hydrogen and carbon,

(b) to find out which of the above two gases the cylinder contained.

Write a concise account of an industrial chemical process which uses **one** of the products of the distillation of **either** coal-tar **or** crude oil (petroleum). (W)

9. By drawing their structural formulae show the differences in structure between ethane and ethylene (ethene). Explain how the bond between a carbon atom and a hydrogen atom in these compounds is formed.

By naming the reagents, stating the conditions, and writing an equation for each reaction, describe how ethylene could be converted into (a) ethane, (b) ethyl chloride (chloro-ethane), (c) ethylene dibromide (1,2 dibromoethane), (d) ethyl alcohol (ethanol). (JMB)

10. What is (a) an exothermic reaction, (b) an endothermic reaction? Give **one** industrial example of each.

$20\,cm^3$ of a mixture of ethene (ethylene) and ethane need $64\,cm^3$ of oxygen for complete combustion (both volumes measured under the same conditions of temperature and pressure). What are the volumes of the two gases?

How could you chemically distinguish between ethane and ethene (ethylene)?

Name an industrially important compound (**one** in each case) which is made from:

(i) ethene;
(ii) acetylene (ethyne).

What kind of reaction takes place in each case? (S)

11. Under suitable conditions it is possible to obtain a hydrocarbon, **X**, directly from ethanol (ethyl alcohol). **X** reacts with bromine in the cold to give a liquid, **Y.** The relative molecular mass of **Y** is 188 and it contains 12·8% carbon, 2·13% hydrogen and 85·1% bromine by mass.

(a) Calculate (i) the empirical formula, (ii) the molecular formula, of the liquid **Y**. Name this liquid.
(b) Name the hydrocarbon, **X**, and give its *structural* formula.
(c) What type of reaction occurs between bromine and the hydrocarbon **X**?
(d) Give **one** large-scale use for **X**.
(e) State briefly how **X** can be obtained from ethanol.
(f) Give **two** other reactions of ethanol. (AEB)

12. A hydrocarbon contains 82·76% by mass of carbon. Find its empirical formula.

If the vapour density of the gas is 29, what is its molecular formula? Write structural formulae for **two** different compounds which have this molecular formula.

Complete the following statement: When two compounds have the same empirical and molecular formulae but different structural formulae they are said to exhibit *?* and the different forms are called *?* (JMB)

13. Describe how you would obtain a sample of almost pure ethanol (ethyl alcohol) from a sugar.

Outline the steps by which ethanol is manufactured from petroleum oil. Name all the chemical processes involved.

Describe **one** chemical and **one** physical test to distinguish between pure water and pure ethanol. State the result of each test on each of these substances. (JMB)

14. (a) Explain the meaning of each of the following statements:

(i) the alkanes are a series of hydrocarbons with the general formula C_nH_{2n+2}
(ii) there are two isomeric compounds each of which has the molecular formula C_4H_{10}
(iii) polythene is an addition polymer of ethylene (ethene).

(b) An organic compound decolorises bromine without any evolution of hydrogen bromide. What conclusion can you draw concerning the nature of the organic compound and of the reaction taking place? Illustrate your answer by reference to a suitable compound. Write an equation for the reaction of the compound with bromine.

(c) Fractional distillation and cracking are two processes used in the petroleum and petrochemicals industries. Explain briefly what each of these processes involves.

L (NUFFIELD)

Chemical energy may be regarded as a rather particular kind of potential energy. The possession of chemical energy by a substance is not obvious until a change occurs in which it is transferred; the form in which it is transferred is usually **heat**. The study of the heat changes accompanying chemical reactions is called **thermochemistry.**

Some substances possess more chemical energy than others. When concentrated sulphuric(VI) acid is added to water, **heat is evolved:** the reaction is said to be **exothermic.** The diluted acid *seems* to possess more energy than the two cold liquids from which it is formed, simply because it is hot. However, when the heat energy has been dissipated to the surroundings as the dilute acid cools to room temperature, it is obvious that the dilute acid contains *less* chemical energy than did the concentrated acid and water at the outset.

Solid ammonium nitrate(V) stirred in water dissolves with the **absorption of heat** energy: the process of solution is said to be **endothermic.** When the solution is allowed to warm up to room temperature, it is clear that aqueous ammonium nitrate(V) contains *more* chemical energy than the solid ammonium nitrate(V) and water from which it is formed.

In some cases energy must be supplied initially to start a reaction which then proceeds with the evolution of heat. You provide this **activation energy** every time you light a fire or a gas cooker in your home. The energy changes involved in an exothermic and in an endothermic reaction, both having activation energies, are shown in Fig. 20.1 and Fig. 20.2. Energy changes involved in the familiar hydrogen–oxygen reaction are discussed in detail in section 9.6 and it would be wise to read or revise this section before proceeding further.

The aims of this Chapter are to show how heat changes accompanying chemical reactions can be measured and to indicate how such measurements, in conjunction with our knowledge of atomic structure and bonding (Chapter 18), can give an insight into the fundamental changes occurring in chemical reactions.

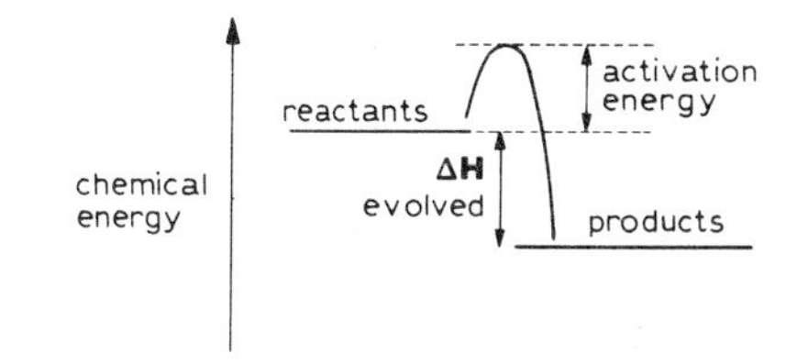

Fig. 20.1 exothermic reaction with activation energy

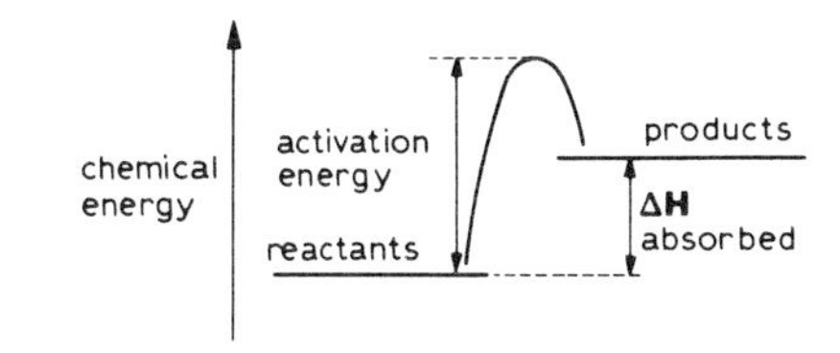

Fig. 20.2 endothermic reaction with activation energy

20.1 expression of energy quantities in chemistry

The units most commonly employed by the chemist for measurement of energies are the joule (J) and kilojoule (kJ); the calorie (cal) and kilocalorie (kcal) may also be encountered.

The joule is defined in terms of the quantity of mechanical work it can do (see also Fig. 2.5): one joule of work is done when a force of one newton moves its point of application through one metre in the direction of the force.

In terms of basic units, $1\,N = 1\,kg\,m\,s^{-2}$

$$1\,\text{cal} = 4{\cdot}187\,\text{J} \approx 4{\cdot}2\,\text{J}$$

$$1\,\text{J} = 0{\cdot}2388\,\text{cal} \approx 0{\cdot}24\,\text{cal}$$

$$1\,\text{kJ} = 1000\,\text{J}$$
$$1\,\text{kcal} = 1000\,\text{cal}$$

$$1\,\text{J} = 1\,\text{kg}\,\text{m}^2\,\text{s}^{-2}$$

The calorie is usually defined in terms of the heating capability of a quantity of energy: one calorie will raise the temperature of one gram of water through 1 °C under strictly defined conditions. The joule and kilojoule are the units of energy used throughout this book, but you will undoubtedly encounter statements of energy in calories and kilocalories in earlier publications. The appropriate conversion factors are given opposite.

20.2 factors affecting the quantity of energy exchanged in a reaction – use of conventions

1. quantity of substance

Fig. 20.3 'put some more coal on'

'Put some more coal on the fire.' The combustion of 2 kg of coal liberates twice as much heat energy as the combustion of 1 kg of coal. The chemist describes an amount of substance in moles rather than in kilograms or grams, which are mass units. The most convenient way of doing this is by using an equation. Simplifying the chemistry of the example to that of a fire of pure carbon, we would write:

$$C + O_2 \longrightarrow CO_2; \quad \Delta H = -393{\cdot}5\,\text{kJ}$$

$C + O_2$

$\Delta H = -393{\cdot}5\,\text{kJ}$

CO_2

Fig. 20.4 exothermic reaction, ΔH negative

This means 'when 12 g of carbon (1 mole) burns in 32 g of oxygen (1 mole) forming 44 g of carbon dioxide (1 mole), 393·5 kJ of energy is released as heat'. ΔH represents the change in 'heat content', or chemical energy, possessed by the substances present; since 44 g of carbon dioxide contains 393·5 kJ of chemical energy less than is contained in 12 g of carbon and 32 g of oxygen, the change is in this case an *energy loss from the substances* and ΔH has a **negative** value. A **positive** value for ΔH would describe an **endothermic** reaction.

2. nature of the substance

Fig. 20.5 not such good coal

'I don't think this coal is as hot as the last load we had.' Different substances react giving different energy changes. Of course the chemist specifies the reacting substances and the products when he writes an equation.

At this point we ought to return to the very important principle stated in section 9.5:

when bonds are broken, energy is required
when bonds are formed, energy is released

A reaction in which relatively few, or relatively weak, bonds are broken and relatively many, or relatively strong, bonds are formed, will be **exothermic.** The hydrogen–oxygen reaction exemplifies this:

H—H H—H
O—O → H—O—H H—O—H

(3 weaker bonds broken) (4 stronger bonds formed)

reaction exothermic

3. physical state of the reactants and products

It was recognised in section 3.7 that the same quantity of the same substance contains more energy in the gaseous state than in the liquid state and more in the liquid than in the solid state. Thus 'H_2O' does not define the chemical energy possessed by water relative to its elements, whereas the symbols $H_2O(g)$, $H_2O(l)$ and $H_2O(s)$ do this, the letters in brackets referring to gas, liquid and solid respectively.

$$H_2(g) + \tfrac{1}{2}O_2(g) \longrightarrow H_2O(g); \quad \Delta H = -243\,\text{kJ}$$
$$H_2(g) + \tfrac{1}{2}O_2(g) \longrightarrow H_2O(l); \quad \Delta H = -287\,\text{kJ}$$

$$H_2(g) + \frac{1}{2}O_2(g) \rightarrow H_2O(s); \quad \Delta H = -293\,kJ$$

The differences between these figures are equal to the latent heats of vaporisation and fusion.

Symbols for physical states of substances should always be included in thermochemical equations. If the physical state is not specified in an equation, you should assume that the substances are in the state stable at 298 K and 101 325 Pa.

4. *allotropes of elements*

Some elements exist in two different forms in the same physical state: a well-known example of such allotropes is that diamond and graphite are both forms of solid carbon. Since the bonding in graphite is stronger than that in diamond, the figures below should not surprise you. It is clearly necessary in such cases to specify which allotrope is used.

$$C(graphite) + O_2(g) \rightarrow CO_2(g); \quad \Delta H = -393{\cdot}5\,kJ$$
$$C(diamond) + O_2(g) \rightarrow CO_2(g); \quad \Delta H = -395{\cdot}4\,kJ$$

5. *standard reference state of temperature and pressure*

Liquid water (1 mole) at 373 K contains more energy than liquid water at 273 K. For this reason, standardised conditions of temperature and pressure should be used in referring to the precise values of energy changes in reactions:

$$H_2(g) + \frac{1}{2}O_2(g) \rightarrow H_2O(l); \quad \Delta H^{\ominus} = -287\,kJ$$

The quantity, 287 kJ, refers to the energy released in converting hydrogen and oxygen at 298 K and 101 325 Pa to water at 298 K and 101 325 Pa, even though in the intermediate stages of the reaction the temperature or pressure (or both) will doubtless rise well above these values. The superscript $^{\ominus}$ is used to denote these standard conditions in advanced work; in this book, the symbol ΔH without the superscript will be used and reference to standard conditions is assumed.

20.3 measurement of heats of reaction

The apparatus used to hold substances for which a heat of reaction is to be measured is called a **calorimeter**. Calorimeters vary in design according to whether gases, liquids, solids or combinations of these are to be used and whether considerable variations in temperature or pressure are expected or not. All calorimeters have two common features:

a) they must minimise heat losses to the surroundings and are therefore insulated,

b) the heat capacity of the calorimeter must either be known or, for approximate work, be so small as to be negligible.

If these requirements are fulfilled, the quantities of substance used in the calorimeter and the temperature change in the reaction can be used to calculate the heat energy transferred.

A 'polythene' bottle makes a good, simple calorimeter for approximate work; its heat capacity is low enough to be neglected and very little heat is transferred to the outer surface of the bottle during the normal time for an experiment. The low specific heat capacity and insulating properties of polythene may be explained by reference to its molecular structure (see section 19.9(5))–the transfer of molecular vibrations through such a structure is as unlikely as the transfer of mechanical energy through a rubber nail.

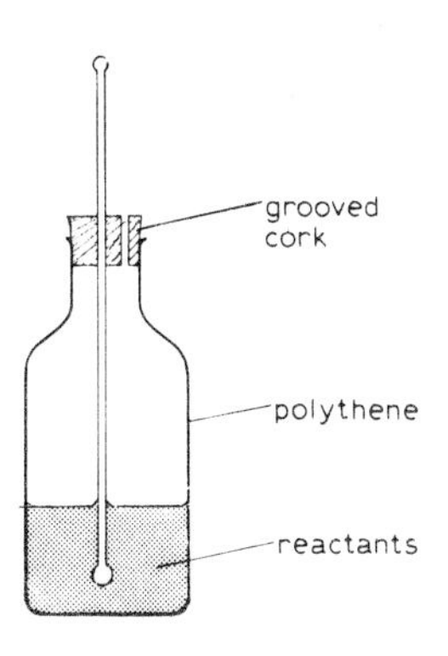

Fig. 20.6 'polythene' bottle calorimeter

20.4 heat of combustion and its measurement

The heat of combustion of a substance is defined as the heat evolved when one mole of a substance burns completely in excess oxygen. Examples are:

$$H_2(g) + \tfrac{1}{2}O_2(g) \longrightarrow H_2O(l); \quad \Delta H = -287\,kJ$$
$$C(graphite) + O_2(g) \longrightarrow CO_2(g); \quad \Delta H = -394\,kJ$$
$$C_2H_2(g) + 2\tfrac{1}{2}O_2(g) \longrightarrow 2CO_2(g) + H_2O(l); \quad \Delta H = -1310\,kJ$$

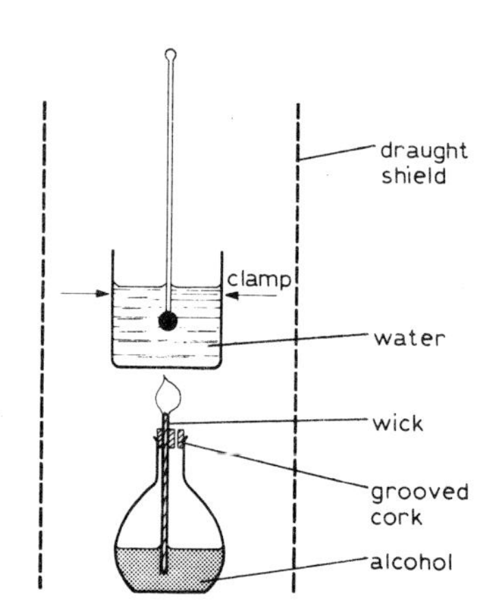

Fig. 20.7 heat of combustion apparatus

The heats of combustion of three alcohols may be compared by the procedure described in the following experiments; it is designed to illustrate how heats of combustion are measured, rather than to give accurate results. The procedure is simple, the experiment brief and the calculation ignores certain heat losses which would need to be considered to obtain an accurate result.

$250\,cm^3$ of water is delivered from a measuring cylinder into a shiny metal can. The mass of the alcohol lamp is determined with the lamp about half-full; the apparatus is then assembled as shown in Fig. 20.7. After noting the temperature of the water, the lamp is lit and the water stirred for about ten minutes, when its temperature will have risen by about 30°C. The highest temperature achieved is noted, the lamp is extinguished and the final mass of the lamp and contents is determined.

The mass of alcohol used in combustion is calculated and it is assumed that the heat evolved has all been used in heating the water. This heat is calculated from the rise in temperature of the known mass of water.

Typical experimental results are shown in the following table.

alcohol used	= methanol, CH_3OH
initial mass of lamp	= 73·46 g
final mass of lamp	= 71·25 g
mass of water used	= 250 g
initial temperature of water	= 26·0°C
final temperature of water	= 54·3°C
M_r for CH_3OH	= 32
mass of methanol used	= 2·21 g
amount of methanol used	$= \frac{2·21}{32}$ mol
	= 0·069 mol
temperature rise of water	= 28·3°C
heat transferred to water	$= 250 \times 28·3 \times 4·2 \times 10^{-3}$ kJ
	= 29·7 kJ
heat of combustion of methanol	$= \frac{29·7}{0·069}$ kJ mol^{-1}
	= **430 kJ mol^{-1}**

The values obtained for the heats of combustion of three alcohols are:

methanol	430 kJ mol^{-1}	
ethanol	910 kJ mol^{-1}	(note the approximate con-
propan-1-ol	1380 kJ mol^{-1}	stancy of the differences)

These experimental results are about two-thirds of the accepted values. What approximations in the experiment account for the discrepancy? How might the apparatus and calculation be modified to give more accurate results?

The approximately constant differences between the results shown above may be accounted for as follows:

$$C_2H_5OH + 3O_2 \longrightarrow 2CO_2 + 3H_2O$$
$$CH_3OH + 1\tfrac{1}{2}O_2 \longrightarrow CO_2 + 2H_2O$$

difference: $$—CH_2— + 1\tfrac{1}{2}O_2 \longrightarrow CO_2 + H_2O$$

Comparison of the result (shown at the top of the next page) with the differences between the heats of combustion for adjacent members of the series of alcohols quoted above offers an interesting experimental test of the principle first stated in section 9.5, that energy is required to break bonds and is released when bonds are formed.

bonds broken (energy required)	bonds formed (energy released)
H–C(–H) (structural: C with two H) $(O{=}O) \times 1\frac{1}{2}$	O—C—O H—O—H
$349 + (416 \times 2) + (496 \times 1\frac{1}{2})$	$(748 \times 2) + (466 \times 2)$
$1181 + 744$	$1496 + 932$
1925 kJ	2428 kJ

$$\text{net energy released} = (2428 - 1925)\,\text{kJ} \approx \mathbf{500\,kJ}$$

bond	energy/kJ mol^{-1}
C—C	349
C—H	416
O=O	496
C=O	748
H—O	466

20.5 Hess's Law

Carbon burns in a limited supply of oxygen to form carbon monoxide:

$$C(graphite) + \tfrac{1}{2}O_2(g) \rightarrow CO(g); \quad \Delta H = -111\,kJ$$

Carbon monoxide will burn in excess oxygen forming carbon dioxide:

$$CO(g) + \tfrac{1}{2}O_2(g) \rightarrow CO_2(g); \quad \Delta H = -284\,kJ$$

The total heat liberated in the **two-stage** conversion of 1 mole of graphite to carbon dioxide is thus 111 + 284 = 395 kJ.

If 1 mole of graphite is converted **directly** to carbon dioxide by burning it in excess oxygen, the heat liberated is 395 kJ:

$$C(graphite) + O_2(g) \rightarrow CO_2(g); \quad \Delta H = -395\,kJ$$

The principle illustrated by this example is stated in a general form as **Hess's Law of Constant Heat Summation:**

The heat change in a reaction depends only on the initial and final states of the system and is independent of the stages by which the reaction occurs, provided that heat is the only form of energy entering or leaving the system.

Hess's Law is a particular case of the Law of Conservation of Energy. The carbon example is represented in Fig. 20.8; $\Delta H_2 + \Delta H_3$ must be the same as ΔH_1. If this were not so, it would be possible to construct a cycle which would generate energy from nothing.

Hess's Law is useful in calculating reaction heats which are difficult to measure, and in obtaining theoretical values for reaction heats for reactions which do not occur.

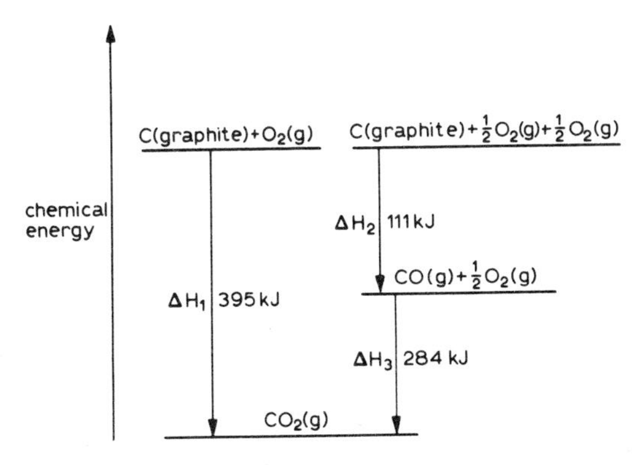

Fig. 20.8 illustration of Hess's law

Example 1: The heats of combustion of graphite and diamond are 393·5 kJ mol^{-1} and 395·4 kJ mol^{-1} respectively. What is the heat change in the conversion of one mole of graphite into diamond?

It is necessary first to express the information given, and the question, in the form of equations:

$$C(graphite) + O_2(g) \rightarrow CO_2(g); \quad \Delta H = -393{\cdot}5\,kJ$$
$$C(diamond) + O_2(g) \rightarrow CO_2(g); \quad \Delta H = -395{\cdot}4\,kJ$$
$$C(graphite) \rightarrow C(diamond); \quad \mathbf{\Delta H = ?}$$

Energy diagrams are then constructed as in Fig. 20.9; it is often best to start the diagram with the quantity to be found (a), then add in the information (b). The algebraic signs are best omitted from the diagram altogether, so long as it is remembered that an *exothermic* process involves a *loss* of chemical energy from the substances, i.e. a vertical *drop* in the diagram. They are included in the final calculation:

$$\Delta H = -393{\cdot}5 + 395{\cdot}4\,kJ$$
$$= +1{\cdot}9\,kJ$$

When one mole of graphite is converted into diamond, 1·9 kJ are absorbed.

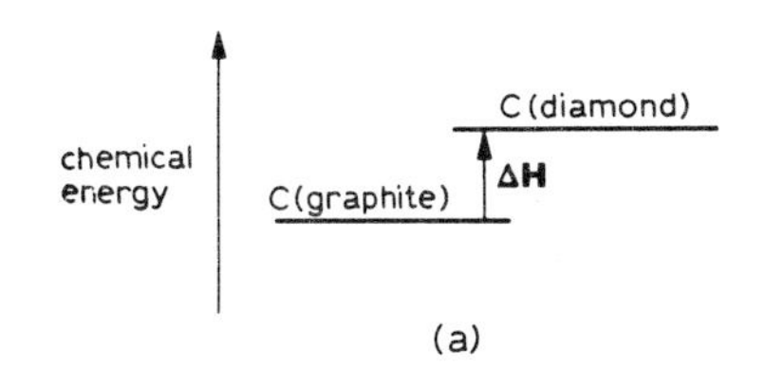

(a)

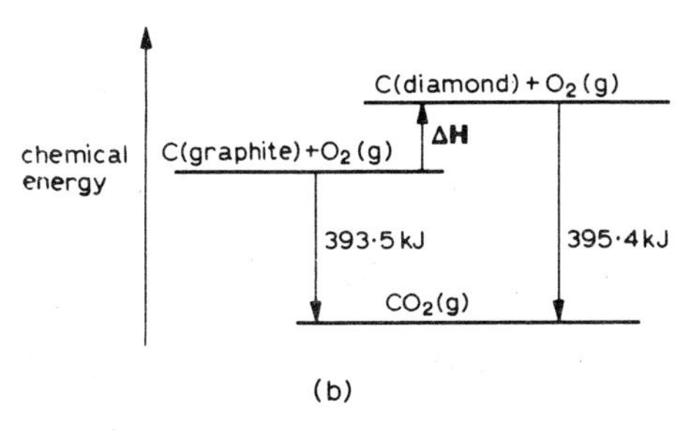

(b)

Fig. 20.9

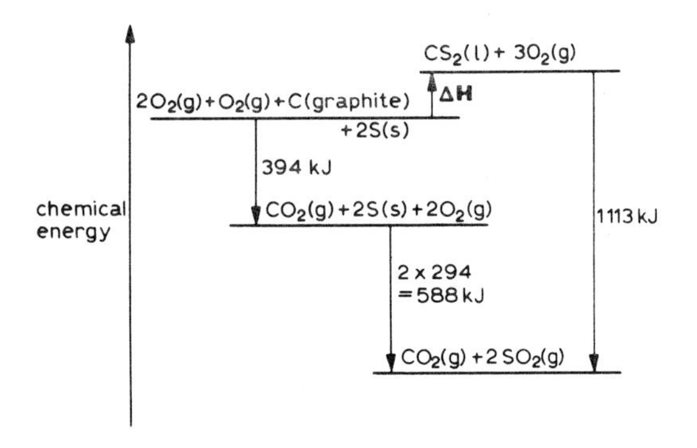

Fig. 20.10

Example 2: The heats of combustion of graphite, sulphur and carbon disulphide are 394, 294 and 1113 kJ mol^{-1} respectively. What is the heat change when one mole of carbon disulphide is formed from its elements?

$$C(graphite) + O_2(g) \rightarrow CO_2(g); \quad \Delta H = -394\,kJ$$
$$S(s) + O_2(g) \rightarrow SO_2(g); \quad \Delta H = -294\,kJ$$
$$CS_2(l) + 3O_2(g) \rightarrow CO_2(g) + 2SO_2(g); \quad \Delta H = 1113\,kJ$$
$$C(graphite) + 2S(s) \rightarrow CS_2(l); \quad \mathbf{\Delta H = ?}$$

$$\Delta H = -394 - 588 + 1113\,kJ$$
$$= +131\,kJ$$

The heat absorbed is 131 kJ

20.6 heat of neutralisation

Neutralisation of an acid by a base in aqueous solutions is an exothermic process. Water is always formed in the reaction and **heat of neutralisation** is defined as the **heat evolved when an acid and base react to form one mole of water** (under conditions such that dilution of the resulting solution gives no further heat change). Heats of neutralisation can be determined using a simple polythene bottle calorimeter, as described in section 20.3. Exactly 50 cm^3 each of 2 M solutions of acids and alkalis are used in each experiment (except for sulphuric(VI) acid, for which a 1 M solution is used–*why?*) The initial temperatures of the reactants are noted, as is the highest temperature attained on mixing; heat lost to the calorimeter is neglected. Typical results for 2 M hydrochloric acid and 2 M sodium(I) hydroxide are given below:

initial temperature acid = 18 °C
initial temperature alkali = 19 °C
highest temperature attained = 32 °C

The dilute aqueous solutions may be assumed to have the same densities and specific heat capacities as water. The heat of neutralisation is calculated as follows:

heat gained by acid solution $= 50 \times 4{\cdot}2 \times 14\,J$
$= 2940\,J$
heat gained by alkali $= 50 + 4{\cdot}2 \times 13\,J$
$= 2730\,J$
total heat gained by solution $= (2940 + 2730)\,J$
$= 5670\,J$
amount of acid used $= \frac{50}{1000} \times 2\,mol$
$= 0{\cdot}1$ mole
amount of alkali used $= \frac{50}{1000} \times 2\,mol$
$= 0{\cdot}1$ mol
therefore amount of water formed $= 0.1$ mol
therefore heat of neutralisation $= 5670 \times 10\,J$
$= 56700\,J$ or $56{\cdot}7\,kJ$

the heat of neutralisation is 57 kJ

Heats of neutralisation of various acid–alkali pairs are shown in Fig. 20.11. The first striking feature is that the **heat of neutralisation of any strong** (completely ionised) **acid by any strong base is constant** at about 57 kJ.

The neutralisation of hydrochloric acid by sodium(I) hydroxide and by potassium(I) hydroxide may be written:

$$H^+(aq) + Cl^-(aq) + Na^+(aq) + OH^-(aq) \rightarrow H_2O(l) + Na^+(aq) + Cl^-(aq)$$
$$H^+(aq) + Cl^-(aq) + K^+(aq) + OH^-(aq) \rightarrow H_2O(l) + K^+(aq) + Cl^-(aq)$$

Eliminating the 'spectator ions', both equations reduce to:

$$H^+(aq) + OH^-(aq) \rightarrow H_2O(l); \quad \Delta H = -57\,kJ$$

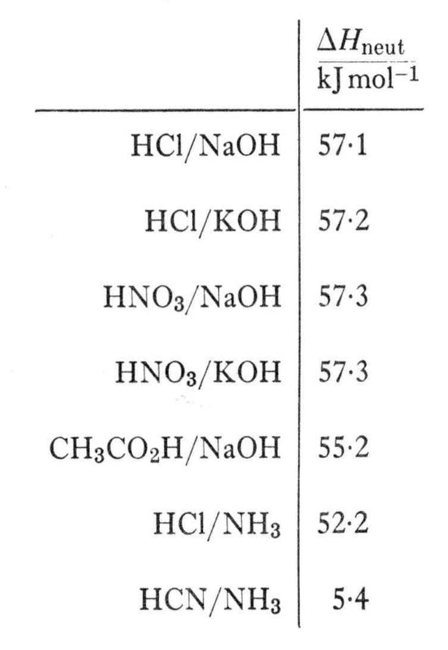

	ΔH_{neut} / kJ mol^{-1}
HCl/NaOH	57·1
HCl/KOH	57·2
HNO_3/NaOH	57·3
HNO_3/KOH	57·3
CH_3CO_2H/NaOH	55·2
HCl/NH_3	52·2
HCN/NH_3	5·4

Fig. 20.11 heats of neutralisation

Ethanoic acid (acetic acid, a *weak* acid) is only *partially ionised*; some energy is absorbed in the process:

$$CH_3CO_2H(aq) \rightarrow CH_3CO_2^-(aq) + H^+(aq)$$

This ionisation goes to completion as the $H^+(aq)$ ions are removed by $OH^-(aq)$ ions, forming water.

Aqueous ammonia is a *weak* (*partially ionised*) alkali; as it is neutralised by an acid, the endothermic ionisation goes to completion:

$$NH_3(aq) + H_2O(l) \rightarrow NH_4^+(aq) + OH^-(aq)$$

Thus the heats of neutralisation of weak acids and weak bases are different from the constant value of $57\,kJ\,mol^{-1}$ for strong acids and bases.

20.7 heat of solution

The process of solution may be exothermic (e.g. sulphuric(VI) acid) or endothermic (e.g. ammonium nitrate(V)). The following simple experiment illustrates an important feature of the heat change when a substance is dissolved in water.

If about six pellets of sodium(I) hydroxide are stirred with $25\,cm^3$ of water in a small beaker, a rise in temperature is observed. When all the solid has dissolved, addition of a further $10\,cm^3$ of water produces another temperature rise. Subsequent additions of small volumes of water produce successively smaller rises in temperature.

The heat of solution of a substance in water is the **heat change when one mole of the substance is dissolved in a volume of water such that addition of more water to the solution produces no further measurable heat change.** The necessary quantity of water is symbolised by aq and the fully hydrated ions in the resulting solution are symbolised by writing (aq) after the formula:

e.g. $NaOH(s) + aq \rightarrow Na^+(aq) + OH^-(aq); \quad \Delta H = -43\,kJ$

20.8 heat of solution of anhydrous and hydrated salts

The heat changes in the interconversion of copper(II) sulphate(VI)-5-water (the blue crystals) and white anhydrous copper(II) sulphate(VI) are described qualitatively in section 5.3. The nature of the water of crystallisation in hydrated salts is also discussed in section 8.16. In this section and the next we shall seek structural reasons for the heat changes on the hydration and the dissolving of salts.

$100\,cm^3$ of cold water is placed in each of two beakers A and B. Addition of 5 g of copper(II) sulphate(VI)-5-water to A produces a *fall* in temperature of about 0·5 °C; addition of 5 g of anhydrous copper(II) sulphate(VI) to B produces a *rise* in temperature of about 4 °C. The experiment may be repeated using zinc(II) sulphate(VI)-7-water and anhydrous zinc(II) sulphate(VI): again the **solution of the hydrated salt is endothermic and that of the anhydrous salt is exothermic.**

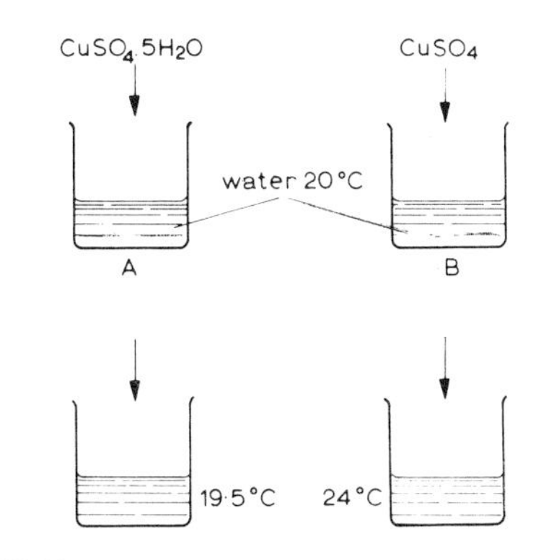

Fig. 20.12

The energy diagram relating the heats of hydration and of solution for copper(II) sulphate(VI) is shown in Fig. 20.13. It is now necessary to explain the structure and bonding which account for these heat changes.

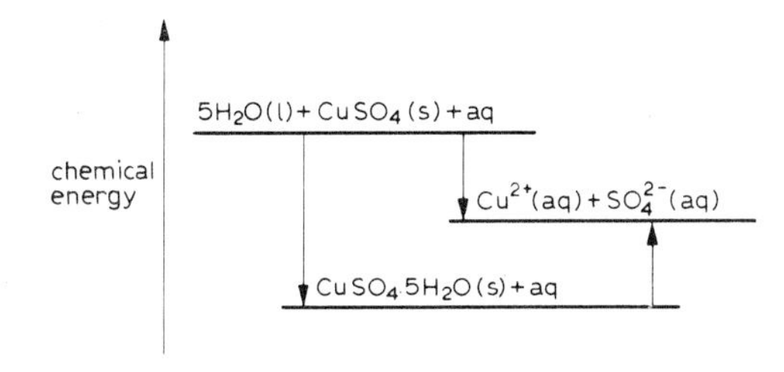

Fig. 20.13

20.9 nature of bonding in hydrated salts

X-ray examination of hydrated salts shows that most (in some cases all) of the water of crystallisation is bonded to the cation (+), not the anion (−). That not bonded to the cation is held elsewhere in the crystal lattice.

e.g. $CuSO_4.5H_2O$ is $[Cu(H_2O)_4]^{2+}SO_4^{2-}.H_2O$

$ZnSO_4.7H_2O$ is $[Zn(H_2O)_6]^{2+}SO_4^{2-}.H_2O$

We may see why this is so if we remember that the molecule of water is dipolar, having two lone pairs of electrons on the oxygen atom (section

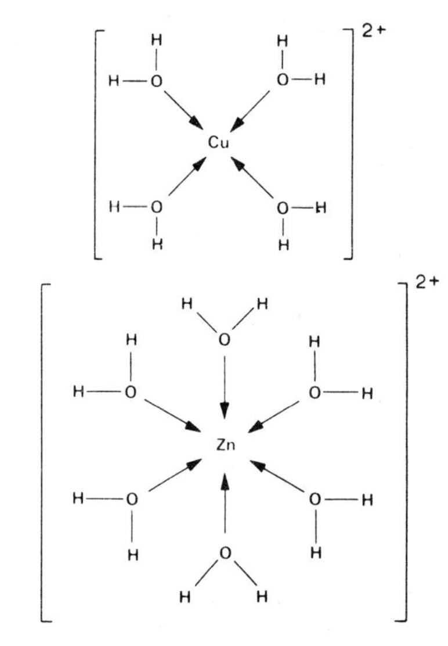

Fig. 20.14 hydrated cations

18.20). One lone pair can be donated to a cation, forming a coordinate bond (section 18.22) between the cation and the water molecule. Four such bonds are formed with the copper(II) ion in $[Cu(H_2O)_4]^{2+}$ and six with the zinc(II) ion in $[Zn(H_2O)_6]^{2+}$.

Remembering the principle that energy is released when bonds are formed, it should be clear why the hydration of an anhydrous salt is an exothermic process.

When an ionic solid dissolves in water, energy must be supplied to separate the ions, just as would be necessary if the solid were sublimed:

e.g. $(Zn^{2+} + SO_4^{2-})(s) \longrightarrow Zn^{2+}(g) + SO_4^{2-}(g)$; **$\Delta H$ positive**

When the resulting ions are hydrated, bonds are formed and energy is released:

e.g. $Zn^{2+}(g) + SO_4^{2-}(g) + aq \longrightarrow Zn^{2+}(aq) + SO_4^{2-}(aq)$; **$\Delta H$ negative**

When the anhydrous salt dissolves in water, the extensive hydration of the ions releases more energy than is absorbed in separating the ions and **the overall dissolving process is exothermic.**

When the hydrated salt dissolves in water, the processes are similar:

$([Zn(H_2O)_6]^{2+} + SO_4^{2-} + H_2O)(s) \longrightarrow [Zn(H_2O)_6]^{2+}(g) + SO_4^{2-}(g) + H_2O(g)$; **$\Delta H$ positive**

$[Zn(H_2O)_6]^{2+}(g) + SO_4^{2-}(g) + aq \longrightarrow Zn^{2+}(aq) + SO_4^{2-}(aq)$; **$\Delta H$ negative**

The process of hydration is much less extensive in this case, since the cation is already largely hydrated. **The overall process of solution is endothermic.**

The argument outlined above would predict that other dipolar molecules bearing lone pairs of electrons should react with anhydrous salts in a way similar to water. The exothermic reaction between anhydrous salts and ammonia described in section 24.7 shows that this is indeed the case.

revision summary: Chapter 20

exothermic reaction: heat evolved
products contain less chemical energy than reactants
ΔH negative

endothermic reaction: heat absorbed
products contain more chemical energy than reactants
ΔH positive

heat of combustion: heat evolved when 1 mole of substance is completely burned in oxygen

Hess's Law: heat change in reaction is dependent only on initial and final states of system, independent of stages by which reaction brought about

heat of neutralisation: heat evolved when acid and base react to form one mole of water

heat of solution: (in water) heat change when 1 mole of substance dissolves in volume of water such that further dilution causes no measurable heat change

questions 20

1. From your general experience state whether the following processes are exothermic or endothermic: (a) the burning of a candle, (b) the melting of ice, (c) the compression of air (e.g. in the barrel of a bicycle pump), (d) the evaporation of surgical spirit (e.g. when your arm is cleaned prior to an injection), (e) the rubbing together of two sticks.

2. Express the following statements as equations:

(a) When 32 g of sulphur burn in air 297 kJ of heat energy is evolved.

(b) When 6 g of graphite is converted to diamond 1·9 kJ of heat energy is absorbed.

(c) When one mole of methane (CH_4) burns to produce only carbon dioxide and water, 882 kJ of heat energy is evolved.

(d) When one mole of hydrogen chloride is formed from its elements 92·3 kJ of heat energy is evolved.

3. Given the information

$$C(s) + O_2(g) \longrightarrow CO_2(g); \quad \Delta H = -394\,kJ,$$
$$C(s) + \tfrac{1}{2}O_2(g) \longrightarrow CO(g); \quad \Delta H = -111\,kJ,$$

calculate the heat of the reaction

$$CO(g) + \tfrac{1}{2}O_2(g) \longrightarrow CO_2(g).$$

4. The heat of reaction of hydrogen with oxygen is given by:

$$H_2(g) + \tfrac{1}{2}O_2(g) \longrightarrow H_2O(l); \quad \Delta H = -286\,kJ.$$

What is the heat change when:

(a) 4 g of hydrogen is burned in 32 g of oxygen,

(b) 4 g of hydrogen is burned in 100 g of oxygen,

(c) 1 g of hydrogen is burned in excess oxygen,

(d) excess hydrogen is added to 80 cm^3 (at s.t.p.) of air and the mixture exploded? (Treat air as a mixture of nitrogen and oxygen in the ratio 4 : 1 by volume; hydrogen does not burn in nitrogen.)

5. If the heats of combustion of ethane, ethene (ethylene) and hydrogen are respectively 1542 kJ, 1387 kJ and 286 kJ, evolved in each case, calculate the heat change when one mole of ethene is converted to ethane by reaction with hydrogen, stating clearly whether heat is evolved or absorbed.

6. $C + O_2 \longrightarrow CO_2; \Delta H = -394\,kJ$
and $N_2 + \tfrac{1}{2}O_2 \longrightarrow N_2O; \Delta H = +82\,kJ$.

What is the heat change for $C + 2N_2O \longrightarrow CO_2 + 2N_2$?

Is heat evolved or absorbed?

7. Calculate the heat of the reaction: $2Al + Cr_2O_3 \longrightarrow 2Cr + Al_2O_3$ given that the heats of combustion of aluminium and chromium are respectively 835 kJ and 564 kJ (evolved in both cases).

8. If the heats of combustion of benzene (C_6H_6) and ethyne (acetylene) are respectively 3273 kJ and 1305 kJ (evolved in both cases), what is the heat change of the polymerisation reaction $3C_2H_2 \longrightarrow C_6H_6$? (*Benzene and ethyne both burn in oxygen to produce carbon dioxide and water only.*)

9. (a) Explain why the heat of neutralisation of sodium(I) hydroxide by hydrochloric acid is the same as the heat of neutralisation of potassium(I) hydroxide by nitric(V) acid.

$$NaOH + HCl \longrightarrow NaCl + H_2O; \quad \Delta H = -57{\cdot}5\,kJ$$
(heat given out)

$$KOH + HNO_3 \longrightarrow KNO_3 + H_2O; \quad \Delta H = -57{\cdot}5\,kJ$$
(heat given out)

(b) Given M sodium(I) hydroxide solution and 0·5 M sulphuric(VI) acid describe how you would prepare solutions of (i) sodium(I) sulphate(VI), (ii) sodium(I) hydrogensulphate(VI).

Write equations for the two reactions concerned, indicating the heat change you would expect to occur in each case per mole of sulphuric(VI) acid used.

Describe a simple test, other than the use of indicators, which would enable you to distinguish between solutions of the two salts. Give the result of the test in each case.

(c) Describe carefully how you would obtain a dry crystalline specimen of sodium(I) sulphate(VI) from its solution.

(JMB)

10. When excess of aqueous ammonia is added to aqueous copper(II) sulphate(VI) a deep blue solution results.

(a) What is the formula for the ion responsible for the deep blue colour?

(b) State and explain the type of chemical bond which joins the ammonia molecule to the cation in this ion.

(c) Do you think copper and methane will join together to form a similar ion? Explain your answer.

11. (a) Describe how you would attempt to find by experiment the quantity of heat liberated when one gram of the liquid methanol, CH_4O, is burnt in air.

(b) The table shows the heat liberated when 1·0 g of each of three alcohols is burnt in air.

	Heat evolved
Methanol, CH_4O	22·6 kJ
Ethanol, C_2H_6O	29·7 kJ
Propanol, C_3H_8O	33·4 kJ

For each of these alcohols, calculate the heat of combustion in kJ per mole.

(c) From your results in (b), estimate the heat of combustion of butanol, $C_4H_{10}O$.

(d) The substance dimethyl ether has the same molecular formula, C_2H_6O, as ethanol, but its heat of combustion is different.

Suggest a reason for this difference.

L (NUFFIELD)

carbon and silicon

Carbon has been known from antiquity. Man must have found charcoal (an impure form of carbon) in his fires at a very early date. Cave painters and bronze founders used it in their crafts in prehistoric times. Diamonds, too, were known to the ancients, but it is unlikely that any relation between this and charcoal was recognised. Indeed, the Greeks thought of diamonds as ice permanently frozen by the intense cold of the far North.

21.1 different forms of carbon: allotropes

We now recognise that carbon, even though it is an element, exists in a number of different forms. To prove that a substance is pure carbon, it is necessary to burn it in pure oxygen.

i) If **carbon dioxide** is formed, the substance may be pure carbon, but it could also be a hydrocarbon or other carbon-bearing compound:

e.g. $$CH_4 + 2O_2 \rightarrow CO_2 + 2H_2O$$

ii) If carbon dioxide **only** is formed, the substance may be pure carbon, or another oxide of carbon:

e.g. $$C_{12}O_9 + 7\tfrac{1}{2}O_2 \rightarrow 12CO_2$$

iii) Only if 12g of the substance yield **exactly 44g of carbon dioxide** can the substance be pure carbon:

$$\underset{12g}{C} + O_2 \rightarrow \underset{44g}{CO_2}$$

(12g of $C_{12}O_9$ yields 22g of CO_2)

Try to design an apparatus which could be used to test whether or not an unknown solid is pure carbon.

Such experiments reveal the existence of three different forms of solid carbon:

a) diamond,
b) graphite,
c) amorphous carbon: a term covering charcoal, coke and other solids.

These different forms of the same element in the same physical state are called **allotropes**. The difference between the three lies in their crystal structures, which in turn cause differences in properties.

21.2 diamonds

Although diamonds are so well known, they are extremely rare. Less than ten tons come each year on to the international market, nearly all from southern Africa. Here they occur in decomposed volcanic material, which is broken up by a jet of water and washed over greased tables, to which the diamonds adhere.

The **structure** of diamond is based on the tetrahedral arrangement of four covalent bonds about a carbon atom. Each bond is joined to another carbon atom; the second atom is joined to four other atoms (the original atom and three more) and so on. The resulting arrangement is shown in Fig. 21.1; the whole diamond is one 'giant molecule'. Since the carbon–carbon bond is very strong, and the bonds extend through the crystal, a diamond exhibits the following properties:

a) *hardness:* it is the hardest substance known, and can be cut only by other diamonds. Skilled men can cleave the diamond along planes between the atoms, using a hammer and chisel. This produces the gemstones. Much larger quantities of industrial (small or discoloured) diamonds are made into saws or grinding wheels by embedding them in steel or resin. Diamonds can be made synthetically for this, using graphite and a catalyst at very high temperatures and pressures.

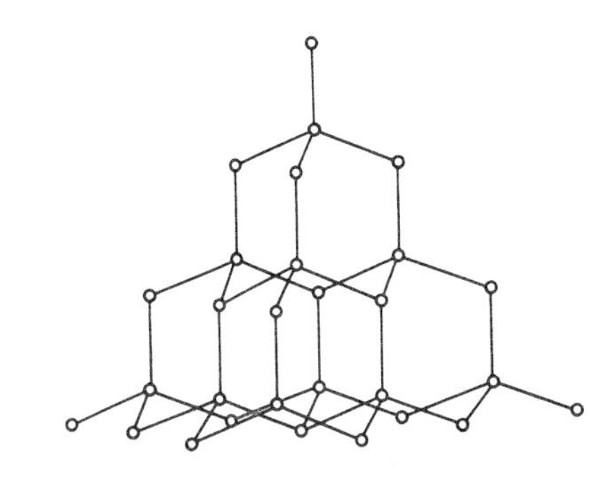

Fig. 21.1 diamond lattice

Fig. 21.2 gemstones

Fig. 21.3 grinding wheel

b) *resistance to oxidation and melting:* diamond saws can be used at very high temperatures; they do not corrode, even after long periods.

c) *brilliance:* diamonds are used in jewellery, as the high refractive index of the solid leads to the light entering the stone, after internal reflections, being emitted in narrow-angled beams–the characteristic flashing effect of a gem.

21.3 graphite

Graphite is the other crystalline form of carbon, having a regular layer structure, as shown in Fig. 21.4.

Within each layer the atoms are joined by very strong covalent bonds in a hexagonal arrangement, each atom being about 0·14 nm from its three nearest neighbours. The next layer is 0·34 nm away and the forces holding adjacent layers together are weaker than covalent bonds. The overall strength of the bonding in graphite is, however, greater than that in diamond (see section 20.2(4)).

Because of the weak forces between adjacent layers, the layers slide over one another readily and graphite is consequently used as a lubricant.

Graphite was originally mined in the Lake District, where the mines were heavily guarded; graphite was then of great military importance, being the only known lubricant for the boring of cannon. Now, with greater demands and resources, a synthetic route has been discovered for its manufacture using coke and a silicon(IV) oxide catalyst in an electric furnace. Graphite is often used when high temperatures are employed (because of its high melting point) or in abrasive conditions (it does not pick up sand and grit).

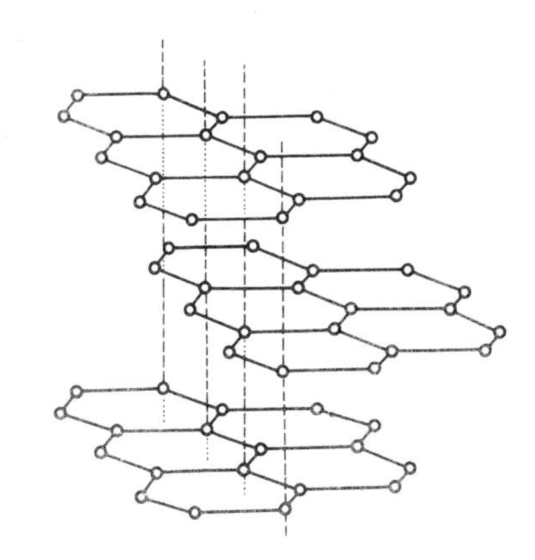

Fig. 21.4 graphite lattice

As graphite slides across a surface, microscopic flakes are left behind as a black mark. Pencils contain graphite to act in this way, together with a little baked clay in all but the softest 'lead'. Otherwise the 'lead' wears too rapidly.

Graphite is a conductor of electricity, since the electrons between the layers are not tightly held by atoms. It is used in electric motors for high-speed (and hence high-temperature) contacts, for dry batteries and as a corrosion-resistant electrode. As the only conducting non-metal, it is invaluable as an electrode at which chlorine is produced.

The carbon nucleus is able to slow down neutrons and graphite is therefore used to moderate, or control, nuclear reactions in atomic piles.

21.4 amorphous carbon

There are various forms of amorphous carbon, all of which probably contain small regions (perhaps only a few atoms across) in which the graphite structure occurs. Each region is local and at angles to neighbouring regions, so that the bulk of the substance is not crystalline.

The largest in tonnage production is probably

i) *coke*, which is formed by heating coal in large ovens in the absence of air until all the volatile compounds in the coal are driven off. It is used in a variety of industrial processes, from the relatively small-scale production of graphite to the reduction of millions of tons of iron ore and other metal ores (see Chapter 26). It is also widely used as a fuel. Next in commercial importance comes

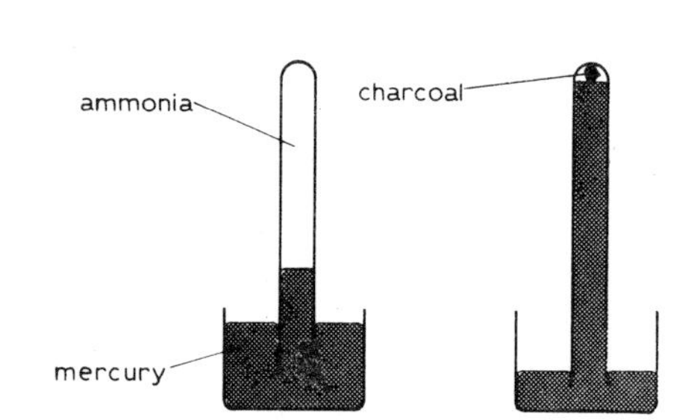

Fig. 21.5 ammonia absorbed by charcoal

ii) *lampblack*, produced by the combustion of any conveniently cheap oil fraction in a limited supply of air. It is a fine black powder – you can obtain a similar product by holding a cold surface just above a candle flame. The commercial product is mainly used as a 'filler': in car-tyres (for strength) and in the floors of public buildings, where its electrical conductivity is sufficient to suppress sparks which might cause a fire. It is also mixed with oil to make printer's ink and with wax to make shoe polish.

iii) *wood charcoal* is of greater chemical interest. It is produced by heating wood in a limited supply of air; the tars and aqueous solutions which distil off are valuable for the chemicals they contain, while the charcoal remains as a solid residue.

Fig. 21.6 gas-mask

Charcoal is full of microscopic holes; it is porous and has a very large surface area. The surface is very active in **adsorbing** gases, which means that molecules of gases become attached to the surface of the charcoal. It is particularly efficient at adsorbing gases and vapours which can also be liquefied easily.

If a piece of wood charcoal is heated and allowed to cool in a vacuum or under a covering of fine sand, the surface layer is cleaned ready for adsorption: it is said to be activated. If such a piece is dropped into a gas jar of air and bromine, the brown colour of the bromine rapidly disappears. Ammonia or sulphur dioxide trapped over mercury are similarly adsorbed, as shown in Fig. 21.5. This principle is used when activated charcoal is incorporated in gas masks, where it is effective against all common poisonous gases except carbon monoxide.

Fig. 21.7 sugar boiling pan

Charcoal is also effective at removing coloured material from aqueous solutions. If a dilute solution of litmus is boiled with a little powdered charcoal, the colour disappears from the solution; on filtering the suspension, a clear colourless liquid is obtained. The coloured pigment may be recovered from the charcoal by washing it with ethanol. Brown sugar is refined to white by this method, though in this process **animal charcoal** is generally used rather than wood charcoal. It is formed by burning bones and consists of finely deposited charcoal on the mineral of the bone, calcium(II) phosphate(V).

21.5 chemical reactions of carbon

Amorphous carbon is more reactive than the crystalline forms and will react with a few substances. It will combine with the more reactive metals (e.g. calcium, aluminium) at white heat to form carbides and with fluorine at red heat to form tetrafluoromethane, CF_4.

The best-known reaction of carbon is the **combustion** of coke in air. In a limited supply of air, as in the middle of a fire, **carbon monoxide** is formed:

$$C + \frac{1}{2}O_2 \rightarrow CO$$

but in a plentiful supply of air **carbon dioxide** is formed:

$$C + O_2 \rightarrow CO_2$$

Thus fires should always be well ventilated to avoid the poisonous monoxide.

The energy changes associated with these reactions are discussed in section 20.5; the energy available from the processes enables carbon to reduce oxides of metals below zinc in the reactivity series (see sections 27.2 and 27.3) to the metals.

21.6 preparation of carbon monoxide

The experiments in this section must be conducted in a properly ventilated fume-cupboard.

$$Mg + \frac{1}{2}CO_2 \rightarrow MgO + \frac{1}{2}C$$
$$Zn + CO_2 \rightarrow ZnO + CO$$
$$Fe + CO_2 \rightarrow \text{no reaction}$$

The passage of air over heated coke is not a practical method of preparation, since the main impurity, nitrogen, is impossible to remove. Partial reduction of carbon dioxide, as opposed to oxidation of carbon, is a possible method. Metals from a narrow range in the reactivity series will do this; magnesium is too reactive and produces carbon; iron is not sufficiently reactive to reduce carbon dioxide at all; but zinc, at the temperature of a bunsen flame, is just right. It is instructive to design an apparatus for the preparation of carbon monoxide from carbon dioxide and heated zinc powder; the apparatus should include a method of removing carbon dioxide, which has not been reduced, from the gas formed.

A more usual laboratory preparation is to dehydrate methanoic (formic) acid, H_2CO_2. The apparatus shown in Fig. 21.8 is used, with concentrated sulphuric(VI) acid as the dehydrating agent. The reaction can easily be controlled by adding sulphuric(VI) acid from a tap-funnel.

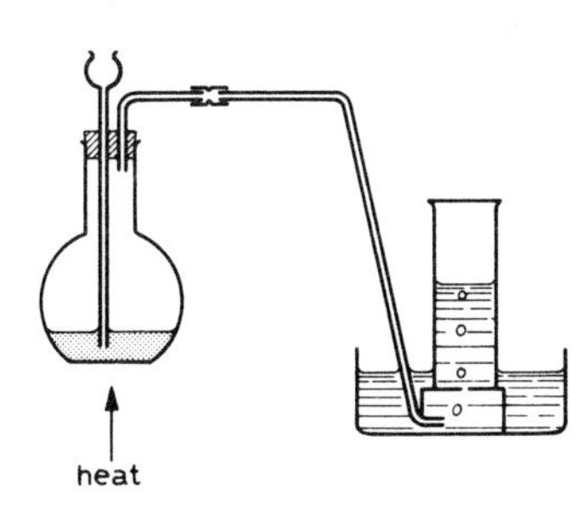

Fig. 21.8 preparation of carbon monoxide

21.7 properties of carbon monoxide

1. physical properties

Observation of the preparation shows that carbon monoxide is colourless, and not more than sparingly soluble in water. It is also odourless, but this must not be tested!

Since its relative molecular mass (28) is very close to the average for air (28·8), it has a density similar to that of air and therefore cannot be collected by displacement of air (see section 15.7). It is liquefied only with difficulty – at atmospheric pressure it must be cooled to $-192°C$ to affect this.

2. poisonous nature

A notorious chemical property of carbon monoxide is that it is poisonous, combining strongly with the haemoglobin of the blood. The carboxyhaemoglobin formed is bright pink and people severely poisoned show this colour in the face. The formation of carboxyhaemoglobin blocks the other biochemical reactions connected with respiration described in section 7.8. The danger is compounded because the gas is not detectable by odour; the first symptom is dizziness, by which time it may be too late for resuscitation. The treatment for suspected carbon monoxide poisoning is to remove the patient to fresh air and administer artificial respiration, using oxygen if available.

	volume %
limit for safe working	1×10^{-4}
headache, dizziness within 1 hour	4×10^{-4}
fatality in 1 hour	4×10^{-3}

Fig. 21.9 poisonous limits for carbon monoxide

Carbon monoxide is emitted in large quantities by badly adjusted car engines and concentrations above the safe working limit have been recorded in many major cities. Fortunately many soil bacteria are able to convert the monoxide to the dioxide.

3. reducing properties

Carbon monoxide has no acidic or basic properties under laboratory conditions. Most of its other chemical reactions depend on its reducing properties. A jar of the gas burns with a characteristic bright blue flame, to produce carbon dioxide:

$$CO + \tfrac{1}{2}O_2 \rightarrow CO_2$$

The carbon dioxide formed can be detected using limewater. This blue flame is seen above coke fires, where it is generated by a process shown in Fig. 21.10.

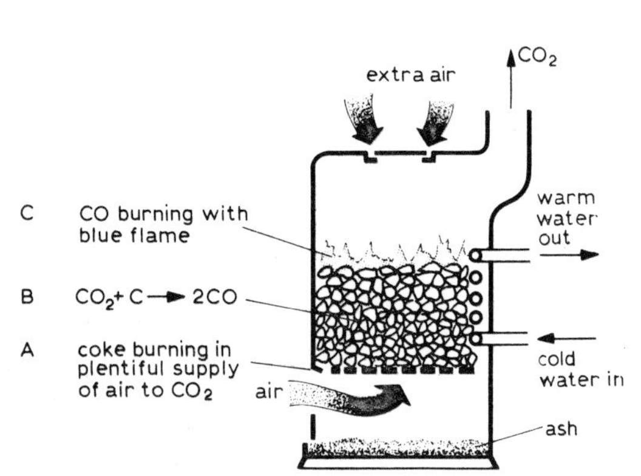

Fig. 21.10 formation of carbon monoxide in coke fire

Zone A: plentiful air supply, fairly hot, temperature maintained by the reaction:

$$C(s) + O_2(g) \rightarrow CO_2(g); \quad \Delta H = -394\,kJ$$

Zone B: little air available, some heat produced in zone A used in the process:

$$C(s) + CO_2(g) \rightarrow 2CO(g); \quad \Delta H = +173\,kJ$$

Zone C: plentiful air supply, relatively cool but hot enough to ignite carbon monoxide from zone B:

$$CO(g) + \tfrac{1}{2}O_2(g) \rightarrow CO_2(g); \quad \Delta H = -284\,kJ$$

Carbon monoxide is a good reducing agent for metal oxides since it is deficient in oxygen with respect to the dioxide. If carbon monoxide prepared as previously described is passed over heated black copper(II) oxide, the pink colour of copper soon appears:

$$CuO + CO \rightarrow Cu + CO_2$$

Similarly, yellow or orange lead(II) oxide is reduced by carbon monoxide to silvery globules of lead:

$$PbO + CO \rightarrow Pb + CO_2$$

Iron oxides and oxides of metals below iron in the reactivity series are readily reduced; zinc oxide is reduced at high temperatures:

$$ZnO + CO \underset{400\,°C}{\overset{800\,°C}{\rightleftharpoons}} Zn + CO_2$$

Reduction of magnesium oxide is possible, reversibly, at 2000°C. These reactions are used for the commercial preparation of metals, the carbon monoxide being generated *in situ* by burning coke (see Chapter 26). About one thousand million tons of coke are used each year in this way.

Carbon monoxide is also generated industrially by the reaction of steam with oil at 900 °C (formerly steam and coke were used). The reaction produces hydrogen and alkanes of low relative molecular mass as well as carbon monoxide. The gas mixture produced can be used either for the production of organic chemicals, for example, methanol:

$$CO + 2H_2 \xrightarrow[\text{oxides}]{\text{Cr/Al}} CH_3OH$$

or, by reaction with more steam, to generate hydrogen for ammonia synthesis (see section 24.11):

$$CO(g) + H_2O(g) \xrightarrow[500\,°C]{Fe_2O_3} CO_2(g) + H_2(g)$$

The removal of carbon dioxide from the gas mixture is described in section 8.26. Although a number of industrial processes yield carbon dioxide as a by-product, this reaction is probably its chief commercial source.

21.8 preparation of carbon dioxide

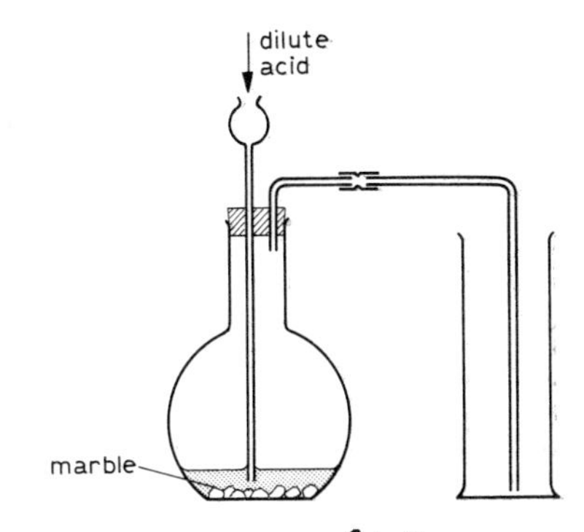

Fig. 21.11 preparation of carbon dioxide

The laboratory preparation of carbon dioxide depends on the action of acids on carbonates, which is fully discussed in Chapter 11. For the purposes of preparation, calcium(II) carbonate in the form of marble lumps is suitable: this gives a steady stream of gas lasting for a long time. Any dilute acid which forms a soluble calcium(II) salt may be used. (*Which acid does this condition exclude?*) For instance,

$CaCO_3$	+	$2HCl$	→	$CaCl_2$	+ H_2O +	CO_2
calcium(II) carbonate		hydrochloric acid		calcium(II) chloride		carbon dioxide

This reaction is quite general for carbonates (see Chapter 11):

carbonate + acid → carbon dioxide + water

$$CO_3^{2-}(s) + 2H^+(aq) \rightarrow CO_2(g) + H_2O(l)$$

The reaction is used in food chemistry, where a mixture of sodium(I) hydrogencarbonate and tartaric acid is used as baking powder. These two substances do not react when dry, but they dissolve and react as soon as water is added, forming carbon dioxide, which gives a light and fluffy texture to cakes.

sodium(I) hydrogencarbonate + tartaric acid → sodium(I) tartrate + carbon dioxide + water

$$HCO_3^-(aq) + H^+(aq) \rightarrow CO_2(g) + H_2O(l)$$

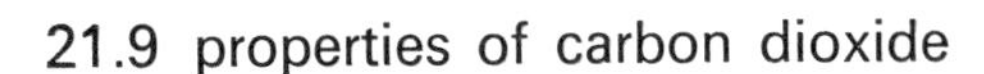

21.9 properties of carbon dioxide

1. *physical properties*

Direct observation of the preparation shows that it is colourless and nearly odourless–some people can detect a faint sweet smell. It is sparingly soluble in water, with a solubility small enough to enable it to be collected over water. This solubility increases with pressure, as does the solubility of all gases, and this is used industrially to manufacture soda-water (pure aqueous carbon dioxide) and fizzy drinks (which have flavouring and colour added). The relative molecular mass of carbon dioxide is 44 (half as much again as air) and its density is therefore high. A jar of the gas can be 'poured', displacing the less dense air. Because of its high density, the gas can be collected by upward displacement of air (see section 15.7). Carbon dioxide liquefies under pressure alone at room temperature (50 atmospheres at 15 °C). The liquid, however, is unstable at all pressures below 5 atmospheres, whatever the temperature. Consequently, when carbon dioxide is cooled, it solidifies directly at −78°C to a white crystalline substance commonly known as 'dry ice'. Small quantities can probably be obtained from your local ice-cream factory and you can do many interesting low-temperature experiments with it. Gloves should be worn when handling it. Try the effect of dipping familiar objects into a bath of 'dry ice' in paraffin. (*No flames*; keep paraffin off your skin.)

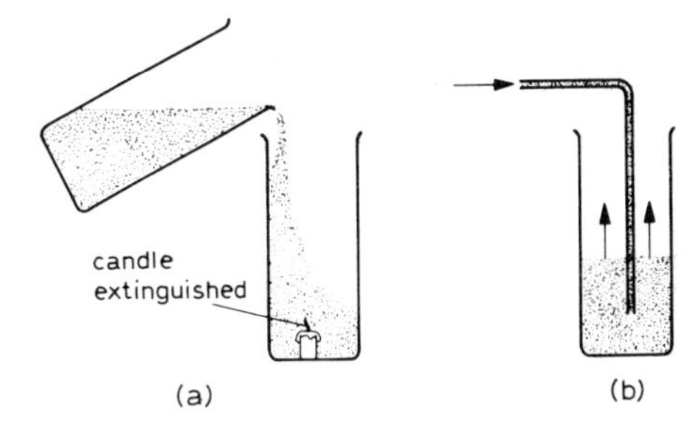

Fig. 21.12 high density of carbon dioxide enables it to be (a) poured in air, (b) collected by upward displacement of air

2. *as a weak acid*

Carbon dioxide is a weakly acidic gas, and if a test-tube full of gas is shaken with a little litmus solution, the litmus turns red–though not the bright red given by more strongly acidic solutions. A solution of carbon dioxide in water contains some carbonate and hydrogen ions, loosely referred to as carbonic acid:

$$CO_2(aq) + H_2O(l) \rightleftharpoons CO_3^{2-}(aq) + 2H^+(aq)$$

Since it is an acid, it will react with alkalis:

$$CO_2(g) + OH^-(aq) \rightarrow HCO_3^-(aq)$$
$$HCO_3^-(aq) + OH^-(aq) \rightarrow CO_3^{2-}(aq) + H_2O(l)$$

3. *stability towards oxidation and reduction*

Carbon dioxide is the product of complete combustion of many fuels, including the food we eat. This suggests that it is a stable compound with respect to oxidation or reduction. It cannot, in fact, be oxidised at all, and only the most vigorous reducing agents attack it. If burning magnesium is plunged into a jar of carbon dioxide, it does continue to burn, but in a spluttering fashion. When the residue is washed with acid, the white magnesium(II) oxide dissolves, leaving black specks of carbon:

$$Mg(s) + \frac{1}{2}CO_2(g) \longrightarrow MgO(s) + \frac{1}{2}C(s)$$

Other examples of its reduction have been referred to in sections 21.6 and 21.7.

4. *uses of carbon dioxide*

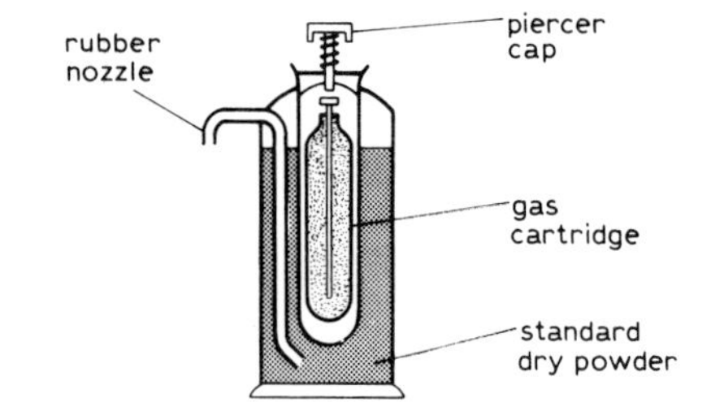

Fig. 21.13 carbon dioxide fire extinguisher, 'dry type'

Because carbon dioxide is unreactive, non-inflammable, dense and non-poisonous, it is very suitable for use in **fire extinguishers**. There are two basic types of extinguisher:

i) *dry type*

This contains carbon dioxide under pressure, and possibly a powder which is blown out with the gas as the pressure is released. It is suitable only for small fires, since it has no great cooling effect; it can, however, be used several times. It is especially suitable for oil fires. Extinguishers based on tetrachloromethane are also used, but give toxic fumes which are dangerous in a confined space.

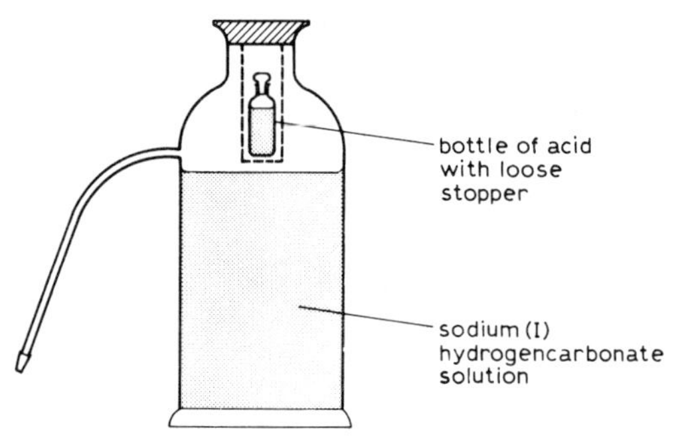

Fig. 21.14 carbon dioxide fire extinguisher, 'wet type'

ii) *wet type*

This contains a solution of sodium(I) hydrogencarbonate with a small phial of acid and frequently a foaming agent. Wet extinguishers should not be used on oil fires and once discharged must be refilled. They can be used on large fires, since the water cools the contents of the fire below the temperature of ignition.

Large quantities of carbon dioxide are now used to make urea, by heating it under pressure with ammonia. Urea is a **fertiliser** and a constituent of some types of thermosetting or 'unbreakable' plastics.

$$CO_2 + 2NH_3 \longrightarrow \underset{\text{urea}}{CO(NH_2)_2} + 2H_2O$$

21.10 the carbon dioxide cycle

Carbon dioxide is an important constituent of the atmosphere, although only present in a concentration of 0·03% by volume. As yet, the amount produced by man's activities is not a serious atmospheric pollutant. Although thousands of millions of tons of the gas are released from chimneys each year, this is still a small amount compared with the natural turnover by vegetation. Although carbon dioxide does not trap the sun's high-energy radiation directly, it does absorb heat re-radiated by the earth. Thus it is possible that an increase in the percentage of carbon dioxide in the atmosphere would raise the average temperature of the earth. Ice-caps would melt and seas would rise. However, this could be self-regulating, as the increased radiation would evaporate more water and produce more cloud, thus cutting down the amount of radiation reaching the surface of the earth. This type of control is called 'positive feedback'.

The circulation of carbon and carbon dioxide in the atmosphere is summarised in Fig. 21.15, which includes references to the Chapters in which the various processes are discussed in detail.

21.11 similarities and differences between silicon and carbon

Silicon is the element below carbon in the Periodic Table; the two elements

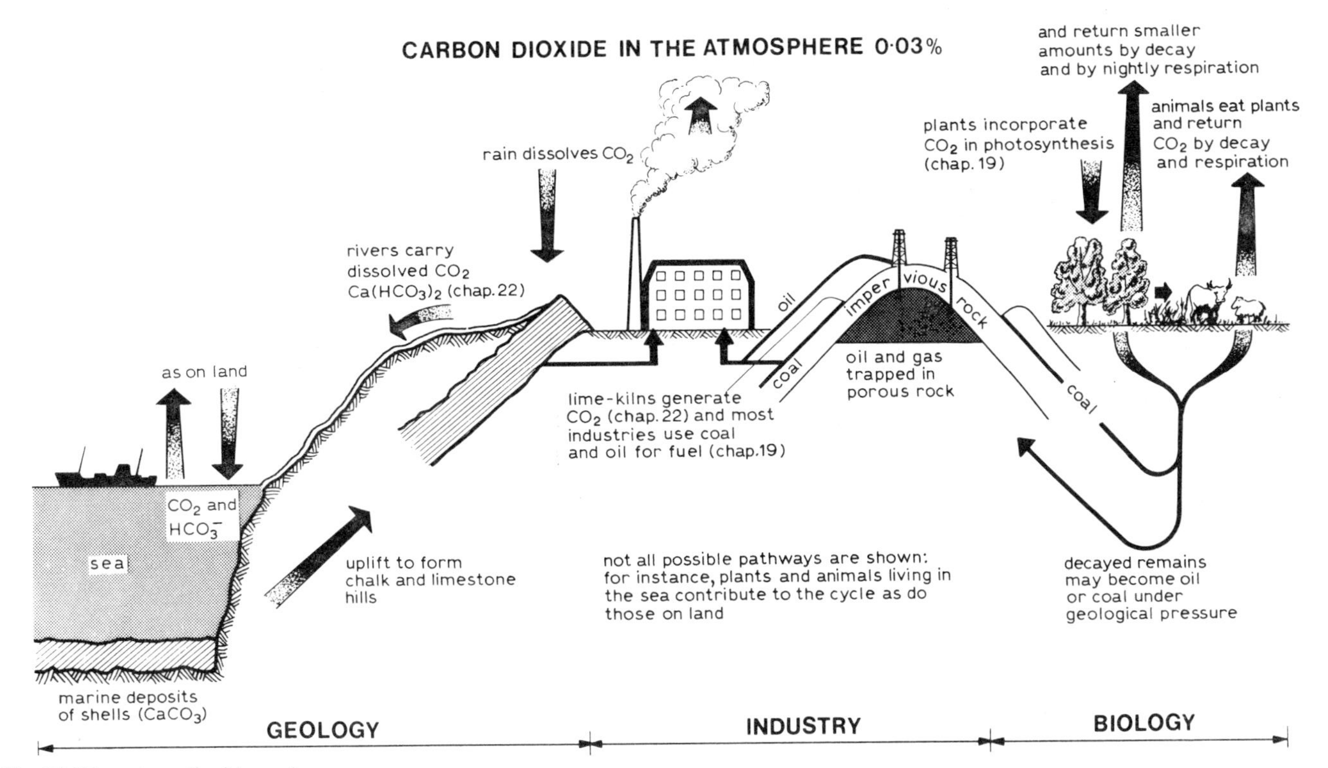

Fig. 21.15 carbon dioxide cycle

show some similarities. They each have four electrons in the highest energy level of their atoms and have a common valency of four. Silicon forms a gaseous inflammable hydride (silane, SiH_4; compare methane, CH_4) and a liquid chloride (silicon tetrachloride, $SiCl_4$; compare tetrachloromethane CCl_4). The element itself crystallises with the diamond structure, but since Si–Si bonds are weak compared to C–C bonds, a silicon crystal is not as hard as diamond. When a silicon crystal is treated with traces of impurity, it takes on special conducting properties and is used in transistors and similar electric devices.

21.12 silicon(IV) oxide (silica)

Silicon(IV) oxide, SiO_2, resembles carbon dioxide in its formula and in being a weak acid. However, carbon dioxide is a molecular compound (see section 18.15), existing in individual molecules, O=C=O, whereas silicon(IV) oxide, or silica, has a non-molecular (giant) structure. In one crystalline form (quartz), each silicon atom is surrounded by four oxygen atoms each shared by another silicon atom, this pattern being repeated throughout the structure: the result is a hard substance with a high melting-point. One common and slightly impure form of quartz is sand, which contains traces of iron oxide which are usually responsible for the colour. Sand grains in wind and water are common agents of geological erosion.

21.13 glass

Sand is used for making glass. The sand is heated with a basic mixture of sodium(I) and calcium(II) carbonates together with some scrap glass to assist melting. Carbon dioxide is displaced and sodium(I) and calcium(II) silicates(IV) are formed; the glass also contains unchanged silica.

$$SiO_2 + Na_2CO_3 \rightarrow Na_2SiO_3 + CO_2$$
$$SiO_2 + CaCO_3 \rightarrow CaSiO_3 + CO_2$$

essentially:
$$SiO_2 + CO_3^{2-} \rightarrow SiO_3^{2-} + CO_2$$

The fact that curved surfaces are formed when glass is broken is evidence that **glass is a mixture**; a pure crystalline substance cleaves in straight

Fig. 21.16 glass-making furnace

Fig. 21.17 glassblowing

planes. Glass has widespread use in windows, chemical apparatus and decoration. Its constitution and properties can be varied easily (since it is a mixture) and the table below gives typical compositions used for different purposes:

type of glass	composition		special property	use
soda glass	Na_2SiO_3 $CaSiO_3$ SiO_2	28% 26% 46%	clear, low m.p.	windows
lead glass	K_2SiO_3 $PbSiO_3$ SiO_2	21% 44% 35%	very clear, high refractive index	decorative glass, jewellery
borosilicate glass	Na_2SiO_3 $Al_2(SiO_3)_3$ B_2O_3 SiO_2	9% 6% 12% 73%	small thermal expansion	ovenware, laboratory equipment
bottle glass	Na_2SiO_3 $CaSiO_3$ SiO_2	34% 10% 46%	cheap (impure sand used)	bottles

Toughened glass is produced by heat treatment, rather than by use of a special composition.

Clay is essentially silicon(IV) oxide in combination with aluminium(III) oxide, alkali metal oxides and water, forming aluminosilicates. The widespread occurrence of sand and clay makes silicon the most abundant element other than oxygen in the earth's crust (see Fig. 5.18). Pottery is made by shaping, drying and finally firing clay.

revision summary: Chapter 21

carbon:
- *allotropes:* diamond, graphite, amorphous carbon
- *properties:* non-metal, reductant

carbon monoxide:
- *preparation:* hot zinc on carbon dioxide
  $Zn + CO_2 \rightarrow ZnO + CO$
  conc. H_2SO_4 on methanoic (formic) acid
  $H_2CO_2 \rightarrow H_2O + CO$
- *collection:* over water
- *properties:* burns with a blue flame (test for the gas)
  poisonous
  neutral
  reductant

carbon dioxide:
- *preparation:* dil. acid (not H_2SO_4) on $CaCO_3$
  $CO_3^{2-} + 2H^+ \rightarrow H_2O + CO_2$
- *collection:* over water, or
  upward displacement of air
- *dry by:* conc. H_2SO_4
- *tests:* will not burn or support combustion
  turns limewater milky
- *properties:* dense, weakly acidic gas

silicon: formally similar to carbon

silicon(IV) oxide (silica): weakly acidic, involatile
used in glass-making

questions 21

1. Animals exhale carbon dioxide. What other processes contribute to the presence of carbon dioxide in the atmosphere? Which natural process regenerates oxygen to the atmosphere?

Lithium(I) hydroxide is used to remove carbon dioxide from the atmosphere of spacecraft. What reaction is taking place? What advantage might the lithium compound have over the corresponding sodium compound?

2. Describe how you could test the following statements, making clear what you would do and what you would expect to see.
(a) Carbon dioxide is an acidic oxide.
(b) The product of the combustion of carbon monoxide is carbon dioxide.
(c) Carbon dioxide is denser than air.

3. Mention **three different** chemical changes as a result of which carbon dioxide is produced. What reagents are used respectively (a) to detect, (b) to absorb carbon dioxide?

When carbon dioxide is passed into a strongly heated combustion tube packed with charcoal, an inflammable gas is produced; name this gas, and give an equation for the reaction. Assuming you have the facilities for producing a steady stream of this gas and you also have a supply of hydrogen available, outline experiments you would make to show (i) one similarity and (ii) one difference between this gas and hydrogen. (O)

4. By means of a labelled diagram and an equation, show how a sample of carbon dioxide can be made and collected in the laboratory.

Without giving any details of the apparatus used, describe briefly how you could convert carbon dioxide into pure carbon monoxide.

A journalist in a motoring magazine wrote, 'On a busy roadway, the proportion of carbon monoxide has varied from 6 parts per million to 180 parts per million.'
(a) At what time of day would you expect the concentration of carbon monoxide to be high?
(b) By what reaction is the carbon monoxide formed?
(c) What is the effect of carbon monoxide on blood, and why does this make the gas so poisonous? (JMB)

5. In what forms does carbon occur naturally? Give some indication of their order of abundance.

By what processes do the percentages of carbon dioxide and oxygen in the air remain approximately constant?

6. a) How could you obtain a pure, dry specimen of the form of carbon which is obtained by dehydrating sugar?
b) How could you obtain carbon monoxide from a gas stream containing carbon monoxide and carbon dioxide?
c) How could you demonstrate that carbon monoxide is a reducing agent? (C, part question)

7. Carbon monoxide may be prepared by the action of sulphuric(VI) acid on **either** methanoic (formic) acid **or** oxalic acid.

Choose **one** of these two methods, give a diagram of a suitable apparatus for carrying out the experiment and write the equation for the reaction. State:
(a) what concentration of sulphuric(VI) acid is used;
(b) if it is necessary to warm the mixture;
(c) the name of the product or products of the reaction other than carbon monoxide;
(d) how the main impurity in the carbon monoxide can be removed. (AEB)

8. The atomic number and relative atomic mass of carbon are 6 and 12 respectively.
(a) Give a labelled diagram showing the structure of a carbon atom.
(b) Explain what is meant by a covalent bond. Give an electronic diagram for **one** named carbon compound.

Diamond, graphite and the charcoal obtained from sugar are all pure forms of carbon.
(c) How are the atoms arranged in graphite and in diamond? Show how this knowledge enables us to understand the physical characteristics of these substances.
(d) Briefly state how it could be shown that graphite and the charcoal obtained from sugar are chemically identical. (Experimental details are **not** required.) (AEB)

9. Carbon, atomic number 6, and silicon, atomic number 14, are elements in the same group of the periodic table. Give the electronic configuration for an atom of each element, and state in which group the elements occur.

Both carbon and silicon form dioxides. Give **one** similarity and **one** difference between these oxides.

Name **three** major raw materials used in the manufacture of glass. (JMB)

10. Write balanced equations for the following reactions. Calculate the volumes of gaseous products and reactants (at s.t.p.), given the quantities specified.
(a) lead(II) oxide + carbon monoxide ⟶ lead + carbon dioxide (4·46 g)
(b) magnesium + carbon dioxide ⟶ magnesium(II) oxide + carbon (4 g)
(c) zinc + carbon dioxide ⟶ zinc(II) oxide + carbon monoxide (26 g)

What can you deduce about the reactivity of lead, magnesium and zinc, from a comparison of these three reactions?

11. What volume of air (20% by volume of which is oxygen) would be needed for the complete combustion of the following? (All volumes are measured at the same temperature and pressure.)
a) 10 cm^3 of carbon monoxide
b) 10 cm^3 of a mixture of equal volumes of carbon monoxide and hydrogen (water gas)
c) 10 cm^3 of a mixture containing 2/3 nitrogen and 1/3 carbon monoxide by volume (producer gas)
d) 10 cm^3 of methane, CH_4 (natural gas)
e) 10 cm^3 of a mixture containing 50% hydrogen, 15% carbon dioxide, 35% methane (town gas)

22.1 occurrence of carbon

The compounds of carbon are to be seen in every inhabited part of the earth, forming the basis for all living material (see Chapter 19). Yet in fact this layer of life is thinner in proportion to the size of the earth than is the layer of paint on a cricket ball. Carbon makes up less than 0·1% of the earth's crust. On a planet such as ours, rich in oxygen, we may expect to find much of the carbon in an oxidised form. This is true: although animals and plants attract our attention, nearly all of the carbon on the earth is in one of two forms:
i) *carbon dioxide* in the atmosphere (0·03% by volume, or $2{\cdot}35 \times 10^{15}$ kg in all);
ii) *carbonates*, formed by reaction between acidic carbon dioxide and basic metal oxides (a total of $6{\cdot}7 \times 10^{19}$ kg).

22.2 occurrence of carbonates

Fig. 22.1 calcite crystals

Most of the massive deposits are of **calcium(II) carbonate**, formed about 100–400 million years ago from small marine animals. As these animals died, their shells of calcium(II) carbonate fell to the sea bed, building up deposits up to six hundred metres in thickness. As these deposits became compacted, they formed chalk rock, which was uplifted by earth movements to form, for example, the Downs of South-East England. In **chalk**, which is soft, shells can still be seen easily using a hand-lens. Older deposits which have been subjected to greater heat and compression, are in the form of **limestone**, which is a harder and more compact rock than chalk and contains fewer visible fossils. Limestone forms many of the upland areas of the Pennines, and also some of the world's most spectacular mountains in the Alps and Himalayas. When limestone is heated by volcanic action it may melt and re-crystallise as beautifully clear crystals of **calcite**, or as finer crystals in blocks of **marble**, prized for sculpture. Chalk, limestone, calcite and marble are all naturally occurring forms of calcium(II) carbonate.

Many other carbonates occur naturally, and these are listed in the table below.

mineral	formula	remarks
dolomite	$CaCO_3.MgCO_3$	named after the Dolomite Alps, Italy
magnesite	$MgCO_3$	worked for its magnesium content
limonite	$FeCO_3$	worked in the Midlands for iron
calamine	$ZnCO_3$	found in veins and do not occur in massive deposits, as do those above
malachite	$CuCO_3.Cu(OH)_2$	
witherite	$BaCO_3$	

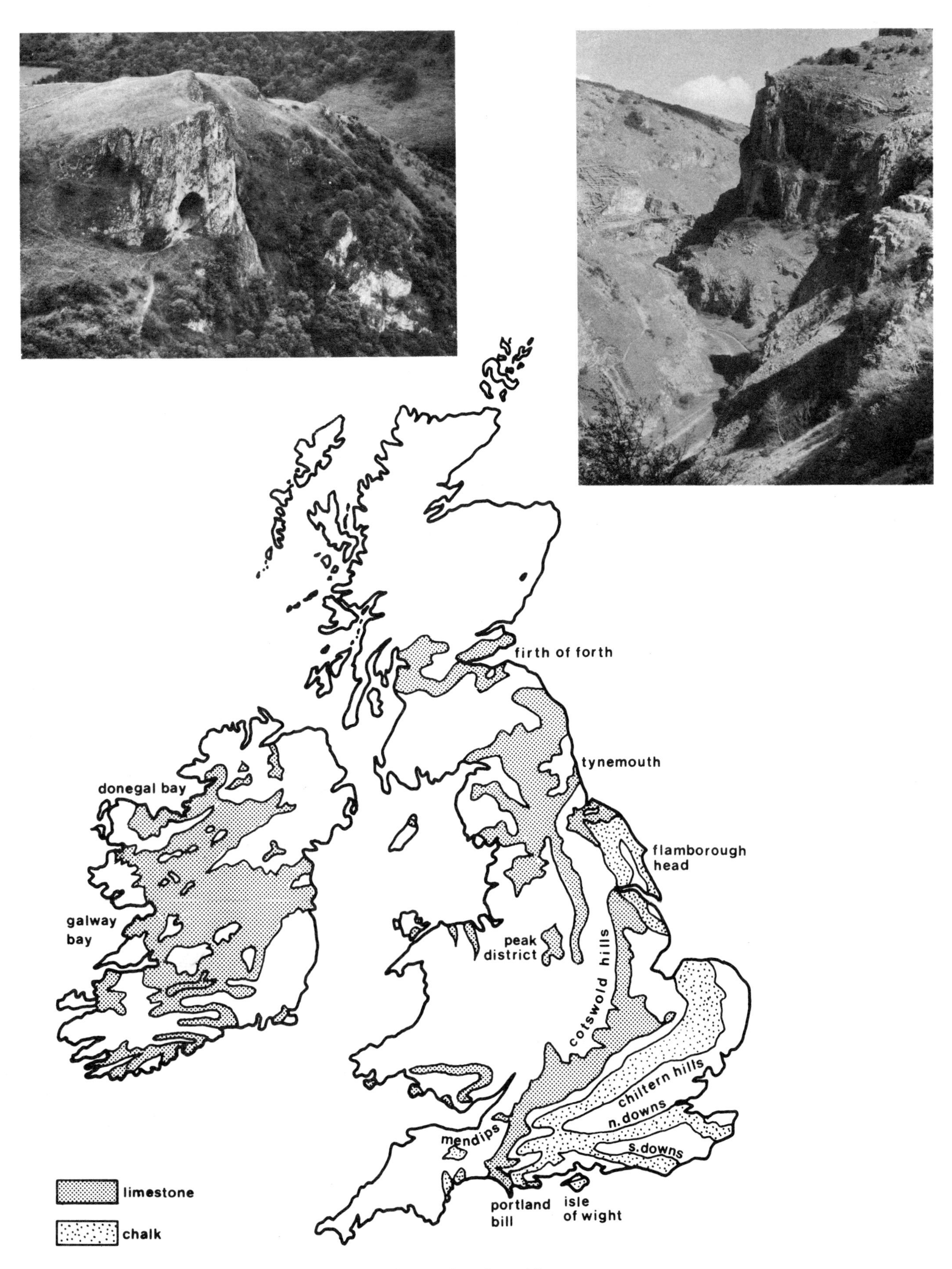

Fig. 22.2 Geological map of Great Britain showing carbonate deposits and limestone scenery

22.3 properties of a typical carbonate

Sodium(I) and potassium(I) carbonates, Na_2CO_3 and K_2CO_3, are exceptional in that they are stable to heat and soluble in water, unlike all other common carbonates, which decompose on heating and are insoluble in water. A carbonate other than that of sodium or potassium may consequently be obtained from a solution containing the metal ion by precipitation using aqueous sodium(I) carbonate. Often, however, precipitates obtained in this way are basic carbonates, containing hydroxide ions; this is due to the alkaline nature of aqueous sodium(I) carbonate (see section 16.7).

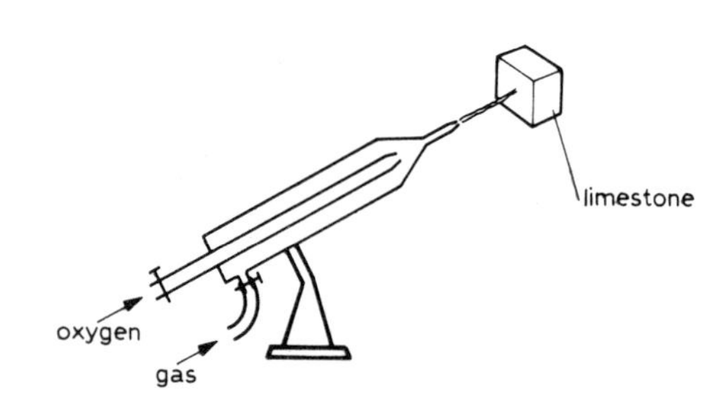

Fig. 22.3 blowpipe for decomposition of limestone

Limestone (calcium(II) carbonate) has properties more typical of carbonates in general. If a few limestone chips are placed on an iron tray and heated with an intense flame (from a blowpipe or oxy-town gas torch), the limestone crumbles into powder. This white powder gives off a brilliant white light as it is heated in the flame. The reaction occurring is:

calcium(II) carbonate ⟶ calcium(II) oxide + carbon dioxide
(*limestone*) (*quicklime*)

$$CaCO_3(s) \longrightarrow CaO(s) + CO_2(g)$$

The white light is the original '*limelight*' used in stage lighting.

Other carbonates of less reactive metals decompose at lower temperatures. For example, lead(II) carbonate decomposes to lead(II) oxide at about 200 °C, and copper(II) carbonate to copper(II) oxide at about 80 °C. Lead(II) carbonate is white and copper(II) carbonate is blue-green; what colour would you expect them to turn on heating? How could you show that carbon dioxide is evolved in the process? It is a simple matter to test your predictions.

22.4 a brief study of calcium(II) oxide ('quicklime')

If the calcium(II) oxide obtained as described above is allowed to cool and a little water is then added, a great deal of heat is evolved. Some of the water boils and clouds of steam are seen, indicating that an exothermic reaction is taking place. The reaction is:

calcium oxide + water ⟶ calcium hydroxide

$$CaO(s) + H_2O(l) \longrightarrow Ca(OH)_2(s)$$

or

$$O^{2-}(s) + H_2O(l) \longrightarrow 2OH^-(s)$$

The vigour of this reaction and the physical movement associated with it have led to the calcium(II) oxide being called **'quicklime'** ('*quick*' meaning '*alive*'). The hydroxide, which has had its 'thirst' for water slaked, is called **'slaked lime'.** If the slaked lime formed as described is stirred with excess water, a creamy suspension is formed **('milk of lime')** which, on filtering, yields a clear colourless solution. This is dilute aqueous calcium(II) hydroxide, or **'limewater',** the well-known laboratory test reagent. Being alkaline, it reacts with the acidic gas carbon dioxide to form the salt calcium(II) carbonate. This reaction has already been referred to as a test for carbon dioxide–when carbon dioxide is bubbled through limewater a white precipitate of calcium(II) carbonate is formed:

$$Ca(OH)_2(aq) + CO_2(g) \longrightarrow CaCO_3(s) + H_2O(l)$$

The action of carbon dioxide on sodium(I) hydroxide is described in section 16.5(6).

22.5 preparation of hydrogencarbonates

Hydrogencarbonates are the acid salts corresponding to carbonates; an acid salt is made by adding extra acid to an aqueous solution of the normal salt (see section 12.19). In this case, the acid is carbonic acid, which can be obtained by passing carbon dioxide gas into water:

$$CO_2(g) + H_2O(l) \longrightarrow H_2CO_3(aq)$$

When excess carbon dioxide is bubbled through limewater the precipitate of calcium(II) carbonate does not remain, but slowly dissolves as the soluble hydrogencarbonate is formed:

$$CO_3^{2-}(s) + H_2CO_3(aq) \rightarrow 2HCO_3^-(aq)$$

Most acid salts decompose when heated, to form the normal salt and the acid. Hydrogencarbonates are no exception and, like the parent carbonic acid, decompose at a very low temperature. If the solution of calcium(II) hydrogencarbonate is evaporated, the cloudy white precipitate of calcium(II) carbonate returns. Only sodium(I), potassium(I) and other group I hydrogencarbonates can be obtained as solids, and even these decompose below 100 °C.

$$Ca(HCO_3)_2 \rightleftharpoons CaCO_3 + H_2O + CO_2$$

or

$$2HCO_3^-(aq) \rightleftharpoons CO_3^{2-}(s) + H_2O(l) + CO_2(g)$$

Notice that this reaction is *reversible*; the direction it takes is determined by whether the system is hot or cold.

All the reactions can be summarised in the flow diagram in Fig. 22.5.

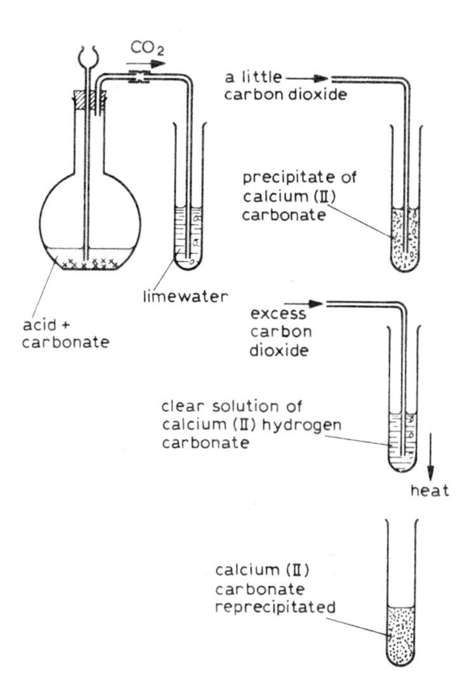

Fig. 22.4 action of carbon dioxide on aqueous calcium(II) hydroxide (limewater)

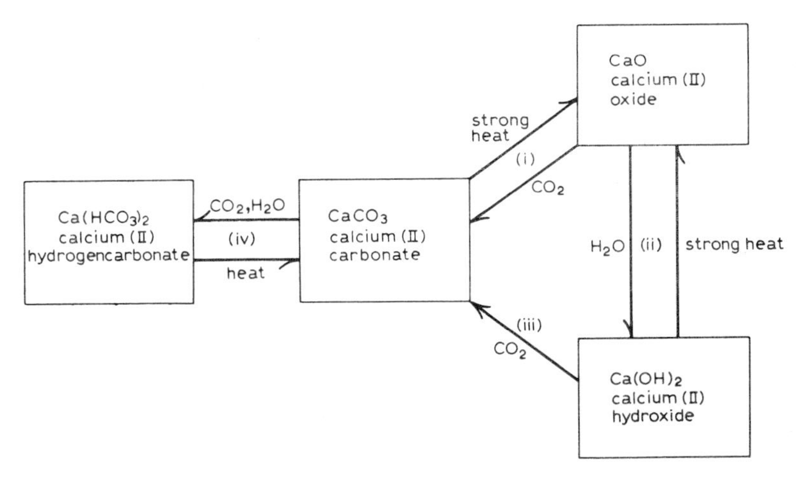

Fig. 22.5 interconversion of calcium (II) compounds

22.6 applications of the foregoing reactions

The conversions (*i*) to (*iv*) shown in Fig. 22.5 are the basis of all concrete and mortar construction; they play a substantial part in the evolution of our landscape and are involved in the behaviour of cleaning materials.

1. mortar

Limestone (or chalk) is burnt in a kiln to produce calcium(II) oxide. Since this reaction is reversible, a draught of air is maintained through the kiln to remove carbon dioxide and prevent any recombination (*reaction* (*i*)). The calcium(II) oxide produced is slaked (*reaction* (*ii*)) and mixed with sand and water. This mixture sets, initially by loss of water, but later by taking up carbon dioxide from the atmosphere. Long crystals of calcium(II) carbonate are formed (*reaction* (*iii*)) which give strength to the mortar.

Fig. 22.6 lime-kiln

2. cement

Cement is used as a building material, often mixed with sand or gravel to set as concrete. Again the limestone is 'burnt' (*reaction* (*i*)) but this time mixed with clay. The acidic oxides of silicon and aluminium in the clay combine with the basic calcium(II) oxide to form a mixture of calcium(II) silicate and aluminate(III).

The setting action is similar to that of mortar, but the product is stronger and more resistant to weathering.

Fig. 22.7 reinforced concrete bridge

3. *as a flux*

Limestone or dolomite ($CaCO_3.MgCO_3$), is added to blast furnaces in the smelting of iron, where it is decomposed by heat into calcium(II) and magnesium(II) oxides. The sand (silicon(IV) oxide) often present in the iron ore is acidic and combines with these basic oxides to form calcium(II) silicate(IV). This melts at the temperature of the furnace and is run off as a slag.

4. *in agriculture*

Slaked lime, produced by reactions (*i*) and (*ii*) shown in Fig. 22.5, is spread on the fields in powder form. Most cereals grow best at a pH of about 7 (see section 11.8), but heavy use of ammonium fertilisers, or poor drainage, encourages acidity. Slaked lime, or calcium(II) hydroxide, is the cheapest alkali available to correct this. The fields are treated in winter, and the effect is like that of a heavy local snowfall!

5. *in geology*

Limestone is calcium(II) carbonate; rain-water is a dilute aqueous solution of carbon dioxide, some of this carbon dioxide being derived from the air, but rather more from the decomposing vegetation overlying the limestone. These two react as in the laboratory experiment (*reaction* (*iv*) shown in Fig. 22.5), and the limestone dissolves to form calcium(II) hydrogencarbonate. This process is fastest along the cracks or joints in the rock, and the surface often develops little pits and fissures which grow into large pot-holes. Many of these can be seen in Pennine regions, and one of them, Gaping Ghyll, swallows a stream to become England's highest waterfall–underground! **Chalk** has fewer joints, and rainfall sinks less spectacularly, although there are small swallow-holes at North Mimms and in the Mole Valley.

Once underground, the water continues to promote dissolving, forming caves and passages until a system many miles long may be formed–30 to 40 miles in Kentucky and Yugoslavia.

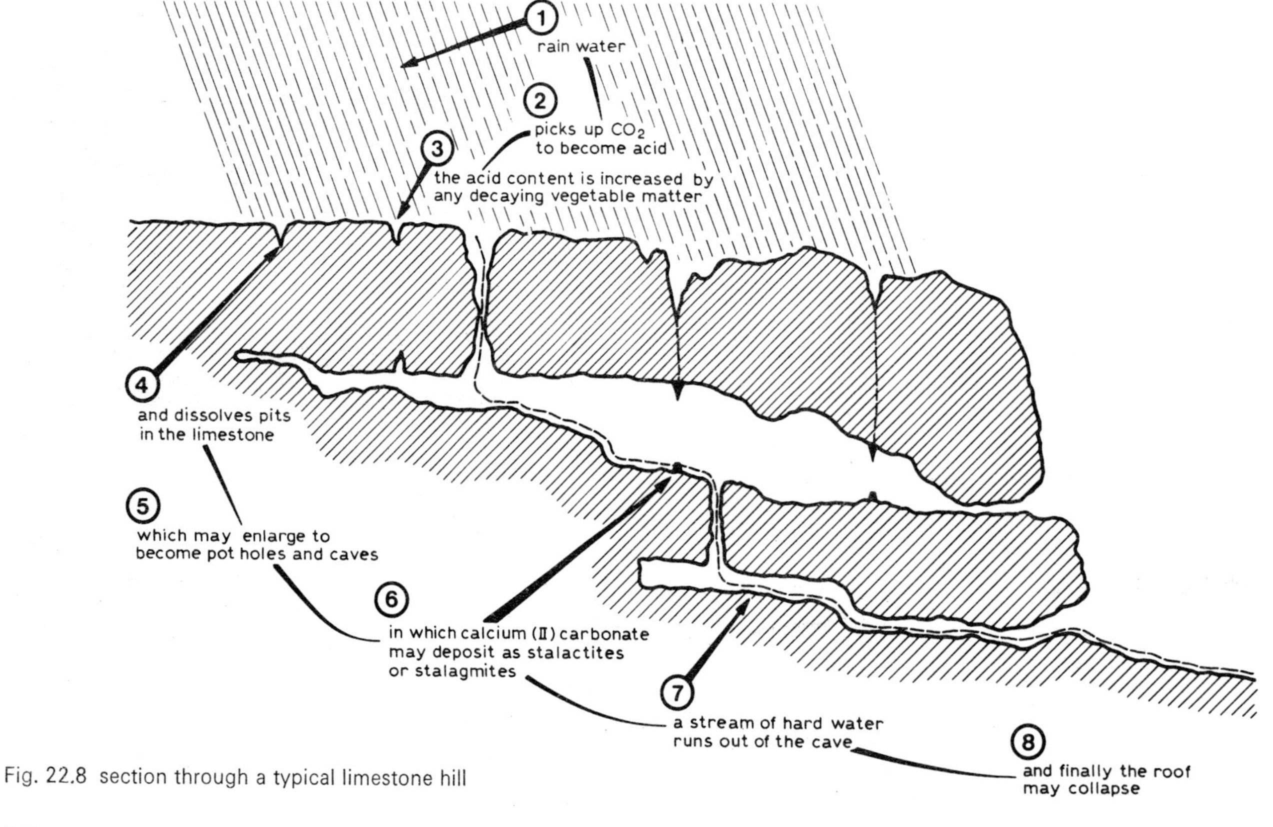

Fig. 22.8 section through a typical limestone hill

When the water containing calcium(II) hydrogencarbonate drips from the roof of a cave which has a current of air flowing, the water evaporates. By the reverse of *reaction (iv)*, calcium(II) carbonate must now be formed. This builds up as more drips flow over the original deposit to form the famous **stalactites** (rock icicles) of tourist caves. Further evaporation may take place on the floor as drips fall on to it from a stalactite, and a column of calcium(II) carbonate builds up from the floor to become a **stalagmite** (stala**C**tites hang from the **C**eiling; stala**G**mites rise from the **G**round). In many show caves, whole curtains and columns have been formed, coloured by the presence of other metal salts.

Finally, as caves become larger, their roofs can collapse, forming the gorges typical of limestone areas.

Fig. 22.9 stalagmites and stalactites

22.7 impurities in water supplies

When water engineers are planning a supply for a large city, there are many things that they must bear in mind, which may be summarised under the three headings: quantity, quality and cost. In Britain, with a cool climate and regular rainfall, the problem of quantity has not yet been severe. Lowland Britain (the South-East) has underground storage of large capacity in the chalk rock to offset its lower rainfall; highland Britain has a sufficiently high rainfall to justify the construction of small storage reservoirs. Desert countries of the Middle East, however, are forced to rely on distillation of sea-water, which requires large amounts of energy and so increase the cost of the water supply.

As for quality, even rain-water contains dissolved impurities. *Carbon dioxide* has been mentioned as one example; *sulphur dioxide* and small *dust* particles are also important. As rain-water runs over the soil and into rivers it picks up small particles of *soil* (which give a colour), *bacteria* (which may be harmful) and *acids* (from the decomposition of plant material). This surface water will also dissolve small amounts of sparingly soluble calcium(II) sulphate(VI), if it passes over such a deposit, and calcium(II) hydrogencarbonate as described in the previous section. Underground water will contain even more of these *minerals*. The engineer must plan to remove all of these, as well as any *sewage* or *industrial waste*–although water containing such wastes is usually avoided.

The purification process is outlined in Fig. 22.10. At the end of the process it can be seen that the remaining impurities consist of traces of chlorine, a little sterilised organic matter and relatively large quantities of calcium(II) salts which have not been removed.

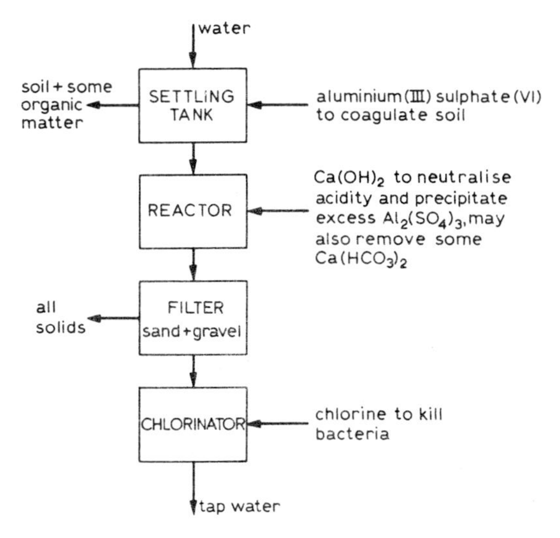

Fig. 22.10 water purification

22.8 effect of dissolved calcium(II) salts in tap water

If you live in the South of England, try using soap to make a lather, first in tap water, then in distilled water. In other parts, where tap water contains only small quantities of dissolved calcium(II) salts, use a dilute solution of calcium(II) sulphate(VI) instead of tap water. You will notice that the **water containing calcium(II) ions lathers with difficulty,** forming a greyish scum first.

Soap consists principally of the salt sodium(I) stearate, $Na^+C_{17}H_{35}CO_2^-$, and owes its properties to the action of the stearate ion, as is described later. Unfortunately, calcium(II) stearate is insoluble, and consequently calcium(II) ions in the water react with stearate ions from soap, precipitating a scum of calcium(II) stearate:

$$Ca^{2+}(aq) + 2C_{17}H_{35}CO_2^-(aq) \longrightarrow Ca(C_{17}H_{35}CO_2)_2(s)$$

The calcium(II) ions are usually derived from either the sulphate(VI) or the hydrogencarbonate. Any water containing calcium(II) ions will not make a lather with soap, and for this reason is called **hard water.**

Fig. 22.11 water resources of some of England's major cities and towns

Hard water has other **disadvantages.** If calcium(II) ions are present with hydrogencarbonate ions, calcium(II) carbonate is precipitated on heating the water (*reaction (iv)*, Fig. 22.5; cf. stalactites):

$$Ca(HCO_3)_2(aq) \longrightarrow CaCO_3(s) + H_2O(l) + CO_2(g)$$

This accumulates in kettles as 'kettle fur' and more seriously in the tubes of industrial steam boilers as 'scale'. Time and money must be spent in removing this. In addition, some of the materials used in the dyeing and tanning industries also form insoluble calcium(II) salts; the water used in these industries must therefore be free from such salts.

However, hard water does have some positive **advantages.** It is more pleasant to taste and is thought to contribute calcium(II) salts necessary for bone and other growth. For this reason, brewing industries are situated in hard water areas so that the yeast grows more vigorously. In older houses, with lead piping, the sulphate(VI) ions often present in hard water reprecipitate any lead(II) ions (which may be formed in solution by the action on the pipe of traces of acid) as lead(II) sulphate(VI), preventing cumulative poisoning by lead.

$$Pb^{2+}(aq) + SO_4^{2-}(aq) \longrightarrow PbSO_4(s)$$

22.9 removal of hardness from water supplies

1. boiling

Calcium(II) carbonate is precipitated when a solution of the hydrogencarbonate is boiled. This, then, is one way of removing calcium(II) ions from solution. When this is done, the water is said to be softened. However, only calcium(II) hydrogencarbonate is removed by this method; calcium(II) sulphate(VI) remains. For this reason, hardness due to hydrogencarbonate is called **temporary hardness** and that due to calcium(II) sulphate(VI) or other soluble calcium(II) salts is called **permanent hardness.**

2. *addition of alkali*

Another method of removing temporary hardness is to add an alkali, frequently calcium(II) hydroxide. The acidity of the hydrogencarbonate is neutralised and calcium(II) carbonate precipitates:

$$HCO_3^-(aq) + OH^-(aq) \rightarrow CO_3^{2-}(aq) + H_2O(l)$$
$$Ca^{2+}(aq) + CO_3^{2-}(aq) \rightarrow CaCO_3(s)$$

For this method to be effective, the exact concentration of calcium(II) hydrogencarbonate in the water must be known as it is essential that the exact quantity of calcium(II) hydroxide to neutralise this must be added–and no more. Can you see why?

3. *addition of a soluble carbonate*

Permanent hardness can be removed only by precipitating calcium(II) ions as the carbonate or other insoluble compound. These methods automatically remove temporary hardness as well. In an older method, still used domestically, sodium(I) carbonate-10-water is added to the water (as 'washing soda' or 'bath salts'). All calcium(II) ion is precipitated:

$$Ca^{2+}(aq) + CO_3^{2-}(aq) \rightarrow CaCO_3(s)$$

4. *ion-exchange*

A more recently discovered method is that of ion-exchange, used extensively in industry. Some natural alumino-silicates (*zeolites*) and some synthetic

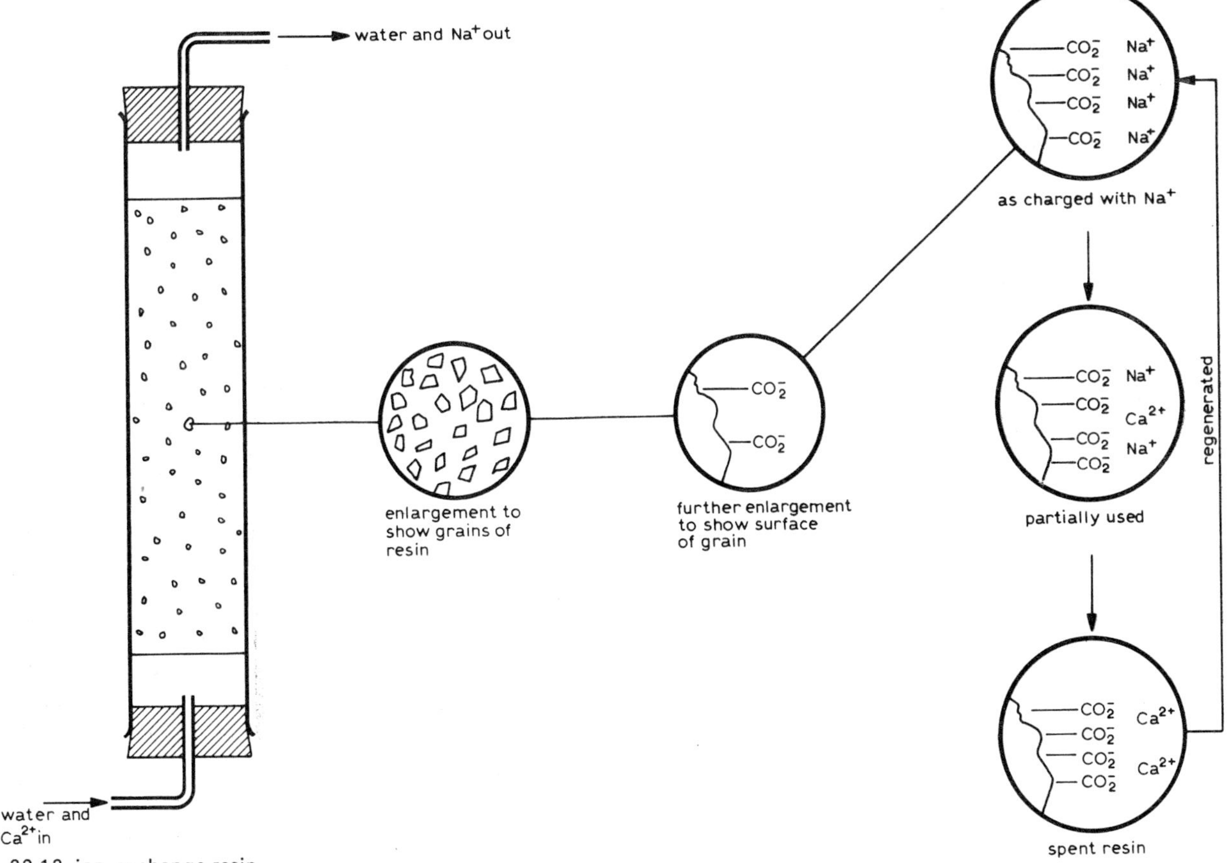

Fig. 22.12 ion-exchange resin

resins consist of a polymeric framework with negatively charged ions (anions) attached to the framework by covalent bonds. These anions attract positively charged ions (cations) from solutions, calcium(II) and other **divalent cations being more strongly attracted** than sodium(I) and other monovalent ions.

If hard water is passed through a glass tube (or column) packed with this resin (charged with sodium(I) ions), the sodium(I) ions in the resin are replaced by calcium(II) ions, thus removing calcium(II) ions from the water. Of course, the resulting water contains sodium(I) ions, but these have no adverse effect on soap: the water is effectively 'softened'. The resin can subsequently be **re-charged** by passing saturated brine (aqueous sodium(I) chloride) over it, when the high concentration of sodium(I) ions is sufficient to displace the calcium(II) ions and regenerate the resin.

Similar resins, an anionic one charged with H^+ ions and a cationic one charged with OH^- ions, can be used to remove all minerals from water. The product is sold and used in medicine, or for car batteries, as purified water. It still contains traces of organic matter which can be removed only by distillation. An example of the use of such a pair of resins to remove sodium(I) chloride from water is shown in Fig. 22.13. One mole of water is formed for each mole of sodium(I) chloride removed.

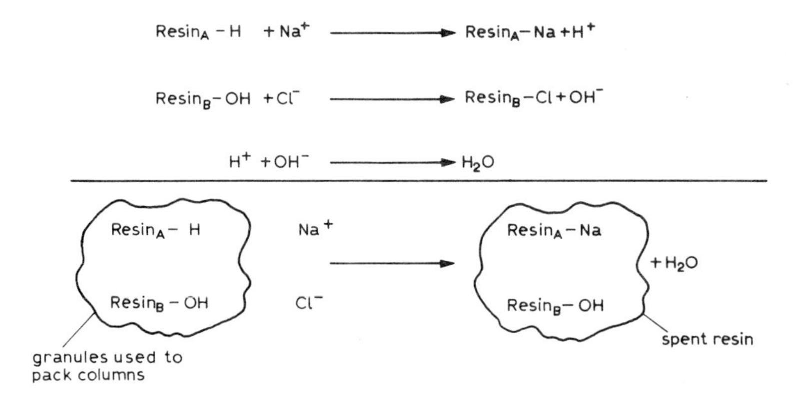

Fig. 22.13 'equations' for resins removing Na^+ and Cl^- from water

22.10 measurement of degree of hardness in water

A convenient reaction for estimating hardness is that with soap. A standard soap solution, Clark's soap, is added from a burette to $25cm^3$ of hard water. The volume of soap solution required before a lather is formed corresponds to total hardness, both temporary and permanent. The titration is repeated using a $25cm^3$ sample of water which has been boiled and cooled. The volume of soap solution required to form a lather with this sample corresponds to the permanent hardness only (the temporary hardness having been removed by boiling). The difference between the two volumes corresponds to the temporary hardness in the water. Nowadays, other methods, including spectrometry, may be used.

22.11 molecular action of soaps and detergents

All soaps and detergents are found to have a similar chemical structure–a long chain consisting of carbon and hydrogen atoms, with a 'head' consisting of an ionic group. These are two typical structures:

i) *a typical soap*

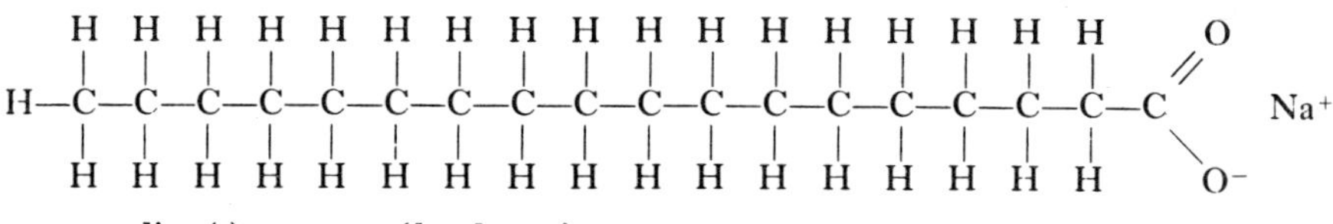

sodium(I) *stearate: 'hard soap'*

Softer soaps may have shorter chains of carbon atoms, or K^+ instead of Na^+.

ii) *a typical detergent*

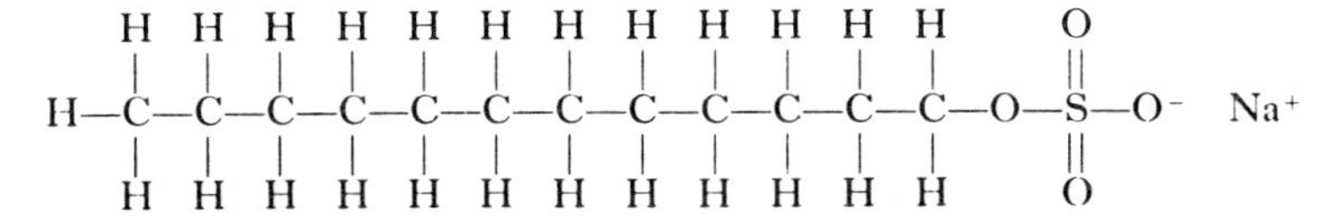

Other detergents may contain different carbon chains but all are based on sulphur-bearing anionic groups.

The ionic 'head', like other ionic substances, will dissolve readily in water. However, the long chain of carbon and hydrogen atoms is similar to, and will dissolve in, greases and fats. The presence of these two functional groups in the soap or detergent molecule enables the grease (and dirt which sticks to the grease) to be dispersed in water. The process is represented in Fig. 22.14.

Soap is prepared by heating vegetable oils and fats together with sodium(I) hydroxide. From the reaction mixture, the soap is precipitated using concentrated brine. Detergents are based on petrochemicals and are usually obtained by polymerisation of alkenes, followed by the action of sulphuric(VI) acid. The calcium(II) salts of detergents are soluble; detergents do not form a scum and therefore lather readily in hard water. Detergents were first introduced in hair shampoos; the formation of a scum in freshly washed hair would be unwelcome and unsightly.

Commercial detergents also contain sodium(I) phosphate(V) as an alkali to attack grease, sodium(I) peroxoborate(III) (perborate) as a mild bleach, fluorescent agents for whitening and perfume.

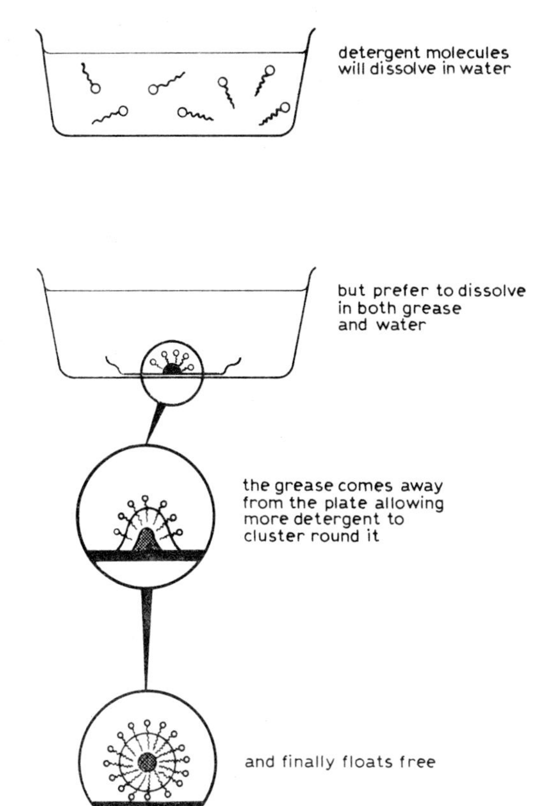

Fig. 22.14 detergents acting on grease on plate

revision summary: Chapter 22

carbonates: all except those of Na^+ and K^+ are insoluble and decompose on heating, to the oxide + CO_2
all effervesce with acids
$CO_3^{2-} + 2H^+ \rightarrow H_2O + CO_2$

products from limestone:

$$CaCO_3(s) \xrightarrow{heat} CaO(s) + CO_2(g)$$
$$CaO(s) + H_2O(l) \rightarrow Ca(OH)_2(s)$$
sparingly soluble

$$Ca(OH)_2(aq) + CO_2(g) \rightarrow CaCO_3(s) + H_2O(l)$$
$$CaCO_3(s) + H_2CO_3(aq) \rightarrow Ca(HCO_3)_2(aq)$$
$$Ca(HCO_3)_2(aq) \xrightarrow{heat} CaCO_3(s) + H_2O(l) + CO_2(g)$$

hard water: water which does not readily give a lather with soap due to $Ca^{2+}(aq)$ and $Mg^{2+}(aq)$ in the water

temporary hardness: due to $Ca(HCO_3)_2$, $Mg(HCO_3)_2$ – removed by boiling

permanent hardness: due to other Ca^{2+}, Mg^{2+} salts; not removed by boiling

removal of hardness: temporary only – boiling
addition of $Ca(OH)_2$
both types – addition of $CO_3^{2-}(aq)$
ion-exchange

soaps and detergents:	molecules contain long hydrocarbon chain and ionic 'head' form link between grease and water Ca^{2+}, Mg^{2+} form insoluble compounds with soaps form soluble compounds with detergents

questions 22

1. Explain why:
(a) caves and underground streams are frequently found in chalk and limestone areas, but rarely elsewhere;
(b) stalactites have been observed growing from the underside of concrete motorway bridges;
(c) lead pipes are dangerous in soft water areas.

2. Write brief notes stating whether you would rather live in a soft water or a hard water area, giving your reasons.

3. From this list of calcium(II) salts–chloride, sulphate(VI) carbonate, hydrogencarbonate, oxide, hydroxide–choose one which
a) is soluble in water,
b) decomposes below 100 °C,
c) reacts violently with water,
d) exists in equilibrium with the oxide at 900 °C,
e) forms a hydrate unchanged on exposure to the atmosphere,
f) is responsible for permanently hard water,
g) is responsible for temporarily hard water,
h) is deliquescent.

4. The following experiments were carried out with sodium(I) hydrogencarbonate, $NaHCO_3$;
Expt. 1. 4·20 g of sodium(I) hydrogencarbonate were heated to constant mass in a crucible. The white powder which remained weighed 2·65 g.
Expt. 2. The gas evolved when 0·02 gram formula (formula-weight) of sodium(I) hydrogencarbonate was treated with excess dilute hydrochloric acid was collected in a gas syringe. The volume of the dry gas corrected to s.t.p. was 448 cm^3.
($H = 1$; $C = 12$; $O = 16$; $Na = 23$.)
(a) Use the above information to answer the following:
(i) What fraction of a gram-formula of sodium(I) hydrogencarbonate was used in Expt. 1?
(ii) What was the white powder which remained in Expt. 1?
(iii) What fraction of a gram-formula of this white powder was formed?
(iv) What gas was evolved in Expt. 2?
(v) What fraction of a gram-molecule of this gas was collected in Expt. 2?
(b) Write equations for the reactions occurring in the above experiments and in each case show how the experimental results agree with the equations. (W)

5. (a) Calcium(II) carbonate, present in rocks or soil, is one of the causes of hardness in water. Explain why this is so.
(b) Explain the use of (i) calcium(II) hydroxide, and (ii) sodium(I) carbonate, in the softening of hard water.
(c) Why does the presence of dissolved sodium(I) carbonate not make water hard?
(d) A copper boiler used in the preparation of distilled water is encrusted with a layer of white scale caused by the hardness in the water used. Explain how this scale was formed from the hard water.
If supplies of dilute sulphuric(VI), hydrochloric and nitric(V) acids were available, which of these acids would you use to remove the scale from the boiler? Give the reasons for your choice. (C)

6. (a) Some calcium(II) hydroxide was well shaken with water and the mixture filtered. Into the filtrate was bubbled carbon dioxide until no further change could be observed. Separate portions of this final liquid were (i) boiled in a beaker, (ii) shaken with a solution of soap (sodium(I) stearate). State what could be observed at each stage of this sequence of operations, and explain each observation.
(b) Approximately 0·2 g calcium(II) sulphate(VI) is sufficient to saturate 100 g water at room temperature. Starting from marble (calcium(II) carbonate), describe how you would prepare a reasonably pure sample of calcium(II) sulphate(VI).
(c) If you were told that magnesium compounds closely resembled the corresponding compounds of calcium, what would you expect to happen if portions of a solution of magnesium(II) chloride were mixed with (i) sodium(I) carbonate solution, (ii) sodium(I) hydrogencarbonate solution? Briefly state your reasons. (O)

7. Name **two** substances which are obtained commercially from sea water.
Sea water is very hard. Name **one** substance which causes hardness in sea water.
Explain briefly what happens when soap is mixed with sea water. (JMB)

8. Describe in detail an experiment by which you could determine the mass of oxygen which combines with 1 g of hydrogen to form water.
What is meant by the term *hardness of water*? Describe **two** methods by which all the calcium ions present in a sample of water can be removed. (AEB)

9. Describe the manufacture of quicklime from limestone. (A **simple** diagram is wanted.)
Give **two** large-scale uses for lime, either quicklime or slaked lime (hydrated lime). How would you distinguish between quicklime and slaked lime? Starting from slaked lime, how would you prepare (i) calcium(II) hydrogencarbonate solution; (ii) anhydrous calcium(II) chloride? (C)

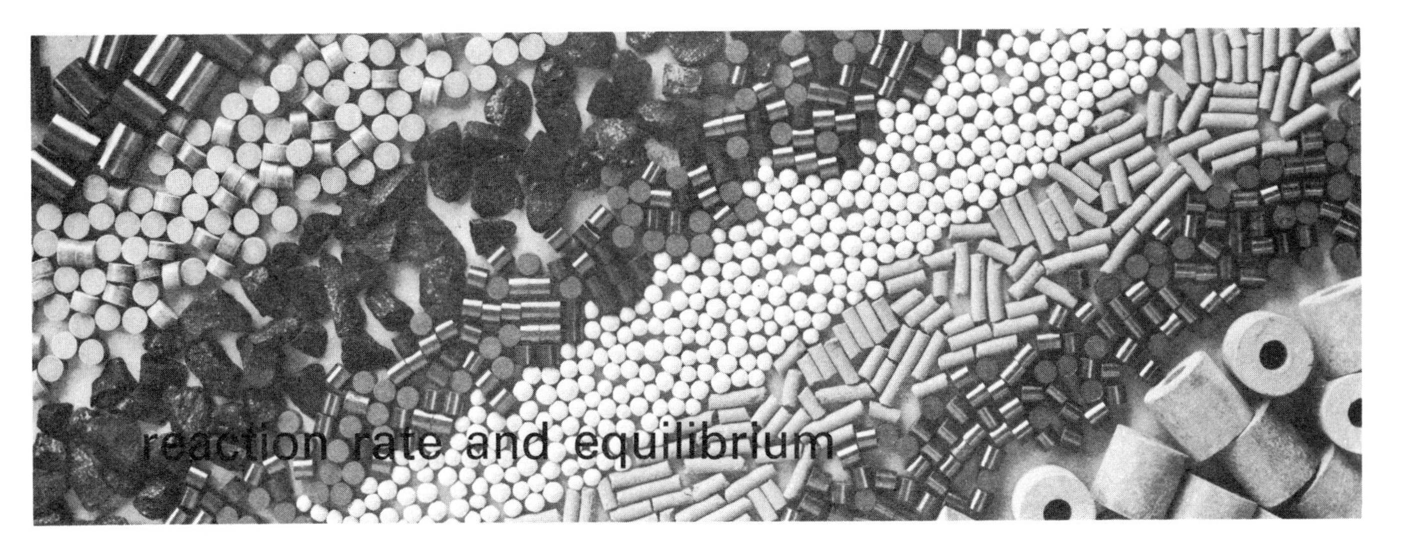

23.1 chemical reactions proceed at different speeds

Some reactions are so fast as to appear instantaneous; some are so slow that no observable change can be detected except after a long period of time; some take place at intermediate speeds. Can you think of examples of each of these three types, either from previous work or everyday experience?

1. fast reactions

A dense white precipitate of silver(I) chloride appears as soon as aqueous solutions of sodium(I) chloride and silver(I) nitrate(V) are mixed.

Addition of an acid to blue litmus solution reddens it immediately, without any observable transition through intermediate shades of purple.

2. slow reactions

One cannot see a change occurring when iron rusts or when a newspaper or a piece of white fabric becomes yellow on prolonged exposure to air and sunlight. The change is taking place very slowly and its effect is noticeable after a long interval of time.

3. reactions of intermediate speed

Coal burning in a fire grate is, fortunately, neither instantaneous nor very slow. A steady supply of heat energy is available for a period of a few hours.

Zinc reacts with dilute hydrochloric acid liberating hydrogen. The gas is evolved at a convenient rate, enabling several jars to be collected in a few minutes.

23.2 meaning of the term 'concentration'

Before the factors affecting the speeds of reactions can be understood, it is necessary to be clear about the meaning of the term 'concentration'.

$1 cm^3$ of concentrated nitric(V) acid is placed in a test-tube. $2 cm^3$ of concentrated nitric(V) acid is added to $1 dm^3$ of water in a large flask. If a scrap of copper is added to each, the acid in the test-tube gives an immediate reaction, but no change at all is observed in the large flask (see Fig. 23.1).

The test-tube contains only half as much acid as is present in the flask; the important difference is that in the test-tube, the acid is in a much more 'concentrated' form, meaning that the amount of acid per unit volume is higher in this case than in the case of the acid in the flask. **Concentration means amount of substance per unit volume;** amount of substance is expressed in moles and volume in units either of cubic decimetres (dm^3) or

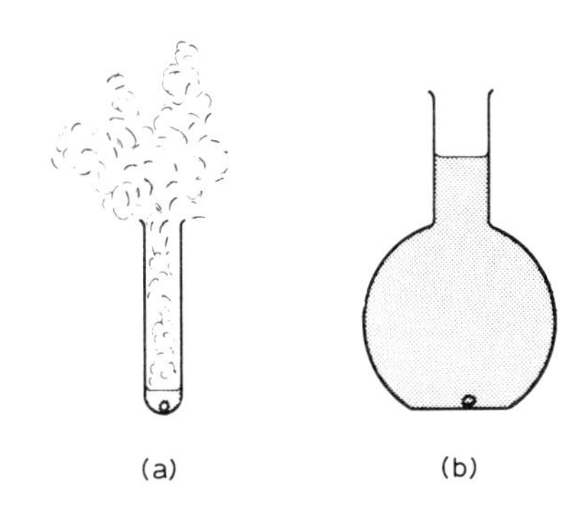

Fig. 23.1 copper and (a) concentrated, (b) dilute, nitric(V) acid

litres (l). Concentration is therefore expressed in the units **mol dm^{-3}** or mol l^{-1}.

23.3 rate of a chemical reaction

The rate of a reaction may be compared with the rate of production of articles in a factory. In a small factory eight workers might each produce 50 cardboard boxes in a working day. The rate of production of boxes is

$$\frac{\text{number of boxes produced}}{\text{time taken}} = \frac{400 \text{ boxes}}{1 \text{ day}} = 400 \text{ box day}^{-1}$$

The manager might decide that this is not a sufficiently high rate of production. By doubling the number of workers he can double the rate of production:

$$\text{rate} = \frac{800 \text{ boxes}}{1 \text{ day}} = 800 \text{ box day}^{-1}$$

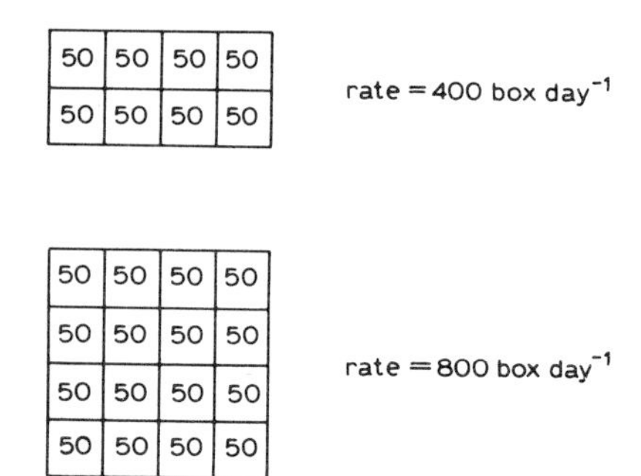

Fig. 23.2 cardboard box production figures

This is one way in which we can express the rate of a chemical reaction, in terms of the amount of product B formed in unit time.

$$A \longrightarrow B$$

If 6 mol of B are formed in one minute, we would express the rate of reaction as:

$$\frac{6 \text{ mol}}{60 \text{ s}} = 0{\cdot}1 \text{ mol s}^{-1}$$

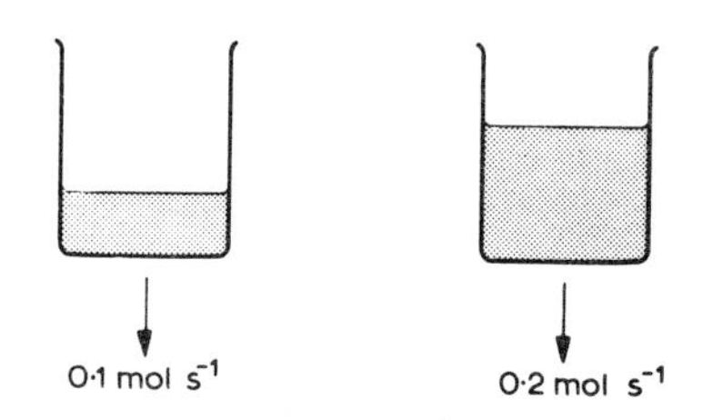

Fig. 23.3 doubling the scale doubles the rate of production

If much more starting material A is used (in a larger reaction vessel) it is possible to double the rate of production of B (Fig. 23.3). Thus when the rate of a reaction is expressed in terms of the *amount* of B formed, the *rate depends on the scale* on which the reaction is carried out, just as the rate of production of boxes depends on the number of workers.

Factory management is not quite so simple as has been represented above. Let us suppose that the Factory Act laid down that each worker requires a floor area of 5 m^2. The manager needs to double the floor area used when he doubles his number of workers. Now the two rates of production of boxes per unit floor area become identical:

for eight workers: $\dfrac{400 \text{ boxes}/40 \text{ m}^2}{1 \text{ day}} = \mathbf{10 \text{ box m}^2\text{day}}$ 1

for sixteen workers: $\dfrac{800 \text{ boxes}/80 \text{ m}^2}{1 \text{ day}} = \mathbf{10 \text{ box m}^{-2}\text{day}^{-1}}$

This result is obvious if each of the squares in Fig. 23.2 is taken to represent 5 m^2.

The quantity 10 box m^{-2}day^{-1} tells us something more meaningful about box production than the earlier statement. Similarly, in the reaction:

$$A \longrightarrow B$$

the rate of production of B can be doubled by using twice as much A, but if this doubles the volume occupied, the *rate of production of B* ***as measured by its concentration*** *is independent of the scale* of the reaction.

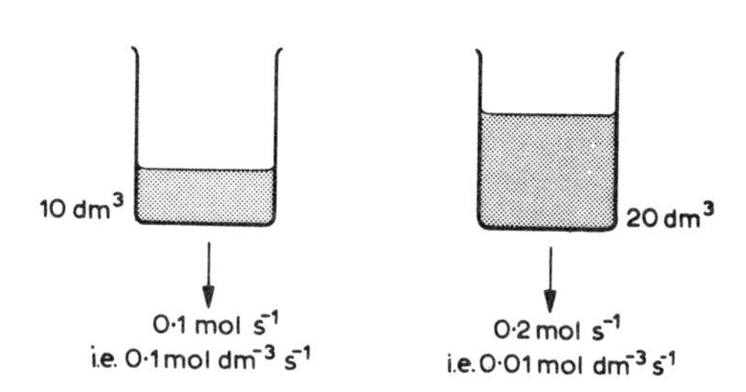

Fig. 23.4 doubling the scale does not affect the rate of increase of concentration

In a volume of 10 dm^3, if 6 mol B are produced in 1 minute,

$$\text{rate of increase of concentration of B} = \frac{6 \text{ mol}/10 \text{ dm}^3}{60 \text{ s}}$$
$$= \mathbf{0{\cdot}01 \text{ mol dm}^{-3}\text{s}^{-1}}$$

If the scale of the reaction is doubled, then
in a volume of 20 dm^3, 12 mol B are produced in 1 minute,

$$\text{rate of increase of concentration of B} = \frac{12 \text{ mol}/20 \text{ dm}^3}{60 \text{ s}}$$
$$= \mathbf{0{\cdot}01 \text{ mol dm}^{-3}\text{s}^{-1}}$$

It is important to distinguish between a rate which is measured in terms of

amount ($mol\,s^{-1}$), which depends on the scale of the reaction, and one which is measured in terms of **concentration** ($mol\,dm^{-3}s^{-1}$), which is independent of the scale. In this Chapter we will usually refer to rates measured by increases in concentration of product.

The concentration of a substance is often denoted by a square bracket:

$$[B] = 0{\cdot}6\,mol\,dm^{-3}$$

Less commonly (and less usefully), **mass concentration** may be used. Hence if the molar mass of B is 60, its mass concentration equivalent to the above concentration would be:

$$0{\cdot}6\,mol\,dm^{-3} \times 60\,g\,mol^{-1}$$
$$= 36\,g\,dm^{-3}$$

23.4 measurement of the rate of a reaction

In order to measure the rate of a reaction, the increase in either amount (mol) or, better, concentration ($mol\,dm^{-3}$) of a product, in a given time, must be determined by analysis. Such techniques are beyond the scope of this book, but the following simple experiment, illustrating the speed at which hydrogen gas is produced, may be useful. It must be stressed that this experiment measures the rate as a volume of gas produced in a given time and therefore it does not give the rate of the reaction within the strict meaning of the term. The results are nevertheless significant.

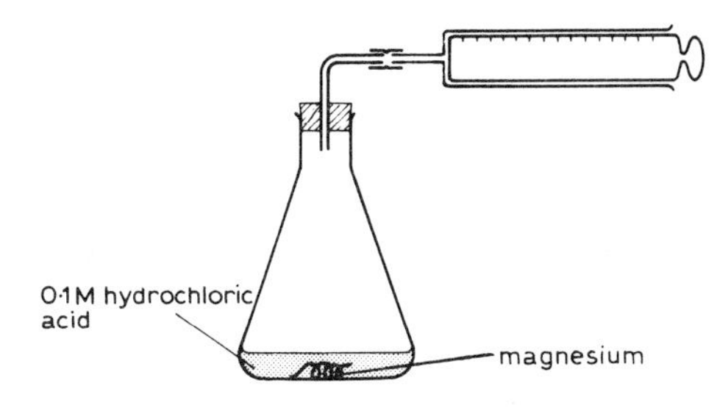

Fig. 23.5 measurement of a simple reaction rate

Exactly one metre of magnesium ribbon is cleaned with emery paper and its mass is determined. The length corresponding to a mass of 0·008 g is calculated; this length is cut off and added to about 2 cm^3 of 0·1 M hydrochloric acid in the apparatus shown in Fig. 23.5. *Immediately* on addition of the magnesium, the cork should be inserted and the time noted. Measurements of the volume of gas are taken after 15 s, 30 s, 45 s, 60 s and after 2, 3 4, 5 and 6 minutes. The results are most clearly represented graphically, as shown in Fig. 23.6.

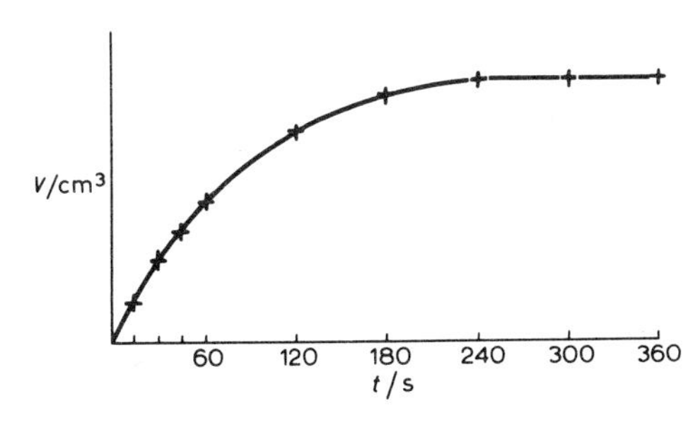

Fig. 23.6 volume of hydrogen produced (V) plotted against time (t)

The rate of the reaction is high at the outset and *decreases as the reaction proceeds*. The reaction is complete after about four minutes. Why does the rate change in this way during the course of the reaction?

23.5 factors which influence the rate of a reaction

1. *state of division of reactants*

a) Two beakers are each half-filled with dilute hydrochloric acid. A *lump* of marble is added to one – bubbles of gas are produced slowly from the lump of marble and rise to the surface of the liquid. *Powdered* chalk is added to the other – bubbles of gas are produced rapidly and the mixture may even froth over the top of the beaker. Marble and chalk are both forms of calcium(II) carbonate and the gas produced is carbon dioxide.

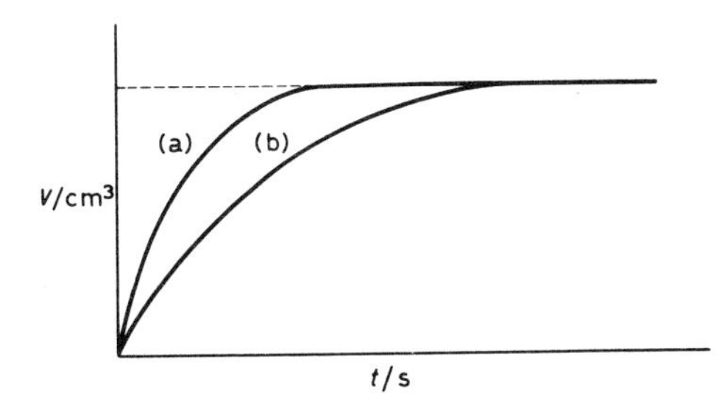

Fig. 23.7 reaction 'rates' using (a) magnesium powder, (b) magnesium ribbon

b) *Solid* mercury(II) chloride and *solid* potassium(I) iodide are ground together in a mortar using a pestle. A pink colour appears slowly. If *solutions* of mercury(II) chloride and potassium(I) iodide are mixed, a salmon-pink precipitate is produced immediately. The coloured product in each case is mercury(II) iodide. Even after grinding the solids, the particle size is still very many thousands of times greater than in solution, in which the particles have sizes comparable with those of atoms.

c) If the experiment described in section 23.4 is repeated using magnesium *powder* instead of magnesium *ribbon*, the rate of reaction is at all times higher and the reaction is completed in a shorter time (see Fig. 23.7).

A reaction involving a solid takes place more rapidly if the particles are small than if the particles are large. Fig. 23.8 illustrates this; in (*a*), 20 particles out of the total of 36 particles present are exposed to the second reacting substance, whereas in (*b*), 32 of the 36 particles are exposed.

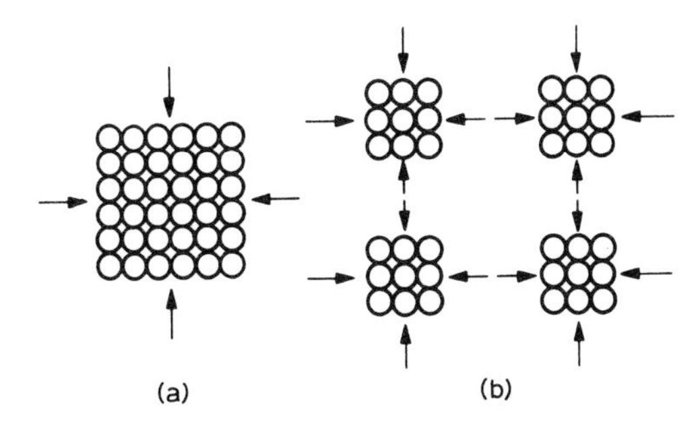

Fig. 23.8 small particles (b) of solid react more rapidly than large particles (a)

2. *concentration of reacting substances (reactants)*

a) A piece of granulated zinc is placed in each of two beakers. Dilute hydrochloric acid is added to one and concentrated hydrochloric acid is added to the other. Bubbles of gas (hydrogen) are produced at a much greater rate in the case of the concentrated acid.

b) A piece of magnesium ribbon, ignited and plunged into a jar containing air, burns with a bright white flame. A similar piece of magnesium ribbon, ignited and plunged into a jar containing pure oxygen, burns with a much brighter white flame and the reaction ceases in a shorter time.

The concentration of oxygen in pure oxygen is five times that in air (Chapter 6). The flame is brighter since the same quantity of energy is being released in a shorter time.

c) A small quantity of manganese(IV) oxide is sprinkled into two beakers containing respectively '20-volume' and '1-volume' aqueous hydrogen peroxide. The rate of production of gas is much higher from the '20-volume' solution.

d) Identical flasks A and B are charged with the following solutions:

A: water (250 cm³)
0·1 M aqueous potassium(I) iodate(V) (10 cm³)
0·2 M hydrochloric acid (15 cm³)
starch solution (15 cm³)

B: water (240 cm³)
0·1 M aqueous potassium(I) iodate(V) (20 cm³)
0·2 M hydrochloric acid (15 cm³)
starch solution (15 cm³)

10 cm³ of 0·1 M aqueous sodium(I) sulphate(IV) is added to each flask and a stop-clock is started immediately on addition. A deep blue colour appears suddenly in flask A after about 30 seconds and in flask B after about 15 seconds.

Note that the total volume of solution in the two flasks is the same (300 cm³). In flask B the **concentration** of iodate(V) ion is twice that in flask A.

Sulphate(IV) ion reduces iodate(V) ion to iodide ion (for a treatment of oxidation–reduction reactions see Chapter 14):

$$IO_3^-(aq) + 6H^+(aq) + 6e^- \longrightarrow I^-(aq) + 3H_2O(l)$$
$$[SO_3^{2-}(aq) + H_2O(l) \longrightarrow SO_4^{2-}(aq) + 2H^+(aq) + 2e^-] \times 3$$
$$\mathbf{IO_3^-(aq) + 3SO_3^{2-}(aq) \longrightarrow I^-(aq) + 3SO_4^{2-}(aq)}$$

This is a fairly slow reaction and when all the sulphate(IV) ion has been used up, excess iodate(V) ion reacts with the iodide ion formed, producing iodine:

$$IO_3^-(aq) + 6H^+(aq) + 5e^- \longrightarrow \tfrac{1}{2}I_2(aq) + 3H_2O(l)$$
$$[I^-(aq) \longrightarrow \tfrac{1}{2}I_2(aq) + e^-] \times 5$$
$$IO_3^-(aq) + 5I^-(aq) + 6H^+(aq) \longrightarrow 3I_2(aq) + 3H_2O(l)$$

The iodine then reacts with starch giving the blue colour.

The important reaction is that shown in bold type above; it is completed in flask B in half the time taken in flask A, because the concentration of iodate(V) ion in flask B is doubled.

Reactions proceed more rapidly if the concentrations of reactants are increased.

3. *pressure*

This factor is only applicable to reactions involving gases. It is found that **increase of pressure increases the rate of a reaction between gases.**

If the pressure on a fixed amount of a gas is increased at constant temperature then its volume must be reduced (Chapter 3). Since the amount of gas present does not change, then the **concentration** of the gas (amount of substance per unit volume) must increase. The effect of increasing the pressure on a gaseous reactant is to increase its concentration, which in turn increases the rate of the reaction.

4. *temperature*

A small cross is made in the centre of a piece of white paper. 10 cm³ of 0·2 M aqueous sodium(I) thiosulphate(VI) is placed in a 100 cm³ conical flask, a thermometer is placed in the solution and the flask is warmed until the temperature is just above 20 °C. 5 cm³ of 2 M hydrochloric acid is added, a stop-clock is started and the temperature of the mixture is noted. The mixture is gently swirled and the flask is then placed on the paper above the cross. A pale yellow suspension appears and the liquid slowly becomes opaque. When the cross (viewed through the liquid) is no longer visible, the stop-clock is stopped and the time taken is noted. The experiment is repeated four times, heating the sodium(I) thiosulphate(VI) solution before the addition of acid to a little over 30 °C, 40 °C, 50 °C and 60 °C, so that the reaction can be started as near to these temperatures as possible. A typical set of results is shown below.

initial temperature $\theta/°C$	time taken t/s	$\frac{1}{t}/s^{-1}$
19	34	0·029
28	26	0·038
38	18	0·056
47	10	0·10
57	6	0·17

The reaction taking place is:

$$S_2O_3^{2-}(aq) + 2H^+(aq) \rightarrow S(s) + SO_2(aq) + H_2O(l)$$

What is actually being measured is the time taken for a concentration of sulphur to develop sufficient to mask the cross. Since the rate of the reaction can be expressed as

$$\text{rate} = \frac{\text{concentration of product formed}}{\text{time taken}}$$

the rate of this reaction is proportional to the *reciprocal* of the time taken for the cross to disappear.

The rate of the reaction increases with increase of temperature. This is a general rule for all chemical reactions; the rate of a chemical reaction is approximately doubled by a rise in temperature of about 10C°.

5. *catalysis*

The catalytic effect of manganese(IV) oxide in the production of oxygen is described in section 7.2 and that of platinum in the reaction between hydrogen and oxygen in section 9.6(3). **A catalyst is a substance which alters the rate of a reaction without itself being consumed or changed in chemical composition at the end of the reaction.** The possible mechanisms by which different catalysts act are described in the appropriate sections (e.g. 9.6(3)). Usually a catalyst increases the rate of a reaction, but in some cases (negative catalysts) the effect can be to reduce the rate.

6. *light*

If silver(I) bromide, precipitated from aqueous solutions of silver(I) and bromide ions, is exposed to sunlight for a short time, or diffused daylight for a longer time, the precipitate darkens in colour. The same effect may be achieved much more rapidly by burning magnesium close to a sample of freshly precipitated silver(I) bromide.

The factors mentioned in this section may be summarised as follows:

	rate increased by	*rate reduced by*
1.	small particles (solids)	large particles (solids)
2.	increase in concentration of reactants	decrease in concentration of reactants
3.	increase of pressure (gases)	decrease of pressure (gases)
4.	increase of temperature	decrease of temperature
5.	catalyst	negative catalyst
6.	light (limited application)	dark (limited application)

23.6 collision theory of chemical reactions

In general, reactions can occur only when particles of the reactants collide with each other. If there is a significant *activation energy* necessary before reaction can proceed (e.g. to break bonds in the reactant molecules), the collisions must be sufficiently energetic to provide this activation energy.
i) Small particles, having a large surface area, maximise frequency of collisions with a second reactant.
ii) Increase in concentration of a reactant increases the likelihood of collisions with particles of a second reactant; this also holds for increase of pressure of a gas.
iii) Increase of temperature increases the kinetic energy of particles; this increases both the number of collisions per second with particles of a second reactant, and the average energy of such collisions.
iv) A catalyst can act by increasing the concentration of reactants near its surface and by decreasing the activation energy of a reaction.
Increased frequency of collisions, *increased energy* of collisions and *reduction in activation energy*, will all have the effect of increasing the rate of a reaction.

23.7 reversible reactions

1) When hydrated copper(II) sulphate(VI) is heated, it changes colour from blue to white and steam is given off:

$$CuSO_4.5H_2O(s) \longrightarrow CuSO_4(s) + 5H_2O(g)$$

If water is added to cold anhydrous copper(II) sulphate(VI), the reverse reaction occurs:

$$CuSO_4(s) + 5H_2O(l) \longrightarrow CuSO_4.5H_2O(s)$$

This is an example of a **reversible reaction** – a reaction which can be made to go in either direction by changing the conditions. The two changes above can be written as one reversible change:

$$CuSO_4.5H_2O(s) \rightleftharpoons CuSO_4(s) + 5H_2O(l)$$

2) About 10 cm³ of aqueous bromine is poured into a beaker standing on a piece of white paper. By means of a dropping pipette, a small volume of aqueous sodium(I) hydroxide is added and the mixture is swirled gently; the orange solution becomes colourless. If a small volume of dilute hydrochloric acid is added, the orange colour is restored. This procedure may be repeated many times – addition of acid always produces the orange colour, addition of alkali removes it. The equation for the reaction is:

$$\underset{(orange)}{Br_2(aq)} + \underbrace{2OH^-(aq) \rightleftharpoons Br^-(aq) + BrO^-(aq) + H_2O(l)}_{(colourless)}$$

Addition of acid causes the reaction to proceed to the left, addition of alkali causes the reaction to proceed to the right.
3) If the above experiment is repeated using aqueous potassium(I)

chromate(VI) instead of aqueous bromine, addition of acid causes the yellow solution to turn orange and subsequent addition of alkali restores the yellow colour:

$$\underset{\text{(yellow)}}{2CrO_4^{2-}(aq)} + 2H^+(aq) \rightleftharpoons \underset{\text{(orange)}}{Cr_2O_7^{2-}(aq)} + H_2O(l)$$

Addition of acid causes the reaction to proceed to the right, addition of alkali causes the reaction to proceed to the left.

4) Aqueous ammonia smells strongly of ammonia gas; the aqueous solution also reacts with aqueous copper(II) ions to produce the deep blue compound which is formed from anhydrous copper(II) ions and ammonia gas (see section 24.7). Thus *aqueous ammonia contains ammonia molecules.*

The solution conducts electricity, which indicates that it contains ions. If it is added to aqueous iron(III) ions, a red-brown precipitate of iron(III) hydroxide is obtained:

$$Fe^{3+}(aq) + 3OH^-(aq) \longrightarrow Fe(OH)_3(s)$$

Aqueous ammonia contains hydroxide ions.

These facts can be accounted for by assuming a *reversible acid–base reaction* to occur when ammonia dissolves in water:

$$NH_3(aq) + H_2O(l) \rightleftharpoons NH_4^+(aq) + OH^-(aq)$$

23.8 equilibrium in chemical reactions

Consider a reversible change between two substances:

$$A \rightleftharpoons B$$

If we start with pure A, its rate of change to B will be high at first, but will fall as B is produced, because the concentration of A will fall. The concentration of B is initially zero, but it gradually increases; hence the rate of conversion of B to A gradually increases. At a certain point, the *rate of production of B and the rate of its conversion to A will be equal* (point X in Fig. 23.9). At this point no further change is apparent, as both A and B are being formed and decomposed at the same rate; the composition of the mixture will remain constant, even though the individual particles are changing. This is described as a situation of **dynamic equilibrium.**

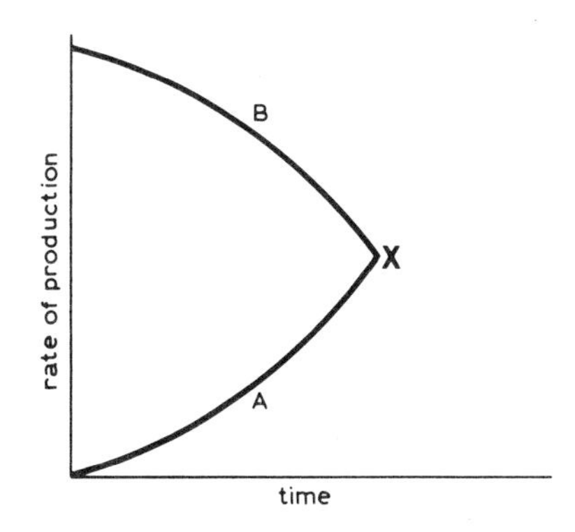

Fig. 23.9 dynamic equilibrium—note that at X the rates are not zero

In a reversible reaction, when the rate of the forward reaction is equal to the rate of the reverse reaction, the system is said to be in a state of dynamic equilibrium.

The reversible reactions mentioned in the preceding section are all examples of dynamic equilibrium. When acid is added to aqueous bromine, the equilibrium

$$Br_2(aq) + 2OH^-(aq) \rightleftharpoons Br^-(aq) + BrO^-(aq) + H_2O(l)$$

is disturbed and the proportion of bromine in the mixture is increased. Addition of alkali disturbs the equilibrium in favour of a high proportion of bromide and bromate(I) ions. Both in acid and alkali, however, all of the species represented in the equation are present, in greater or lesser amounts.

23.9 factors affecting the composition of equilibrium mixtures

The effect of a constraint, or external influence, upon the composition of a system in equilibrium, is described by **Le Chatelier's principle:**

If a constraint is applied to a system in equilibrium, the reaction proceeds in such a way as to reduce the effect of the constraint.

Three important constraints which can influence the composition of equilibrium mixtures are:

i) *pressure:* for reactions involving gases, if the number of moles of gas in the reactants and products is different,

ii) *temperature:* for exothermic or endothermic changes,

iii) *change in concentration* of reactants or products.

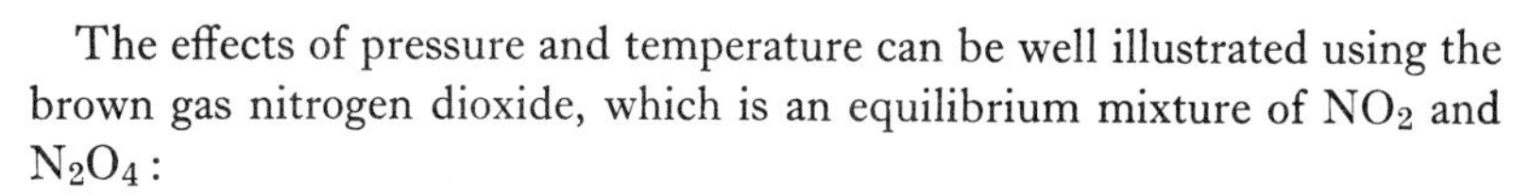

The effects of pressure and temperature can be well illustrated using the brown gas nitrogen dioxide, which is an equilibrium mixture of NO_2 and N_2O_4:

$$\underset{\text{colourless}}{N_2O_4} \rightleftharpoons \underset{\text{brown}}{2NO_2}$$

To understand how changes in pressure and temperature affect this equilibrium, try to work out whether conversion of dinitrogen tetraoxide (N_2O_4) to nitrogen dioxide (NO_2) will

a) increase or decrease the pressure of the gas mixture,
b) evolve or absorb heat energy, and therefore
c) increase or decrease the temperature of the gas mixture.

1. pressure

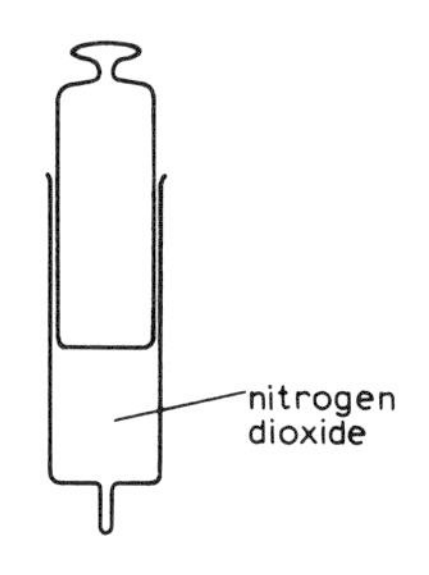

Fig. 23.10 effect of pressure on composition of equilibrium mixture
$N_2O_4(g) \rightleftharpoons 2NO_2(g)$

The end of a 10 cm^3 transparent plastic syringe is sealed by holding the tip in a flame for a few seconds, then pressing it on a metal surface. The syringe is filled with the brown gas generated from copper and concentrated nitric(V) acid and the plunger is inserted (Fig. 23.10). The syringe is laid on a piece of white paper and the plunger is pushed in *rapidly*. The colour momentarily deepens, then goes paler. The plunger is *rapidly* pulled back; the colour momentarily becomes paler still, then deepens to its original intensity. These operations can be repeated many times, until the important sequence of changes may be demonstrated beyond argument.

When the gas mixture is compressed, its volume is reduced; before any change in the composition of the mixture occurs, the brown colour is intensified, as the same number of moles of nitrogen dioxide is present in a smaller volume. The explanation of the mixture then becoming paler can only be that the composition of the mixture is changed in the direction

$$2NO_2 \rightarrow N_2O_4$$

This is consistent with Le Chatelier's Principle. Increase of the pressure *on* the gas mixture causes an increase in pressure *of* the gas mixture. By proceeding in the direction shown, the reaction forms *fewer moles* of gas in the mixture, thus *reducing its pressure* and opposing the effect of the constraint.

The changes observed on withdrawing the plunger can be explained in similar terms.

Change in pressure will affect the composition of an equilibrium mixture only if the reaction involves a change in the number of moles of gas present in the mixture. Which of the following equilibria will be affected by a change in pressure?

$$H_2(g) + I_2(g) \rightleftharpoons 2HI(g)$$
$$2SO_2(g) + O_2(g) \rightleftharpoons 2SO_3(g)$$
$$PCl_5(g) \rightleftharpoons PCl_3(g) + Cl_2(g)$$
$$N_2(g) + 3H_2(g) \rightleftharpoons 2NH_3(g)$$
$$CaCO_3(s) \rightleftharpoons CaO(s) + CO_2(g)$$

2. temperature

Two thin-walled glass bulbs ('Dumas bulbs' suffice well) are filled with nitrogen dioxide gas mixture and are sealed. One bulb is immersed in hot water, the other in ice-cold water (Fig. 23.11). After about ten seconds, withdrawal of the bulbs shows that the colour of the gas in the hot bulb has intensified and that in the cold bulb has become paler. The bulbs may be reversed and the effect repeated.

Since the volume is kept constant, increase in temperature will cause an increase in the pressure of the gas; we would expect from (1) above that this would *decrease* the intensity of colour–in fact the *reverse effect* is observed. Change in temperature must therefore have another effect on the composition of the equilibrium mixture, independent of the change in pressure it causes.

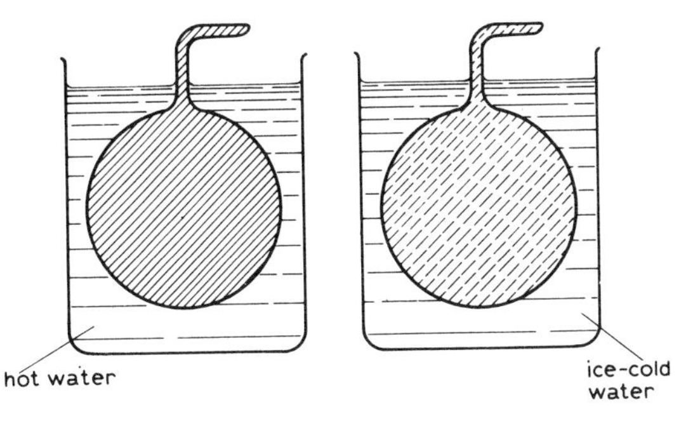

Fig. 23.11 effect of temperature on composition of equilibrium mixture
$N_2O_4(g) \rightleftharpoons 2NO_2(g)$

$$O_2N{-}NO_2 \rightarrow O_2N \quad NO_2$$

The above change involves breaking a bond and will therefore be endothermic: the energy necessary for the change is supplied at the expense of the kinetic energy, and therefore the temperature, of the system.

When the temperature of the system is raised, the reaction proceeds in the direction

$$N_2O_4 \longrightarrow 2NO_2$$

which *absorbs energy*, *reduces the temperature* of the system and hence opposes the constraint.

What will be the effect on the following equilibria of increasing the temperature of the system?

$$N_2(g) + O_2(g) \rightleftharpoons 2NO(g); \quad \Delta H = +180\,kJ$$
$$N_2(g) + 3H_2(g) \rightleftharpoons 2NH_3(g); \quad \Delta H = -92\,kJ$$

3. concentration

The action of acid and alkali on aqueous bromine, described in section 23.7(2), illustrates this effect well.

$$\underset{(orange)}{Br_2(aq)} + \underbrace{2OH^-(aq) \rightleftharpoons Br^-(aq) + BrO^-(aq) + H_2O(l)}_{(colourless)}$$

Addition of alkali increases the concentration of $OH^-(aq)$ in the mixture. The reaction opposes this constraint by reducing $[OH^-(aq)]$ and it can do this only by forming bromide and bromate(I); the orange colour of the bromine is discharged.

Addition of acid has the effect of reducing the concentration of $OH^-(aq)$ and the orange colour of the bromine reappears.

4. catalysts do not affect composition of equilibrium mixtures

If a catalyst increases the rate of a change $A \longrightarrow B$ it will also increase the rate of the reverse change $B \longrightarrow A$. A catalyst therefore has **no effect** on the composition of an equilibrium mixture, though its presence will reduce the time taken for the equilibrium to be established.

23.10 thermal dissociation and thermal decomposition

When solid ammonium chloride is heated in a dry test-tube, a ring of white solid appears on the cool upper part of the tube. Between the original solid and this new deposit, the contents of the tube appear colourless and transparent. If the experiment is repeated incorporating a glass wool plug and two pieces of litmus paper, as shown in Fig. 23.12, both pieces of litmus paper change colour. The colour changes indicate the formation of the gases hydrogen chloride and ammonia, which are partially separated as a consequence of their different rates of diffusion through the plug.

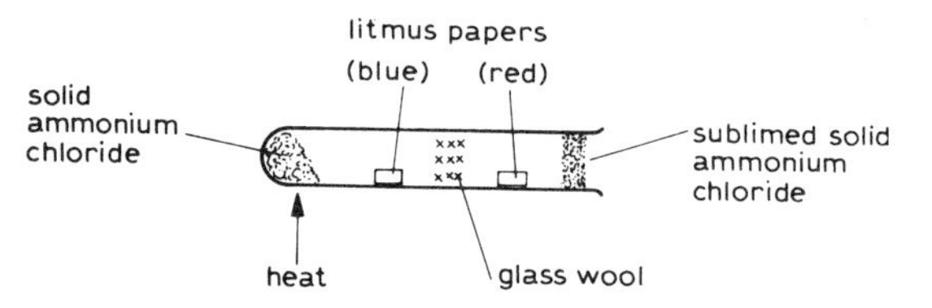

Fig. 23.12 thermal dissociation of ammonium chloride

Heating ammonium chloride converts it to the gases ammonia and hydrogen chloride. On cooling the gas mixture, ammonium chloride is re-formed:

$$NH_4Cl(s) \underset{\text{cool}}{\overset{\text{heat}}{\rightleftharpoons}} NH_3(g) + HCl(g)$$

This is an example of **thermal dissociation**, in which a change produced on heating is reversed on cooling. Another example is afforded by calcium(II) carbonate:

$$CaCO_3(s) \underset{\text{cool}}{\overset{\text{heat}}{\rightleftharpoons}} CaO(s) + CO_2(g)$$

When a substance is broken down on heating and the products do *not* recombine on cooling, the change is called a **thermal decomposition.** An example is the action of heat on ammonium dichromate(VI):

$$(NH_4)_2Cr_2O_7(s) \longrightarrow N_2(g) + Cr_2O_3(s) + 4H_2O(g)$$

Under no conditions can nitrogen gas, chromium(III) oxide and water be made to combine to form ammonium dichromate(VI).

revision summary: Chapter 23

concentration:	amount of substance per unit volume, mol dm^{-3}
rate of reaction:	in elementary work, indicated by rate of increase of concentration of product B, units mol dm^{-3}s^{-1}
reaction rate increased by:	fine subdivision of particles (solids) concentration of reactants pressure (gases) temperature (rate about doubled by 10 °C rise) catalyst light
dynamic equilibrium:	in a reversible reaction, when the rate of the forward reaction is equal to the rate of the reverse reaction the system is said to be in a state of dynamic equilibrium
Le Chatelier's Principle:	if a constraint is applied to a system in equilibrium, the reaction proceeds in such a way as to oppose the constraint *constraints:* pressure temperature concentration

questions 23

1. A boy added a certain mass of powdered calcium(II) carbonate to 50 cm^3 of M hydrochloric acid at room temperature and noted the *time* taken for all the carbonate to dissolve. State in each of the following cases whether this *time* would have increased or decreased had the same mass of carbonate been added to
(a) 100 cm^3 of 0·5 M hydrochloric acid.
(b) 25 cm^3 of 2·0 M hydrochloric acid.
(c) 50 cm^3 of M hydrochloric acid (warmed after measurement). (W)

2. The following statements are made in a text-book.
The rates of most chemical reactions are approximately doubled by raising the temperature at which the reactions are carried out by 10 °C.
The rate at which a chemical substance reacts is directly proportional to its concentration.
(i) Describe the experiments you would carry out to test the truth of these two statements when applied to **either** the reaction between a metal and a dilute acid **or** the decomposition of hydrogen peroxide catalysed by manganese(IV) oxide (manganese dioxide).
(ii) Explain simply, in terms of the ions or molecules present, why the rate of a reaction is increased both by raising the temperature and also by increasing the concentration of the reagents. (C)

3. The following results were obtained from a series of four experiments in which identical portions of powdered chalk were added to aqueous hydrochloric acid at room temperature.
In each case, the total volume of carbon dioxide evolved was measured and the time which had elapsed from the start to when the reaction effectively ceased was noted.

experiment	*acid used*	*volume of CO_2*	*time*
1	20 cm^3 M HCl	225 cm^3	10 min
2	30 cm^3 M HCl	260 cm^3	2 min 50 s
3	40 cm^3 M HCl	260 cm^3	2 min 20 s
4	60 cm^3 0·5 M HCl	260 cm^3	5 min 10 s

(i) Why was the volume of gas collected in experiment 1 different from that in the other experiments?
(ii) What information does the result of Experiment 3 provide about the amounts of acid being used in Experiments 2 and 4?
(iii) What do you notice about the actual amounts of acid being used in Experiments 2 and 4?
(iv) What principle do the results of Experiments 2 and 4 illustrate? (W)

4. Explain what is meant by (i) reversible reaction, (ii) chemical equilibrium.
Indicate **three** reversible reactions by appropriate equations. Choose **one** of these and describe in detail experiments by which the reaction can be shown to be reversible. (Mere statement of formation of a different substance is insufficient; some description or identification of the substance is necessary.) (AEB)

5. The equation for the reaction by which ammonia is manufactured is

$$N_2 + 3H_2 \rightleftharpoons 2NH_3$$

(a) What would be the effect on the equilibrium concentration of ammonia of
(i) increasing the pressure?
(ii) increasing the nitrogen concentration?
(b) The equilibrium concentration of ammonia increases as the temperature is lowered.
Is heat evolved or absorbed when ammonia is formed?
(c) Why is a catalyst used in this reaction? (JMB)

6. (a) The formation of methanol (methyl alcohol) from hydrogen and carbon monoxide can be represented by

$$CO + 2H_2 \rightleftharpoons CH_3OH; \Delta H = +91\,kJ\,mol^{-1}$$

What mass of hydrogen would react to cause a heat change of 91 kJ?
(b) What would be the effect on the equilibrium concentration of methyl alcohol in this endothermic reaction if
(i) the temperature was increased,
(ii) the pressure was increased,
(iii) the hydrogen concentration was increased?
(Atomic mass: H = 1·0) (JMB)

7. (a) State Le Chatelier's principle (law).
(b) Sulphuric(VI) acid is manufactured by the Contact process which depends on the reaction

$$2SO_2 + O_2 \rightleftharpoons 2SO_3; \quad \Delta H = -190\,kJ$$

Use Le Chatelier's principle to design conditions which lead to the maximum possible yield of sulphur(VI) oxide, giving your reasoning in full.
What conditions are actually used? Give reasons for the conditions chosen.

8. Explain the terms 'reversible reaction', 'chemical equilibrium' and 'shift of an equilibrium'.
At high temperatures the following reaction occurs:

$$PCl_3(g) + Cl_2(g) \rightleftharpoons PCl_5(g); \quad \Delta H = -93\,kJ$$

What would be the effect on the position of equilibrium if
(a) the pressure was increased,
(b) the temperature was increased,
(c) more chlorine was added,
(d) a catalyst was added?

9. Consider the reaction: $ICl_3 + Cl_2 \rightleftharpoons ICl_5$ at equilibrium. ICl_3 is a brown liquid, Cl_2 is a green gas, ICl_5 is a yellow solid. The forward reaction is exothermic.
State and explain what you would see if
(a) the temperature was increased,
(b) the pressure was reduced,
(c) more chlorine was added. (Sc., part question)

10. Design an experiment to find out whether the reaction between nitrogen oxide (nitric oxide) and oxygen is or is not reversible.

nitrogen compounds

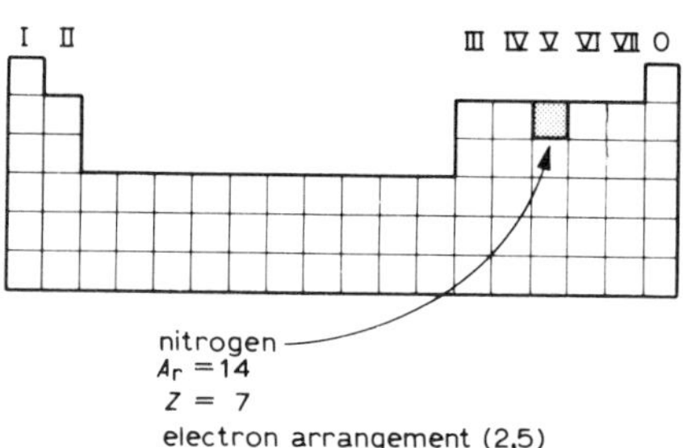

Fig. 24.1 place of nitrogen in Periodic Table

	kJ mol^{-1}
H—H	437
F—F	150
O=O	496
N≡N	950

Fig. 24.2 bond energies

The element nitrogen is essential for life; it is an important constituent of *aminoacids*, which are the compounds of which *proteins* are made. Proteins, in turn, form the material of *cell tissue*. The chemistry of the element is therefore of great importance, but it poses some acute problems.

The place of nitrogen in the Periodic Table is shown in Fig. 24.1. It is a **non-metal** and it forms a gaseous compound with hydrogen; some of its oxides are acidic. The element itself is a well known gas which is the main constituent of air; analysis in the mass spectrometer shows that it consists of diatomic molecules, N_2.

The bond energies joining the atoms in the diatomic molecules, H_2, F_2, O_2 and N_2 are shown in Fig. 24.2. This energy must be supplied to break the bonds in the molecules when the element reacts. The high bond energy of nitrogen explains its very unreactive nature; although there is no shortage of nitrogen in nature (it is all around us in the air) there are few processes which can supply sufficient energy to convert the element nitrogen into compounds reactive enough to be absorbed by plants and animals. This Chapter will be concerned with such reactions and with the nature and use of the important compounds of nitrogen.

24.1 a nitrogen compound obtained from natural products

If a small quantity of a natural product is heated in a small test-tube with sodalime (a solid alkali), a gas is evolved which turns moist red litmus paper blue. Substances which might be investigated in this way include hair, wool, finger nail clippings, meat, fish and cheese. The alkaline gas is **ammonia** and its pungent smell is much in evidence in maggot farms, where plant and animal tissues are caused to decay. The gas may be obtained in a purer form by heating an ammonium salt with an alkali, though the nitrate(V) should not be used as it can explode on heating. In the experiments which follow, the gas may be generated conveniently in this way, or by heating concentrated aqueous ammonia. It should be dried using calcium(II) oxide.

Since ammonia can be obtained so readily from natural substances it is likely to be closely related to important substances in living materials; the composition of ammonia is worth investigating.

24.2 burning ammonia in air and in oxygen

A lighted splint applied to a jar of ammonia is extinguished and the gas does not burn readily. However, if a jar of ammonia is inverted over a Bunsen flame, a dull orange flame can be seen to flicker fleetingly – the gas can be made to burn in air.

If dry ammonia is passed into an atmosphere of oxygen using the apparatus

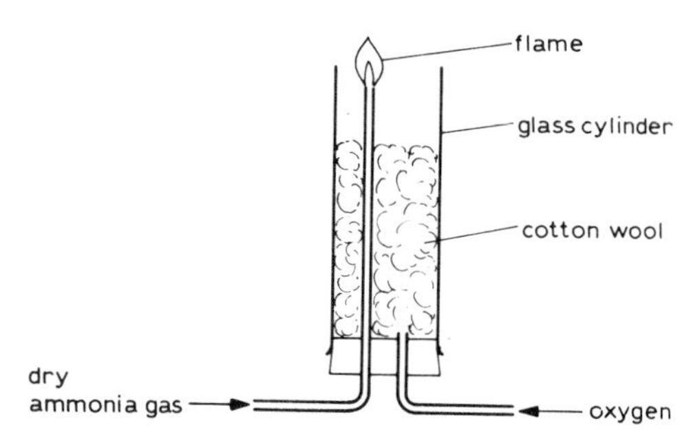

Fig. 24.3 burning ammonia in oxygen (the cotton wool diffuses the oxygen supply)

shown in Fig. 24.3, a lighted splint applied to the stream of ammonia causes the gas to burn with a yellow flame.

Substances which burn in oxygen can often remove oxygen from easily decomposed compounds. The action of ammonia on hot copper(II) oxide gives valuable information concerning the elements present in the gas.

24.3 the elements present in ammonia

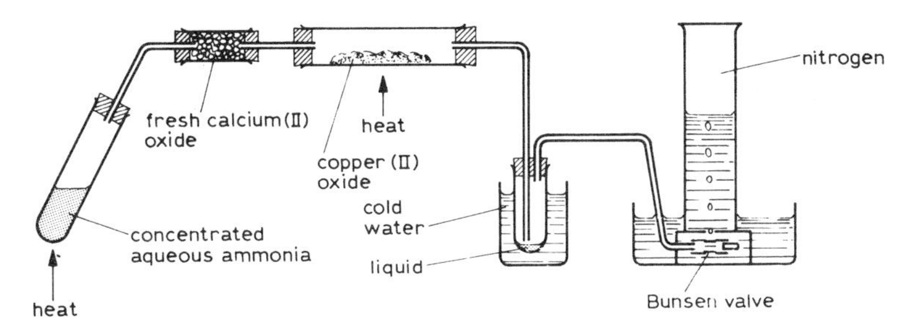

Fig. 24.4 oxidation of ammonia

A slow stream of dry ammonia is passed over hot copper(II) oxide, using the apparatus shown in Fig. 24.4. The delivery tube entering the water in the trough is closed with a Bunsen valve; as will be seen later, ammonia is very soluble in water and the Bunsen valve (Fig. 24.5) allows gas to escape from the tube, but prevents water from sucking back into the apparatus.

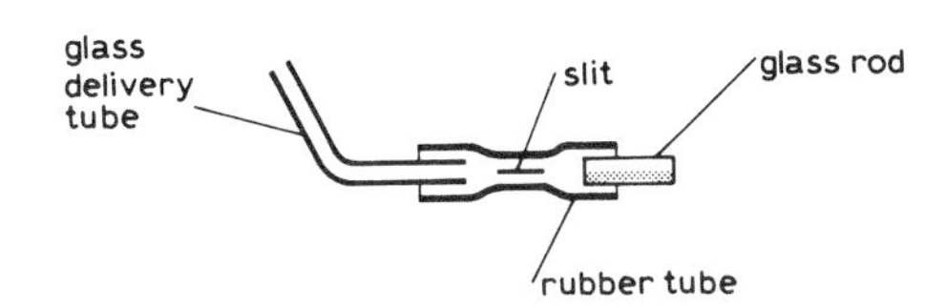

Fig. 24.5 a Bunsen valve

The copper(II) oxide changes colour to pink; clearly the product is copper – ammonia has removed oxygen from the oxide; what are the other products of the reaction?

The liquid which collects in the cooled receiver may be pale blue in colour, due to traces of impurity carried over from the combustion tube. This should not obscure its identity; it freezes at 0 °C and boils at 100 °C. **Ammonia must contain** the **hydrogen** necessary for the formation of this water.

The gas which collects in the jar is colourless, insoluble in water, does not burn and does not support combustion. It responds to none of the tests for the well-known gases and this is normally considered sufficient to enable us to identify it as the unreactive gas, **nitrogen.**

Ammonia contains the elements nitrogen and hydrogen. Whether it contains any other elements is still open to question. Can the gas be synthesised from nitrogen and hydrogen only? The synthesis is possible, but difficult to reproduce on the laboratory scale. It is an important industrial process and it is described in section 24.11. We may assume in view of this that **ammonia contains nitrogen and hydrogen only.**

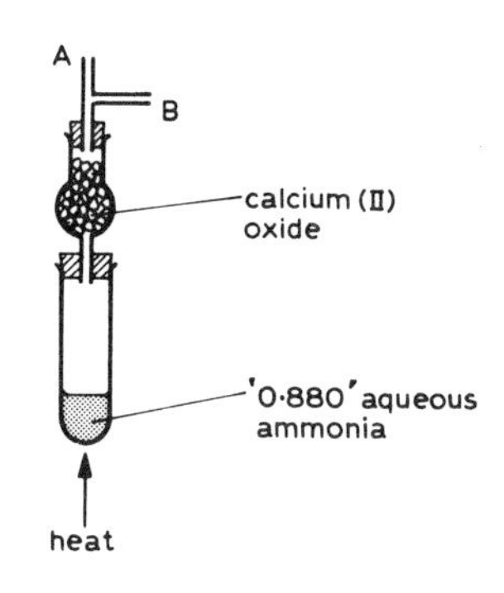

Fig. 24.6

24.4 the formula for ammonia

Dry ammonia is generated in the apparatus shown in Fig. 24.6. A controlled stream of gas is obtained from arm A of the T-piece by closing arm B momentarily with the forefinger. By connecting arm A with A in Fig. 24.7, and controlling the passage of gas using the finger as described, it is possible to introduce exactly 40 cm³ of gas into syringe 1 of the assembly shown.

In order to obtain good results in this experiment the apparatus must be completely dry and the iron wool must be free from oxide. The apparatus must contain no air and must therefore be flushed with dry nitrogen immediately before use.

Syringes 1 and 2 are connected and ammonia is passed back and forth over heated iron until no further increase in volume is observed. The volume increases to 80 cm³; at this stage the ammonia has been decomposed into its elements, nitrogen and hydrogen.

The residual gases are transferred to syringe 2 and this is connected with syringe 3. The gases are passed back and forth over hot copper(II) oxide

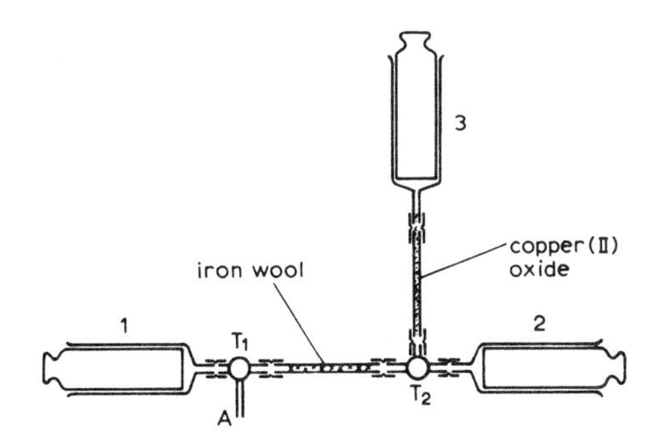

Fig. 24.7 determination of the formula for ammonia (plan view of gas-syringe apparatus)

until no further diminution in volume occurs. The residual gas, which may be assumed to be nitrogen, occupies 20 cm^3.

$$40\,cm^3 \text{ ammonia} \rightarrow 20\,cm^3 \text{ nitrogen} + 60\,cm^3 \text{ hydrogen}$$
$$2 \text{ volumes ammonia} \rightarrow 1 \text{ volume nitrogen} + 3 \text{ volumes hydrogen}$$

by application of *Avogadro's law*:

$$2 \text{ moles ammonia} \rightarrow 1 \text{ mole nitrogen} + 3 \text{ moles hydrogen}$$
$$2N_xH_y \rightarrow N_2 + 3H_2$$

From this is follows that **the formula for ammonia is NH_3.**

$$2NH_3(g) \rightarrow N_2(g) + 3H_2(g)$$

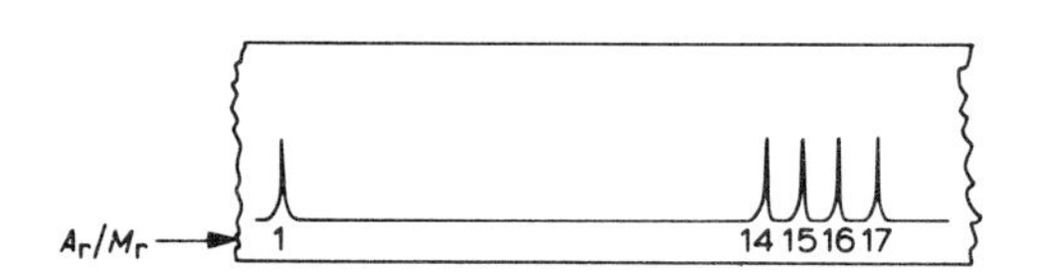

Fig. 24.8 mass spectrograph for ammonia

The mass spectrograph for ammonia is given in Fig. 24.8. It confirms the formula for ammonia and tells us further that the three hydrogen atoms are joined separately to the nitrogen atom. From the electronic structure of the nitrogen atom (2,5) we can deduce that **the ammonia molecule contains a lone pair of electrons**–a diagram of its molecular structure is shown in Fig. 18.35. The lone pair of electrons is very significant in the chemistry of ammonia, as it is responsible for the basic properties of the gas.

24.5 preparation and collection of dry ammonia

Ammonia is generated conveniently in the laboratory by heating any ammonium salt with any alkali; a slightly damp mixture of ammonium chloride and calcium(II) hydroxide serves well:

$$NH_4^+(s) + OH^-(s) \rightarrow NH_3(g) + H_2O(g)$$

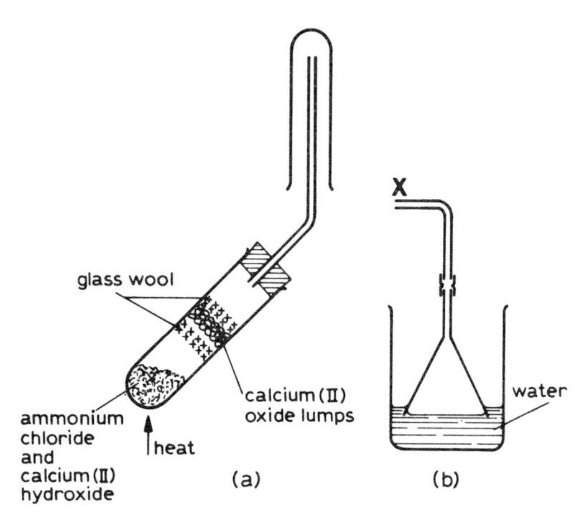

Fig. 24.9 small-scale preparation of (a) dry ammonia gas, (b) aqueous ammonia

The gas is freed from water vapour by passing it over lumps of calcium(II) oxide; the reasons for this choice are discussed in section 24.7. The high solubility in water and low density of the gas (section 24.6) determine that it should be collected by **downward displacement of air**; an apparatus appropriate for the collection of boiling-tubes of the gas is shown in Fig. 24.9. A piece of moist red litmus paper held at the mouth of the collecting tube indicates when the tube is full. The apparatus may be modified as shown for the preparation of an aqueous solution of ammonia; the delivery tube can be attached at X by rotating the reaction tube through 180°.

24.6 physical properties of ammonia

The gas is colourless and has an intensely pungent smell–it attacks the eyes severely. It is easy to liquefy, either by pressure or by cooling it to 239 K–can you explain why?

If a tube of the gas is uncorked with its mouth below the surface of water, water rapidly fills the tube and the solution becomes alkaline. The great solubility of ammonia in water can be demonstrated by using the gas in the Fountain Experiment (see section 15.6). As ammonia dissolves in water, the density of the liquid falls to a minimum value of 0·880 g cm^{-3}–such a saturated solution is often called '0·880 ammonia solution'.

The gas is much less dense than air, having a relative vapour density of 8·5 as compared with 14·4 for air.

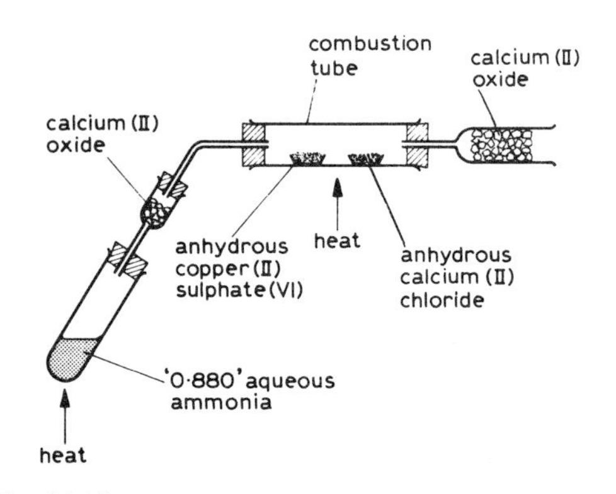

Fig. 24.10

24.7 a drying agent for ammonia

Concentrated sulphuric(VI) acid *cannot* be used to dry ammonia gas, for obvious reasons. For less obvious reasons the other common drying agent, anhydrous calcium(II) chloride, *cannot* be used either, as the following experiment shows.

Dry ammonia is passed over anhydrous copper(II) sulphate(VI) and anhydrous calcium(II) chloride as shown in Fig. 24.10. The copper(II) salt changes colour to a deep blue; no change is observable in the calcium(II) salt until the two boats in which the solids are held are withdrawn from the tube – both boats are hot. An *exothermic reaction* has occurred in both cases. If the solid residues are warmed in test-tubes, ammonia gas is evolved in each case. This shows that both salts react with ammonia:

$$\underset{\text{white}}{Cu^{2+}(s)} + 4NH_3(g) \longrightarrow \underset{\text{deep blue}}{[Cu(NH_3)_4]^{2+}(s)}$$

$$\underset{\text{white}}{Ca^{2+}(s)} + xNH_3(g) \longrightarrow \underset{\text{white}}{[Ca(NH_3)x]^{2+}(s)}$$

Clearly, calcium(II) chloride cannot be used to dry ammonia gas – it will react with ammonia in preference to water, as ammonia will donate its lone pair of electrons to the cation Ca^{2+} more readily than will water.

Calcium(II) oxide can be used to dry ammonia. When calcium(II) oxide absorbs water it is converted to the hydroxide:

$$O^{2-}(s) + H_2O(l) \longrightarrow 2OH^-(s)$$

Here water is acting as an acid, donating a proton to the oxide ion. Since water is more acidic than ammonia, it will be absorbed by oxide ion in preference to ammonia.

24.8 reactions of ammonia

1. with air or oxygen

Ammonia burns, with difficulty when ignited in air, and more easily in oxygen (see section 24.2); the products are nitrogen and steam:

$$4NH_3(g) + 3O_2(g) \longrightarrow 2N_2(g) + 6H_2O(g)$$

2. catalytic oxidation of ammonia

Ammonia can be oxidised to nitrogen oxide at about 600 °C (873 K) in the presence of a platinum or copper catalyst:

$$4NH_3(g) + 5O_2(g) \longrightarrow 4NO(g) + 6H_2O(g)$$

If a red-hot spiral of copper wire is suspended from a glass rod in an atmosphere of ammonia and air, just above the surface of a concentrated aqueous solution of ammonia (Fig. 24.11), the spiral continues to glow brightly showing that an exothermic reaction is occurring at its surface. Brown fumes may be seen in the flask as the nitrogen oxide formed combines spontaneously with more oxygen, to form nitrogen dioxide:

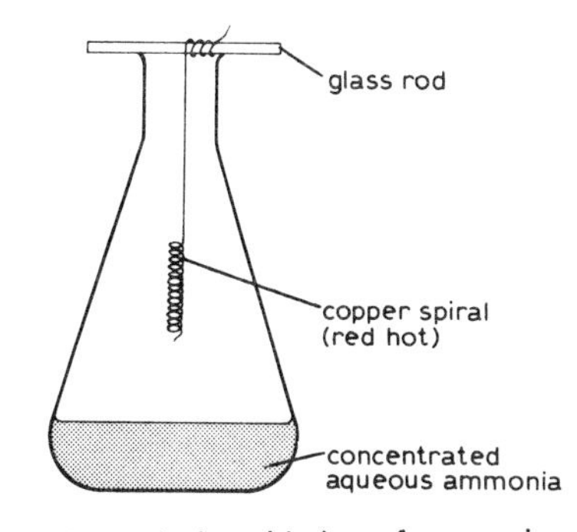

Fig. 24.11 catalytic oxidation of ammonia

$$2NO(g) + O_2(g) \longrightarrow 2NO_2(g)$$

The products of the catalytic oxidation of ammonia depend on the relative concentrations of ammonia and oxygen in the gas mixture; it may be necessary to dilute the ammonia solution and bubble oxygen through it in order to see the brown product.

Nitrogen dioxide dissolves in water and in the presence of oxygen the resulting solution can be oxidised to nitric(V) acid:

$$4NO_2(g) + O_2(g) + 2H_2O(l) \longrightarrow 4HNO_3(aq)$$

The above series of reactions forms the basis of the commercial preparation of nitric(V) acid.

3. with chlorine

If a stream of ammonia is passed into an atmosphere of chlorine, using the apparatus shown in Fig. 24.3, the ammonia ignites spontaneously and burns with a yellow flame, forming nitrogen and hydrogen chloride:

$$2NH_3(g) + 3Cl_2(g) \longrightarrow N_2(g) + 6HCl(g)$$

If the supply of chlorine is switched off, the yellow flame continues for a short time, then it disappears and dense white fumes are formed. Excess ammonia reacts with hydrogen chloride produced in the first reaction,

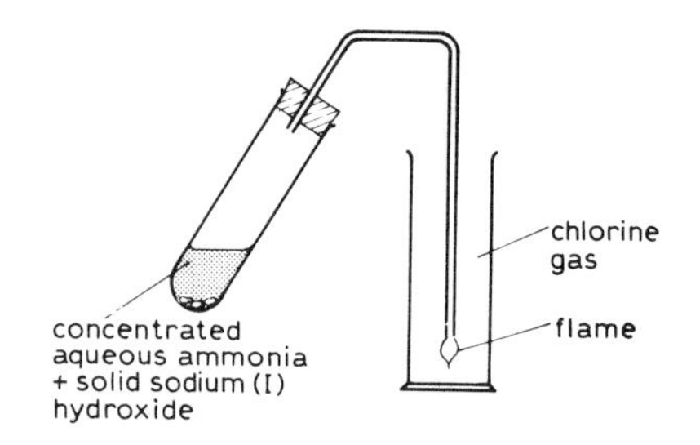

Fig. 24.12

forming ammonium chloride. This is an acid-base (*proton transfer*) reaction:

$$NH_3(g) + HCl(g) \rightarrow (NH_4^+ + Cl^-)(s)$$

These successive reactions can be demonstrated neatly, using the apparatus of Fig. 24.12. For success, the ammonia must be issuing rapidly from the tube before it is lowered quickly into the jar, which must be *full* of chlorine.

4. as a reducing agent

Reactions (1), (2) and (3) involve breaking the relatively weak N–H bonds in ammonia and forming stronger bonds in the products; the reactions are *exothermic*. Some of the energy released may be utilised in removing oxygen from oxides – ammonia can act as a ***reductant***, as exemplified by the reaction described in section 24.3:

$$2NH_3(g) + 3CuO(s) \rightarrow 3Cu(s) + N_2(g) + 3H_2O(g)$$

5. as a base

Ammonia molecules will accept protons, because of the **lone pair of electrons** in the molecule of ammonia (see section 18.19). Thus ammonia gas acts as a **base**. With hydrogen chloride gas, white fumes of solid ammonium chloride are formed:

$$NH_3(g) + HCl(g) \rightarrow (NH_4^+ + Cl^-)(s)$$

A similar reaction occurs when ammonia dissolves in water; in this case water acts as an acid:

$$NH_3(g) + H_2O(l) \rightleftharpoons NH_4^+(aq) + OH^-(aq)$$

24.9 reactions of aqueous ammonia ('ammonium hydroxide')

Aqueous ammonia can behave in these distinct ways:

i) *as an alkali:* it will neutralise acids to form salts,

$$OH^-(aq) + H^+(aq) \rightarrow H_2O(l)$$

but its heat of neutralisation is lower than that for strong bases (Fig. 20.11).

ii) *as a precipitant for hydroxides:* addition of aqueous ammonia to aqueous solutions containing certain cations, precipitates insoluble hydroxides (section 12.29(d)):

$$Pb^{2+}(aq) + 2OH^-(aq) \rightarrow Pb(OH)_2(s)$$

However, aqueous ammonia will *not* precipitate calcium(II) hydroxide from an aqueous solution containing calcium(II) ions, whereas aqueous sodium(I) hydroxide, of the same concentration, will.

iii) *as a complexing agent*, or ligand: certain hydroxide precipitates redissolve in excess aqueous ammonia (section 12.29(4)):

$$Cu(OH)_2(s) + 4NH_3(aq) \rightarrow \underset{\text{deep blue}}{[Cu(NH_3)_4]^{2+}(aq)} + 2OH^-(aq)$$

Addition of alcohol to the deep blue solution precipitates a deep blue solid, which can be shown to be identical with the solid obtained in the experiment described in section 24.7 using ammonia gas. The formation of such complexes results from the formation of **coordinate bonds** between ammonia molecules and cations in the solution; the lone pair of electrons is **donated** to the cation (Fig. 24.13).

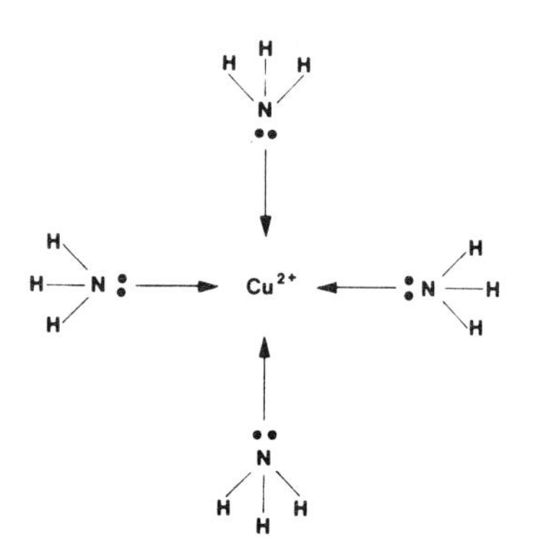

Fig. 24.13

Careful consideration of reactions (i), (ii) and (iii) indicates that aqueous ammonia is only **partially ionised** and consists of an equilibrium mixture of unionised molecules and ions. It is said to be a **weak base.**

$$NH_3(aq) + H_2O(l) \rightleftharpoons NH_4^+(aq) + OH^-(aq)$$

24.10 ammonium salts

The direct formation of ammonium chloride from the gases ammonia and hydrogen chloride affords interesting information about the relative rates of diffusion of the two gases, which can in turn be related to their relative molecular masses. If cotton wool plugs are soaked in concentrated aqueous ammonia and concentrated hydrochloric acid respectively and placed simultaneously in the apparatus shown in Fig. 24.14, the ring of white solid ammonium chloride forms in the position shown.

The ammonium cation, NH_4^+, present in ammonium salts is readily decomposed on heating, hence the easy thermal dissociation of ammonium chloride (section 23.10). Many other ammonium salts sublime, notably ammonium carbonate, which sublimes at room temperature and is therefore used in 'smelling salts'. All ammonium salts are soluble in water, due to the low binding forces between the ions in the solid crystals.

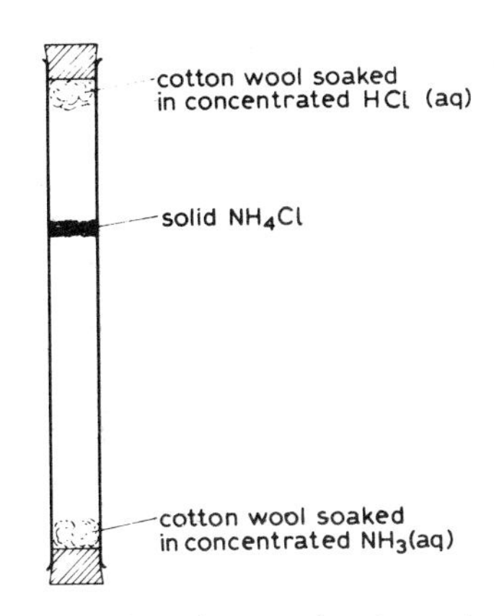

Fig. 24.14 plan view–the tube must be clamped horizontally

24.11 manufacture of ammonia: the Haber Process

All ammonium salts and all nitrates(v) are water-soluble and sufficiently reactive to be absorbed by plants and utilised in the synthesis of aminoacids and proteins. The supply of ammonium salts and nitrates(v) from natural deposits is not inexhaustible and a method had to be found for converting the inactive nitrogen in the air into such compounds. The solution to the problem emerged with the development of the **Haber Process,** which is a method of synthesising ammonia from nitrogen and hydrogen. The route from these elements to nitrogenous fertilisers is shown in Fig. 24.15.

The process illustrates the application of fundamental principles of chemistry, to maximise the yield of desired product in the equilibrium mixture produced in the reversible reaction:

$$N_2(g) + 3H_2(g) \rightleftharpoons 2NH_3(g); \quad \Delta H = -92\,kJ$$

Application of Le Chatelier's Principle (section 23.9) indicates that the conditions which will produce the maximum yield of ammonia in the equilibrium mixture are (*i*) high pressure, (*ii*) low temperature. However, a low temperature slows down the *rate at which equilibrium is attained* and it is found to be more economical to employ a temperature of about 500 °C, producing a reduced yield of ammonia rapidly. Even at this temperature, the rate at which equilibrium is established must be increased by using a catalyst, usually of finely divided iron.

The gases are compressed to about 200 atmospheres and are passed over the heated catalyst; the issuing gases are cooled, still under pressure, until the ammonia liquefies. The unchanged nitrogen and hydrogen are then recirculated.

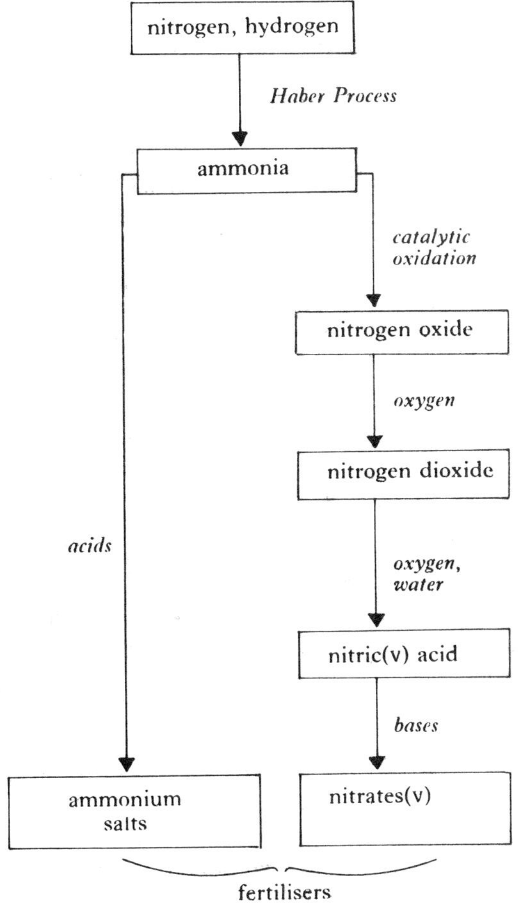

Fig. 24.15

24.12 preparation of nitric(V) acid

Nitric(V) acid is a volatile liquid acid and can therefore be prepared by the action of the less volatile concentrated sulphuric(VI) acid on a nitrate(v). In this reaction, nitrate(v) ion acts as a base, accepting a proton from sulphuric(VI) acid:

$$NO_3^-(s) + H_2SO_4(l) \longrightarrow HSO_4^-(s) + HNO_3(g)$$

At the temperature at which the reaction is carried out, the nitric(v) acid escapes as vapour and is subsequently condensed to a liquid product on cooling. Since nitric(v) acid is very corrosive, an all-glass apparatus must be used, as shown in Fig. 24.16. The liquid product is yellow, due to the presence of dissolved nitrogen dioxide, formed by thermal decomposition of some of the nitric(v) acid during the preparation. The pure acid is a colourless fuming liquid, boiling-point 86 °C (359 K) and density $1{\cdot}5\,g\,cm^{-3}$.

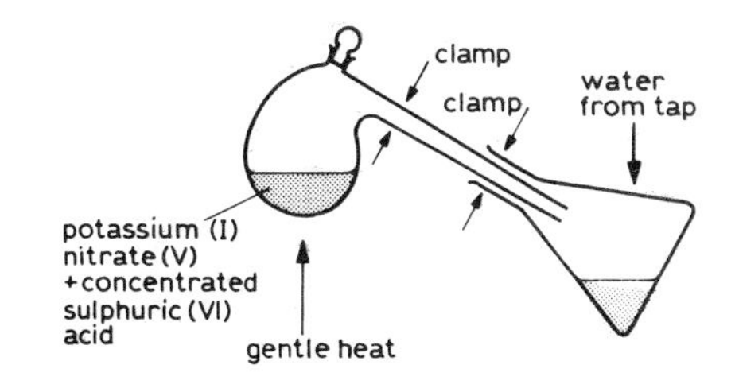

Fig. 24.16 preparation of nitric(V) acid

24.13 reactions of nitric(V) acid

Before embarking on a detailed study of this very important compound, it is advisable to see how it is related to other nitrogen compounds. Nitrogen exists in a variety of oxidation states, from −3 in ammonia to +5 in nitric(v) acid; the principal compounds of nitrogen, showing their interconversion by oxidation (o) and reduction (r), are depicted in Fig. 24.17.

$NH_3 \leftrightarrow NH_4^+$ *−3*

o ⇓⇑ *r*

N_2 *0*

o ⇓⇑ *r*

NO *+2*

o ⇓⇑ *r*

$HNO_2 \leftrightarrow NO_2^-$ *+3*

o ⇓⇑ *r*

$NO_2 \leftrightarrow N_2O_4$ *+4*

o ⇓⇑ *r*

$HNO_3 \leftrightarrow NO_3^-$ *+5*

Fig. 24.17 principal compounds of nitrogen and their oxidation states

Nitric(v) acid is readily reduced, usually to either nitrogen dioxide or nitrogen oxide, and is a **powerful oxidising agent (oxidant).**

1. *thermal decomposition of nitric(V) acid*

A little pure nitric(v) acid is placed in the bottom of a test-tube, which is then two-thirds filled with *dry* glass beads or pieces of broken pot. The delivery tube should be arranged to dip just below the surface of the water in the beaker, as shown in Fig. 24.18. The beads are heated strongly, then the flame is directed on to the acid, which vaporises and passes over the hot solid. Dense brown fumes are evident in the tube, but the gas collected over water is colourless; it rekindles a glowing splint and proves to be oxygen.

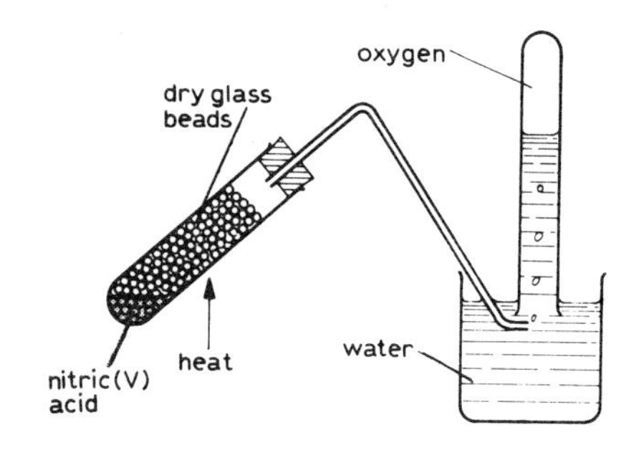

Fig. 24.18 thermal decomposition of nitric(V) acid

$$4HNO_3(g) \xrightarrow{\text{heat}} \underset{\text{(brown)}}{4NO_2(g)} + O_2(g) + 2H_2O(g)$$

The brown nitrogen dioxide is freely soluble in water, giving a mixture of nitrate(v) and nitrate(III) ions:

$$NO_2 + H_2O \longrightarrow NO_3^- + 2H^+ + e^-$$
$$NO_2 + e^- \longrightarrow NO_2^-$$
$$\mathbf{2NO_2(g) + H_2O(l) \longrightarrow \underset{\text{nitrate(v) ion}}{NO_3^-(aq)} + \underset{\text{nitrate(III) ion}}{NO_2^-(aq)} + 2H^+(aq)}$$

This reaction is an example of **disproportionation,** in which a substance in one oxidation state (in this case +4) undergoes oxidation and reduction, forming substances in higher and lower oxidation states (in this case +5 and +3).

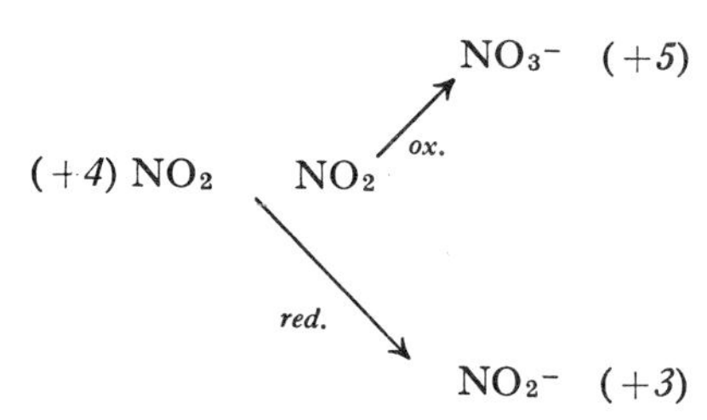

The instability of nitric(v) acid relative to nitrogen dioxide illustrated in this experiment means that nitric(v) acid is a powerful oxidant.

2. *action of nitric(V) acid on non-metals*

Non-metals are oxidised by concentrated nitric(v) acid *to their highest oxides*; carbon to carbon dioxide, sulphur to sulphur(VI) oxide or sulphate(VI) ion, phosphorus to phosphate(v) ion; in each case the reduction of the nitric(v) acid is indicated by the appearance of brown fumes.

a) *carbon*

Warm charcoal gives an immediate reaction with concentrated nitric(v) acid, but the issuing gases fail to turn limewater milky. Can you explain this? If the experiment is repeated using the apparatus shown in Fig. 24.19, the limewater does turn milky, indicating that the carbon has been oxidised to carbon dioxide:

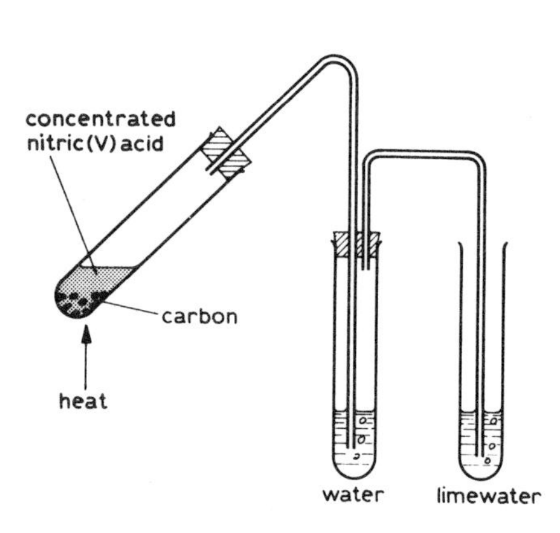

Fig. 24.19 identification of oxidation product of carbon in reaction with nitric(V) acid

$$C(s) + 4HNO_3(l) \longrightarrow CO_2(g) + 4NO_2(g) + 2H_2O(l)$$

b) *sulphur*

Brown fumes are evolved if concentrated nitric(v) acid is added to warm sulphur. If the residue is diluted and filtered, addition of aqueous barium(II) chloride to the filtrate gives a white precipitate, indicating the presence of sulphate(VI) ion:

$$S + 4H_2O \rightarrow SO_4^{2-} + 8H^+ + 6e^-$$
$$[NO_3^- + 2H^+ + e^- \rightarrow NO_2 + H_2O] \times 6$$

reducing to:

$$\mathbf{S + 6NO_3^- + 4H^+ \rightarrow SO_4^{2-} + 6NO_2 + 2H_2O}$$

subsequently:

$$Ba^{2+}(aq) + SO_4^{2-}(aq) \rightarrow BaSO_4(s)$$

c) *phosphorus*

2 or 3 cm^3 of pure nitric(V) acid is boiled in a large flask until the vapour fills the flask. The source of heat is removed and a deflagrating spoon containing red phosphorus is lowered into the vapour (Fig. 24.20). Reaction occurs with a spectacular orange-red glow, which is especially impressive if carried out in a darkened room.

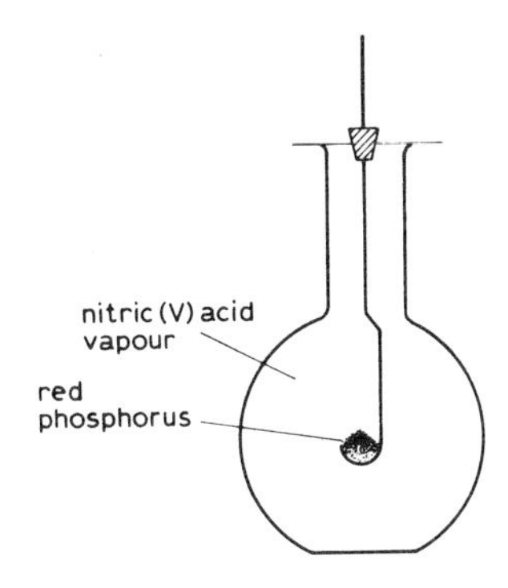

Fig. 24.20 action of phosphorus on nitric(V) acid

$$P + 4H_2O \rightarrow PO_4^{3-} + 8H^+ + 5e^-$$
$$[NO_3^- + 2H^+ + e^- \rightarrow NO_2 + H_2O] \times 5$$

reducing to:

$$\mathbf{P + 5NO_3^- + 2H^+ \rightarrow PO_4^{3-} + 5NO_2 + H_2O}$$

3. oxidation of compounds by nitric(V) acid

a) *sawdust*

If a little pure nitric(V) acid is added to hot sawdust on a sand-tray, the sawdust ignites and copious brown fumes of nitrogen dioxide are evolved. Sawdust contains much carbon and hydrogen and these are oxidised to carbon dioxide and water. *Several seconds may elapse before ignition occurs*; the reaction has an '*induction period*'. Concentrated nitric(V) acid attacks wood, paper, corks and rubber by oxidising these materials in a manner similar to its attack on sawdust.

b) *hydrogen sulphide*

A few drops of concentrated nitric(V) acid, added to a jar of hydrogen sulphide, produces a precipitate of sulphur and the familiar brown fumes:

$$H_2S \rightarrow S + 2H^+ + 2e^-$$
$$[NO_3^- + 2H^+ + e^- \rightarrow NO_2 + H_2O] \times 2$$

reducing to:

$$\mathbf{H_2S(g) + 2HNO_3(l) \rightarrow S(s) + 2NO_2(g) + 2H_2O(l)}$$

c) *iodide ion*

If concentrated nitric(V) acid is added to an aqueous solution of potassium iodide, iodine is liberated. The temperature at which the reaction is carried out gives an illustration of the effect of temperature on the rate of a reaction (see also section 23.5(4)). At 20 °C a yellow solution is formed, at 40 °C a brown solution and at 80 °C purple iodine vapour fills the flask:

$$I^- \rightarrow \tfrac{1}{2}I_2 + e^-$$
$$NO_3^- + 2H^+ + e^- \rightarrow NO_2 + H_2O$$
$$\mathbf{I^-(aq) + NO_3^-(aq) + 2H^+(aq) \rightarrow \tfrac{1}{2}I_2(g) + NO_2(g) + H_2O(l)}$$

4. oxidation of metals by nitric(V) acid

Only in the case of magnesium and very dilute nitric(V) acid is hydrogen liberated from the acid by a metal:

$$Mg(s) + 2H^+(aq) \rightarrow Mg^{2+}(aq) + H_2(g)$$

In all other cases, the metal is oxidised to its aquated cation by the nitrate(V) ion rather than by $H^+(aq)$. The reduction product of the nitric(V) ion depends

i) on the concentration of the acid,
ii) on the metal employed.

Most metals reduce concentrated nitric(v) acid to **nitrogen dioxide,**

e.g. copper: $Cu \rightarrow Cu^{2+} + 2e^-$

$$[NO_3^- + 2H^+ + e^- \rightarrow NO_2 + H_2O] \times 2$$

$$\mathbf{Cu(s) + 2NO_3^-(aq) + 4H^+(aq) \rightarrow Cu^{2+}(aq) + 2NO_2(g) + 2H_2O(l)}$$

Consequently nitric(v) acid will dissolve metals below hydrogen in the electrochemical series; only gold and platinum will not react.

Dilute nitric(v) acid is reduced (when hot) by metals fairly low in the electrochemical series to **nitrogen oxide,**

e.g. copper: $[Cu \rightarrow Cu^{2+} + 2e^-] \times 3$

$$[NO_3^- + 4H^+ + 3e^- \rightarrow NO + 2H_2O] \times 2$$

$$\mathbf{3Cu(s) + 2NO_3^-(aq) + 8H^+(aq) \rightarrow 3Cu^{2+}(aq) + 2NO(g) + 4H_2O(l)}$$

More reactive metals may give N_2O, N_2 or even NH_4^+.

5. as a strong acid

Dilute nitric(v) acid is completely ionised and behaves as a strong monobasic acid. Such a solution does not have oxidising properties unless it is hot, or in contact with a powerful reductant.

24.14 manufacture and uses of nitric(V) acid

Nitric(v) acid is manufactured by the catalytic oxidation of ammonia (section 24.8(2)) obtained from the Haber Process (section 24.11). It is used for making fertilisers and explosives (e.g. TNT) and is a valuable reagent in the dyestuff and pharmaceutical industries.

24.15 nitrates(V)

a) All nitrates(v) are soluble in water.
b) All nitrates(v) react with concentrated sulphuric(vi) acid on heating, liberating nitric(v) acid.
c) All nitrates(v) decompose on heating, the extent of the decomposition depending on the position of the metal in the electrochemical series, e.g.

$$KNO_3 \rightarrow KNO_2 + \tfrac{1}{2}O_2 \quad (\textit{also Na})$$

$$Mg(NO_3)_2 \rightarrow MgO + 2NO_2 + \tfrac{1}{2}O_2 \quad (\textit{also Al, Zn, Fe, Sn, Pb, Cu})$$

$$AgNO_3 \rightarrow Ag + NO_2 + \tfrac{1}{2}O_2 \quad (\textit{also Hg})$$

Ammonium nitrate(v) decomposes explosively on heating, giving dinitrogen oxide:

$$NH_4NO_3(s) \rightarrow N_2O(g) + 2H_2O(g)$$

d) Nitrates(v) cannot be identified by a precipitation reaction, as can chlorides or sulphates(vi), since they are all soluble. Their presence is detected by the **brown ring test:**

To an aqueous solution of the suspected nitrate(v), dilute sulphuric(vi) acid and aqueous iron(ii) sulphate(vi) are added, followed by concentrated sulphuric(vi) acid, carefully, so that it forms a dense lower layer in the test-tube. If a nitrate(v) is present, a brown ring forms at the junction between the liquids.

The brown substance is formed by combination of iron(ii) ions with nitrogen oxide which is obtained by reduction of nitrate(v) ions; it may be formulated as $[Fe(NO)]^{2+}$:

$$NO_3^- + 4H^+ + 3e^- \rightarrow NO + 2H_2O$$

$$[Fe^{2+} \rightarrow Fe^{3+} + e^-] \times 3$$

$$NO_3^- + 4H^+ + 3Fe^{2+} \rightarrow NO + 3Fe^{3+} + 2H_2O$$

then: $Fe^{2+} + NO \rightarrow [Fe(NO)]^{2+}$

overall:

$$\mathbf{NO_3^-(aq) + 4H^+(aq) + 4Fe^{2+}(aq) \rightarrow [Fe(NO)]^{2+}(aq) + 3Fe^{3+}(aq) + 2H_2O(l)}$$

The concentrated sulphuric(VI) acid absorbs water from the aqueous solution above it; in doing so it *concentrates* the dilute nitric(V) acid in the aqueous solution and *heats it up*, enabling it to oxidise iron(II) ions and in turn be reduced to nitrogen oxide.

24.16 the element nitrogen

Nitrogen is a colourless, odourless gas which is so unreactive that it is of little interest. It neither burns, nor will it support combustion–except that of magnesium and calcium, which supply sufficient energy to break the strong bond holding the nitrogen atoms in the molecule. The product is magnesium or calcium nitride:

$$3Mg(s) + N_2(g) \rightarrow Mg_3N_2(s)$$

On boiling the resulting solid with aqueous alkali, ammonia is evolved and can be identified by its action on litmus paper:

$$Mg_3N_2(s) + 6H_2O(l) \rightarrow 3Mg(OH)_2(s) + 2NH_3(g)$$

This series of reactions provides the only positive test for nitrogen.

Repeated sparking of a mixture of nitrogen and oxygen provides sufficient energy to convert it to nitrogen oxide:

$$N_2(g) + O_2(g) \rightleftharpoons 2NO(g)$$

This was the first step in an older process for the manufacture of nitric(V) acid, but it has now been superseded by the catalytic oxidation of ammonia.

Under appropriate conditions (section 24.11), nitrogen will combine with hydrogen forming ammonia.

Nitrogen may be prepared by heating ammonium nitrate(III); this is a redox reaction in which ammonium ion and nitrate(III) ion both form nitrogen:

$$NH_4^+ \rightarrow \tfrac{1}{2}N_2 + 4H^+ + 3e^-$$

$$NO_2^- + 4H^+ + 3e^- \rightarrow \tfrac{1}{2}N_2 + 2H_2O$$

$$\mathbf{(NH_4^+ + NO_2^-)(s) \rightarrow N_2(g) + 2H_2O(l)}$$

The gas is obtained industrially by the fractional distillation of liquid air (see section 6.2); it is used in the manufacture of ammonia; in the liquid form as a coolant; and as an unreactive gas to provide an inert atmosphere for reactions which must be carried out in the absence of oxygen.

24.17 the nitrogen cycle

The importance of nitrogen in living materials has already been emphasised. There are some natural processes by which nitrogen is conserved in nature, but the cycle is imperfect–especially when it is interfered with by civilised man, who conveys nitrogen-rich sewage to places where it cannot readily be returned to the soil. The changes involved in the natural cycle and the artificial processes added by man are summarised in Fig. 24.21.

revision summary: Chapter 24

ammonia:	*preparation:*	heat ammonium salt with alkali
		$NH_4^+(s) + OH^-(s) \rightarrow NH_3(g) + H_2O(g)$
	dry:	CaO (**not** H_2SO_4 or $CaCl_2$)
	collection:	downward displacement of air
	tests:	turns moist red litmus paper blue
		fumes with hydrogen chloride

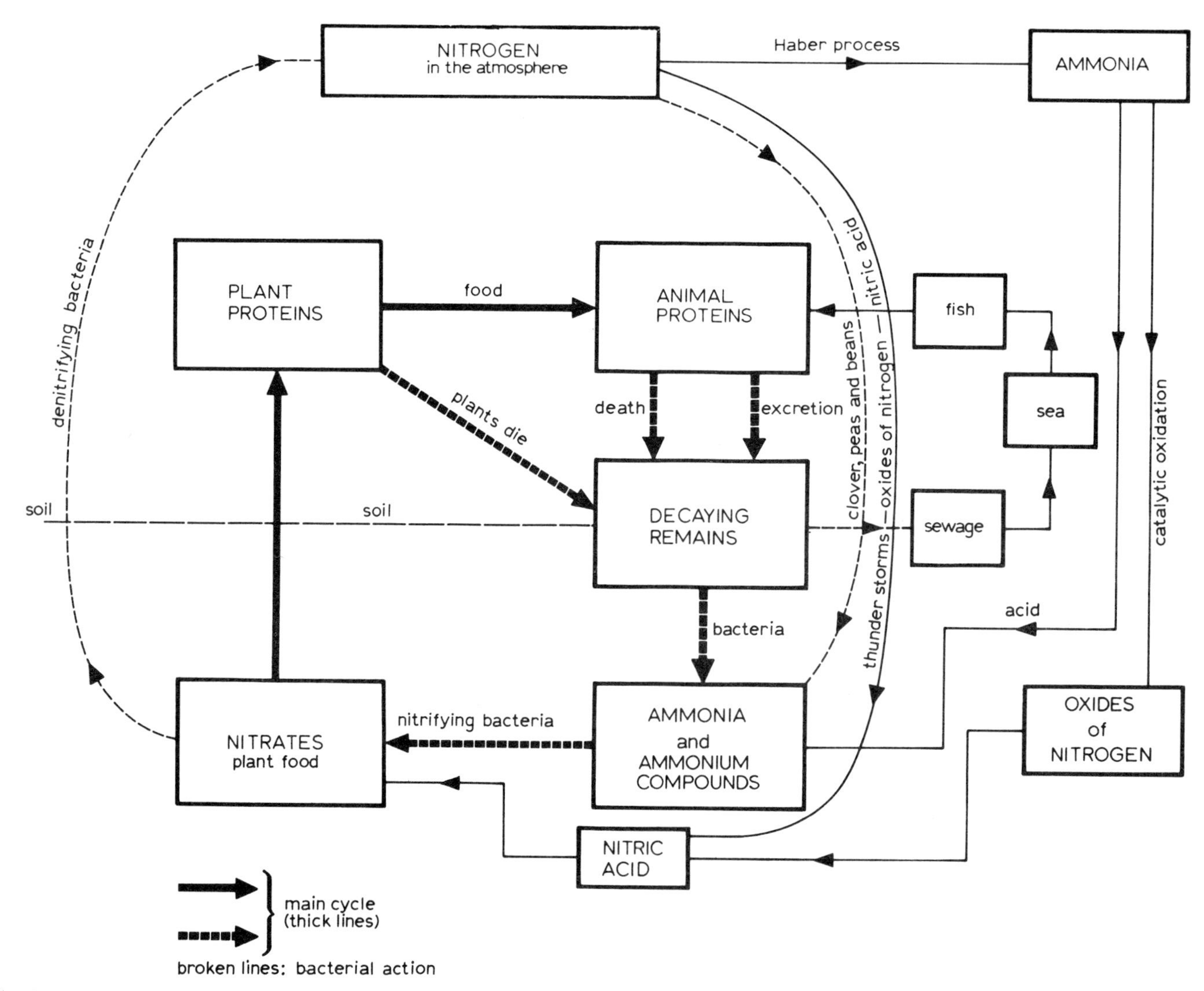

Fig. 24.21 the nitrogen cycle

	properties:	very soluble burns in oxygen and chlorine reductant only common alkaline gas
aqueous ammonia:		weak base $NH_3(aq) + H_2O(l) \rightleftharpoons NH_4^+(aq) + OH^-(aq)$
manufacture of ammonia:		Haber Process $N_2(g) + 3H_2(g) \rightleftharpoons 2NH_3(g)$ high pressure, intermediate temperature, catalyst
nitric(v) acid:	*preparation:*	conc. H_2SO_4 heated gently with nitrate(v) $NO_3^-(s) + H_2SO_4(l) \rightarrow HSO_4^-(s) + HNO_3(g)$
	properties:	thermal decomposition $\rightarrow NO_2, O_2$ (dilute) strong, monobasic acid oxidant when cold, concentrated ($\rightarrow NO_2$) or hot, dilute (usually $\rightarrow NO$)
	manufacture:	catalytic oxidation of ammonia

questions 24

1. Outline briefly how a sample of nitrogen can be obtained from (a) the air, (b) an ammonium salt. What difference would there be between the two samples so prepared?

Describe in outline the industrial process by which nitrogen is converted into ammonia. The reaction involved in this process is reversible. Indicate **three** ways in which the manufacturer makes sure he gets the best possible yield of ammonia.

State **two** industrial uses of ammonia. (S)

2. Describe and give the equation for a reaction by which ammonia can be prepared in the laboratory.

Ammonia is lighter than air, very soluble in water, forming an alkaline solution, and it reacts with calcium(II) chloride. How are these facts taken into account when collecting a dry sample of the gas in the laboratory?

A bottle in the laboratory is labelled 'Liquid Ammonia ·880'. What is the significance of the figures, and why are the words misleading?

Explain what happens when ammonia is passed over heated copper(II) oxide. (S)

3. (a) Give in outline a method by which hydrogen is manufactured on a large scale, and state the conditions under which hydrogen reacts with nitrogen in the commercial synthesis of ammonia.

(b) By what reactions could you obtain (i) ammonia from ammonium chloride, (ii) nitrogen from ammonia?

Draw a labelled diagram of the apparatus you would use to prepare and collect **either** dry ammonia using reaction (i), or moist, but otherwise pure, nitrogen using reaction (ii). (C)

4. Give two physical properties of ammonia gas.

(a) When ammonia gas is brought into contact with gently heated copper(II) oxide, the ammonia is oxidised to nitrogen; write an equation for this reaction. Sketch an apparatus which you would use to carry out the reaction and collect substances produced; give one chemical reaction of each of the products.

(b) What happens when jars of ammonia and hydrogen chloride are brought together and the gases allowed to mix? State the visual change, name what is formed and give an equation for the reaction. Comment upon the type of chemical bonding existing in each of the substances concerned.

(c) What happens when a small quantity of ammonium chloride is placed at the bottom of a six-inch test tube and heated strongly? (O)

5. Describe, by means of labelled diagrams and brief explanations, how you would prove that ammonium chloride contains nitrogen by preparing from it a sample of the gas. State why none of the other reagents you employ in your experiments should contain nitrogen.

Contrast what is seen in each of the following pairs of experiments:

(a) Solid samples of sodium(I) chloride and ammonium chloride are heated in separate test-tubes.

(b) Experiment (a) is repeated using lead(II) nitrate(V) and potassium(I) nitrate(V).

(c) An aqueous solution of ammonia is added in turn to separate solutions of iron(II) and iron(III) salts. (W)

6. (a) Describe in outline the industrial synthesis of ammonia from nitrogen and hydrogen.

(b) Give the equation and conditions for a reaction in which ammonia is converted into nitrogen oxide and indicate briefly the industrial importance of this reaction.

(c) If you were provided with 100 cm^3 of 2M ammonia solution, describe how you would obtain from it a crystalline sample of ammonium sulphate(VI).

Calculate the maximum mass of ammonium sulphate(VI) that could be formed from this solution.

What is the importance of ammonium sulphate(VI)? (C)

7. (a) What effect does ammonia solution have on

(i) iron(III) chloride solution,

(ii) copper(II) sulphate(VI) solution,

(iii) precipitated silver(I) chloride?

(b) Ammonia is oxidised to nitrogen when in contact with hot copper(II) oxide; write an equation for this reaction and calculate what volume of nitrogen (at s.t.p.) could be produced by the oxidation of 6 g of ammonia. (O)

8. A farmer, wishing to buy fertiliser, finds that he can buy a ton of ammonium sulphate(VI) for the same cost as a ton of sodium(I) nitrate(V), both in a reasonable state of purity. Which is the more economical source of nitrogen?

Why are nitrogenous fertilisers added to the soil?

Describe briefly one method whereby the nitrogen of the air can be converted into a nitrogenous fertiliser.

In what way, other than by using chemical fertilisers, can the farmer replenish the nitrogen content of his soil?

9. Describe how nitric(V) acid is manufactured from ammonia.

State **two** commercial uses of nitric(V) acid.

Give **two** chemical reactions in which nitric(V) acid acts as an oxidising agent.

If an acid contains 2·13% hydrogen, 29·79% nitrogen and 68·08% oxygen, what is its empirical formula? (AEB)

10. Describe and explain what you would observe during the following experiments.

(a) (i) Lead(II) nitrate(V) is heated.

(ii) Ammonium chloride is heated.

(b) (i) Sodium(I) hydroxide solution is added to a solution of lead(II) nitrate(V) until no further change occurs.

(ii) Sodium(I) hydroxide solution is warmed with ammonium chloride.

(c) Solutions of lead(II) nitrate(V) and ammonium chloride are mixed. More water is added, the mixture boiled, and allowed to cool. (AEB)

11. The nitrate(V) of an unknown metal decomposes when heated to form oxygen and a white solid only. The white solid evolves nitrogen dioxide when treated with hydrochloric acid. The atomic number of the element lies between 31 and 38 (*inclusive*).

From your knowledge of the general behaviour of nitrates(V) and the periodic table deduce (giving your reasons) the name of the metal.

sulphur compounds

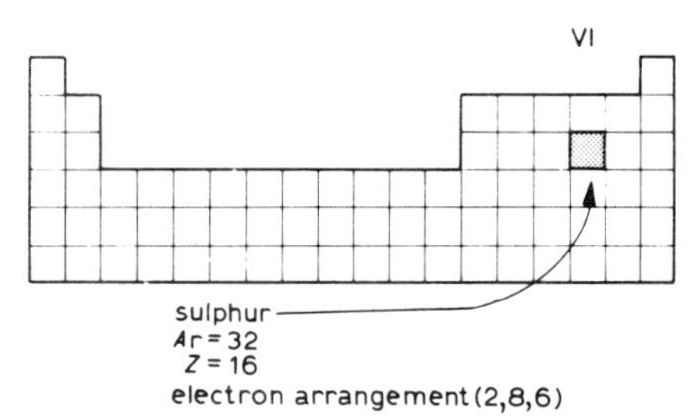

Fig. 25.1 place of sulphur in Periodic Table

	M_r
Cl_2	71
Br_2	160
I_2	254
S_n	?

Fig. 25.2

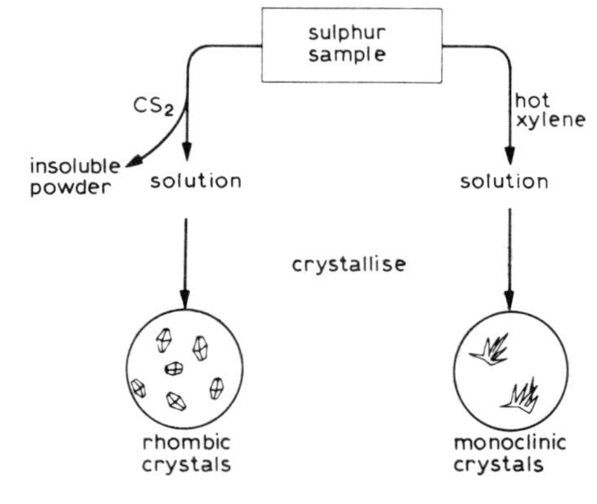

Fig. 25.3

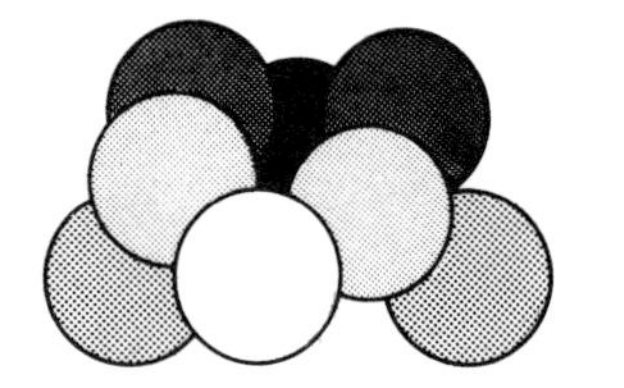

Fig. 25.4 S_8 molecule, showing puckered ring

Sulphur is a substance which has been known for many thousands of years as a yellow material found in nature, in and around volcanoes. It has long been recognised as an element and its position in the Periodic Table is shown in Fig. 25.1; this position reminds us that it is a **non-metal** – it will form salts with metals, a gaseous hydride, acidic oxides and a liquid chloride (see section 7.10). The arrangement of electrons in the sulphur atom indicates that sulphur is likely to form ions, S^{2-}, and compounds in which the sulphur atom forms two covalent bonds; it is also likely to form covalent bonds with oxygen and it can exist in various states of oxidation. Of its many compounds, sulphuric(VI) acid is one of the most useful chemical substances known.

25.1 physical forms of the element

1. *solid, liquid and gaseous phases*

If crushed roll sulphur is heated gently and evenly, it melts to a pale yellow liquid at about 114 °C. Further heating of the liquid causes it eventually to boil to vapour at 444 °C. These observations indicate that sulphur is a **molecular** solid and that the molecules must contain more than one or two atoms – compare these physical properties with those of chlorine, bromine and iodine (see sections 15.12 and 15.15): which of these halogens does sulphur most closely resemble in its physical properties? Can you make an intelligent guess for the value of n in S_n, by reference to the values in Fig. 25.2?

2. *solid sulphur*

The yellow powder found in volcanoes, or obtained from condensing sulphur vapour on a cold surface, is called **'flowers of sulphur'.** It is undoubtedly one substance, since it burns to form only one product, sulphur dioxide. But only part of it dissolves in the solvent carbon disulphide; this part recrystallises in glossy yellow diamond-shaped crystals **(rhombic sulphur)** leaving an insoluble powder. Even more surprisingly, use of hot xylene as a recrystallising solvent leads to different crystals, which are yellow but needle-shaped **(monoclinic sulphur).**

The problem is now easily solved using X-ray analysis of the crystals. Rhombic sulphur contains rings of eight sulphur atoms arranged in a crystal lattice. If this is heated to 96 °C, it slowly changes to monoclinic sulphur, which contains the same rings of eight atoms arranged differently in the lattice. 96 °C is called the **transition temperature;** any crystals of sulphur prepared below this temperature have the rhombic shape, but those prepared above it have the monoclinic shape. The insoluble powdery form of sulphur is **amorphous,** having no structural regularity.

This ability of an element to exist in different physical forms within the same physical state (solid) is called allotropy; the different forms are called allotropes.

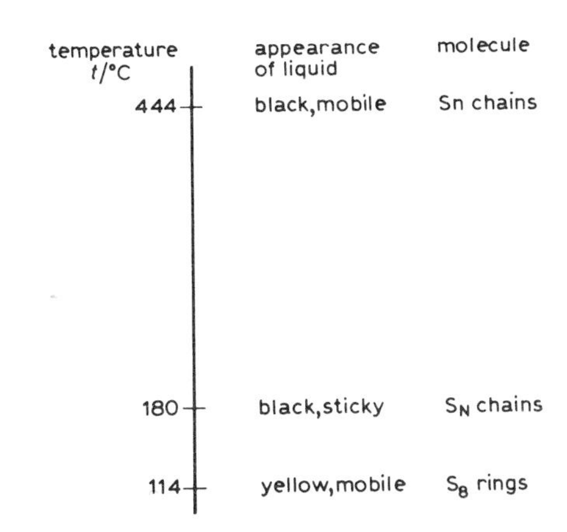

Fig. 25.5 composition of liquid sulphur ($N \approx 1000$, $n \approx 4$–10)

3. liquid sulphur

If sulphur is heated slowly from its melting-point at 114 °C it changes from a mobile, amber liquid to a dark, sticky mass at 180 °C. This becomes *less* viscous as the temperature rises to the boiling-point at 444 °C.

The different physical properties of liquid sulphur indicate changes in the **molecular state** of sulphur as it is heated; the surprising increase in viscosity on heating from 114 °C to 180 °C argues an increase in intermolecular bonding. The effect of heat is to disrupt the S_8 rings; the resulting S_8 *chains* combine to form chains containing hundreds of sulphur atoms, the chains being of maximum length at about 180 °C. Further heating above 180 °C breaks down the chains, with a corresponding decrease in the viscosity of the liquid. The composition of sulphur vapour varies from S_8 to S_2 units.

4. plastic sulphur

If the brown mobile liquid obtained by heating sulphur to between 200 °C and 300 °C is poured into cold water, a rubbery solid is formed (plastic sulphur). This can be drawn into strands of considerable tensile strength. Plastic sulphur contains long helical (or spiral) chains of sulphur atoms, with eight atoms to each cycle of the helix. Plastic sulphur slowly becomes brittle on standing, as it reverts to rhombic sulphur.

5. a simple model for sulphur molecules

The bond length S–S in sulphur molecules is about 2×10^{-10} m (2·0 Å); the bond angle is about 110° (*why?*). Fortunately, most pipe-cleaners are about 16 cm long and can therefore be bent easily into eight lengths of 2 cm each, forming angles of 110°. The construction of a number of such simple models enables the stability of S_8 rings and their conversion into spiral chains to be readily appreciated.

25.2 reactions of sulphur

1. with air and oxygen

Sulphur burns in air with a blue flame and does so more brightly in oxygen, forming sulphur dioxide:

$$S(s) + O_2(g) \rightarrow SO_2(g)$$

Rhombic and monoclinic sulphur can be shown to be allotropic forms of the same element by this reaction: 32 g of each form will give exactly 64 g of sulphur dioxide.

2. with metals

Powdered zinc and sulphur react violently when ignited (see section 2.4); iron and sulphur ignite when heated strongly and hot copper foil glows red-hot in sulphur vapour. In each case a sulphide is formed:

$$Zn(s) + S(s) \rightarrow ZnS(s)$$
$$Fe(s) + S(s) \rightarrow FeS(s)$$
$$Cu(s) + S(g) \rightarrow CuS(s)$$

3. with hydrogen

If hydrogen is bubbled through boiling sulphur, the issuing gas smells of bad eggs, due to partial reaction forming hydrogen sulphide:

$$H_2(g) + S(l) \rightleftharpoons H_2S(g)$$

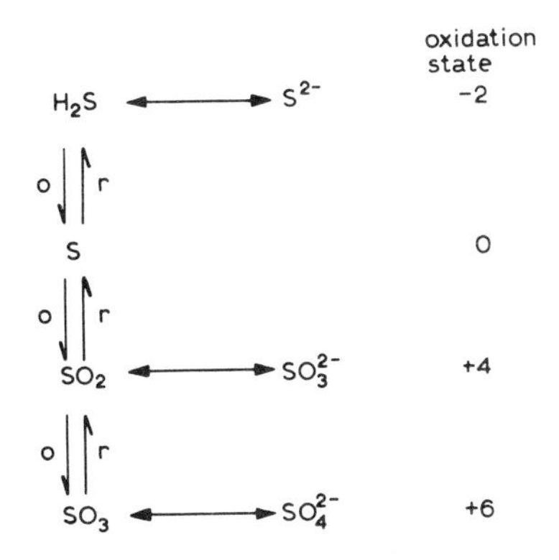

Fig. 25.6 oxidation states of common sulphur compounds

4. oxidation states of sulphur

The most common compounds of sulphur, showing their oxidation states (which vary from -2 to $+6$) are shown in Fig. 25.6. The interconversions are shown as oxidations (o) or reductions (r).

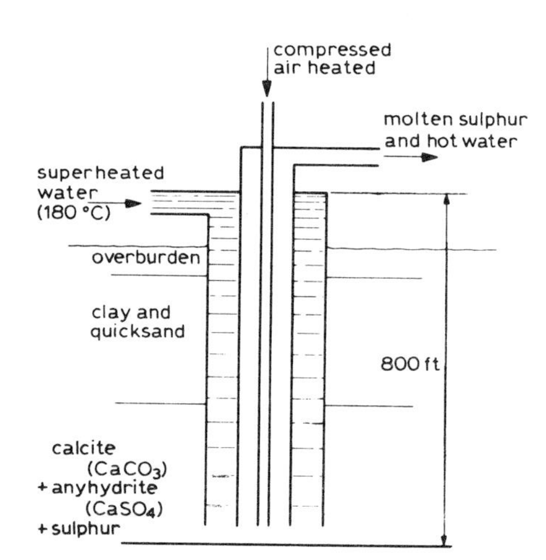

Fig. 25.7 Frasch process

25.3 extraction of naturally occurring sulphur

Sulphur occurs naturally in underground beds in Texas and Poland. It is overlain by quicksand and porous rock and cannot be mined conventionally; it is extracted most cheaply by a special process called the **Frasch process.** A hole is bored down to the sulphur, in which coaxial tubes are laid, as shown in Fig. 25.7. These carry superheated water at 180°C under pressure, to melt the sulphur; compressed air, to force the molten sulphur to the surface; and finally the ascending froth of sulphur, water and air. The product is run into reservoirs to settle and set. Recently, the technology needed for transporting sulphur as a hot liquid has been mastered. In this form it is easy to handle and shipments of it from the USA are displacing pyrites in Europe as a source of sulphur.

25.4 preparation and collection of hydrogen sulphide

Like oxide ions, sulphide ions are quite strong bases; they will accept protons from acids to form the gas hydrogen sulphide:

$$S^{2-}(s) + 2H^+(aq) \longrightarrow H_2S(g)$$

Many metal sulphides will react with dilute hydrochloric acid, liberating hydrogen sulphide in this way; in some cases (e.g. aluminium(III) sulphide), water itself will suffice as the acid:

$$S^{2-}(s) + 2H_2O(l) \longrightarrow H_2S(g) + 2OH^-(aq)$$

Iron(II) sulphide is a convenient sulphide to use in the laboratory preparation of the gas; bearing in mind how iron(II) sulphide is likely to be made, which other gas would you expect to be present as impurity in the hydrogen sulphide so formed? An apparatus suitable for the collection of jars of the gas is shown in Fig. 25.8, though **Kipp's apparatus** (Fig. 8.14) charged with iron(II) sulphide and hydrochloric acid may be used in preference, especially in view of the odious and poisonous nature of the gas. The gas is collected by **upward displacement of air** and the jar can be shown to be full when lead(II) acetate paper, held as shown, turns black–the usual test for the gas.

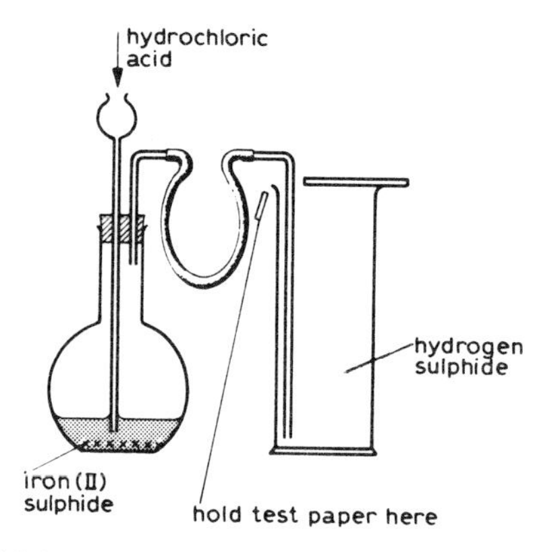

Fig. 25.8 preparation and collection of hydrogen sulphide

25.5 physical properties of hydrogen sulphide

The gas is colourless and smells strongly of bad eggs–this is as well, for it is intensely poisonous and the nauseating smell can act as a warning. An atmosphere polluted by hydrogen sulphide can be freshened by sprinkling dilute aqueous bromine on the floor (see section 25.6(2)). It is fairly soluble in water giving a weakly acidic solution. It is denser than air (R.V.D. 17, compared with 14·4 for air).

25.6 reactions of hydrogen sulphide

1. oxidation

If a lighted splint is plunged into a jar of hydrogen sulphide, the splint is extinguished but the gas burns in air with a blue flame, leaving a pale yellow deposit of sulphur (Fig. 25.9):

$$H_2S(g) + \tfrac{1}{2}O_2(g) \longrightarrow H_2O(g) + S(s)$$

Hydrogen sulphide ignited in excess oxygen is oxidised further, to sulphur dioxide:

$$H_2S(g) + 1\tfrac{1}{2}O_2(g) \longrightarrow H_2O(g) + SO_2(g)$$

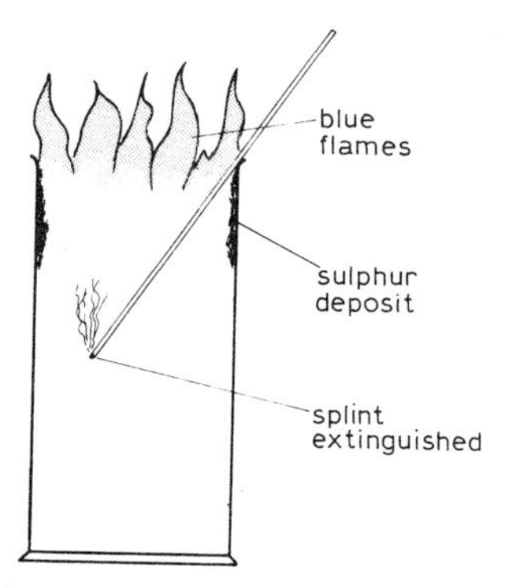

Fig. 25.9 hydrogen sulphide burns, but does not support combustion

2. as a reducing agent (reductant)

Because of its ready oxidation, hydrogen sulphide is a good reductant and its action as a reducing agent is made apparent by the formation of a deposit of sulphur. The half-equation for the oxidation of hydrogen sulphide:

$$H_2S \longrightarrow 2H^+ + S + 2e^-$$

may be combined with the half-equation for the following reductions, which it causes to occur.

a) *aqueous iron*(III) *ions to iron*(II) *ions*,

$$Fe^{3+}(aq) + e^- \longrightarrow Fe^{2+}(aq)$$

(note the difference between this action and the precipitation of insoluble sulphides of other cations, part (4) of this section)

b) *aqueous bromine to bromide ions*,

$$\underset{\text{orange}}{\tfrac{1}{2}Br_2(aq)} + e^- \longrightarrow \underset{\text{colourless}}{Br^-(aq)}$$

c) *aqueous acidic manganate*(VII) *to manganese*(II) *ions*,

$$\underset{\text{purple}}{MnO_4^-(aq)} + 8H^+(aq) + 5e^- \longrightarrow \underset{\text{colourless}}{Mn^{2+}(aq)} + 4H_2O(l)$$

d) *aqueous acidic dichromate*(VI) *to chromium*(III) *ions*,

$$\underset{\text{orange}}{Cr_2O_7(aq)} + 14H^+(aq) + 6e^- \longrightarrow \underset{\text{green}}{2Cr^{3+}(aq)} + 7H_2O(l)$$

3. as a weak acid

Hydrogen sulphide is moderately soluble in water and the dissolved hydrogen sulphide ionises slightly in two stages:

$$H_2S(aq) \rightleftharpoons H^+(aq) + HS^-(aq)$$
$$HS^-(aq) \rightleftharpoons H^+(aq) + S^{2-}(aq)$$

The equilibrium mixture contains a very low concentration of sulphide ions and this may be further reduced by adding a dilute solution of a strong acid. This method of reducing $[S^{2-}(aq)]$ can be used to control the precipitation of sulphides (see part (4) of this section).

Clearly, the above equilibria are displaced in favour of sulphide ions when alkali is present, and hydrogen sulphide is freely soluble in aqueous alkali:

$$H_2S(g) + 2OH^-(aq) \longrightarrow 2H_2O(l) + S^{2-}(aq)$$

4. as a precipitant for sulphides

a) In the presence of very dilute nitric(V) acid, only the cations shown in Fig. 25.10 form sulphides sufficiently insoluble to be precipitated by the low sulphide ion concentration in aqueous hydrogen sulphide, e.g.:

$$Hg^{2+}(aq) + S^{2-}(aq) \longrightarrow HgS(s)$$

b) In neutral or alkaline solutions, hydrogen sulphide will precipitate FeS (black), NiS (black) and ZnS (white) in addition to those shown in Fig. 25.10. NH_4^+, Ca^{2+}, K^+ and Na^+ give no sulphide precipitates under any conditions. This information, taken in conjunction with the action of aqueous sodium(I) hydroxide and aqueous ammonia (see section 12.29), can be valuable in identifying cations in aqueous solution.

Hg^{2+} black
Pb^{2+}, Ag^+, Cu^{2+} } brown
Cd^{2+} yellow

Fig. 25.10 colour of sulphides precipitated by acidic $H_2S(aq)$

25.7 preparation of sulphur dioxide

Sulphur dioxide can be obtained by the oxidation of sulphur or sulphides, but this is not a convenient method of production for laboratory use. It may be generated either by the action of warm dilute hydrochloric acid on a sulphate(IV):

$$SO_3^{2-}(s) + 2H^+(aq) \longrightarrow SO_2(g) + H_2O(l)$$

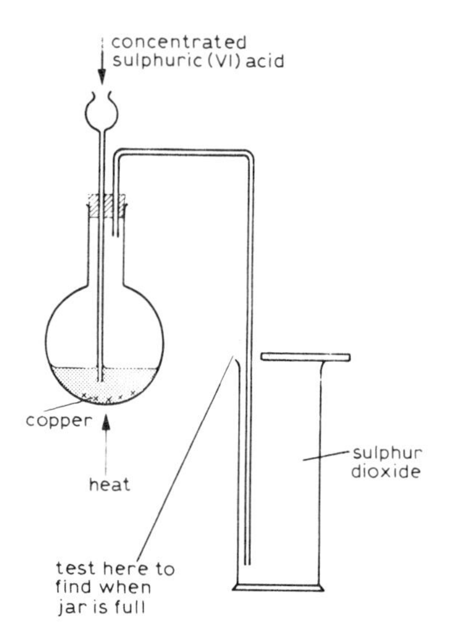

Fig. 25.11 preparation and collection of sulphur dioxide

or by the reduction of hot concentrated sulphuric(VI) acid using copper:

$$Cu \rightarrow Cu^{2+} + 2e^-$$
$$SO_4^{2-} + 4H^+ + 2e^- \rightarrow SO_2 + 2H_2O$$
$$\mathbf{Cu + SO_4^{2-} + 4H^+ \rightarrow Cu^{2+} + 2H_2O + SO_2}$$

A convenient apparatus for preparing the gas by this method and collecting it by **upward displacement of air** is shown in Fig. 25.11. The black residue in the flask indicates that the sulphuric(VI) acid is converted to other reduction products, notably sulphide ion.

25.8 physical properties of sulphur dioxide

The gas is colourless and has a sharp, choking smell. It is poisonous and constitutes an important and dangerous atmospheric pollutant–many fuels contain sulphur compounds which are oxidised to sulphur dioxide. It is easily liquefied under pressure (*why?*) and it can be generated conveniently in the laboratory from pressurised siphons of the liquid. The gas is denser than air (R.V.D. 32 compared with 14·4 for air) and is moderately soluble in water.

25.9 reactions of sulphur dioxide

1. as a weak acid

An aqueous solution of sulphur dioxide ionises partially to produce a weakly acidic solution. It may be considered to form the weak acid sulphuric(IV) acid, which subsequently ionises slightly:

$$SO_2(g) + H_2O(l) \rightleftharpoons \underset{\text{sulphuric(IV) acid}}{H_2SO_3(aq)}$$
$$H_2SO_3(aq) \rightleftharpoons H^+(aq) + \underset{\text{hydrogensulphate(IV) ion}}{HSO_3^-(aq)}$$
$$HSO_3^-(aq) \rightleftharpoons H^+(aq) + \underset{\text{sulphate(IV) ion}}{SO_3^{2-}(aq)}$$

Sodium(I) sulphate(IV) crystals may be obtained in the following way. An aqueous solution of sodium(I) hydroxide is divided into two equal parts. The first part is saturated with sulphur dioxide, producing a solution of sodium(I) hydrogensulphate(IV):

$$SO_2(g) + OH^-(aq) \rightarrow HSO_3^-(aq)$$

The second part of the alkali is added to this:

$$HSO_3^-(aq) + OH^-(aq) \rightarrow SO_3^{2-}(aq) + H_2O(l)$$

Evaporation and cooling yields crystals of $Na_2SO_3.7H_2O$.

2. as a reductant

Sulphur dioxide neither burns nor supports combustion, though oxygen oxidises it in the presence of a catalyst to sulphur(VI) oxide:

$$2SO_2(g) + O_2(g) \rightleftharpoons 2SO_3(g)$$

It is readily oxidised by chemical oxidants to sulphate(VI) and is therefore a useful reductant:

$$SO_2(g) + 2H_2O(l) \rightarrow SO_4^{2-}(aq) + 4H^+(aq) + 2e^-$$

The aqueous solution behaves similarly:

$$SO_3^{2-}(aq) + H_2O(l) \rightarrow SO_4^{2-}(aq) + 2H^+(aq) + 2e^-$$

The following oxidants are reduced by sulphur dioxide:

a) *aqueous acidic manganate*(VII) (see Fig. 25.12),

$$\underset{\text{purple}}{MnO_4^-(aq)} + 8H^+(aq) + 5e^- \rightarrow \underset{\text{colourless}}{Mn^{2+}(aq)} + 4H_2O(l)$$

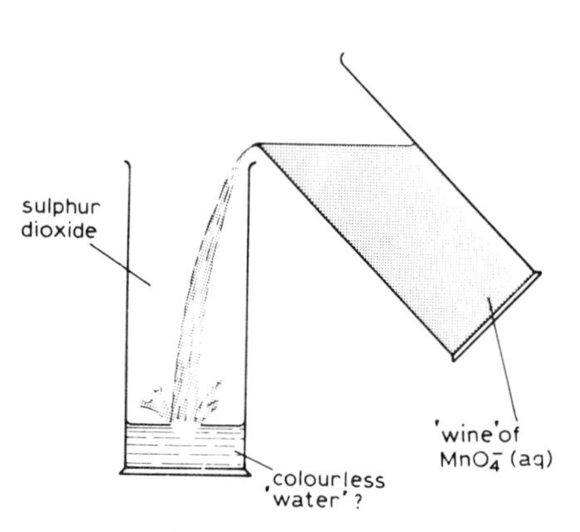

Fig. 25.12 'wine into water'

b) *aqueous acidic dichromate*(VI),

$$Cr_2O_7^{2-}(aq) + 14H^+(aq) + 6e^- \longrightarrow 2Cr^{3+}(aq) + 7H_2O(l)$$
orange green

c) *aqueous bromine,*

$$\frac{1}{2}Br_2(aq) + e^- \longrightarrow Br^-(aq)$$
orange colourless

d) *warm aqueous iron*(III) *ions,*

$$Fe^{3+}(aq) + e^- \longrightarrow Fe^{2+}(aq)$$

3. reduction of sulphur dioxide

Powerful reducing agents will convert sulphur dioxide to sulphur.
a) Burning magnesium continues to burn in the gas:

$$2Mg(s) + SO_2(g) \longrightarrow 2MgO(s) + S(s)$$

b) Gas jars of hydrogen sulphide and sulphur dioxide when mixed give a precipitate of sulphur:

$$(H_2S \longrightarrow 2H^+ + S + 2e^-) \times 2$$
$$SO_2 + 4H^+ + 4e^- \longrightarrow 2H_2O + S$$
$$\mathbf{SO_2(g) + 2H_2S(g) \longrightarrow 3S(s) + 2H_2O(l)}$$

Sulphur dioxide, obtained from roasting sulphur or sulphide ores in air, is used in the manufacture of sulphuric(VI) acid. It is also used as a bleaching agent and as a preservative for certain foods.

25.10 manufacture of sulphuric(VI) acid: the Contact Process

Sulphuric(VI) acid is so widely used that it is produced in larger quantities than almost any other chemical substance. The steps in the manufacturing process are:

i) conversion of sulphur or sulphide ore to sulphur dioxide,
ii) catalytic oxidation of sulphur dioxide to sulphur(VI) oxide,
iii) reaction of sulphur(VI) oxide with water to give sulphuric(VI) acid.

Sulphur dioxide is obtained either by burning sulphur or by roasting sulphur ores in air,

e.g. iron(II) disulphide: $4FeS_2(s) + 11O_2(g) \longrightarrow 2Fe_2O_3(s) + 8SO_2(g)$

Copper pyrites, $CuFeS_2$, is another ore used for this purpose. It is more expensive as a source of sulphur dioxide than is sulphur, but the copper metal obtained from the resulting oxide is valuable and renders the process economically viable. The gas obtained from roasting sulphur-bearing ores requires rigorous purification before it is passed over the catalyst in the next stage of the process.

This stage gives the name to the method of manufacture–the **Contact Process.** The sulphur dioxide formed in the first stage is mixed with oxygen, with which it combines slowly and reversibly:

$$2SO_2(g) + O_2(g) \rightleftharpoons 2SO_3(g); \quad \Delta H = -190\,kJ$$

Conditions favouring a high equilibrium yield of sulphur(VI) oxide are high pressure and low temperature (see Le Chatelier's Principle, section 23.9). In practice, a temperature of 400 °C–500 °C is employed, as lower temperatures render the attainment of equilibrium uneconomically slow. Even at this temperature, equilibrium is established sufficiently rapidly only in contact with a catalyst of platinum or vanadium(V) oxide. If these conditions are employed, compression of the gas mixture is found to be unnecessary.

The formation of sulphur(VI) oxide may be illustrated on the laboratory scale using the apparatus shown in Fig. 25.13. The whole apparatus must be *thoroughly dry*, corks and rubber bungs should not be employed at the exit from the catalyst chamber and crushed ice–*not* a freezing mixture–should

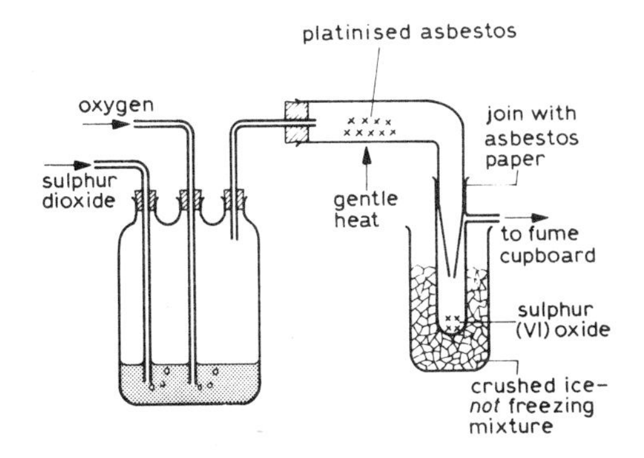

Fig. 25.13 preparation of sulphur(VI) oxide

surround the receiver. (A freezing mixture liquefies excess sulphur dioxide, in which the sulphur(VI) oxide dissolves.) White needle-like crystals of sulphur(VI) oxide are formed in the cooled receiver.

If the white fumes issuing from the catalyst chamber are bubbled through water, they pass through unchanged. The white fumes are, however, efficiently absorbed when bubbled through concentrated sulphuric(VI) acid; the product is oleum, $H_2S_2O_7$:

$$SO_3(g) + H_2SO_4(l) \longrightarrow H_2S_2O_7(l)$$

Sulphur(VI) oxide is absorbed in this way in the industrial process and the product is subsequently diluted (with dilute sulphuric(VI) acid) to produce pure sulphuric(VI) acid:

$$H_2S_2O_7(l) + H_2O(l) \longrightarrow 2H_2SO_4(l)$$

25.11 physical properties of sulphuric(VI) acid

The pure acid is a colourless, viscous liquid, density about $1{\cdot}8\,g\,cm^{-3}$. It causes serious burns in contact with the skin.

25.12 reactions of sulphuric(VI) acid

1. with water

Concentrated sulphuric(VI) acid reacts exothermically with water to such a degree that a small quantity of water added to the acid is likely to boil and eject the contents from the tube: when diluting sulphuric(VI) acid, the acid should therefore always be added to the water.

$$H_2SO_4(l) \rightleftharpoons H^+(aq) + HSO_4^-(aq)$$
$$HSO_4^-(aq) \rightleftharpoons H^+(aq) + SO_4^{2-}(aq)$$

In dilute aqueous solution, the acid is completely ionised and it is therefore termed a **strong acid.** Reactive metals liberate hydrogen from it and it evolves carbon dioxide from carbonates:

e.g.
$$Mg(s) + 2H^+(aq) \longrightarrow Mg^{2+}(aq) + H_2(g)$$
$$CO_3^{2-}(s) + 2H^+(aq) \longrightarrow H_2O(l) + CO_2(g)$$

Concentrated sulphuric(VI) acid is a powerful dehydrating agent and is used to dry gases – but not ammonia or hydrogen sulphide (*why not?*). It owes its use in the 'brown ring' test for nitrates(V) (section 24.15) to its dehydrating properties.

2. as a dehydrating agent

a) Crystals of copper(II) sulphate(VI)-5-water rapidly change colour from blue to white when dropped into concentrated sulphuric(VI) acid, as the anhydrous salt is formed.

b) Sugar treated with concentrated sulphuric(VI) acid changes colour slowly, through brown, to black. The product is a very pure form of carbon – 'sugar charcoal':

$$C_{12}H_{22}O_{11}(s) \longrightarrow 12C(s) + 11H_2O \quad \text{(absorbed by acid)}$$

c) Carbon monoxide (section 21.6) is obtained by the dehydrating action of concentrated sulphuric(VI) acid on sodium(I) methanoate (formate):

$$H_2CO_2(s) \longrightarrow CO(g) + H_2O \quad \text{(absorbed by acid)}$$

3. as a powerful, non-volatile acid

Concentrated sulphuric(VI) acid displaces less powerful or more volatile acids from their salts:

e.g.
$$Cl^-(s) + H_2SO_4(l) \longrightarrow HCl(g) + HSO_4^-(s)$$
$$NO_3^-(s) + H_2SO_4(l) \longrightarrow HNO_3(g) + HSO_4^-(s)$$

4. as an oxidant

Hot, concentrated sulphuric(VI) acid is a powerful oxidant. It is usually reduced to sulphur dioxide in so acting, but powerful reductants may convert it to hydrogen sulphide.

e.g.

a) preparation of sulphur dioxide (section 25.7)

b) oxidation of a bromide to bromine:

$$(Br^- \longrightarrow \tfrac{1}{2}Br_2 + e^-) \times 2$$
$$SO_4^{2-} + 4H^+ + 2e^- \longrightarrow SO_2 + 2H_2O$$
$$\mathbf{SO_4^{2-} + 4H^+ + 2Br^- \longrightarrow Br_2 + 2H_2O + SO_2}$$

c) oxidation of an iodide to iodine:

$$(I^- \longrightarrow \tfrac{1}{2}I_2 + e^-) \times 8$$
$$SO_4^{2-} + 10H^+ + 8e^- \longrightarrow H_2S + 4H_2O$$
$$\mathbf{SO_4^{2-} + 10H^+ + 8I^- \longrightarrow 4I_2 + 4H_2O + H_2S}$$

25.13 uses of sulphuric(VI) acid

Sulphuric(VI) acid is used in the manufacture of fertilisers (both nitrogenous and phosphate), paints, fibres and films (e.g. Rayon), detergents, plastics and dyestuffs, and extensively in the fine chemical industry. The consumption of sulphuric(VI) acid has been used as a guide to the economic prosperity of a nation; the figures for consumption in the United Kingdom shown in Fig. 25.14 tell their own tale.

1961	2662
1962	2731
1963	2881
1964	3136
1965	3305
1966	3118
1967	3182
1968	3282
1969	3235

Fig. 25.14 sulphuric(VI) acid; UK production in thousands of tons

25.14 sulphates(VI)

All sulphates(VI), except those of Pb^{2+} and Ba^{2+} (Ca^{2+} sparingly), are soluble in water. Aqueous solutions of sulphates(VI) give a white precipitate with aqueous barium(II) chloride even in the presence of acid:

$$Ba^{2+}(aq) + SO_4^{2-}(aq) \longrightarrow BaSO_4(s)$$

This is the usual test for a sulphate(VI). Some sulphates(VI) decompose on heating,

e.g. $$CuSO_4(s) \longrightarrow CuO(s) + SO_3(g)$$

If the hydrated crystals are heated, concentrated sulphuric(VI) acid ('oil of vitriol') can be distilled from them. Hydrated copper(II) sulphate(VI) was called 'Blue Vitriol'; zinc(II) sulphate(VI)-7-water, 'White Vitriol'. 'Green Vitriol' undergoes autoxidation on heating:

$$2FeSO_4.7H_2O(s) \longrightarrow 14H_2O(g) + Fe_2O_3(s) + SO_2(g) + SO_3(g)$$

revision summary: Chapter 25

allotropes:		different physical forms of the same element in the same physical state
sulphur:	*allotropes:*	rhombic $\overset{96\,°C}{\rightleftharpoons}$ monoclinic (both S_8 rings) also plastic sulphur (helical chains)
	occurrence:	native, extracted by Frasch process
	properties:	yellow solid, m.p. 114°C, b.p. 444°C
	reactions:	burns in air $\longrightarrow SO_2$; heated with metals $\longrightarrow$ sulphides; hydrogen $\longrightarrow H_2S$ } i.e. a non-metal
hydrogen sulphide:	*preparation:*	hydrochloric acid on iron(II) sulphide $S^{2-}(s) + 2H^+(aq) \longrightarrow H_2S(g)$
	collection:	over water or upward displacement of air

	test:	lead(II) ethanoate (acetate) paper turns black
	reactions:	burns in air reductant weak dibasic acid precipitant for sulphides
sulphur dioxide:	*preparation:*	dilute acid on a sulphate(IV), or copper on hot, conc. H_2SO_4
	collection:	upward displacement of air
	test:	smell, acid dichromate(VI) paper ⟶ green
	reactions:	weak dibasic acid reductant
sulphuric(VI) acid:	*manufacture:*	Contact Process
	reactions:	(*dilute*) strong dibasic acid (*cold, conc.*) dehydrating agent (*hot, conc.*) non-volatile acid oxidant

questions 25

1. What do you understand by allotropy? Describe the appearance of the **two** chief allotropes of sulphur and how you would prepare a sample of **one** of them. Outline how it could be shown that each form of sulphur was composed of the element only.

What volume of oxygen would be required completely to oxidise 10 dm^3 of hydrogen sulphide to sulphur dioxide and water, all volumes being measured at s.t.p.?

2. (a) Sulphur and hydrogen combine to form hydrogen sulphide. Is the hydrogen oxidised or reduced in this process?

(b) Sulphur combines with oxygen to form sulphur dioxide. Is the sulphur oxidised or reduced in this reaction?

(c) Sulphur dioxide and hydrogen sulphide react together– give the equation for this reaction.

(d) State whether the sulphur dioxide is oxidised or reduced in reaction (c). (JMB)

3. (a) Describe how to prepare and collect sulphur dioxide.

(b) Describe what you would see and say what is formed
- (i) when sulphur dioxide is passed into a jar of moist hydrogen sulphide,
- (ii) when a jar of sulphur dioxide is placed mouth downwards in a solution of sodium(I) hydroxide.

What does reaction (i) show concerning the nature of sulphur dioxide?

(c) Outline briefly the stages by which sulphur is converted into sulphuric(VI) acid by the contact process. Give equations and the conditions of reaction, but **no** details of the industrial plant used. (C)

4. Outline the manufacture of sulphuric(VI) acid from a rich sulphur-bearing ore such as iron(II) disulphide (iron pyrites). No description of technical plant is required, but the essential conditions for each stage of the process should be given.

Assuming (i) that this ore was pure and that its formula was **FeS_2 and** (ii) that each reaction stage was completed, calculate **the mass** of sulphuric(VI) acid obtainable from 600 tons of iron **pyrites.**

What is the action of sulphuric(VI) acid upon (a) powdered crystals of copper(II) sulphate(VI)-5-water, (b) magnesium? Specify the necessary conditions. (O)

5. Hydrogen sulphide is usually prepared in the laboratory by the action of hydrochloric acid on iron(II) sulphide (ferrous sulphide).
- (a) Give a diagram of an apparatus for preparing hydrogen sulphide by use of which a supply of the gas is available whenever required. Explain briefly how the apparatus works.
- (b) What is the approximate concentration of the acid used?
- (c) Give the equation for the reaction.
- (d) Indicate how one main impurity can be removed from the gas.
- (e) Calculate the volume at s.t.p. of hydrogen sulphide which can be obtained by using 11 g of iron(II) sulphide.

Give **two** reactions in which hydrogen sulphide acts as a reducing agent, explaining why hydrogen sulphide is considered to be a reducing agent in each reaction. (AEB)

6. (a) Give a reaction by which hydrogen sulphide can be conveniently prepared in the laboratory. **Neither** diagrams **nor** method of collection are required.

Describe the reactions of hydrogen sulphide with **three** of the following:
- (i) air **or** oxygen,
- (ii) chlorine,
- (iii) zinc(II) sulphate(VI) solution,
- (iv) concentrated nitric(V) acid.

Why is it important to remove hydrogen sulphide from town gas or coal gas?

(b) The compound sulphur dichloride dioxide (formula SO_2Cl_2) reacts readily with cold water to give a solution containing sulphuric(VI) acid and hydrochloric acid. Write the equation for this reaction and calculate the volume of 1 M sodium(I) hydroxide needed to neutralise the solution formed when one tenth of a mole of sulphur dichloride dioxide reacts completely with water. (C)

Although plastics are finding an increasing use in everyday life, the new materials have a long way to go before they catch up with the traditional metals. In 1968, about 19 million tons of all types of plastic materials were produced in the world compared with 1000 million tons of iron and steel and 30 million tons of other metals. The sources as well as the production of plastics are limited; it has been estimated that there is only twenty years' supply of oil for plastics, at present rates of consumption, compared with 150 years' supply of iron ore. After this time, recycling methods will have to be introduced if any new objects are to be made from these materials. Compounds of metals are also widely used in our present civilisation, and their chemistry and uses are summarised in this Chapter.

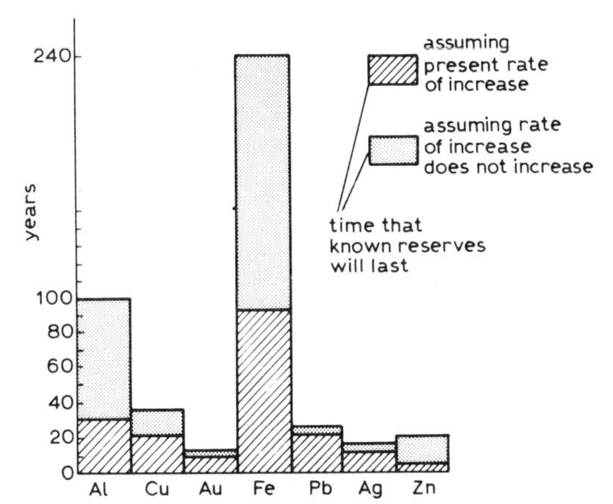

Fig. 26.1 world resources of metals

aluminium

26.1 occurrence and extraction

Aluminium occurs extensively in the earth's crust as aluminosilicates (section 21.13). However, the cheapest source for extraction is the higher-grade ore *bauxite*, $Al_2O_3.2H_2O$, which occurs in the USA, Jamaica, Southern Europe and West Africa. Most extraction processes for metals include purification of the ore, reduction of the ore to the crude metal and refining this product to the pure metal. In the case of aluminium, the ore is purified by dissolving it in aqueous sodium(I) hydroxide (in which some impurities, e.g. iron(III) oxide, are insoluble) and subsequently reprecipitating the hydroxide. At this stage other impurities remain in solution, e.g. silica as silicates. This purification process utilises the *amphoteric* nature of aluminium(III) oxide and hydroxide–see section 26.4. The precipitated hydroxide is then heated to give the oxide.

Aluminium is high in the reactivity series: electrolysis must be used to reduce the oxide to the metal. Unfortunately, the oxide is covalent and therefore a non-electrolyte; it must be dissolved in the molten ionic substance, sodium(I) hexafluoroaluminate(III) (cryolite), to obtain a conducting melt. Cryolite enjoys the dubious distinction of being the first ore to be completely exhausted in all its known sources. The electrolysis is carried out using a p.d. of about 6 V and a current of 100 000 A; the graphite lining of the cell acts as the cathode and a graphite block is used as the anode. The passage of the current provides sufficient heat to keep the electrolyte molten. Aluminium and oxygen are discharged in preference to sodium and fluorine and only small quantities of synthetic cryolite need be added with fresh charges of oxide.

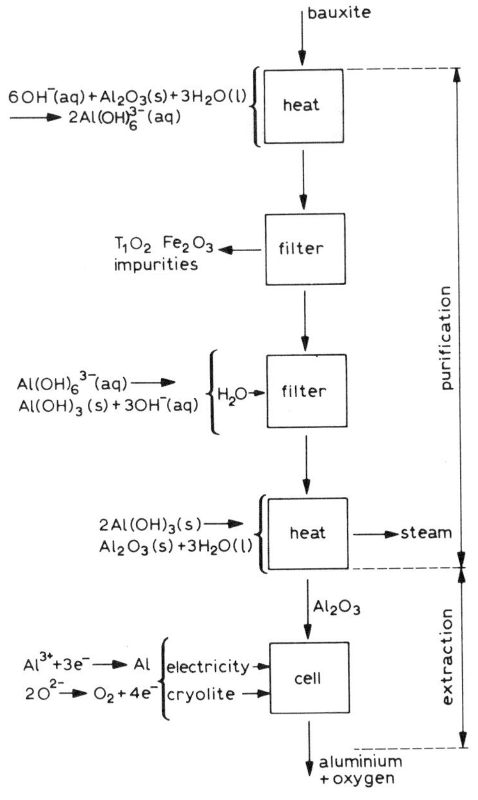

Fig. 26.2 extraction of aluminium–flow sheet

26.2 properties and reactions of aluminium

Aluminium is a typical metal in its physical properties: it is malleable and

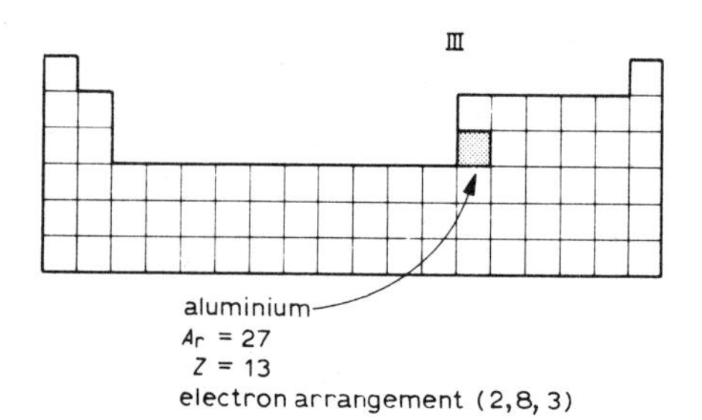

Fig. 26.3 position of aluminium in Periodic Table

K
Na
Ca
Mg
(Al)
Zn
Fe
Sn
Pb
H
Cu

Fig. 26.4 position of aluminium in reactivity series

ductile, and can therefore be machined and rolled. It conducts electricity well, and because of its low density (2700 kg m^{-3}) is a better conductor per unit mass than copper. It melts at 660° C (933 K) and its tensile strength is 6·3 kg mm^{-2}. This last value is raised considerably by alloying with magnesium (4%), or silicon (7%), or copper (4%), or magnesium and manganese (total 2%). Its position in the Periodic Table and its simple atomic properties are shown in Fig. 26.3; it has a valency of III only and the ion Al^{3+} is so small that the intense field around it produces a high degree of covalency in most aluminium compounds (see section 27.5).

From its position in the reactivity series (Fig. 26.4) it would appear to be a very reactive metal–yet it is apparently not attacked by air. This stability depends on the formation of a thin but strongly adhering layer of covalent aluminium(III) oxide on the surface of the metal; it is this layer which is resistant to attack. If aluminium foil is warmed with dilute hydrochloric acid, very little happens for a few minutes, while the oxide film is attacked:

$$Al_2O_3(s) + 6H^+(aq) \longrightarrow 2Al^{3+}(aq) + 3H_2O(l)$$

Soon a violent reaction develops, with the evolution of hydrogen, as the exposed metal is oxidised by the acid:

$$Al \longrightarrow Al^{3+} + 3e^-$$
$$3H^+ + 3e^- \longrightarrow 1\tfrac{1}{2}H_2$$
$$\mathbf{Al(s) + 3H^+(aq) \longrightarrow Al^{3+}(aq) + 1\tfrac{1}{2}H_2(g)}$$

The oxide film can be removed by rubbing the metal foil with cotton wool soaked in aqueous mercury(II) chloride–'whiskers' of aluminium(III) oxide grow from the exposed metal as it is attacked by air, and the foil becomes very hot. The oxide film can be thickened commercially by anodic oxidation; the metal is made the anode of a cell in which oxygen is liberated, e.g. in the electrolysis of chromic(VI) acid. The layer may be developed to a thickness of 3 nm, and can be made to incorporate dyes for coloured kitchenware. The oxide layer produces difficulties in welding aluminium, which has to be done in an inert atmosphere to prevent oxidation.

Other acids have very little effect on aluminium but alkalis dissolve it rapidly to yield hydrogen and sodium(I) hexahydroxoaluminate. For this reason sodium(I) carbonate-10-water (washing-soda) should not be used to clean aluminium pans.

$$Al \longrightarrow Al^{3+} + 3e^-$$
$$Al^{3+} + 6OH^- \longrightarrow [Al(OH)_6]^{3-}$$

$$Al + 6OH^- \longrightarrow [Al(OH)_6]^{3-} + 3e^-$$
$$(H_2O + e^- \longrightarrow \tfrac{1}{2}H_2 + OH^-) \times 3$$
$$\mathbf{Al(s) + 3H_2O(l) + 3OH^-(aq) \longrightarrow [Al(OH)_6]^{3-}(aq) + 1\tfrac{1}{2}H_2(g)}$$

The great strength of the aluminium–oxygen bond means that much energy is released in its formation. Thus if a mixture of aluminium powder and iron(III) oxide is ignited in a Battersea crucible using a fuse, a violently exothermic reaction occurs and molten iron is produced:

$$2Al(s) + Fe_2O_3(s) \longrightarrow Al_2O_3(s) + 2Fe(l)$$

This is the basis of the **Thermite** Process used for welding iron.

26.3 uses of aluminium

The uses of aluminium depend on its properties of

i) resistance to corrosion,
ii) high conductivity,
iii) high tensile strength (in alloys),
iv) low density,
v) non-toxic nature of the metal and its compounds.

Thus it is used for many cooking utensils [(i), (iv) and (v)]; for structural

Fig. 26.5 uses of aluminium

materials (usually as the copper alloy) where weight must be kept low, as in aircraft and ships [(i), (iii) and (iv)]; for grid cables [(ii), (iii) and (iv)]; and more recently for car engines (as the silicon alloy). The speed of supersonic aircraft is limited by the low melting-point of aluminium. 10 million tons of aluminium were produced in 1970 throughout the world, making it second to iron in importance.

26.4 aluminium(III) compounds

1. aluminium(III) hydroxide

This is the white precipitate formed when aqueous sodium(I) hydroxide is added to an aqueous solution of an aluminium(III) salt. Its formation is best understood by representing the hydrated aluminium(III) ion more fully than usual, as $[Al(H_2O)_6]^{3+}$:

$$[Al(H_2O)_6]^{3+} + 3OH^-(aq) \longrightarrow [Al(OH)_3(H_2O)_6](s) + 3H_2O(l)$$

In this process the water coordinating the aluminium(III) cation is acting as an acid, donating protons to hydroxide ions. In excess alkali the same process is carried further as the hydrated hydroxide dissolves, giving a solution containing hexahydroxoaluminate(III) ions:

$$[Al(OH)_3(H_2O)_3](s) + 3OH^-(aq) \longrightarrow [Al(OH)_6]^{3-}(aq) + 3H_2O(l)$$

The amphoteric nature of aluminium(III) oxide can be represented schematically as shown in Fig. 26.6.

It is used as a mordant in dyeing; added to the dye-bath it combines strongly with both the cloth and an appropriate dye, thus fixing the colour on the material. Also, aluminates(III) form a constituent of cement–see section 22.6(2).

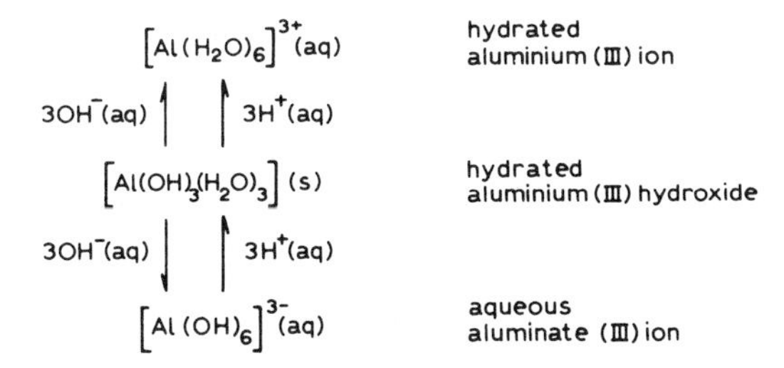

Fig. 26.6 amphoteric nature of aluminium(III) hydroxide

2. aluminium(III) oxide

Aluminium(III) oxide is a white insoluble powder which can be made in the laboratory by heating aluminium(III) hydroxide:

$$2[Al(OH)_3(H_2O)_3](s) \longrightarrow Al_2O_3(s) + 9H_2O(g)$$

It 'ages' to a very unreactive form, but shows the same amphoteric properties as the hydroxide when it is fresh. Its high melting-point (2000°C) leads to its use in refractory linings for kilns. It is also used as an abrasive in emery paper and as an adsorbent in chromatography.

3. aluminium(III) chloride

This may be prepared dry in the laboratory by heating aluminium in chlorine (see Fig. 15.9):

$$Al(s) + 1\tfrac{1}{2}Cl_2(g) \longrightarrow AlCl_3(s)$$

It is made industrially by heating aluminium(III) oxide with coke and chlorine:

$$Al_2O_3(s) + 3Cl_2(g) + 3C(s) \longrightarrow 2AlCl_3(s) + 3CO(g)$$

It is a white covalent substance which sublimes at a low temperature. Both solid and vapour contain the dimer (double molecule) Al_2Cl_6. It dissolves in water with a violent exothermic reaction as the hydrated ion is formed:

$$AlCl_3(s) + 6H_2O(l) \longrightarrow [Al(H_2O)_6]^{3+}(aq) + 3Cl^-(aq)$$

This reaction is not reversible: if an aqueous solution of aluminium(III) chloride (prepared by the usual method, from the metal and aqueous acid) is heated, aluminium(III) oxide is formed by hydrolysis. This is another example of the strength of the aluminium–oxygen bond:

$$2[Al(H_2O)_6]^{3+}(aq) \longrightarrow Al_2O_3(s) + 6H^+(aq) + 9H_2O(l)$$

4. aluminium(III) sulphate(VI)

This is most easily prepared by dissolving the hydroxide in dilute sulphuric(VI) acid:

$$Al(OH)_3(s) + 3H^+(aq) \longrightarrow Al^{3+}(aq) + 3H_2O(l)$$

The salt crystallises with eighteen moles of water per mole of sulphate(VI), $Al_2(SO_4)_3.18H_2O$. Its solutions contain the $[Al(H_2O)_6]^{3+}$ ion, which is effective in coagulating finely divided organic material; thus it is used in water purification (section 22.7), in tanning, and as a substance to promote the clotting of blood (a styptic).

5. tests for aluminium(III) ions

Aluminium(III) ions in solution may be identified by the following tests, taken in conjunction:

i) aqueous sodium(I) hydroxide: white precipitate soluble in excess alkali,
ii) aqueous ammonia: white precipitate, insoluble in excess alkali,
iii) hydrogen sulphide in acid solution: no precipitate.

calcium

26.5 occurrence and extraction

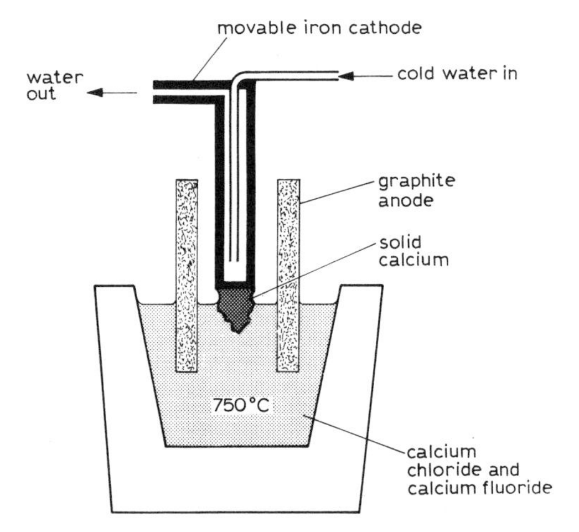

Fig. 26.7 extraction of calcium

Although combined calcium is found abundantly in the earth's crust, the metal itself is too reactive to find general use. It is prepared by the electrolysis of fused calcium(II) chloride, with a little fluoride added to lower the melting-point. The cathode is steel and the anode is graphite:

cathode: $Ca^{2+}(l) + 2e^- \longrightarrow Ca(l)$

anode: $2Cl^-(l) \longrightarrow Cl_2(g) + 2e^-$

26.6 properties and reactions of calcium

Calcium is a white metal, malleable and ductile, with a very low density (1550 kg m^{-3}) and low tensile strength (5·7 kg mm^{-2}). Its melting-point is 810 °C (1083 K). In air it rapidly tarnishes and flakes, forming first the oxide, then the hydroxide and carbonate:

$$Ca \xrightarrow{O_2} CaO \xrightarrow{H_2O} Ca(OH)_2 \xrightarrow{CO_2} CaCO_3$$

The position of calcium in the Periodic Table and its simple atomic properties are shown in Fig. 26.8 and its position in the reactivity series in Fig. 26.9. These positions are consistent with its properties as a reactive metal having a valency of II only and forming predominantly ionic compounds.

The metal itself reacts with cold water forming hydrogen (section 8.12):

$$Ca \longrightarrow Ca^{2+} + 2e^-$$
$$2H_2O + 2e^- \longrightarrow H_2 + 2OH^-$$
$$\mathbf{Ca(s) + 2H_2O(l) \longrightarrow Ca^{2+}(aq) + 2OH^-(aq) + H_2(g)}$$

It also liberates hydrogen from dilute acids, but it is unaffected by alkalis.

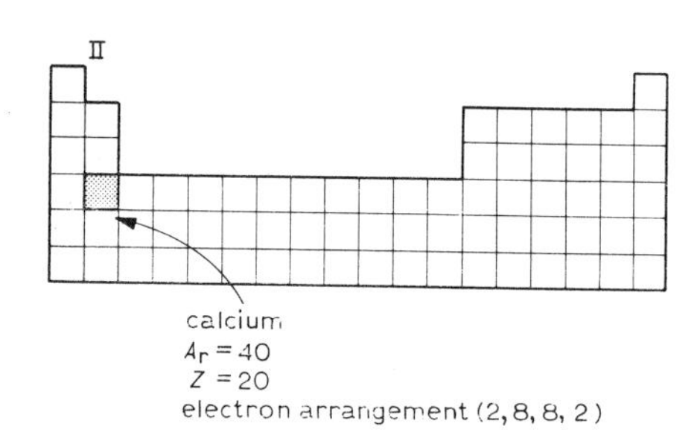

Fig. 26.8 position of calcium in Periodic Table

Its reactivity finds a use when it is added to steel or aluminium alloys to remove the last traces of oxygen. It is also incorporated in alloys (e.g. 3% in lead, for bearings).

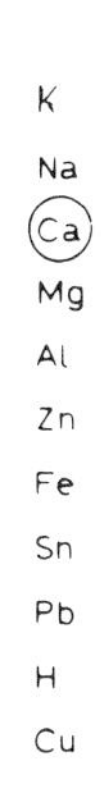

Fig. 26.9 position of calcium in reactivity series

26.7 calcium(II) compounds

1. calcium(II) hydroxide (slaked lime)

This is prepared in the laboratory by precipitation from an aqueous solution containing calcium(II) ions, using aqueous sodium(I) hydroxide. It is sparingly soluble, and therefore is not precipitated by the weaker alkali, aqueous ammonia (see section 24.9).

$$Ca^{2+}(aq) + 2OH^-(aq) \longrightarrow Ca(OH)_2(s)$$

It is a white substance and strongly basic, neutralising and dissolving in dilute acids (Chapter 11). Its relation to other calcium(II) compounds is described in section 22.4.

2. calcium(II) oxide (quicklime)

This strongly basic white powder may be prepared either by burning the metal or heating the hydroxide or carbonate:

$$Ca(s) + \tfrac{1}{2}O_2(g) \longrightarrow CaO(s)$$
$$Ca(OH)_2(s) \rightleftharpoons CaO(s) + H_2O(g)$$
$$CaCO_3(s) \rightleftharpoons CaO(s) + CO_2(g)$$

Its further chemistry and uses are described in Chapter 22.

3. calcium(II) chloride

This may be prepared in the anhydrous form by heating calcium in chlorine; the hydrated crystals may be obtained from the solution resulting from the neutralisation of dilute hydrochloric acid with calcium(II) carbonate.

$$Ca(s) + Cl_2(g) \longrightarrow CaCl_2(s)$$
$$CaCO_3(s) + 2H^+(aq) \longrightarrow Ca^{2+}(aq) + H_2O(l) + CO_2(g)$$

The anhydrous material is a white powder (often fused into lumps) and is used in the laboratory as a drying agent for gases and organic liquids. The crystals obtained from solution are the hexahydrate, which is deliquescent. Although produced in large quantities by the Solvay process, no significant use has been discovered for the salt.

4. calcium(II) sulphate(VI)

This occurs naturally as **anhydrite** ($CaSO_4$) and **gypsum** ($CaSO_4.2H_2O$). It is a white, almost insoluble solid, but it does dissolve sufficiently to cause permanent hardness in water (see Chapter 22). Sulphuric(VI) acid has been made from anhydrite, but this process is no longer important in competition with the Contact Process.

If gypsum is heated carefully to 120°C, the hemihydrate $CaSO_4.\tfrac{1}{2}H_2O$ is formed. Further heating forms the anhydrous salt. The hemihydrate is **'Plaster of Paris'**, and this combines exothermically with water to re-form gypsum. The reaction is accompanied by slight expansion so that, if carried out in a mould, a clear cast is obtained. It is used for plasterboard on walls, plaster casts for broken limbs and for blackboard 'chalk'.

5. tests for calcium(II) ions

Calcium(II) compounds impart a brick-red colour to a flame. An aqueous solution containing calcium(II) ions gives a white precipitate with aqueous sodium(I) hydroxide, insoluble in excess alkali, but gives no precipitate with aqueous ammonia.

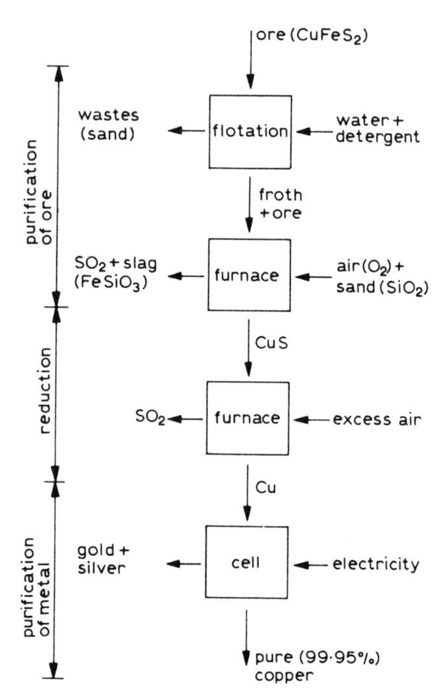

Fig. 26.10 extraction of copper–flow sheet

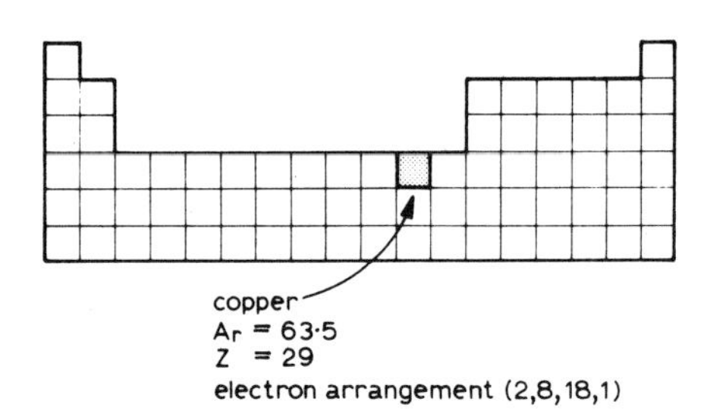

Fig. 26.11 position of copper in Periodic Table

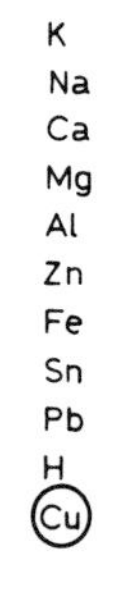

Fig. 26.12 position of copper in reactivity series

copper

Copper is one of the less reactive metals, and it is easily obtained by reduction of its ores. It was known to Bronze Age man as an alloy with zinc (*brass*) or tin (*bronze*). Both pure and alloyed, copper finds extensive uses, and it is third in scale of production (world production: 7 million tons, 1970).

26.8 occurrence and extraction

The principal ore is a sulphide, $CuFeS_2$, although carbonate ores have also been worked. The ore is concentrated by floating it in the froth of a special detergent, and the copper(II) sulphide is separated from the iron(II) compounds as shown in Fig. 26.10. The copper(II) sulphide is then roasted in air to form copper and sulphur dioxide (used for sulphuric(VI) acid production). The reactions occurring in this process are many and complex.

The copper obtained in this way contains appreciable amounts of impurities and is refined electrolytically. The impure copper is made the anode of a cell containing aqueous copper(II) sulphate(VI) and thin sheets of pure copper are used as the cathode.

At the *anode* the copper dissolves:

$$Cu(s) \longrightarrow Cu^{2+}(aq) + 2e^-$$

Copper reforms on the *cathode*:

$$Cu^{2+}(aq) + 2e^- \longrightarrow Cu(s)$$

Metallic impurities more reactive than copper dissolve, but are not discharged,

e.g. iron: $$Fe(s) \longrightarrow Fe^{2+}(aq) + 2e^-$$

Those less reactive do not dissolve; they are deposited as solids on the floor of the cell, as copper dissolves from the anode. Silver and gold are recovered from the 'anode slime' as an important by-product of the process.

26.9 properties and reactions of copper

Copper is typically metallic. It is very malleable and ductile, and can be rolled into sheets or drawn into wire. It is an exceptionally good conductor of electricity, only surpassed by silver. It has a high density ($8920\,kg\,m^{-3}$) and moderate tensile strength ($22{\cdot}5\,kg\,mm^{-2}$). Its melting point is 1080°C (1353 K) and it has a pleasant reddish-gold colour. The place of copper in the Periodic Table and its simple atomic properties are shown in Fig. 26.12. The atom cannot attain a noble gas electron structure by loss of electrons and copper shows variable valency. Since it is below hydrogen in the reactivity series, it is not attacked by water or by dilute acids in the absence of oxidising agents. It is attacked by concentrated nitric(V) acid and by hot concentrated sulphuric(VI) acid, because of their oxidising properties. The chemistry of these reactions is discussed in sections 24.13(4) and 25.7 respectively.

Carbonic acid in rain-water will attack the metal slowly in the presence of air, turning exposed copper surfaces (e.g. roofs) green, as copper(II) carbonate is formed.

The metal reacts with air on heating, forming a black surface layer of copper(II) oxide:

$$Cu(s) + \tfrac{1}{2}O_2(g) \longrightarrow CuO(s)$$

26.10 uses of copper

The uses of copper depend on

i) its high conductivity of heat and electricity
ii) its low reactivity
iii) its decorative appearance

Thus it is used extensively in electrical wiring and switchgear (i), and in heat exchangers and boilers [(i), (ii)]. Alloying improves its strength; *brass* (60% Cu, 40% Zn) is used in many small corrosion-resistant items (e.g. buttons, screws); *bronze* (90% Cu, 10% Sn + Pb) is used for springs and gears. *Coinage metal* is now principally copper: *coinage bronze* is 95% Cu, 5% (Zn + Sn) and *coinage 'silver'* is 75% Cu, 25% Ni.

Fig. 26.13 some uses of copper

26.11 copper as a transition element

Transition elements contain incompletely filled electron energy levels below the highest energy level in the atomic particle, either in their atoms or ions. They occupy the centre block of the Periodic Table, for which the gradation in properties across horizontal rows and down vertical groups does not apply. The transition elements are all metals; they are characterised by variable valency and the formation of coloured salts and many complex ions.

Copper is a transition element (see Fig. 26.11). It forms copper(I) and copper(II) compounds and many of the latter are blue or green, as the hydrated copper(II) ion, $[Cu(H_2O)_4]^{2+}$, is blue. Complex ions containing copper(I) and copper(II) ions are well known. Only the commoner copper(II) compounds are described in the next section.

26.12 copper(II) compounds

1. copper(II) hydroxide

This is a blue solid precipitated from an aqueous solution containing copper(II) ions by aqueous sodium(I) hydroxide:

$$Cu^{2+}(aq) + 2OH^-(aq) \longrightarrow Cu(OH)_2(s)$$

The precipitate contains other anions, but we shall not consider the detail of its composition. It is insoluble in excess sodium(I) hydroxide, but it dissolves in acid and it is also soluble in excess aqueous ammonia, forming a deep blue solution containing the tetraamminecopper(II) ion:

$$Cu(OH)_2(s) + 4NH_3(aq) \longrightarrow [Cu(NH_3)_4]^{2+}(aq) + 2OH^-(aq)$$

The deep blue complex is used for making viscose rayon. In this process it dissolves cellulose, which is regenerated as a fine thread by squirting a jet of the solution into an acid bath.

2. copper(II) oxide

This is a black, insoluble, basic oxide, which may be prepared by heating the hydroxide, carbonate or nitrate(V). It may be reduced to copper by heating it with hydrogen, ammonia or carbon.

3. copper(II) chloride, sulphate(VI) and nitrate(V)

These are all typical salts, crystallising as hydrates from solutions: $CuCl_2.2H_2O$ (green); $CuSO_4.5H_2O$ (blue); $Cu(NO_3)_2.3H_2O$ (deep blue). The sulphate(VI) can be dehydrated by gentle heating to form white anhydrous copper(II) sulphate(VI). The blue is restored by a little water, or aqueous solution.

The salts are soluble and may be prepared by the methods given in Chapter 12, except that copper metal does not react with non-oxidising acids. Copper(II) sulphate(VI) may be prepared from the metal by warming it with dilute sulphuric(VI) acid and hydrogen peroxide; on the commercial scale, air is blown through the acid.

Copper(II) salts are used in several formulations for fungicides, and copper(II) sulphate(VI) is used as the electrolyte in copper-plating.

4. copper(II) sulphide

This is a black substance, insoluble in water, and prepared by precipitation from a solution of aqueous copper(II) ions using hydrogen sulphide:

$$H_2S(aq) \rightleftharpoons 2H^+(aq) + S^{2-}(aq)$$
$$Cu^{2+}(aq) + S^{2-}(aq) \rightarrow CuS(s)$$

5. tests for copper(II) ions

Aqueous ammonia added to an aqueous solution containing copper(II) ions gives a light blue precipitate of copper(II) hydroxide, which redissolves in excess ammonia to give a deep blue solution.

iron

26.13 occurrence and extraction

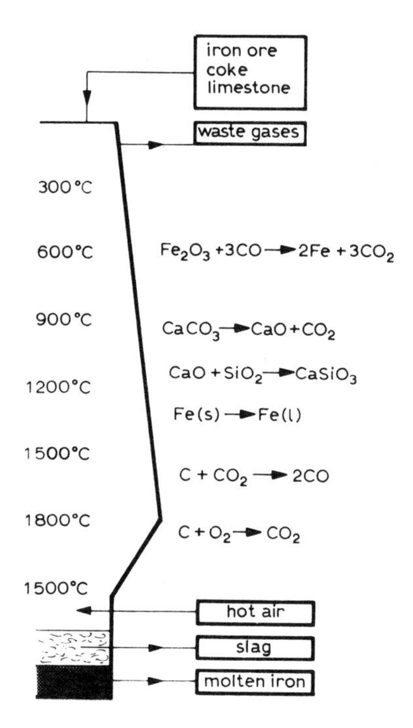

Fig. 26.14 reactions in a blast furnace

Iron is produced on a tremendous scale all over the world. A typical furnace is capable of producing 300 tons of crude pig-iron every four hours. Low-grade ores, such as the iron(II) carbonate found in Britain, are roasted to convert them to the oxide:

$$2FeCO_3(s) + \tfrac{1}{2}O_2(g) \rightarrow Fe_2O_3(s) + 2CO_2(g)$$

The oxide Fe_2O_3 also occurs naturally as **haematite,** or 'red iron ore'. This, or a higher-grade ore, is loaded into the top of a **blast furnace,** with coke and limestone; preheated air is blown in at the bottom. The series of reactions occurring in the furnace is complex, but the essential reactions are outlined below and are summarised in Fig. 26.14. It must be emphasised that neither coke nor limestone feature in the essential reactions; they are merely the raw materials used to produce carbon monoxide and quicklime respectively.

At the bottom of the furnace (about 1500°C) coke is oxidised to carbon dioxide:

$$C(s) + O_2(g) \rightarrow CO_2(g)$$

As this rises up the furnace, it is converted to carbon monoxide:

$$C(s) + CO_2(g) \rightarrow 2CO(g)$$

Near the top of the furnace, the ore is reduced *by the carbon monoxide* at about 600°C:

$$Fe_2O_3(s) + 3CO(g) \rightarrow 2Fe(s) + 3CO_2(g)$$

As the iron passes down the furnace it melts and dissolves carbon; the liquid metal is tapped from the base of the furnace.

Limestone is decomposed half-way down the furnace at about 900°C:

$$CaCO_3(s) \rightarrow CO_2(g) + CaO(s)$$

The acidic oxide silica (SiO_2) present in the iron ore, reacts with the quicklime formed, to produce a slag of calcium(II) silicate(IV):

$$CaO(s) + SiO_2(s) \rightarrow CaSiO_3(s)$$

This slag melts as it passes down the furnace and floats on top of the liquid iron. It is tapped off periodically and it finds some use as a material for road-making.

The whole process is continuous: ore, limestone and coke are continuously fed in at the top and an air blast is maintained at the bottom. The waste gases emerging from the top are fed into a heat exchange system and molten iron and slag are removed periodically from the base of the furnace.

The liquid iron is cast into 'pigs'; the resulting **pig-iron,** or cast-iron, contains silicon, phosphorus and about 4% of carbon, which renders it very brittle. It may be used for simple castings, but most is converted into steel.

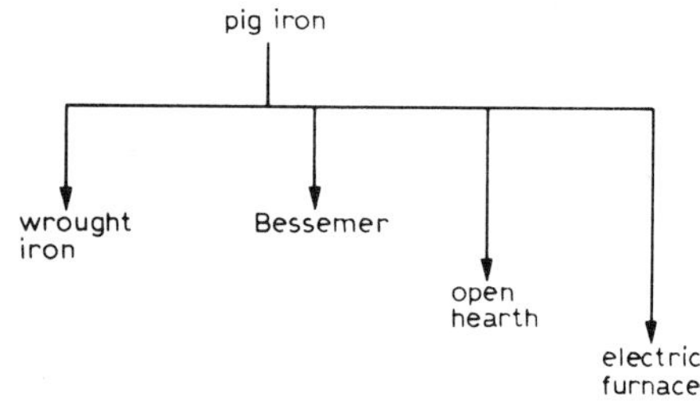

Fig. 26.15 steel manufacture–flow sheet

26.14 production of steel

In a perfect crystal of a metal as shown in Fig. 26.16(a), the atoms are arranged with complete regularity. Such crystals are rare and almost inevitably faults appear in the structure as shown in (b). If a shearing force is applied to such a crystal, the faults shift through the structure and produce permanent deformation (c), much as a heavy carpet can be moved by making a slight ruck at one side and easing it across to the other. The faults in the system can be filled by introducing a small percentage of *bulky* atoms, either in the 'holes' or interstices in the structure, (d), or substituted for metal atoms (e). Such **interstitial** or **substitutional** alloys have a greater strength and toughness than the pure metal, though too high a percentage of the large atoms can lead to cleavage as shown in (f) and brittleness results.

Pig-iron or cast-iron has a high carbon content and exemplifies case (f). Pure wrought iron exemplifies case (b) and this type of iron is useless for constructional purposes, though it has some value in ornamental metalwork. Steels correspond with cases (d) and (e); the principal alloying element is **carbon**, though many other elements are used in addition and give steels of different properties. The compositions and uses of several steels are given in Fig. 26.17.

The usual procedure in making steel is to remove all impurities from pig-iron by oxidation, using air enriched with oxygen at a high temperature, producing wrought iron. The required quantity of carbon and other elements is then added in a carefully controlled way. Heat treatment will also affect the properties of a steel: rapid quenching gives a hard, brittle material, but subsequent heating to 200°C–500°C and cooling gives a strong, tough product–tempered steel.

The **Bessemer** and **Open-Hearth** processes for steel production are shown diagrammatically in Fig. 26.18. Nowadays much steel is regenerated from scrap in an electric furnace; careful selection of the scrap employed can give steels of very precise composition and properties.

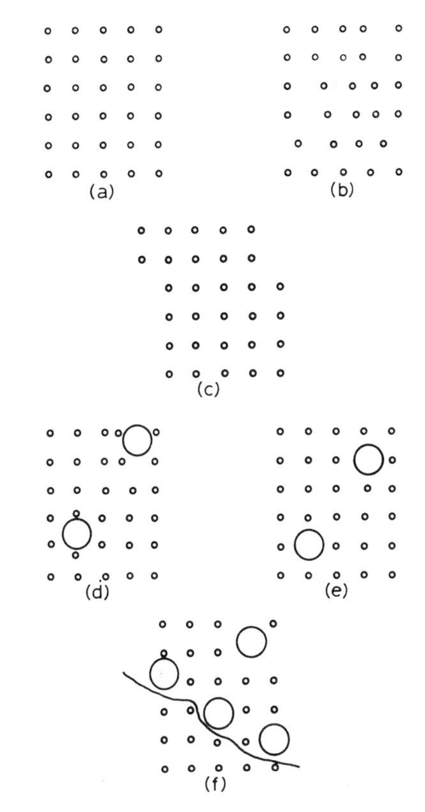

Fig. 26.16 (a) perfect metal structure, (b) structure with 'faults', (c) deformation, (d) interstitial alloy, (e) substitutional alloy, (f) fracture

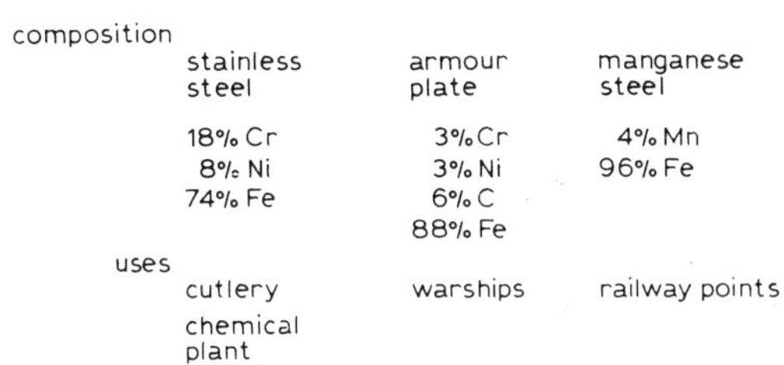

	stainless steel	armour plate	manganese steel
composition	18% Cr 8% Ni 74% Fe	3% Cr 3% Ni 6% C 88% Fe	4% Mn 96% Fe
uses	cutlery chemical plant	warships	railway points

Fig. 26.17 composition and uses of some steels

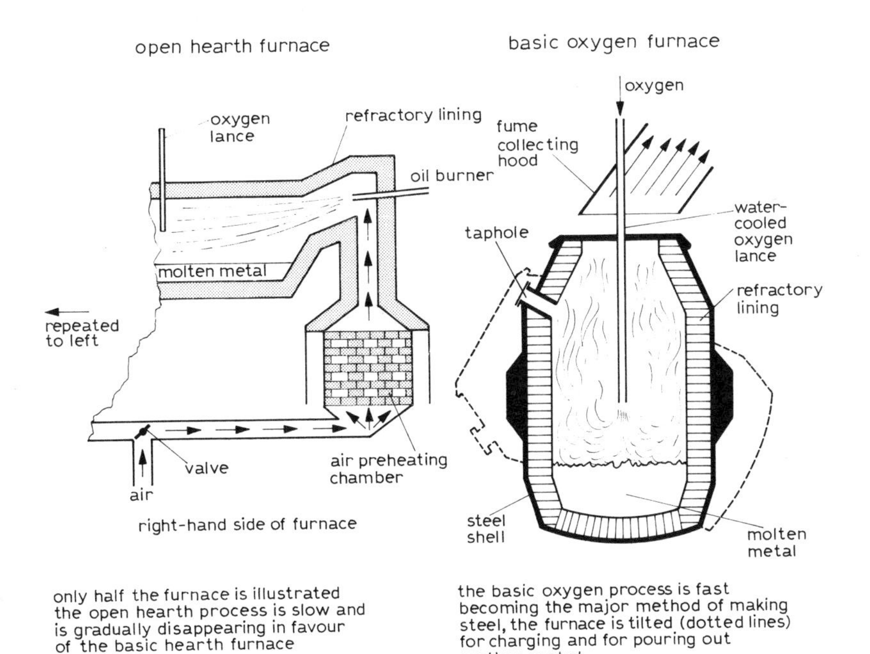

Fig. 26.18

26.15 properties and reactions of iron

Iron is a good conductor, fairly dense (7900 kg m^{-3}), with a melting point of 1530°C. The position of iron in the Periodic Table and its simple atomic properties are shown in Fig. 26.19; Fig. 26.20 shows its position in the reactivity series. Like copper, iron cannot attain a noble gas electron structure by loss of electrons and it shows variable valency. It is a fairly reactive metal;

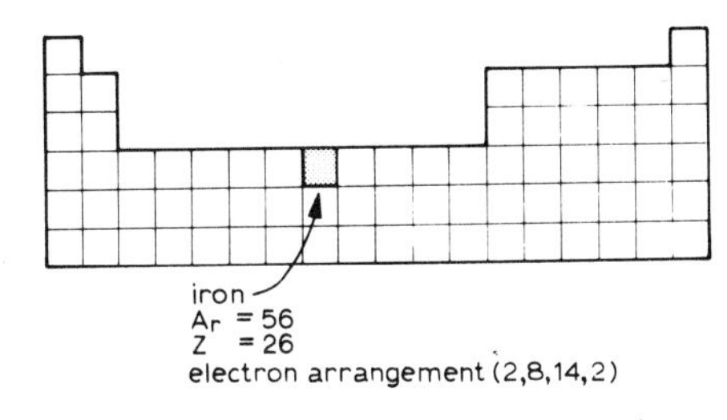

Fig. 26.19 position of iron in Periodic Table

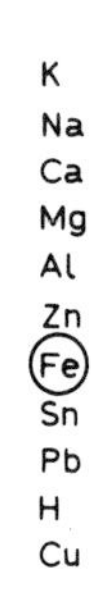

Fig. 26.20 position of iron in reactivity series

it **rusts** in air fairly rapidly, forming hydrated iron(III) oxide, and must be protected by painting or plating (see section 14.25). It displaces hydrogen from steam (section 8.13) and it dissolves in dilute acids to yield hydrogen, though it is unaffected by alkalis.

$$Fe(s) + 2H^+(aq) \rightarrow Fe^{2+}(aq) + H_2(g)$$

Iron does not react with concentrated nitric(V) acid, due to the formation of an oxide layer similar to that formed on aluminium. In this condition it is said to be *passive.*

Iron reacts directly with sulphur on heating, to form black iron(II) sulphide, in a manner very similar to nickel (section 4.5); with chlorine on heating, to form black anhydrous iron(III) chloride (Fig. 15.9); and with oxygen on heating to form the black mixed oxide tri-iron tetraoxide (Fe_3O_4).

26.16 iron(II) compounds

Like copper (see section 26.11), iron is a **transition element**; it forms iron(II) and iron(III) compounds. Aqueous solutions and hydrated crystals of iron(II) compounds are pale green, those of iron(III) compounds are yellow. The oxidation of iron(II) to iron(III) and the reverse reduction are described in section 14.7.

1. iron(II) hydroxide

This is formed as a dull green precipitate when aqueous sodium(I) hydroxide is added to an aqueous solution of an iron(II) salt:

$$Fe^{2+}(aq) + 2OH^-(aq) \rightarrow Fe(OH)_2(s)$$

It darkens very rapidly in air, by partial oxidation to iron(III) hydroxide. It is basic and will not dissolve in excess of either aqueous sodium(I) hydroxide or aqueous ammonia.

2. iron(II) oxide

Iron(II) oxide is unstable in air and although it can be formed in the absence of air by thermal decomposition of certain iron(II) compounds, exposure to air changes it rapidly to iron(III) oxide.

3. iron(II) chloride and iron(II) sulphate(VI)

The preparation of crystalline iron(II) salts presents a slight problem, as hot solutions containing hydrated iron(II) ions are oxidised by air to iron(III) ions. This difficulty is overcome if metallic iron is added to acid of a sufficiently high concentration to yield a saturated solution of the salt without further evaporation of the solution:

$$Fe(s) + 2H^+(aq) \rightarrow Fe^{2+}(aq) + H_2(g)$$

So long as hydrogen is present, no oxidation will occur. Cooling such solutions yields crystals of $FeCl_2.4H_2O$ or $FeSO_4.7H_2O$, according to the acid used.

Heating iron(II) sulphate(VI)-7-water (Green Vitriol) yields a distillate of sulphuric(VI) acid. The salt undergoes autoxidation to iron(III) oxide and sulphur dioxide:

$$2FeSO_4.7H_2O(s) \rightarrow Fe_2O_3(s) + SO_2(g) + SO_3(g) + 14H_2O(g)$$

4. iron(II) sulphide

This is obtained as a black precipitate by the action of hydrogen sulphide on a neutral solution of an iron(II) salt. Addition of acid reverses this reaction and forms a method of preparation of hydrogen sulphide:

$$Fe^{2+}(aq) + H_2S \rightleftharpoons FeS(s) + 2H^+(aq)$$

26.17 iron(III) compounds

1. *iron(III) hydroxide*

This is obtained as a brown precipitate by adding aqueous sodium(I) hydroxide or aqueous ammonia to an aqueous solution containing iron(III) ions:

$$Fe^{3+}(aq) + 3OH^-(aq) \rightarrow Fe(OH)_3(s)$$

It is entirely basic and will not dissolve in excess alkali. Addition of acid reverses its formation:

$$Fe(OH)_3(s) + 3H^+(aq) \rightarrow Fe^{3+}(aq) + 3H_2O(l)$$

2. *iron(III) oxide*

This red-brown powder can be prepared by heating iron(III) hydroxide (or iron(II) sulphate(VI)):

$$2Fe(OH)_3(s) \rightarrow Fe_2O_3(s) + 3H_2O(g)$$

It is used as jeweller's rouge for polishing gems and as a pigment, particularly in coloured glass.

3. *iron(III) chloride*

Anhydrous iron(III) chloride is formed when chlorine is passed over heated iron (see Fig. 15.9). Like aluminium(III) chloride, it is covalent; it forms a dimer Fe_2Cl_6 and it sublimes on heating. It is deliquescent and forms many hydrates.

4. *tests for iron(II) and iron(III) ions*

Aqueous sodium(I) hydroxide or aqueous ammonia give precipitates insoluble in excess alkali, either dull green $Fe(OH)_2$ or red-brown $Fe(OH)_3$. Alternative means of identification, which are useful in cases of red-green colour-blindness (which is surprisingly widespread) are

i) aqueous solutions of iron(III) ions (*but not iron*(II)) give a deep red colour with aqueous thiocyanate:

$$Fe^{3+}(aq) + 6CNS^-(aq) \rightleftharpoons [Fe(CNS)_6]^{3-}(aq)$$

ii) aqueous solutions of iron(III) ions give a deep blue colour or precipitate with aqueous hexacyanoferrate(II):

$$4Fe^{3+}(aq) + 3[Fe(CN)_6]^{4-}(aq) \rightarrow \underset{\text{deep blue}}{Fe_4[Fe(CN)_6]_3(s)}$$

The same deep blue compound is formed if aqueous hexacyanoferrate(III) is added to aqueous iron(II), by exchange between iron(II) and iron(III) with the reagent.

lead

26.18 occurrence and extraction

Lead is a metal which has been known since ancient times. It was mined extensively by the Romans in the Mendip Hills and the Pennines, but all the veins above the water-table are now worked out. Pumping to lower the level of water would be too costly. The principal ore of lead is **galena,** lead(II) sulphide, (PbS). It is concentrated by froth flotation and roasted to the oxide:

$$PbS(s) + 1\tfrac{1}{2}O_2(g) \rightarrow PbO(s) + SO_2(g)$$

The oxide is smelted with carbon and a little limestone, which on heating provides quicklime which removes acidic sandy materials as calcium(II) silicate(IV):

$$2PbO(s) + C(s) \rightarrow 2Pb(s) + CO_2(g)$$
$$CaO(s) + SiO_2(s) \rightarrow CaSiO_3(l)$$

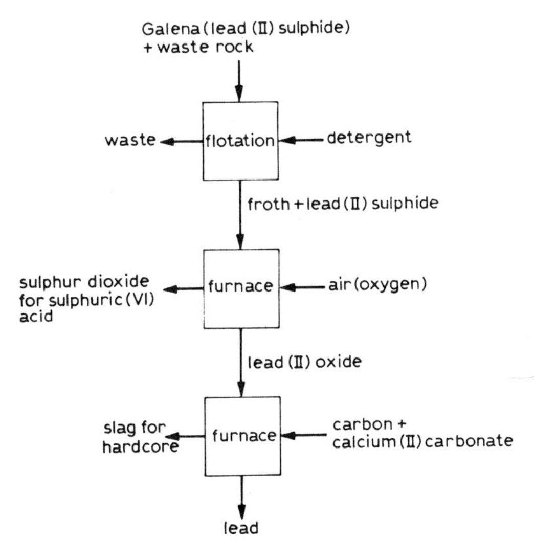

Fig. 26.21 extraction of lead–flow sheet

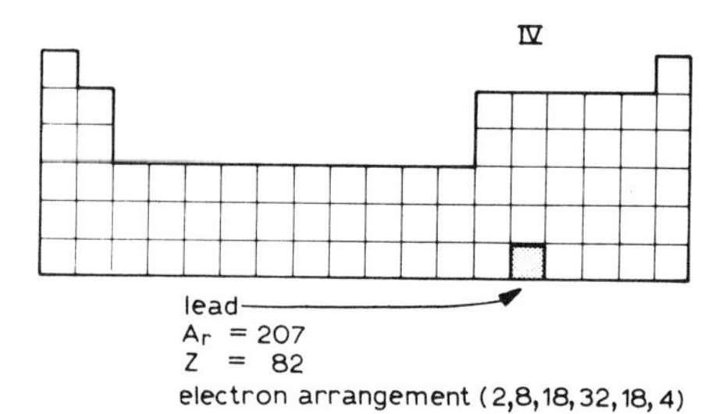

Fig. 26.22 position of lead in Periodic Table

The process is very similar to the extraction of copper or zinc; the essential operations are shown in Fig. 26.21. Recently a combined process has been published which extracts lead and zinc simultaneously from a mixed ore.

26.19 properties and reactions of lead

Lead is celebrated as a heavy metal, and is certainly the densest in common use, at 11 300 $kg\,m^{-3}$. It is also a very soft metal with the low melting-point of 328 °C (601 K), but it is not a good conductor compared with other metals. The tensile strength is poor (1·3 $kg\,mm^{-2}$). The position of lead in the Periodic Table and its simple atomic properties are shown in Fig. 26.22; it shows valencies of II and IV. Its position in the reactivity series is shown in Fig. 26.23.

K
Na
Ca
Mg
Al
Zn
Fe
Sn
(Pb)
H
Cu

Fig. 26.23 position of lead in reactivity series

Although placed just above hydrogen in the reactivity series, its reactions with acids are very slow, unless an oxidising agent is present. It dissolves in concentrated nitric(v) acid:

$$Pb \longrightarrow Pb^{2+} + 2e^-$$
$$(NO_3^- + 2H^+ + e^- \longrightarrow NO_2 + H_2O) \times 2$$
$$\mathbf{Pb(s) + 2NO_3^-(aq) + 4H^+(aq) \longrightarrow Pb^{2+}(aq) + 2H_2O(l) + 2NO_2(g)}$$

It also dissolves in warm dilute nitric(v) acid:

$$(Pb \longrightarrow Pb^{2+} + 2e^-) \times 3$$
$$(NO_3^- + 4H^+ + 3e^- \longrightarrow NO + 2H_2O) \times 2$$
$$\mathbf{3Pb(s) + 2NO_3^-(aq) + 8H^+(aq) \longrightarrow 3Pb^{2+}(aq) + 4H_2O(l) + 2NO(g)}$$

Concentrated hydrochloric and sulphuric(vi) acids attack it slowly, but the latter produces an insoluble coating on the metal which prevents further attack:

$$Pb^{2+}(aq) + SO_4^{2-}(aq) \longrightarrow PbSO_4(s)$$

Concentrated sulphuric(vi) acid is transported and stored in lead containers.

Lead tarnishes slowly in air; on heating in air it forms a variety of oxides, discussed in section 26.21. Neither cold water nor steam attacks the metal.

26.20 uses of lead

The important properties of the metal are

i) softness,
ii) high density,
iii) low melting-point,
iv) resistance to corrosion.

Some care must be exercised in its use because its compounds are poisonous. It has been used in the past for roofing and damp-coursing (i), but is now too expensive for this. It forms water-pipes, cable covers, and is sometimes used in chemical plant [(i), (iv)]. It is used as radioactive shielding, either as the metal or as lead oxide in glass (ii), and in several alloys, particularly those of a low melting- point (iii):

solder:	67% Pb + 33% Sn
type metal:	75% Pb + 25% (Sn, Sb)
accumulator plates:	90% Pb + 10% Sn
bearings:	80% Pb + 20% (Sn, Sb)

The world production of lead was about 4 million tons in 1970.

26.21 oxides of lead

Lead, heated in air at its melting-point, forms the pink monoxide, lead(ii) oxide or litharge. This oxide is also produced on heating lead(ii) hydroxide, carbonate, or nitrate(v):

$$Pb(s) + \tfrac{1}{2}O_2(g) \longrightarrow PbO(s)$$

$$Pb(OH)_2(s) \longrightarrow PbO(s) + H_2O(g)$$
$$PbCO_3(s) \longrightarrow PbO(s) + CO_2(g)$$
$$Pb(NO_3)_2(s) \longrightarrow PbO(s) + 2NO_2(g) + \tfrac{1}{2}O_2(g)$$

When lead(II) oxide is heated in air at 400 °C, dilead(II) lead(IV) oxide–red lead–is formed; the reaction is reversed at about 470 °C:

$$3PbO(s) + \tfrac{1}{2}O_2(g) \rightleftharpoons Pb_3O_4(s)$$

Red lead acts as a mixture of lead(II) and lead(IV) oxides. On the addition of dilute nitric(V) acid, the brown, neutral lead(IV) oxide remains and can be filtered off; the filtrate contains lead(II) ions:

$$Pb_3O_4(s) + 4H^+(aq) \longrightarrow PbO_2(s) + 2Pb^{2+}(aq) + 2H_2O(l)$$

Lead(II) oxide is amphoteric and will dissolve in acid or alkali; lead(IV) oxide dissolves more readily in alkali than in acid. The higher oxide is a good oxidising agent; it will oxidise warm concentrated hydrochloric acid to chlorine:

$$2Cl^- \longrightarrow Cl_2 + 2e^-$$
$$PbO_2 + 4H^+ + 2e^- \longrightarrow Pb^{2+} + 2H_2O$$
$$\mathbf{PbO_2(s) + 4H^+(aq) + 2Cl^-(aq) \longrightarrow Pb^{2+}(aq) + 2H_2O(l) + Cl_2(g)}$$

then $$Pb^{2+}(aq) + 2Cl^-(aq) \longrightarrow PbCl_2(s)$$

All the oxides are reduced to silvery beads of lead if heated with carbon, or in a stream of hydrogen:

$$PbO(s) + H_2(g) \longrightarrow Pb(l) + H_2O(g)$$
$$PbO_2(s) + C(s) \longrightarrow Pb(l) + CO_2(g)$$

The oxides find uses in paints and accumulators (see section 14.22).

26.22 other lead(II) salts

Lead(II) salts are more stable than lead(IV) salts; the most common soluble lead(II) salt is lead(II) nitrate(V). It can be prepared by one of the standard methods for a soluble salt. It decomposes to the oxide on heating:

$$Pb(NO_3)_2 \longrightarrow PbO(s) + 2NO_2(g) + \tfrac{1}{2}O_2(g)$$

Other lead(II) salts, except the acetate, are all insoluble and all white, except the sulphide which is black and the iodide which is yellow. Lead(II) chloride and lead(II) iodide are soluble in hot water and can be recrystallised from it. Lead(II) hydroxide is amphoteric; it will dissolve in dilute acid and in aqueous sodium(I) hydroxide, but not in aqueous ammonia:

$$Pb(OH)_2(s) + 4OH^-(aq) \longrightarrow \underset{\text{hexahydroxoplumbate(II) ion}}{[Pb(OH)_6]^{4-}}$$

26.23 tests for lead(II) ions

A solution containing aqueous lead(II) ions gives a white precipitate with aqueous sodium(I) hydroxide, soluble in excess alkali, a white precipitate with aqueous ammonia, insoluble in excess, and a black precipitate with hydrogen sulphide, even in acid solution.

magnesium

26.24 occurrence and extraction

Magnesium occurs as the solid carbonate **magnesite** ($MgCO_3$), and the double carbonate **dolomite** ($MgCO_3.CaCO_3$), but larger quantities still are available from **sea water.** Although the concentration of magnesium(II) ions is low (0·12%), this is compensated for by the vast volume of the oceans.

Both sources are used for the extraction of magnesium. The process based on the carbonate involves heating to form the oxide and subsequent reduction with silicon, as shown on the next page.

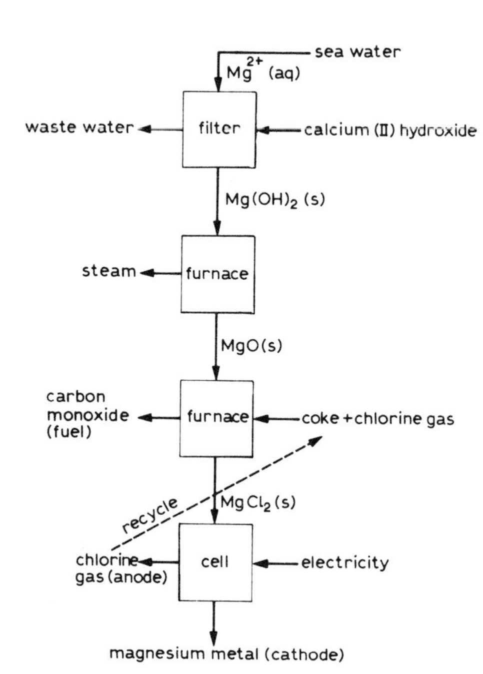

Fig. 26.24 extraction of magnesium from sea water – flow sheet

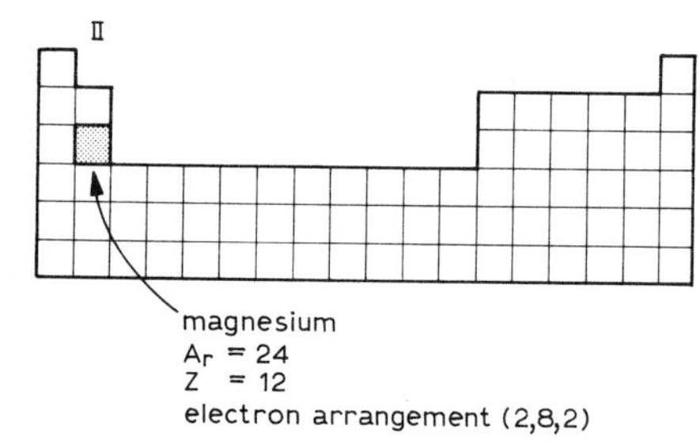

Fig. 26.25 position of magnesium in Periodic Table

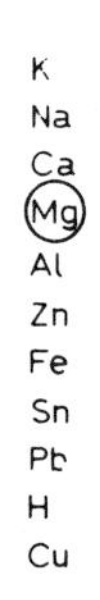

Fig. 26.26 position of magnesium in reactivity series

$$CaCO_3.MgCO_3 \rightarrow CaO, \quad MgO \xrightarrow{Si} CaSiO_3, \quad Mg$$

Magnesium is extracted from sea water by the following electrolytic method.

Magnesium(II) hydroxide is first precipitated from sea water by adding calcium(II) hydroxide; the precipitate is filtered off and heated to form the oxide. This is further heated with coke in a stream of chlorine, to form anhydrous magnesium(II) chloride:

$$MgO(s) + C(s) + Cl_2(g) \rightarrow MgCl_2(s) + CO(g)$$

When this is mixed with a little sodium(I) chloride, it has a sufficiently low melting-point (700 °C) for electrolysis; a carbon anode and steel cathode are used. The magnesium, which is liquid at this temperature and less dense than the electrolyte, floats to the surface.

26.25 properties and reactions of magnesium

Magnesium is a white metal, malleable and ductile, with a very low density ($1500\,kg\,m^{-3}$) and reasonable tensile strength ($9{\cdot}1\,kg\,mm^{-2}$). The tensile strength is improved by alloying (see section 26.14). Its melting-point is 650 °C (923 K). Its position in the Periodic Table and simple atomic properties are shown in Fig. 26.25 – it has a constant valency of II. Its position in the reactivity series is shown in Fig. 26.26.

The metal tarnishes in air, forming a surface layer of oxide which is resistant to further attack. At higher temperatures this is no longer a protection: magnesium catches fire in either air or steam on strong heating, to form the oxide:

$$Mg(s) + \tfrac{1}{2}O_2(g) \rightarrow MgO(s)$$
$$Mg(s) + H_2O(g) \rightarrow MgO(s) + H_2(g)$$

It reacts rapidly with dilute acids, liberating hydrogen:

$$Mg(s) + 2H^+(aq) \rightarrow Mg^{2+}(aq) + H_2(g)$$

Magnesium is the only metal to liberate hydrogen from nitric(V) acid, which it will do if the acid is very dilute. The metal is unaffected by alkalis.

26.26 uses of magnesium

Its very low density and resistance to corrosion in dry conditions indicate a use in lightweight alloys. It is used in interior fittings in aircraft, usually as an alloy (90% Mg, 10% (Al + Zn + Mn)); it is also used for fan-blades and in printing presses to form the moving parts of high-speed machinery.

26.27 magnesium(II) compounds

1. magnesium(II) hydroxide

This is a white solid, precipitated from aqueous solutions of magnesium(II) ions by aqueous sodium(II) hydroxide or aqueous ammonia:

$$Mg^{2+}(aq) + 2OH^-(aq) \rightarrow Mg(OH)_2(s)$$

Being entirely basic, it dissolves in dilute acids but not in alkalis:

$$Mg(OH)_2(s) + 2H^+(aq) \rightarrow Mg^{2+}(aq) + 2H_2O(l)$$

It is used to relieve indigestion by neutralising stomach acidity (see section 11.6).

2. magnesium(II) oxide

This is a white powder which may be obtained by heating the oxide, carbonate or nitrate(V) – or in a less pure state by heating magnesium in air. It is used as a lining for furnaces because of its high melting-point (2500 °C).

3. *magnesium(II) salts*

All magnesium(II) salts are white or colourless and, except the carbonate, soluble. They may be prepared in the hydrated form (chloride $MgCl_2.6H_2O$; nitrate(V) $Mg(NO_3)_2.6H_2O$; sulphate(VI) $MgSO_4.7H_2O$) by the standard methods for soluble salts. The chloride cannot be dehydrated by heating, since magnesium oxide is formed by hydrolysis:

$$Mg^{2+}(aq) + H_2O(l) \longrightarrow MgO(s) + 2H^+(aq)$$

The anhydrous salt is best formed by direct combination of magnesium with chlorine. Magnesium(II) sulphate(VI) is used in medicine as a purgative (Epsom salt) and also in tanning and dyeing.

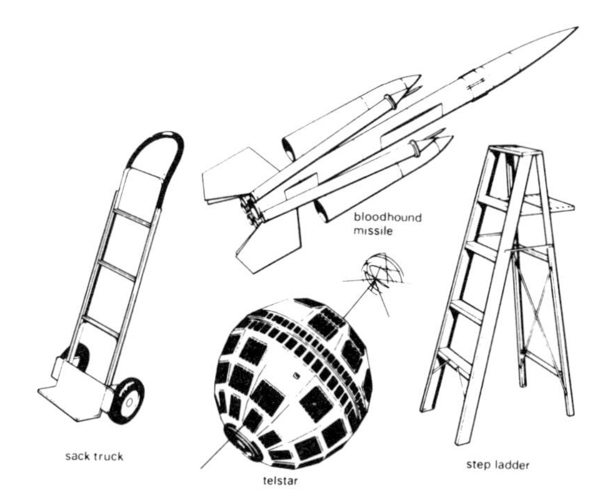

Fig. 26.27 some uses of magnesium

4. *tests for magnesium(II) ions*

An aqueous solution containing magnesium(II) ions gives a white precipitate with both aqueous sodium(I) hydroxide and aqueous ammonia. In neither case does the precipitate dissolve in excess alkali.

zinc

26.28 occurrence and extraction

Zinc, like lead, occurs principally in sulphide ores, e.g. **zinc blende** (ZnS). The extraction process is virtually identical to that for lead (section 26.18). the only major difference being that zinc is volatile. The metal distils and is condensed in chambers beyond the top of the retort in which the oxide is reduced. It is redistilled in an inert atmosphere to obtain a metal of higher purity. A blast furnace method, which has the advantage of being continuous, has recently been developed.

26.29 properties and reactions of zinc

Zinc is reasonably malleable, ductile and tough, and has a moderate density ($7100\,kg\,m^{-3}$), conductivity and tensile strength ($10{\cdot}5\,kg\,mm^{-2}$). The melting-point is quite low (420°C or 693 K). Its position in the Periodic Table and simple atomic properties are shown in Fig. 26.28; it has a constant valency of II and is *not* a typical transition element. Its position in the reactivity series is shown in Fig. 26.29.

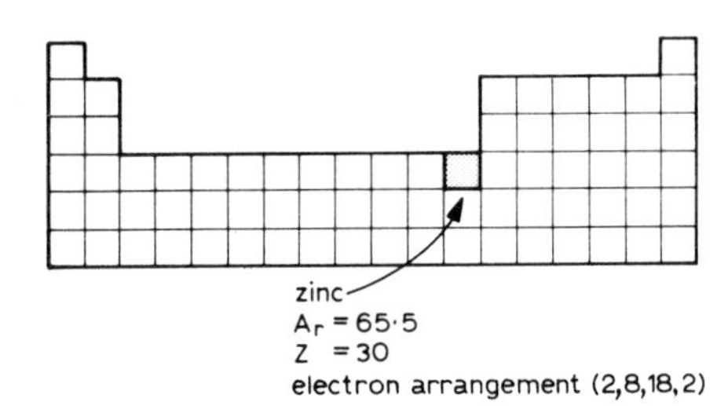

Fig. 26.28 position of zinc in Periodic Table

The metal burns with a green flame when heated in air or steam to form the oxide:

$$Zn(s) + \tfrac{1}{2}O_2(g) \longrightarrow ZnO(s)$$
$$Zn(s) + H_2O(g) \longrightarrow ZnO(s) + H_2(g)$$

Commercial zinc dissolves readily in dilute hydrochloric or sulphuric(VI) acid liberating hydrogen:

$$Zn(s) + 2H^+(aq) \longrightarrow Zn^{2+}(aq) + H_2(g)$$

Like aluminium, zinc dissolves in hot aqueous sodium(I) hydroxide liberating hydrogen:

$$Zn \longrightarrow Zn^{2+} + 2e^-$$
$$Zn^{2+} + 6OH^- \longrightarrow [Zn(OH)_6]^{4-}$$
$$Zn + 6OH^- \longrightarrow [Zn(OH)_6]^{4-} + 2e^-$$
$$2H_2O + 2e^- \longrightarrow H_2 + 2OH^-$$
$$\mathbf{Zn(s) + 4OH^-(aq) + 2H_2O(l) \longrightarrow [Zn(OH)_6]^{4-}(aq) + H_2(g)}$$

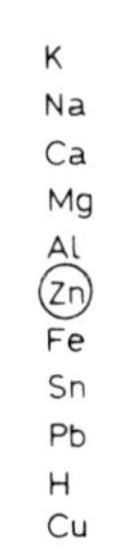

Fig. 26.29 position of zinc in reactivity series

26.30 uses of zinc

Because of its moderate reactivity, zinc is used as a sacrificial coating for iron in **'galvanised iron'**; this is described in detail in section 14.25(3)(b). The coating of zinc is applied either as powder, or by dipping the objects to be coated in molten zinc. This use accounts for about half the world production of 5 million tons (1970).

Zinc forms good castings when alloyed with 5% (Al + Cu + Mn) and is used for carburettors, padlocks and other small articles. It is also a constituent of brass (section 26.10) and forms the casing of dry batteries (section 14.22).

26.31 zinc(II) compounds

1. zinc(II) hydroxide

This white solid is precipitated from aqueous solutions containing zinc(II) ions by aqueous sodium(I) hydroxide or aqueous ammonia:

$$Zn^{2+}(aq) + 2OH^-(aq) \rightarrow Zn(OH)_2(s)$$

It is amphoteric and dissolves readily in excess hydroxide or in dilute acid; it also dissolves in excess aqueous ammonia because of the formation of tetraamminezinc(II) ions:

$$Zn(OH)_2(s) + 2H^+(aq) \rightarrow Zn^{2+}(aq) + 2H_2O(l)$$
$$Zn(OH)_2(s) + 4OH^-(aq) \rightarrow [Zn(OH)_6]^{4-}$$
$$Zn(OH)_2(s) + 4NH_3(aq) \rightarrow [Zn(NH_3)_4]^{2+}(aq) + 2OH^-(aq)$$

2. zinc(II) oxide

This may be prepared in the laboratory by heating the hydroxide, carbonate, or nitrate(V); it is produced commercially by burning zinc. It is yellow when hot and white when cold and like the hydroxide, it is amphoteric. It is used in paints which are not to be affected by atmospheric pollution (zinc(II) sulphide is white, whereas lead(II) sulphide is black), and as a filler for white rubber.

3. zinc(II) chloride, sulphate(VI) and nitrate(V)

These salts are white or colourless and water-soluble; they may be prepared by the standard methods for soluble salts as their hydrates: $ZnCl_2.H_2O$, $ZnSO_4.7H_2O$, and $Zn(NO_3)_2.6H_2O$. The chloride is deliquescent.

4. zinc(II) carbonate and sulphide

These salts are white and insoluble in water. The sulphide is sometimes used as a pigment and the carbonate is used in calamine lotion for the treatment of skin irritation.

5. tests for zinc(II) ions

Aqueous solutions containing zinc(II) ions give white precipitates which re-dissolve in excess alkali, with both aqueous sodium(I) hydroxide and aqueous ammonia.

Thermal decomposition of zinc(II) salts usually gives the oxide, which is readily identifiable as a yellow solid when hot, which turns white on cooling.

questions 26

1. Aluminium, copper, iron, lead, magnesium, sodium, zinc. From the above list of metals select one in each case which

(a) forms an insoluble sulphate(VI)
(b) forms a nitrate(V) which when heated yields oxygen as the only gaseous product
(c) forms both divalent and trivalent ions
(d) forms an amphoteric oxide
(e) is used in the thermit reaction
(f) forms an oxide which is yellow when cold. (JMB)

2. For each of the following substances outline **one** experiment to demonstrate a typical chemical property of a metal or a metallic compound; each experiment should concern a different property which must be stated clearly:

Magnesium ribbon; metallic calcium; copper(II) oxide; copper(II) sulphate(VI) solution.

What are the **two** principal methods used for the extraction of metals not found in the metallic state in nature? Confine your answer to concise statements outlining the methods in general terms.

State which of the following metals are extracted by each of the methods outlined:

Sodium; iron; zinc; aluminium; lead; magnesium. (W)

3. Iron displaces copper from solution according to the equation:

$$Cu^{2+} + Fe \longrightarrow Fe^{2+} + Cu.$$

Devise an experiment whereby you could determine the mass of copper produced by adding a known mass of iron to an excess of a solution of copper(II) sulphate(VI).

In such an experiment, it was found that 1·9 g of iron yielded 2·0 g of a powder which was found to react as follows:
(i) with dilute nitric(V) acid a very pale blue solution was formed,
(ii) with dilute hydrochloric acid hydrogen was evolved.

Comment on these results.
($Fe = 56$, $Cu = 64$) (C)

4. Describe
(a) the reaction of calcium metal with (i) air, (ii) water;
(b) the reaction of calcium(II) oxide with water;
(c) the effect of heat on calcium(II) carbonate.

Show how all these reactions agree with the position of calcium in the electrochemical series.

Starting with calcium(II) chloride describe briefly how you would prepare (d) hydrogen chloride, (e) calcium(II) carbonate. (S)

5. Describe what you would observe and explain simply what would happen in **three** of the following experiments:
(i) a small piece of sodium is carefully placed on a filter paper floating on water;
(ii) sodium, in a deflagrating spoon, is ignited and then plunged into a gas jar of oxygen;
(iii) iron wire is heated at one end until red hot and is immediately plunged into a gas jar of oxygen;
(iv) steam is passed over heated magnesium ribbon in a silica tube. (C)

6. How would you prepare a pure sample of iron(II) sulphate(VI) crystals in the laboratory, starting from iron filings?

A solution contains a mixture of iron(II) and iron(III) chlorides. State what you would see, and explain the reactions that occur when:
(a) chlorine is passed into a portion of the solution until there is no further change;
(b) another portion of the solution is acidified with hydrochloric acid, then zinc is added and the mixture allowed to stand;
(c) excess sodium(I) hydroxide solution is added to the products from (a) and (b). (AEB)

7. Give a brief account of the chemistry of the industrial process for the extraction of aluminium from aluminium(III) oxide. (No technological details are required.)

Mention **three** properties of aluminium that make it a very useful metal.

Describe how you would prepare from aluminium (a) anhydrous aluminium(III) chloride, (b) aluminium(III) hydroxide. (S)

8. Name **one** ore of zinc.
Outline the chemistry for the manufacture of zinc from this ore. Describe the changes which take place when:
(a) zinc(II) carbonate is strongly heated in air;
(b) excess zinc is placed in copper(II) sulphate(VI) solution;
(c) pure zinc is placed in dilute sulphuric(VI) acid and a few drops of copper(II) sulphate(VI) solution are added later.

Give one use of zinc, explaining why it is used for the purpose mentioned. (S)

9. (a) Name and give the formula of an ore of iron. Explain how the metal is extracted from the ore, and describe the reactions which take place in the furnace. What are the **two** functions of carbon in the furnace, and for what reason is limestone added?

(b) The following experiments were carried out starting with some iron filings:
(i) the filings were dissolved in dilute sulphuric(VI) acid,
(ii) more dilute sulphuric(VI) acid and a few drops of concentrated nitric(V) acid were added to the clear solution, which was then heated to boiling.

State all that you would have observed in stages (i) and (ii) and say what colour **each** of the solutions had. Name the salts formed in stages (i) and (ii) respectively and give equations for the reactions which formed them. Explain how you would test the two solutions to prove the presence of the metallic radicals in the salts you name. (C)

10. Explain the following observations; where possible, name the elements and compounds involved and give equations for the reactions.

(a) A white crystalline solid decomposes on heating to give a yellow residue and a mixture of gases which is brown in colour. In this mixture a glowing splint bursts into flame. The residue remains yellow on cooling.

(b) When a white compound is very strongly heated it gives off a gas which turns limewater cloudy and leaves a white residue. When water is added to the cold residue a hissing noise is heard, great heat is evolved and steam is given off. The compound so formed is slightly soluble in water.

(c) When potassium(I) iodide solution is added to a solution of a salt in water, a yellow precipitate is obtained. On boiling the solution, the precipitate dissolves, but on cooling under the tap, glistening yellow crystals appear.

(d) When a blue solution is electrolysed using platinum electrodes, the cathode becomes coated with a reddish-brown deposit and a colourless gas collects at the anode. (AEB)

11. Identify the unnamed substances in the following reaction schemes. Explain your reasonings and write equations for the reactions.

(a) A is a blue solid, which on heating liberates a brown gas and leaves a black residue B. The black residue dissolves in dilute sulphuric(VI) acid and when the resultant solution is made alkaline, a pale blue precipitate C is observed which turns black on boiling.

(b) D is a solid with a metallic lustre. It reacts with dilute sulphuric(VI) acid liberating a gas E and producing a pale green solution F. A dark green precipitate G is observed when the solution is made alkaline. G turns to a brown solid H on standing. (S)

27.1 aids to memory

The emphasis in much of this book has been on the understanding of principles and their application to chemical processes. A good memory is, nevertheless, valuable to the student of chemistry, and some important facts can be memorised easily with a little help.

1. solubility of salts

It is not difficult to remember that all nitrates(v) and all sodium, potassium and ammonium salts are soluble, or that all carbonates are insoluble–with the obvious exceptions of those of sodium, potassium and ammonium.

Pb, Ba, INsol **SU**lphates

'Pub bars in summer' may help you 'remember that lead and barium form insoluble sulphates.

Pb, Ag, Hg(I) INsol **C**h**L**orides

'Publicans against having one in Club'–lead, silver and mercury(I) form insoluble chlorides.

2. electrochemical series

K Ca Na Mg Al Zn Fe Sn,
Pb H Cu Ag

The order of the metals in this series is very important. **'Kind cannibals never must ask zombies for supper; poltergeists hate coffee afterwards'** gets them in the right order.

3. Periodic Table

Li Be B C N O F
Na Mg Al Si P S Cl

		He
Li	F	Ne
Na	Cl	Ar
K	Br	Kr
Rb	I	Xe
Cs		Rn

It is neither necessary nor specially valuable to try to memorise the whole table. **'Little beggar boys catching newts or fishes'** and **'nasty Margaret always sitting peeling spuds clumsily'** give the order for the first two horizontal rows. If about three vertical groups are remembered, one will be doing well. **'He never argues (with) killer X-rays'** helps with the noble gases; **'Little Nan keeps robbing castles'** with the alkali metals; **'foolish clerks brew ink'** with the halogens.

4. industrial electrolyses

V essel
E lectrolyte
C athode
A node
P olarity
P roducts
T emperature
S iting

The important features of electrolytic manufacturing processes to include in a labelled diagram are:

material of vessel, electrolyte, nature of cathode, nature of anode, polarity, products, temperature and siting (of electrodes and products). **'Very earnest chemists and physicists produce top scientists.'**

27.2 the electrochemical series and properties

Strictly speaking, the electrochemical series is an arrangement of the metals in order of their standard electrode potentials, which means that it can be applied with complete justification only when considering how a metal will behave when it is in equilibrium with a molar aqueous solution of its ions,

e.g. $$Zn^{2+}(aq) + 2e^- \rightleftharpoons Zn(s), \quad E^{\ominus} = -0{\cdot}76\,V$$

However, the reactivity of metals towards air, water and acids and the thermal stability of a number of their compounds fall into an order very close to that of the electrochemical series. Thus, if the series is amended slightly with respect to calcium, a table can be constructed which summarises much useful information about metals and their compounds (Fig. 27.1).

	metal			reduction of oxide		heat on	
	heated in air	water or steam	acids	C	H_2	carbonate	nitrate(V)
K	burn to oxide	react with cold water	violent reaction	×	×	soluble, thermally stable	nitrate(III) formed
Na				×	×		
Ca		burn in steam	(rapid)	×	×	insoluble, decompose to oxide on heating	decomposed to oxide
Mg				×	×		
Al				×	×		
Zn			reaction	✓	×		
Fe	slowly oxidised	⇌		✓	⇌		
Sn		not attacked	(slow)	✓	✓		
Pb			attacked only by oxidising acids	✓	✓		
Cu				✓	✓		
Hg	⇌			decomposed by heat alone		do not exist	decomposed to metal
Ag	not oxidised						
Au							
Pt							

Fig. 27.1 electrochemical series and properties

27.3 extraction of metals

The extraction of a metal from its ore is essentially a process of reduction:

$$M^{n+}(s) + ne^- \longrightarrow M(s)$$

1. electrolytic processes

The most reactive metals shown in Fig. 27.1 can be obtained only by reduction at the cathode in the electrolysis of their molten chlorides (potassium–magnesium) or molten fluoride (aluminium).

2. reduction of oxides

Less reactive metals can be obtained by reduction of their oxides, using carbon or carbon monoxide at fairly high temperatures, since both carbon and carbon monoxide react with oxygen to form a stable oxide.

The oxidation

$$\tfrac{2}{3}Fe(s) + \tfrac{1}{2}O_2(g) \longrightarrow \tfrac{1}{3}Fe_2O_3(s); \quad \Delta H = -275\,kJ$$

requires 275 kJ to be supplied to reverse it; this is supplied by the process

$$CO(g) + \tfrac{1}{2}O_2(g) \longrightarrow CO_2(g); \quad \Delta H = -284\,kJ$$

and the reduction of iron ore by carbon monoxide is feasible:

$$\tfrac{1}{3}Fe_2O_3(s) + CO(g) \longrightarrow \tfrac{2}{3}Fe(s) + CO_2(g); \quad \Delta H \textbf{ negative}$$

The reduction of iron ore by carbon is *not* feasible; in the blast furnace, carbon monoxide is the reductant, not carbon; carbon cannot supply sufficient energy for the reduction.

$$\tfrac{1}{2}C(s) + \tfrac{1}{2}O_2(g) \longrightarrow \tfrac{1}{2}CO_2(g); \quad \Delta H = -197\,kJ$$

$$\frac{1}{3}Fe_2O_3(s) + \frac{1}{2}C(s) \longrightarrow \frac{2}{3}Fe(s) + \frac{1}{2}CO_2(g); \quad \Delta H \textbf{ positive}$$

In the above account, all the changes are referred to the standard of half of a mole of oxygen, hence the equations are somewhat unwieldy.

The energy changes quoted are for 298 K – they are, in fact, rather different at high temperatures, but the principle remains the same. Indeed, high temperatures are employed in industrial processes for the extraction of metals by reduction of oxides, in order to obtain favourable values for the energies available from the reductant and required by the oxidant (the ore).

In earlier work we have noticed that a volatile acid can be displaced from its salt by a less volatile acid (e.g. hydrogen chloride and nitric(V) acid displaced by sulphuric(VI) acid). Similarly, if a metal is volatile, a reduction which would not be feasible below its boiling-point, takes place readily above its boiling point: magnesium(II) oxide and zinc(II) oxide can both be reduced by carbon at a temperature high enough to vaporise the metal, although the energy values do not appear favourable:

$$MgO(s) + \frac{1}{2}C(s) \longrightarrow \frac{1}{2}CO_2(g) + Mg(g)$$
$$ZnO(s) + \frac{1}{2}C(s) \longrightarrow \frac{1}{2}CO_2(g) + Zn(g)$$

Such a process is used for zinc; it is not used for magnesium because of the difficulty of condensing magnesium vapour safely.

3. reduction of sulphides

Carbon will not readily extract sulphur from a sulphide ore:

$$C(s) + 2S(s) \longrightarrow CS_2(l); \quad \Delta H = +131\,kJ$$

It is difficult to find a reductant which will pick up sulphur readily from sulphide ores. The best available process is to roast the ore in air and oxidise the sulphur to sulphur dioxide:

$$S(s) + O_2(g) \longrightarrow SO_2(g); \quad \Delta H = -298\,kJ$$

This provides sufficient energy to reverse such processes as

$$Cu(s) + S(s) \longrightarrow CuS(s); \quad \Delta H = -49\,kJ$$

Thus in the case of an unreactive metal (mercury, copper) roasting the sulphide ore produces the metal. Unfortunately, more reactive metals (lead, zinc) combine with oxygen during the process to give the oxide; in these cases roasting the sulphide ore must be followed by reduction of the oxide using carbon.

27.4 economics of industrial processes

To be economical, an industrial process must utilise raw materials which are relatively cheap and readily available (including any sources of energy required); must produce a valuable and useful main product, and should if possible form by-products which are valuable in themselves or which can be re-used in the process. These considerations enable us to interpret certain features of well-known manufacturing processes.

1. siting of the plant

Oil refineries occur in or near ports; plants for the extraction of magnesium near the sea, for aluminium near to sources of cheap hydroelectricity; ironworks have traditionally developed on or near coalfields and sulphuric(VI) acid plants side by side with gas works. These facts exemplify the importance of transportation, availability of raw materials, hydroelectric power and fuel, and the interchange of the waste product in one process which can be utilised as a raw material in another:

$$Fe_2O_3 + 3H_2S \longrightarrow Fe_2S_3 + 3H_2O \quad (\textit{gasworks})$$
$$Fe_2S_3 + 4\tfrac{1}{2}O_2 \longrightarrow Fe_2O_3 + 3SO_2 \quad (\textit{Contact Process})$$

Industry does not have a completely free hand in its choice of site; even when its development in rural areas brings welcome employment, it can be opposed by conservationists both on aesthetic grounds and because of the pollution which it brings.

2. continuity of working

A continuous process is generally more economical than a batch process which requires periodic recharging. In particular, processes involving fluids can be operated on continuous flow lines much more easily than those involving batches of solids. Liquid oil (*hydrocarbons*), liquid sulphur (*Contact Process*), gaseous nitrogen and hydrogen and liquid ammonia (*Haber Process*) all exemplify this; many other processes are now being developed by continuous flow methods.

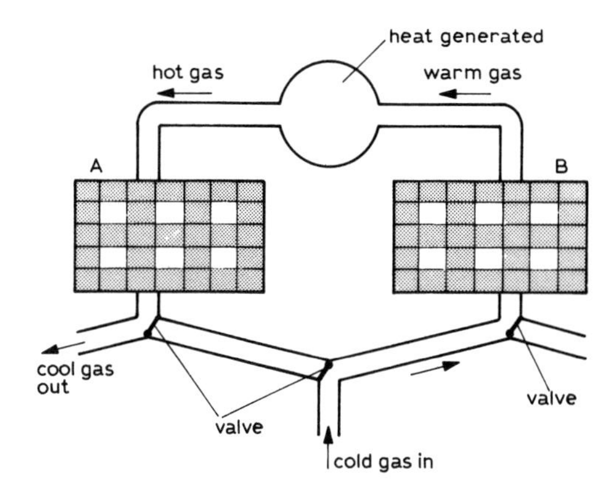

Fig. 27.2 'chequer-brickwork' heat exchange system

3. heat exchange

Many reactions which require a high temperature of working are nevertheless exothermic. When the products are cooled, it is desirable to transfer heat from them to heat up the raw materials for the reaction. This is again most convenient with fluids; waste gases can be passed over a **chequer-brickwork** system to heat up the brickwork (Fig. 27.2) and a valve system can then direct incoming gases over this, while switching the hot waste gases to a second system. Many open-hearth steel works use such a procedure. An alternative heat-exchange system which can be used with liquids employs a **coiled pipe,** as illustrated in Fig. 27.3.

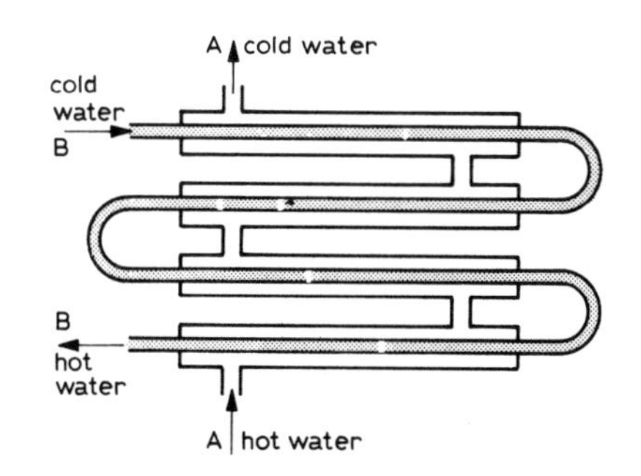

Fig. 27.3 'coiled-pipe' heat exchange system

4. re-use of by-products

Processes which accomplish only a partial conversion of raw materials to the main product (Haber Process, Contact Process, nitric(v) acid manufacture) are rendered economical by **recycling** unchanged reactants–this is especially easy if the substances concerned are fluids.

The **Solvay Process** (section 16.8) illustrates well the efficient re-use of by-products. When the hydrogencarbonate is heated to give the carbonate, carbon dioxide is returned to the carbonator:

$$2HCO_3^-(s) \longrightarrow H_2O(g) + CO_3^{2-}(s) + CO_2(g)$$

The principal source of carbon dioxide is limestone:

$$CaCO_3(s) \longrightarrow CaO(s) + CO_2(g)$$

The quicklime produced in this reaction is utilised to liberate ammonia (a raw material) from ammonium ions in the waste products:

$$O^{2-}(s) + 2NH_4^+(aq) \longrightarrow H_2O(l) + 2NH_3(g)$$

5. valuable by-products

Sodium may be manufactured by the electrolysis of molten sodium(I) chloride **(Downs Process)** or molten sodium(I) hydroxide **(Castner Process).** One of the main advantages of the Downs Process has been that its by-product (chlorine) is more valuable than that of the Castner Process (oxygen). Now that 'tonnage oxygen' is being used to enrich the air blast in steel manufacture, the demand for oxygen has increased and the Castner Process has once more become competitive.

The **petroleum industry** offers perhaps the most conspicuously successful example of the development of a by-product. At one time, the unsaturated gases generated in the 'cracking' plant (section 19.8) were burnt as waste gases. They have now been put to such use that a whole industry–the petrochemicals industry–is based on them; it may even be considered to rival the parent petroleum industry in importance.

27.5 Fajans' rules

Ionic and covalent bonding in compounds is discussed in Chapter 18 and reference is made there to the fact that many covalent bonds are partially polar (ionic) in nature. It is also true that positively and negatively charged ions, in a compound predicted to be purely ionic, may interact with each other to produce a degree of covalency. This feature was not dealt with in Chapter 18, as it is rather beyond the level of treatment appropriate to the main arguments in the Chapter, but it has a place in this retrospective review.

In a purely ionic compound, the cations and anions in the lattice attract each other strongly, but they do not distort the distribution of charge within the ions. If a cation has a very intense charge, it may well attract the electrons in the neighbouring anions towards itself; the anion is then said to be **polarised,** because the distribution of electrons in it is no longer symmetrical (see Fig. 27.4). Such a polarisation leads to the electrons of the anion being partially **shared** by the cation–a degree of **covalency.**

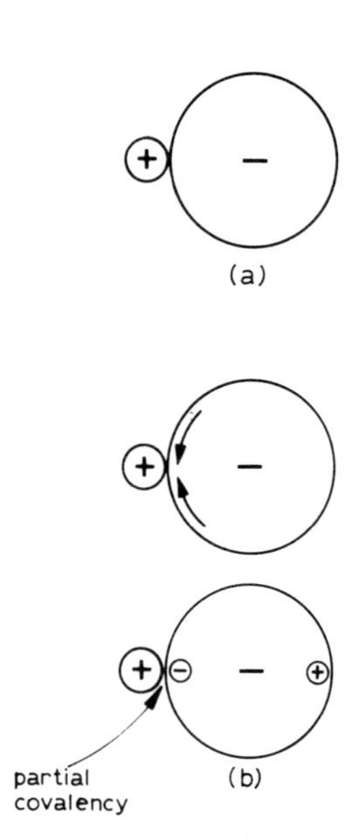

Fig. 27.4 polarisation of an anion by a cation: (a) purely ionic–no polarisation, (b) partly covalent due to polarisation

The characteristics of cations and anions which produce this effect are summarised in **Fajans' rules.** These state:

i) a high degree of covalency will result when a cation has a high polarising power or when an anion has a high polarisibility,
ii) high polarising power results in a cation with a multiple charge and a small radius,
iii) high polarisibility results in an anion with a multiple charge and a large radius.

The halides of silver and lead form a simple illustration of this effect. The ions Ag^+, Pb^{2+}, Cl^-, Br^- and I^- are all colourless. AgCl and $PbCl_2$ are white, which is not surprising. AgBr and $PbBr_2$ are pale yellow and AgI and PbI_2 are quite distinctly yellow–they cannot be purely ionic. As the halide ions increase in size from Cl^- to I^-, they become increasingly *polarisible* by Ag^+ and Pb^{2+}; a successively higher degree of covalency results.

Silver(I) chloride is soluble in aqueous ammonia, which reacts with the silver(I) ion:

$$Ag^+(s) + 2NH_3(aq) \rightleftharpoons [Ag(NH_3)_2]^+(aq)$$

Silver(I) iodide is insoluble in aqueous ammonia–more evidence that the compound is not purely ionic.

Iron(III) chloride and aluminium(III) chloride sublime readily (Fig. 15.9); this is a sure sign of **molecular** composition, resulting from **covalent** bonding, which in turn arises from the high polarising power of the Fe^{3+} and Al^{3+} ions. Aluminium fluoride is the only aluminium halide which can be electrolysed (section 26.1); it is ionic, as the small fluoride ion has a very low polarisibility.

27.6 atomic and ionic radii

It may be appreciated from the previous section that a knowledge of the sizes of atoms and ions can be of great importance in interpreting the chemistry of the elements and their compounds. The tables at the back of the book show the sizes of atoms and ions in terms of atomic and ionic radii. The following simple points may be made about the patterns which emerge:

i) atomic radii decrease across a horizontal row of the Periodic Table;
ii) cations are small and anions large, compared with neutral atoms;
iii) atomic and ionic radii increase down a vertical group of the Periodic Table;
iv) atomic and ionic radii do not vary greatly across a row in the centre block (transition elements).

The chemistry of groups of elements can be interpreted more meaningfully using these patterns: can you see why metallic character increases as

Group I is descended; why carbon, at the top of Group IV, is a non-metal and lead, at the bottom, is a metal; why fluorine atoms have more marked oxidising properties than iodine atoms; why iodide ions are so much more easily oxidised than chloride ions?

The authors hope that you will go on asking and answering questions such as these. The study of chemistry is demanding, but rewarding; this Chapter is intended to give a prospect of wider horizons as well as a retrospect.

questions 27

1. Iron is manufactured by passing a blast of hot air through a mixture of iron ore, coke and limestone.

(a) Explain carefully, with equations, the function of
  (i) the air,
  (ii) the coke,
  (iii) the limestone.

(b) Give the reason why pig-iron is brittle.

(c) State **briefly** what has to be done to pig-iron to convert it into steel.

(d) If you were to build a new factory for manufacturing steel from imported iron ore, suggest two factors which would influence your choice of site. (JMB)

2. The following list is the 'Electrochemical series' of some metals and hydrogen in decreasing order of activity.

Mg Zn Fe Pb (H) Cu Hg Ag
——decreasing activity——→

Choose elements from this list to answer the following questions.

(a) Which metal, other than mercury, could be obtained by the action of heat alone upon its oxide?

(b) Which metal will replace zinc from zinc(II) sulphate(VI) solution? Write an equation for this reaction.

(c) Give two metals which could be deposited at a platinum cathode during electrolysis of solutions of their salts. (JMB)

3. The relative positions of the elements rubidium (Rb), beryllium (Be) and bismuth (Bi) in the electrochemical or metal activity series are given below, with the positions of some more familiar elements. The formulae of three chlorides are also given.

position in the series	chloride
rubidium	RbCl
sodium	
magnesium	
beryllium	$BeCl_2$
iron	
lead	
hydrogen	
bismuth	$BiCl_3$
copper	

Make use of this information to answer the following questions about rubidium, beryllium and bismuth. You may assume that these elements do not show variable valency.

(a) Write formulae for the hydroxides of these three elements. Which one of these hydroxides will be soluble in water?

(b) Give the formulae of the solid compounds you would expect to remain when samples of the three hydroxides in (a) are heated.

(c) How, and under what conditions, would you expect rubidium and beryllium to react with water? Write equations for these reactions. State, with reasons, whether any reaction between bismuth and water is to be expected.

(d) What would you expect to happen when beryllium and bismuth are separately treated with dilute hydrochloric acid?

(e) Which one of the three elements could readily be obtained by heating its oxide in a stream of hydrogen? (C)

4. The atomic and ionic radii of two Periodic Table groups of elements are shown below measured in nanometres (nm):

	atomic radius	ionic radius		atomic radius	ionic radius
A	0·123	0·060	W	0·072	0·136
B	0·157	0·095	X	0·099	0·181
C	0·203	0·133	Y	0·114	0·195
D	0·216	0·148	Z	0·133	0·216
E	0·235	0·169			

(a) Which group contains metals and which non-metals?

(b) Why are the ionic radii of elements A to E smaller than the atomic radii, whereas the ionic radii of elements W to Z are larger than the atomic radii?

(c) Why do both atomic and ionic radii increase down each group?

(d) Which element is the most reactive in the group A to E and which in the group W to Z?

appendices

1 units

The International System of Units (SI) is used in this book. The six dimensionally independent physical quantities encountered, together with their symbols, their units and their symbols, are shown in Fig. A. Italic type is used for the symbol for a physical quantity and Roman type is used for the symbol for a unit. Remember that

physical quantity = number × unit

physical quantity	symbol	unit	symbol
length	l	metre	m
mass	m	kilogram	kg
time	t	second	s
electric current	I	ampere	A
temperature	T	kelvin	K
amount of substance	n	mole	mol

Fig. A basic SI units

Multiples and submultiples of the above units are indicated by prefixes having the meanings shown in Fig. B.

	prefix	symbol		prefix	symbol
10^9	giga	G	$\mathbf{10^{-1}}$	**deci**	d
10^6	mega	M	$\mathbf{10^{-2}}$	**centi**	c
$\mathbf{10^3}$	**kilo**	k	$\mathbf{10^{-3}}$	**milli**	m
			10^{-6}	micro	μ
			10^{-9}	nano	n

Fig. B prefixes for multiples and sub-multiples (those most commonly encountered are shown in bold type)

The physical quantities most commonly encountered in elementary chemistry, together with their symbols and units, are shown in Fig. C. In some cases units other than the coherent SI unit are given where they are in common use and are likely to remain acceptable.

physical quantity	symbol	SI unit	other units
atomic number	Z		
Avogadro constant	L	mol^{-1}	
common temperature	t, θ		°C
concentration of solute B	c_B, $[B]$	$mol\,m^{-3}$	$mol\,dm^{-3}$, $mol\,l^{-1}$
density	ρ	$kg\,m^{-3}$	$g\,cm^{-3}$
electrical charge	Q	(coulomb) C	
electrical potential	V	(volt) V	
electrical potential difference	V	V	
electrode potential (standard)	$E^{\ominus}$	V	
electromotive force	E	V	
electron charge	e	C	
energy	E	(joule) J (i.e. $kg\,m^2s^{-2}$)	kJ
Faraday constant	F	$C\,mol^{-1}$	
force	F	(newton) N (i.e. $kg\,m\,s^{-2}$)	
gas constant (molar)	R	$J\,K^{-1}mol^{-1}$	
half-life (radioactive)	$t_{\frac{1}{2}}$	s	
heat	Q, q	J	kJ
heat capacity (molar)	C	$J\,K^{-1}mol^{-1}$	
latent heat (molar)	L	$J\,mol^{-1}$	
mass concentration of solute B	ρ_B	$kg\,m^{-3}$	$g\,dm^{-3}$, $g\,l^{-1}$
mass number	A		
mass of electron	m_e	kg	
neutron	m_n	kg	
proton	m_p	kg	
molar heat content (enthalpy)	H	J	kJ
molar volume	V_m	m^3mol^{-1}	dm^3mol^{-1}, $l\,mol^{-1}$
neutron number	N		
power	P	(watt) W	
pressure	p	(pascal) Pa (i.e. $N\,m^{-2}$)	kPa
rate of increase of concentration of solute B	r_B	$mol\,dm^{-3}s^{-1}$	$mol\,l^{-1}s^{-1}$
relative atomic mass	A_r		
relative molecular mass	M_r		
resistance	R	(ohm) Ω	
specific heat capacity	c	$J\,kg^{-1}K^{-1}$	
specific latent heat	l	$J\,kg^{-1}$	
volume	V	m^3	cm^3, dm^3, l
weight	W	N	

Fig. C commonly encountered physical quantities

Inevitably, sources of information will for some time continue to use units other than those listed above and a list of some useful conversion factors is printed at the beginning of this book. It should be noted particularly that the unit 'atmosphere' for pressure should be used only when expressing an approximate pressure, e.g. '200 atmospheres' is acceptable, but '1·75 atmospheres' is not. The word 'molar' in Fig. C means 'per unit amount of substance'; it is still also used (strictly speaking, incorrectly) to mean 'having a concentration of 1 mol dm^{-3}'. The word 'specific' means 'per unit mass'.

2 nomenclature

It is recognised that in the early stages of a chemistry course, a name like 'copper(II) sulphate(VI)-5-water' may cause unnecessary perplexity at a stage when 'blue crystals of copper sulphate' describes the solid adequately. Nevertheless, systematic nomenclature is employed throughout this book, on the grounds that material covered in the early Chapters will probably be referred to quite late in a course. It is recommended that the Roman numerals in a name should be accepted without explanation until the point in the course at which their significance can be adequately treated (see section 14.3). After all, if the '(VI)' in 'sulphate(VI)' is a mystery, so is the 'ate' in 'sulphate'.

If trivial names are used in the early stages of a course, they should be chosen with care. Use of 'black copper oxide' and 'red copper oxide' has been known to lead students to believe that there are two kinds of copper, black copper and red copper! 'The black oxide of copper' avoids this confusion. There is much to be said for encouraging students to use the correct names as early as possible in the course – certainly before Chapter 14 is encountered. Dangerous confusion is likely to arise between, e.g. nitric(V) acid (HNO_3) and nitric(III) acid (HNO_2); sulphate(VI) (SO_4^{2-}) and sulphate(IV) (SO_3^{2-}), if the Roman numeral of the name is too long ignored.

A list of names in common use which refer to substances encountered in this book, together with the recommended systematic name, is given in the following table:

COMMON NAME	SYSTEMATIC NAME
acetate	ethanoate, acetate
acetic acid	ethanoic acid, acetic acid
acetone	propanone
acetylene	ethyne
alcohol (ethyl)	ethanol
aluminate ion	aluminate(III) ion
aluminium ion	aluminium(III) ion
aluminium ion (hydrated)	hexaaquaaluminium(III) ion
ammonia	ammonia
ammonium hydroxide	aqueous ammonia
ammonium ion	ammonium ion
anhydrite	calcium(II) sulphate(VI)
argentous ion	silver(I) ion
barium ion	barium(II) ion
bicarbonate, bisulphate, etc.	hydrogencarbonate, hydrogensulphate(VI), etc.
bleaching powder	bleaching powder
bromate ion	bromate(V) ion
calcium ion	calcium(II) ion
calcium carbide	calcium(II) dicarbide
carbon dioxide	carbon dioxide
carbon disulphide	carbon disulphide
carbon monoxide	carbon monoxide
carbon tetrachloride	tetrachloromethane
carbonate ion	carbonate ion
chlorate ion	chlorate(V) ion
chloride ion	chloride ion
chloroform	trichloromethane
chromate ion	chromate(VI) ion
chromic ion	chromium(III) ion
chromium sesquioxide	chromium(III) oxide
cryolite	sodium(I) hexafluoroaluminate(III)
cuprammonium ion	tetraamminecopper(II) ion
cupric ion	copper(II) ion
cuprous ion	copper(I) ion
cyanide ion	cyanide ion
dichromate ion	dichromate(VI) ion
diethyl ether	ethoxyethane
ethylene	ethene
ethylene dibromide	1,2-dibromoethane
ethylene glycol	ethane-1,2-diol
ferric ion	iron(III) ion
ferricyanide ion	hexacyanoferrate(III) ion
ferrocyanide ion	hexacyanoferrate(II) ion
ferrosoferric oxide	iron(II) diiron(III) oxide
ferrous ion	iron(II) ion
fluoride ion	fluoride ion
hydrogen ion (see hydronium ion)	hydrogen ion
hydrogen peroxide	hydrogen peroxide
hydronium or hydroxonium ion	oxonium ion
hydroxyl ion	hydroxide ion
hypobromite ion	bromate(I) ion
hypobromous acid	bromic(I) acid

COMMON NAME	SYSTEMATIC NAME
hypochlorite ion	chloric(I) ion
hypochlorous acid	chloric(I) acid
inert gas	noble gas
iodate ion	iodate(V) ion
iodide ion	iodide ion
iron pyrites	iron(II) disulphide
lead acetate	lead(II) ethanoate (acetate)
lead dioxide	lead(IV) oxide
lead ion	lead(II) ion
lead monoxide	lead(II) oxide
magnesium ion	magnesium(II) ion
manganese dioxide	manganese(IV) oxide
manganous ion	manganese(II) ion
mercuric ion	mercury(II) ion
mercurous ion	mercury(I) ion
nickelous ion	nickel(II) ion
nitrate ion	nitrate(V) ion
nitric acid	nitric(V) acid
nitric oxide	nitrogen oxide
nitride ion	nitride ion
nitrite ion	nitrate(III) ion
nitrogen dioxide (peroxide)	nitrogen dioxide
nitrous acid	nitric(III) acid
nitrous oxide	dinitrogen oxide
olefines	alkenes
orthophosphate ion	phosphate(V) ion
paraffins	alkanes
permanganate ion	manganate(VII) ion
phosphate ion	phosphate(V) ion
phosphoric acid	phosphoric(V) acid
phosphorus	phosphorus
phosphorus pentachloride	phosphorus pentachloride
phosphorus pentoxide	phosphorus(V) oxide

COMMON NAME	SYSTEMATIC NAME
phosphorus trichloride	phosphorus trichloride
phosphorus trioxide	phosphorus(III) oxide
plumbate ion	plumbate(IV) ion
plumbic compounds	lead(IV) compounds
plumbous ion	lead(II) ion
polythene	poly(ethene)
polyvinyl chloride	poly(chloroethene)
potassium ion	potassium(I) ion
quicklime	calcium(II) oxide
red lead	dilead(II) lead(IV) oxide
silica	silicon(IV) oxide
silicate ion	silicate(IV) ion
silicon dioxide	silicon(IV) oxide
silver ion	silver(I) ion
slaked lime	calcium(II) hydroxide
sodalime	sodalime
sodium ion	sodium(I) ion
stannic compounds	tin(IV) compounds
stannous ion	tin(II) ion
strontium ion	strontium(II) ion
sulphate ion	sulphate(VI) ion
sulphite ion	sulphate(IV) ion
sulphur dioxide	sulphur dioxide
sulphur trioxide	sulphur(VI) oxide
sulphuric acid	sulphuric(VI) acid
sulphurous acid	sulphuric(IV) acid
thiocyanate ion	thiocyanate ion
thiosulphate ion	thiosulphate(VI) ion
tin dioxide	tin(IV) oxide
toluene	methylbenzene
vanadium pentoxide	vanadium(V) oxide
zinc ion	zinc(II) ion
zincate ion	zincate(II) ion

answers to numerical questions

Chapter 2, page 15.
8 (a) water; **(b)** 340 J g^{-1}. **9 (a)** 230 J g^{-1}; **(b)** 50 J g^{-1}; **(c)** 200 J g^{-1}. **10 (a)** 60 °C; **(b)** 400 J g^{-1}; **(c)** below 20 °C.

Chapter 3, page 23.
1 808 kPa. **2 (a)** 310 cm^3; **(b)** 120 kPa; **(c)** 215 kPa; **(d)** 109·2 cm^3; **(e)** 120 cm^3; **(f)** 36 kPa; **(g)** 57 °C (330 K); **(h)** $\frac{1}{16}$ cm^3 (0·0625 cm^3); **(i)** 105 cm^3; **(j)** 71·4 cm^3. **4** 200 cm^3.

Chapter 4, page 28.
1 compounds: water, salt, ice, sugar; mixtures: tomato sauce, wood, whisky, paint. **3 (a)** green; **(b)** brown. **5** 94 °C.

Chapter 5, page 34.
2 (b), (c) and **(d)** all possible. **3 (a)** mixtures; **(b)** compounds; **(c)** compounds; **(d)** mixtures; **(e)** mixtures; **(f)** compounds; **(g)** mixtures. **6 (a)** D; **(b)** A; **(c)** B; **(d)** C; **(e)** A; **(f)** E.

Chapter 7, page 47.
4 (i) C; **(ii)** C. **8** D; B; C; A. **11 (c)** and **(d)**. **12** A–metal; B–non-metal; C–non-metal; D–metal; E–metal.

Chapter 8, page 59.
10 1 part hydrogen: 8 parts oxygen. **12 (i)** increase; **(ii)** decrease; **(iii)** increase.

Chapter 9, page 67.
5 NH_3. **6** H_2S. **7** NO. **8** TiO. **9** C_2H_6. **10** N_2O.

Chapter 10, page 89.
1 MgO. **2** I_2O_5. **3** 10 g. **4** CrO_3. **5** O_2. **6** 60%. **7** $H_{12}O_{18}Cr_2S_3$ [$Cr_2(SO_4)_3.6H_2O$]. **8** 351·7, UF_6. **9** X_2O. **10** 2·8 dm^3 H_2; 1·4 dm^3 O_2. **11 (b)** $6{\cdot}02 \times 10^{18}$ molecules. **12 (a)** N; **(b)** 16N; **(c)** $\frac{1}{2}$N; **(d)** $\frac{1}{3}$N. **13 (a)** 1 : 4; **(b)** 8 : 1; **(c)** Mn_2O_3. **14 (a)** 0·56 dm^3; **(b)** 1·12 dm^3. **15 (b)** 35 kg; **(c)** C_3S_2. **17** 64. **18 (b)** SO_3; **(c)** 8 g; **(d)** 24.

Chapter 11, page 98.
3 (a) A; **(b)** B; **(c)** (i) acidic; (ii) acidic, (iii) acidic, (iv) alkaline, (v) acidic. **5** limewater–pH 11; Soap–ph 8; hydrochloric acid–pH 1; lemon juice–pH 5·5; sodium(I) chloride–pH 7. **6 (a)** 3 mole dm^{-3}; **(b)** 0·05 mole dm^{-3}; **(c)** 0·5 mole dm^{-3}; **(d)** 0·4 mole dm^{-3}; **(e)** 0·4 mole dm^{-3}. **7 (i)** (a) 0·05 mole, (b) 3·70 g; **(ii)** (a) 0·125 mole, (b) 4·56 g; **(iii)** (a) 0·002 mole, (b) 0·499 g hydrated; **(iv)** (a) 0·0063 mole, (b) 0·394 g; **(v)** (a) 0·0059 mole, (b) 0·330 g; **(vi)** (a) 0·0034 mole, (b) 0·454 g. **8** 0·054 M (5·72 g dm^{-3} anhydrous). **9 (a)** 0·15 M; **(b)** 40 cm^3. **10** 400 cm^3. **11 (b)** 2 g; **(c)** 0·448 dm^3. **12 (a)** (i) 800 cm^3, (ii) 194 g, (iii) 2.

Chapter 12, page 113.
8 25 g $NaNO_3$; **(i)** 11 g NaCl, 17·5 g $NaNO_3$; **(ii)** 25 g $NaNO_3$, 1 g NaCl. **9 (b)** 1 dm^3. **11 (b)** 37·93 g per 100 g H_2O. **12 (f)** x = 7. **13 (ii)** 208 g; **(iv)** 6 cm^3; **(v)** 3 dm^3; **(vi)** 3; **(vii)** $M_2(SO_4)_3$; **(viii)** 27. **17 (a)** X = Pb_3O_4; **(b)** Y = Al; **(c)** Z = CuO. **18 (a)** 461 g and 233 g resp.; **(b)** $\frac{1}{100}$; **(c)** A – 2·305 g, B = 4·61 g, C = 2·33 g, D = 2·33 g; **(e)** A = neutral, B = neutral, C = neutral, D = acidic; **(f)** alkaline.

Chapter 13, page 128.
2 (i) poor; **(ii)** non-conductor; **(iii)** poor; **(iv)** good; **(v)** medium; **(vi)** medium. **4** 35·5 g. **7** 3·36 dm^3. **8 (b)** 28 cm^3; **(c)** $\frac{1}{400}$. **9 (c)** (i) 193 coulombs, (ii) 96 500 coulombs (1 Faraday), (iii) 193 000 coulombs (2 Faradays).

Chapter 14, page 149.
5 (a) 24 g; **(b)** 7. **12** 101·6 g. **13 (a)** 11·2 dm^3; **(b)** 22·4 dm^3. **17 (a)** A(−) and D(+); **(c)** A. **24 (a)** $\frac{1}{200}$ mole Fe^{3+}(aq.); **(b)** $\frac{1}{400}$ mole Sn^{2+}(aq.); **(c)** 2; **(d)** Sn^{4+}(aq.); **(h)** 0·05 mole.

Chapter 16, page 172.
5 25 cm^3.

Chapter 17, page 180.
9 (i) BaO, SeO_2; **(ii)** 2 and 4.

Chapter 18, page 202.
2 (a) 4; **(b)** 2; **(c)** 6; **(d)** T and V; **(e)** X; **(f)** W. **4 (a)** 17; **(b)** 17; **(c)** 20; **(d)** 18; A : B = 3 : 1. **5 (i)** B; **(ii)** B; **(iii)** A; isotopes. **6 (b)** 20 years. **7 (i)** 2, 6; **(ii)** 2, 8, 8; **(iii)** 2, 8, 8; **(iv)** 2, 8, 8; **(vi)** 2, 8. **9 (b)** 2, 8, 2; **(c)** F^-, Mg^{2+}; **(d)** (i) MgF_2, (ii) NaF; **(g)** (ii) $^{24}_{12}Mg$. **10 (a)** (i) 13, (ii) 27; **(d)** (i) Al(2, 8, 3), F(2, 7), (iii) Al^{3+}, (iv) AlF_3; **(f)** $^{14}_{7}N$. **12 (a)** B; **(b)** D; **(c)** E; **(d)** C; **(e)** A; **(f)** C.

Chapter 19, page 219.
1 160 cm^3. **2 (a)** O_2; **(b)** 10 cm^3 : 20 cm^3 (1 : 2); **(c)** CO_2; **(d)** 10 cm^3; **(e)** H_2O; **(f)** CH_4. **4** 3·5 dm^3 O_2; 2 dm^3 CO_2. **5 (a)** 100 cm^3 O_2; **(b)** 60 cm^3 CO_2. **6** A = C_2H_2; B = C_2H_4; D = CH_2BrCH_2Br; E = polythene. **7** 50 cm^3 CO_2; 100 cm^3 steam. **10** 12 cm^3 C_2H_4; 8 cm^3 C_2H_6. **11 (a)** (i) CH_2Br, (ii) $C_2H_4Br_2$; 1, 2 dibromomethane; **(b)** X = C_2H_4; **(c)** addition; **12** empirical C_2H_5, molecular C_4H_{10}.

Chapter 20, page 229.
1 (a) exo; **(b)** endo; **(c)** exo; **(d)** endo; **(e)** exo. **4** ΔH = **(a)** −572 kJ; **(b)** −572 kJ; **(c)** −143 kJ; **(d)** −0·41 kJ. **5** ΔH = −131 kJ. **6** ΔH = −558 kJ. **7** ΔH = −542 kJ. **8** ΔH = −642 kJ. **11 (b)** methanol 723 kJ mole^{-1}; ethanol 1366 kJ mole^{-1}; propanol 2004 kJ mole^{-1}.

Chapter 21, page 239.
10 (a) 0·448 dm^3 CO, 0·448 dm^3 CO_2; **(b)** 1·86 dm^3 CO_2; **(c)** 8·96 dm^3 CO_2, 8·96 dm^3 CO.

Chapter 22, page 250.
4 (a) (i) $\frac{1}{20}$, (ii) anhydrous Na_2CO_3, (iii) $\frac{1}{40}$, (iv) CO_2, (v) $\frac{1}{50}$.

Chapter 23, page 261.
1 (a) increased; **(b)** decreased; **(c)** decreased. **6 (a)** 3·32 g.

Chapter 24, page 273.
6 (c) 13·2 g. **7 (b)** 3·95 dm^3 **9** HNO_2.

Chapter 25, page 282.
1 15 dm^3. **4** 980 tons. **5 (e)** 2·8 dm^3. **6 (b)** 400 cm^3.

Chapter 26, page 298.
11 (a) A = $Cu(NO_3)_2$, B = CuO, C = $Cu(OH)_2$; **(b)** D = Fe, E = H_2, F = $FeSO_4$(aq.), G = $Fe(OH)_2$, H = $Fe(OH)_3$ or Fe_2O_3.

index

Items in bold type, thus, **29,** form the general introduction to the subject. If subsequent items are marked with an asterisk*, such items are also covered (usually more fully) in this introductory reference.

periodic table in its modern form

PERIODS \ GROUPS	I	II	TRANSITION METALS										III	IV	V	VI	VII	O
1	$_{1}$H																	$_{2}$He
2	$_{3}$Li	$_{4}$Be											$_{5}$B	$_{6}$C	$_{7}$N	$_{8}$O	$_{9}$F	$_{10}$Ne
3	$_{11}$Na	$_{12}$Mg											$_{13}$Al	$_{14}$Si	$_{15}$P	$_{16}$S	$_{17}$Cl	$_{18}$Ar
4	$_{19}$K	$_{20}$Ca	$_{21}$Sc		$_{22}$Ti	$_{23}$V	$_{24}$Cr	$_{25}$Mn	$_{26}$Fe	$_{27}$Co	$_{28}$Ni	$_{29}$Cu	$_{30}$Zn	$_{31}$Ga	$_{32}$Ge	$_{33}$As	$_{34}$Se	$_{35}$Br	$_{36}$Kr
5	$_{37}$Rb	$_{38}$Sr	$_{39}$Y		$_{40}$Zr	$_{41}$Nb	$_{42}$Mo	$_{43}$Tc	$_{44}$Ru	$_{45}$Rh	$_{46}$Pd	$_{47}$Ag	$_{48}$Cd	$_{49}$In	$_{50}$Sn	$_{51}$Sb	$_{52}$Te	$_{53}$I	$_{54}$Xe
6	$_{55}$Cs	$_{56}$Ba	$_{57}$La	↑	$_{72}$Hf	$_{73}$Ta	$_{74}$W	$_{75}$Re	$_{76}$Os	$_{77}$Ir	$_{78}$Pt	$_{79}$Au	$_{80}$Hg	$_{81}$Tl	$_{82}$Pb	$_{83}$Bi	$_{84}$Po	$_{85}$At	$_{86}$Rn
7	$_{87}$Fr	$_{88}$Ra	$_{89}$Ac	↑															

LANTHANIDES

$_{58}$Ce	$_{59}$Pr	$_{60}$Nd	$_{61}$Pm	$_{62}$Sm	$_{63}$Eu	$_{64}$Gd	$_{65}$Tb	$_{66}$Dy	$_{67}$Ho	$_{68}$Er	$_{69}$Tm	$_{70}$Yb	$_{71}$Lu

ACTINIDES

$_{90}$Th	$_{91}$Pa	$_{92}$U	$_{93}$Np	$_{94}$Pu	$_{95}$Am	$_{96}$Cm	$_{97}$Bk	$_{98}$Cf	$_{99}$Es	$_{100}$Fm	$_{101}$Md	$_{102}$No	$_{103}$Lw